Notes on Numerical Fluid Mechanics and Multidisciplinary Design

Volume 130

About this Series

Notes on Numerical Fluid Mechanics and Multidisciplinary Design publishes state-of-art methods (including high performance methods) for numerical fluid mechanics, numerical simulation and multidisciplinary design optimization. The series includes proceedings of specialized conferences and workshops, as well as relevant project reports and monographs.

More information about this series at http://www.springer.com/series/4629

Sharath Girimaji · Werner Haase
Shia-Hui Peng · Dieter Schwamborn
Editors

Progress in Hybrid RANS-LES Modelling

Papers Contributed to the 5th Symposium on Hybrid RANS-LES Methods, 19–21 March 2014, College Station, A&M University, Texas, USA

Springer

Editors
Sharath Girimaji
Aerospace Engineering Department
Texas A&M University
College State, TX
USA

Werner Haase
Werner Haase Aeronautics Consultants
Neubiberg, Bayern
Germany

Shia-Hui Peng
Division of Information and
Aeronautical Systems
Swedish Defence Research Agency FOI
Stockholm
Sweden

Dieter Schwamborn
Center of Computer Applications
in Aerospace Science and Engineering
DLR, Institut für Aerodynamik und
Strömungstechnik
Göttingen
Germany

ISSN 1612-2909 ISSN 1860-0824 (electronic)
Notes on Numerical Fluid Mechanics and Multidisciplinary Design
ISBN 978-3-319-15140-3 ISBN 978-3-319-15141-0 (eBook)
DOI 10.1007/978-3-319-15141-0

Library of Congress Control Number: 2015930203

Springer Cham Heidelberg New York Dordrecht London

Printed on acid-free paper

Springer International Publishing AG Switzerland is part of Springer Science+Business Media
(www.springer.com)

Preface

Motivated by the immediate computational needs of industry, *accuracy-on-demand* turbulence simulation approaches that judiciously combine the advantages of Reynolds-Averaged Navier-Stokes (RANS) and Large Eddy Simulation (LES), broadly classified as hybrid RANS-LES methods, have witnessed rapid development over the last decade. Despite the best efforts of turbulence researchers worldwide, many challenges still remain, especially those pertaining to robustness and applicability to complex engineering flows. It is recognized that further improvement and development are required before hybrid RANS-LES methods can become frontline Computational Fluid Dynamics (CFD) tools for practical applications. The *Symposium on Hybrid RANS-LES Methods* (HRLM) is an important international forum that places specific emphasis on hybrid methods. The main aim of the series of symposia has been to bring together researchers from universities and research institutes, as well as industrial engineers, R&D managers and consultants, to report and discuss the latest developments and applications of advanced turbulence-resolving modelling and simulation methods. The previous HRLM symposia have played an important and unique role in communicating current activities and progress in the field.

This book contains the contributions presented at the 5th Symposium on Hybrid RANS-LES Methods (HRLM-5), which took place in College Station, Texas, USA, 18–21, March 2014. The previous HRLM symposia took place in Stockholm (Sweden, 2005), Corfu (Greece, 2007), Gdansk (Poland, 2009) and Beijing (China, 2011). It is hoped that the book will serve as a useful source of information and inspiration for further advancement of engineering turbulence closure modelling.

The HRLM-5 Symposium included four invited lectures—by K. Hanjalic (Delft University of Technology), P. Spalart (Boeing), V. Yakhot (Boston University) and B. Basara (AVL)—and 43 contributed papers addressing the following topics: *Novel turbulence-resolving simulation and modelling methods, Improved hybrid RANS-LES methods (including DES-type and Embedded LES approaches), Comparative studies of difference modelling methods, Modelling-related numerical issues and Industrial applications.* All the papers included in the present book, 40 in total, have been peer-reviewed.

The HRLM-5 Symposium was co-organized by Texas A&M University and the EU Go4Hybrid Project Consortium. The symposium owes its success to the support of the participants and further, for the publication of the book, to the invited and contributing authors. The Scientific Committee members and a number of external experts served to review the full papers, which has greatly helped in further improving the quality of the book. We are grateful to the following experts for reviewing the full manuscripts appearing in the present HRLM-5 Symposium book: B. Aupoix, B. Basara, R.A. Bauerle, M. Braza, L. Davidson, S. Deck, S. Fu, M.K. Frendi, T. Gatski, M. Germano, K. Hanjalic, S. Jakirlic, J. Kok, D. Laurence, F. Menter, C. Mockett, C. Rumsey, P. Spalart, M. Strelets, F. Thiele, S. Wallin, and V. Yakhot.

Last, but not the least, the editors are grateful for the excellent and dedicated work by the local organizing team consisting of Texas A&M University students. Their effort contributed greatly to making this 5th Symposium a success. Moreover, we wish to express our sincere gratitude to our corporate sponsors: AIRBUS, ANSYS, AVL, CFD Software GmbH and Rolls-Royce Ltd.

November 2014

Sharath Girimaji
Werner Haase
Shia-Hui Peng
Dieter Schwamborn

Contents

Part I
Invited Keynote Papers

PANS Method as a Computational Framework from an Industrial Perspective

B. Basara

Abstract Although Computational Fluid Dynamics (CFD) is routinely used in a wide variety of industries, there are many remaining challenges in physical modelling as well as in numerical methods, which have to be tackled and eventually solved in the near future. Turbulence modelling, especially for industrial CFD, is still one of those open issues. For the purpose of a better and more practical or affordable representation of turbulence in complex flows, the variable resolution methods have emerged as an alternative to a computationally more costly Large Eddy Simulation (LES) method. At present, and among many approaches, the Partial-Averaged Navier Stokes (PANS) approach is one of the most attractive methods for industrial CFD. Therefore, the capabilities of the PANS on a wide range of CFD applications are shown in this paper. The results are presented for simple and well established benchmarks but also for industrial flows in complex geometries. The basic theory and arguments for the usage of this method are given. Besides the present status, the paper also provides some hints for possible improvements and explains some of the on-going activities in this field.

1 Introduction

Ideas about a turbulence closure method which can be used at all levels of scale resolution have been intensively pursued in the last two decades. From the beginning, such 'smart' models, which should provide the optimum solution on any computational mesh, have been very attractive to CFD users, especially to those involved in simulations of complex industrial flows. Presently, a large majority of industrial users tend to ignore the importance of turbulence models as long as calculation results are in a line with the expectations (based on the measurements or just from the experience) and simulations provide the correct behavior or relative differences in the case

B. Basara (✉)
Advanced Simulation Technology, AVL List GmbH,
Hans-List-Platz 1, 8020 Graz, Austria
e-mail: branislav.basara@avl.com

S. Girimaji et al. (eds.), *Progress in Hybrid RANS-LES Modelling*,
Notes on Numerical Fluid Mechanics and Multidisciplinary Design 130,
DOI 10.1007/978-3-319-15141-0_1

of parametric studies. The Reynolds-Averaged Navier-Stokes (RANS) turbulence models have done a good job for decades, but in many applications, the limitations of such an approach, regardless of which RANS turbulence model is used, could be a large source of the calculation error especially when the absolute accuracy is targeted. Indeed, one could recommend the best or the optimum RANS turbulence model for a certain application but there will always be some 'gray zones' in such guidelines especially for the complex industrial flows. On the other hand, the Large Eddy Simulation approach has been more frequently used in the last years but mainly for benchmarks or visibility studies. This is still due to high computational costs. One should also bear in mind that results obtained by the Large Eddy Simulation have to be carefully analyzed in order to conclude if they are trustworthy, which could be a very difficult task for everyday use in industry. And, if the conclusion is that the mesh is too coarse, the new mesh has to be created and calculations must be repeated.

Therefore, there is a clear gap between the RANS and LES simulations which require a deep knowledge in turbulence modelling, both RANS and LES, which is just one of the fields that modern simulation engineers have to cover. Therefore, the models like PANS should be the right answer for the industry. It provides the best possible physical fidelity on any given numerical grid, while varying seamlessly between Reynolds-Averaged Navier-Stokes (RANS) model and Direct Numerical Simulations (DNS). This means that the user makes calculations on the affordable mesh and the model itself combines RANS and DNS by using the arbitrary cutoff length scale concept. The PANS method belongs to the bridging (seamless) variable resolution methods which mean that the basic model is employed in the entire domain. Contrary to this group, zonal methods divide a solution domain into two modeling regions: the RANS turbulence model near the wall and the Large Eddy Simulation (LES) in the rest of the flow domain. An important drawback of zonal methods is a definition of the interface between different modeling zones, especially for complex flows.

In recent years, the bridging methods have become very popular for simulations of complex turbulent flows. Probably, the most attractive bridging method is the Partially-Averaged Navier-Stokes (PANS) formulated by Girimaji [6, 8]. This method is derived from the RANS model equations. It inevitably improves results when compared with its corresponding RANS model if more scales of motions are resolved. The PANS model is used in the industry more than other seamless methods due to its simplicity, robustness and recent theoretical extensions as well as due to the detailed validations on the number of complex cases presented in many publications. It was shown in previous studies that the implied cut-off for the PANS method can be placed in any part of the spectrum including the dissipation range. This is done by varying the unresolved-to-total ratios of kinetic energy (f_k) and dissipation (f_e). In practice, the parameter which determines the unresolved-to-total kinetic energy ratio is defined by using the grid spacing and calculated integral length scale of turbulence. If the resolution parameters, f_k and f_e, are equal unity, the PANS model recovers the RANS model. As Girimaji pointed out in his work, the better the RANS model is, the better the corresponding PANS model will be. On the other side, if the resolution parameters are very low, then the modeling of the small unresolved scales will affect the overall solutions less. The main target of such an approach is to have an optimum turbulence model for any mesh used in calculations. There are different ways to define the resolution parameter f_k and the work presented here will show the differences between various approaches as this is a crucial point to get the PANS method widely used in the industry. We already know that numerical meshes for most industrial applications are usually coarser near the wall to achieve so-called a wall-resolved LES. A similar situation is related to PANS calculations: one could expect that f_k is equal or close to unity which means that the RANS model is used near the wall. This issue is getting more pronounced for separating flows from the curvature rather than from sharp edges. Therefore, we use the PANS variant [3] derived from the four equation near-wall eddy viscosity transport model, namely k-ε-ζ-f turbulence model [11]. As this model represents a practical and accurate RANS choice for a wide range of industrial applications, especially when used in conjunction with the universal wall approach, its PANS variant therefore guarantees that the proper near-wall model is used when f_k is of a higher value. The paper presents a variety of test cases, from simple flow benchmarks to complex industrial flows. The results will show that this modeling approach can be successfully used for complex industrial applications.

2 Computational Method

The partially-Averaged Navier-Stokes equations are written in term of partially averaged or filtered velocity and pressure field, thus

$$\frac{\partial U_i}{\partial t} + U_j \frac{\partial U_i}{\partial x_j} + \frac{\partial \tau\left(V_i, V_j\right)}{\partial x_j} = -\frac{1}{\rho}\frac{\partial p}{\partial x_i} + \nu \frac{\partial^2 U_i}{\partial x_j \partial x_j} \tag{1}$$

where the velocity field is decomposed into two components, the partially filtered component and the sub-filter component as

$$V_i = U_i + u_i \tag{2}$$

The closure for the sub-filter stress can be obtained by using the Boussinesq approximation as

$$\tau\left(V_i, V_j\right) = -2\nu_u S_{ij} + \frac{2}{3} k_u \delta_{ij} \tag{3}$$

where the eddy viscosity of unresolved scales and the resolved stress tensor are equal to

$$\nu_u = c_\mu \frac{k_u^2}{\varepsilon_u} \qquad S_{ij} = \frac{1}{2}\left(\frac{\partial U_i}{\partial x_j} + \frac{\partial U_j}{\partial x_i}\right) \tag{4}$$

and the resolution parameters are

$$f_k = \frac{k_u}{k} \qquad f_\varepsilon = \frac{\varepsilon_u}{\varepsilon} \tag{5}$$

However, values of these parameters can be checked only at the end of calculations but they are needed at the start of calculations and even at the start of every time step. This input unresolved-to-total kinetic energy ratio f_k is therefore based on the grid spacing [3, 7], thus

$$f_k \geq \frac{1}{\sqrt{c_\mu}} \left(\frac{\Delta}{\Lambda}\right)^{2/3} \tag{6}$$

where Δ is the grid cells dimension and $\Lambda = (k^{3/2}/\varepsilon)$ is the integral scale of turbulence, while f_ε was taken to be equal 1. A dynamic parameter f_k changes at each point at the end of every time step, and then it is used as a fixed value at the same location during the next time step. Note that at the end of calculations or at the end of each time step, it must be ensured that

$$\frac{1}{\sqrt{c_\mu}} \left(\frac{\Delta}{\Lambda}\right)^{2/3} \geq \frac{k_u}{k} \tag{7}$$

In other words, the parameter f_k, which is imposed on calculations, should be at the end supported by the mesh. It is also important to know that the PANS method produces the correct production-to-dissipation ratio which changes gradually from RANS value to DNS value as f_k, is reduced from 1 to 0. This is analytically proven in the paper of Girimaji [8]. This is clear evidence that PANS captures all intermediate resolution with precision. The original model of Girimaji [8], which provides the unresolved kinetic energy equations

$$\frac{Dk_u}{Dt} = P_u - \varepsilon_u + \frac{\partial}{\partial x_j}\left[\left(\nu + \frac{\nu_u}{\sigma_{ku}}\right)\frac{\partial k_u}{\partial x_j}\right] \tag{8}$$

and the unresolved energy equation

$$\frac{D\varepsilon_u}{Dt} = C_{\varepsilon 1} P_u \frac{\varepsilon_u}{k_u} - C^*_{\varepsilon 2}\frac{\varepsilon_u^2}{k_u} + \frac{\partial}{\partial x_j}\left[\left(\nu + \frac{\nu_u}{\sigma_{\varepsilon u}}\right)\frac{\partial \varepsilon_u}{\partial x_j}\right] \tag{9}$$

with the new constants, thus

$$C^*_{\varepsilon 2} = C_{\varepsilon 1} + \frac{f_k}{f_\varepsilon}\left(C_{\varepsilon 2} - C_{\varepsilon 1}\right); \sigma_{ku,\varepsilon u} = \sigma_{k,\varepsilon}\frac{f_k^2}{f_\varepsilon} \tag{10}$$

is extended by Basara et al. [3] for more accurate near-wall modelling. This new PANS variant, namely PANS k-ε-ζ-f, solves additional equations, thus

$$\frac{D\zeta_u}{Dt} = f_u - \frac{\zeta_u}{k_u} P_u + \frac{\zeta_u}{k_u} \varepsilon_u (1 - f_k) + \frac{\partial}{\partial x_j}\left[\left(\nu + \frac{\nu_u}{\sigma_{\zeta_u}}\right) \frac{\partial \zeta_u}{\partial x_j}\right] \tag{11}$$

$$L_u^2 \nabla^2 f_u - f_u = \frac{1}{T_u}\left(c_1 + C_2' \frac{P}{\varepsilon}\right)\left(\zeta_u - \frac{2}{3}\right) \tag{12}$$

with the unresolved eddy viscosity now taken as

$$\nu_u = C_\mu \zeta_u \frac{k_u^2}{\varepsilon_u} \tag{13}$$

Constants c_1 and C_2' are taken as proposed in the original model. L_u and T_u are the length and time scales defined by using unresolved kinetic energy respectively. It is clear that for $f_k = 1$, the equation for ζ_u will get its RANS form. Note also that $f_\varepsilon = 1$ implies that $\varepsilon_u = \varepsilon$.

The approach shown above has been proved on large number of cases which include static meshes. The main issue is that the total kinetic energy needed for the integral length scale and for the input of f_k can be accurately calculated only after averaged field is obtained, thus

$$k = k_r + k_u \tag{14}$$

where the resolved turbulence is obtained from

$$k_r = \frac{1}{2}\left(U_i - \overline{U_i}\right)^2 \tag{15}$$

This means that for the calculation of the resolved part, we need the average velocity which is impractical for the moving meshes (e.g. engines) or cases with transient boundaries.

Nevertheless, it was shown in the past that for the static mesh cases, the PANS method presented above can produce practical, accurate and reliable results. In order to avoid averaging the velocity field and calculation the resolved kinetic energy, Basara and Girimaji [9] derived the additional equation for so called the scale supplying variable which actually represents the modelled resolved kinetic energy. This equation is derived by using the PANS basic principles and following the similar procedure as done for the unresolved kinetic energy shown above. The equation can be written as

$$\frac{Dk_{ssv}}{Dt} = \frac{k_{ssv}}{k_u}(P_u - \varepsilon_u) + \frac{\partial}{\partial x_j}\left[\left(\nu + \frac{\nu_u}{\sigma_{ku}}\right) \frac{\partial k_{ssv}}{\partial x_j}\right] \tag{16}$$

Note that the total kinetic energy is now calculated as

$$k = k_{ssv} + k_u \tag{17}$$

The results are shown in the next section.

There are other possibilities to predict f_k. Recently Foroutan and Yavuzkurt [5] proposed the formulation for f_k which ensures that the predicted value is between 0 and 1, thus

$$f_k = 1 - \left[\frac{\left(\frac{\Lambda}{\Delta}\right)^{2/3}}{0.23 + \left(\frac{\Lambda}{\Delta}\right)^{2/3}}\right]^{4.5} \tag{18}$$

Values of f_k predicted with Eq. (18) are lower in general than those obtained with Eq. (6). In any case it must be checked that Eq. (7) is satisfied especially in the near wall region.

It is also worth of revisiting the work of Speziale [15], see also Hussaini et al. [10], who provided a possible path in deriving the f_k value in the form of

$$f_k = \left[1 - \exp\left(-\beta\Delta/\Lambda_k\right)\right]^n \tag{19}$$

where β and n are constants.

This modelling approach is further enhanced by including the hybrid wall treatment. This entails combining integration up to the wall with wall functions. Smoothing functions which blend two formulations together are known by different names: automatic, hybrid or compound wall treatments (see [2, 14]). The blending formula for the quantities specified at the cell next to the wall is given as

$$\phi = \phi_\nu e^{-\Gamma} + \phi_t e^{-1/\Gamma} \tag{20}$$

where 'ν' is the viscous and 't' the fully turbulent value of the variables: wall shear stress, production and dissipation of the turbulent kinetic energy. The function Γ is given as

$$\Gamma = \frac{0.01\left(y^+\right)^4}{1 + 5y^+} \tag{21}$$

This ensures that the optimum RANS approach is used for $f_k = 1$ and any distance from the wall (any y^+ values).

The concept presented in this section implies that there are regions with spatial or temporal variations in filter width. Hence, Wallin and Girimaji [16] and Wallin et al. [17], have introduced commutation residual terms in the resolved momentum and unresolved kinetic energy equations in the context of the PANS approach (see also [9]). It is easy to implement proposed modifications in the computer code and there are no additional penalties in calculations, see original references.

This brief review shows the present status and the focus of on-going activities related to the PANS method.

2.1 Numerical Method

The PANS model is implemented into the commercial CFD code AVL FIRE. All dependent variables, such as momentum, pressure, density, turbulence kinetic energy, dissipation rate, and passive scalar are evaluated at the cell center. The cell-face based connectivity and interpolation practices for gradients and cell-face values are introduced to accommodate an arbitrary number of cell faces. A second-order midpoint rule is used for integral approximation and a second order linear approximation for any value at the cell-face. The convection is solved by a variety of differencing schemes (upwind, central differencing, MINMOD, and SMART). The rate of change is discretized by using implicit schemes, namely Euler implicit scheme and three time level implicit scheme of second order accuracy. The overall solution procedure is iterative and is based on the Semi-Implicit Method for Pressure-Linked Equations algorithm (SIMPLE). For the solution of a linear system of equations, a conjugate gradient type of solver and algebraic multi-grid are used.

3 Results and Discussion

The predictive performances of the PANS k-ε-ζ-f model are shown on few examples below. Note also that these results are extracted here from various systematic studies just to emphasize certain important points in calculations specific to the PANS usage.

3.1 Channel Flow

PANS calculations for the channel flow with the $Re_\tau = 650$ based on the wall friction velocity u_τ, the channel half width δ and kinematic viscosity ν is analyzed in the work of Basara et al. [3] in all detail. Direct Numerical Simulation data of Iwamoto et al. [12] was used for comparisons. Instantaneous iso-surfaces of the second invariant of the velocity gradient as computed by PANS on two meshes are shown in Fig. 1 (left), (right). Following Eq. (6) the ratio between unresolved-to-total kinetic energy f_k will be decreased with the grid refinement and consequently finer structures are captured. The present standard procedure with the PANS method is to use Eq. (6) for f_k which provides typical variable instantaneous values as shown in Fig. 2 (left). However, using the most recent development described by Eqs. (16, 17), predicted f_k will vary less than with the standard approach as shown in Fig. 2 (right). However, the averaged velocity is close to each other and a very good agreement with DNS data is obtained, see Fig. 3 (left).

It is useful to know that fluctuating f_k helps clearly to capture instantaneous flow and the smooth f_k could lead to the steady flow without predicting any small structures. But even 'smooth' f_k values if they are low, as shown in Fig. 2 (right) will lead to the fluctuating velocities. Note also that the condition given by Eq. (7) must be satisfied. This new method is only introduced here in order to give the answer at the

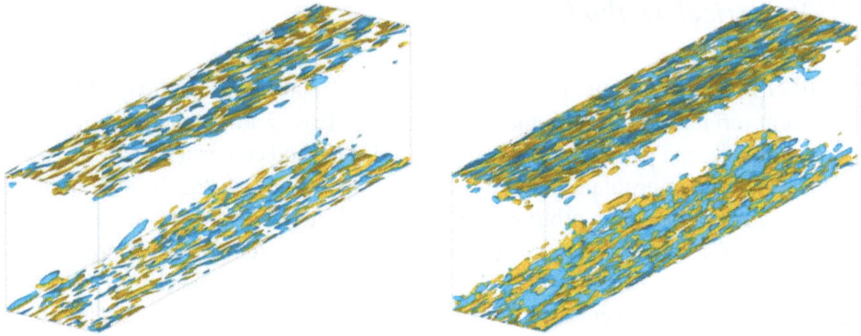

Fig. 1 An instantaneous isosurface of the second invariant of the velocity gradient as predicted by the PANS k-ε-ζ-f on different grids

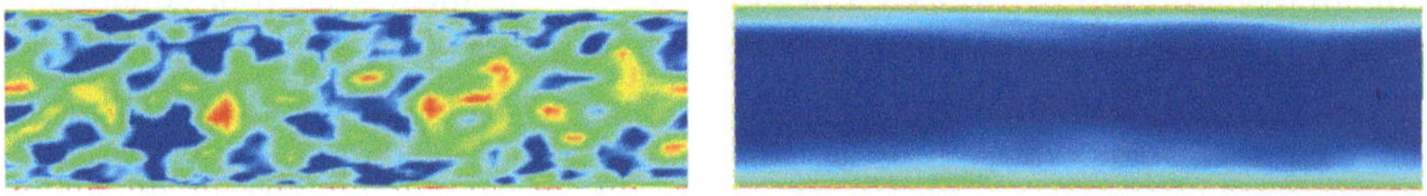

Fig. 2 Unresolved-to-total kinetic energy computed by Eqs. (6) and (14–15) (*left*) and by Eqs. (16–17) (*right*)

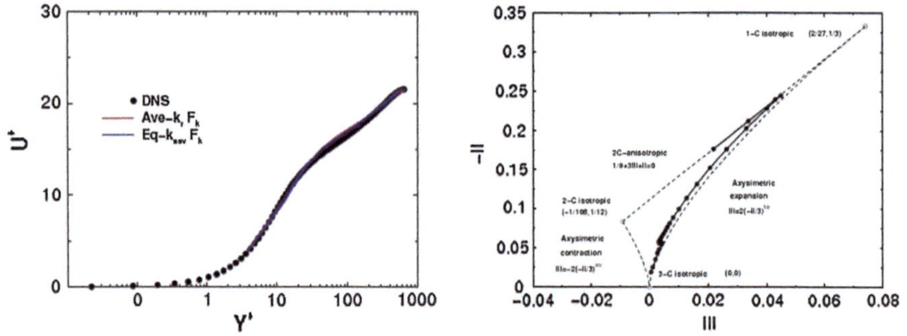

Fig. 3 Channel flow: computed mean velocity (*left*) and the predicted invariant map (*right*)

first place what can be done for the moving geometries or calculations which have transient boundaries. Otherwise, one has to apply cycle to cycle averaging which is not practical for general applications. Or, one has to start with the constant f_k and then values should be corrected for the next cycle depending of the maximum values obtained in the present cycle. This is possible to make but it would not lead to the optimum use of computer resources. However, it is important to say, that the standard method explained above is proved on many simple and complex applications.

In order to close the channel flow performance assessment, invariants of Reynolds stress anisotropy on the so-called Lumley invariant map, as predicted by the standard approach, is shown in Fig. 3 (right).

Fig. 4 Mesh sequence after the adaptive grid refinement (**a**), the predicted velocity magnitude on the coarse mesh (**b**) and on the fine mesh (**c**)

The behavior of the invariants in the wall-normal direction provides the strongest indication of the fidelity of the model calculation.

3.2 Square Cylinder

Data for the vortex shedding flow around a square cylinder at Re = 21,400 is available in the ERCOFTAC classic database. The PANS predictions are well described in the work of Basara and Pavlovic (2010). A direct effect of the automatic mesh refinement is shown here as well. Cells, which have f_k values larger than 0.5, are refined. Only two refinements starting on the mesh containing 30,000 cells are applied. Meshes created with the refinement have the size 408,000 and 696,700 cells. The extract of the finest mesh is shown in Fig. 4a, and the velocity magnitude on the second mesh is shown in Fig. 4b and on the finest mesh in Fig. 4c. It is easy to observe that smaller structures are captured with the mesh refinement.

Predicted time-averaged velocity and time-averaged turbulent kinetic energy are shown in Fig. 5 (left), (right). Results are very much improved with the mesh refinement. It is the same effect as by using meshes refined in advance, but with the adaptive

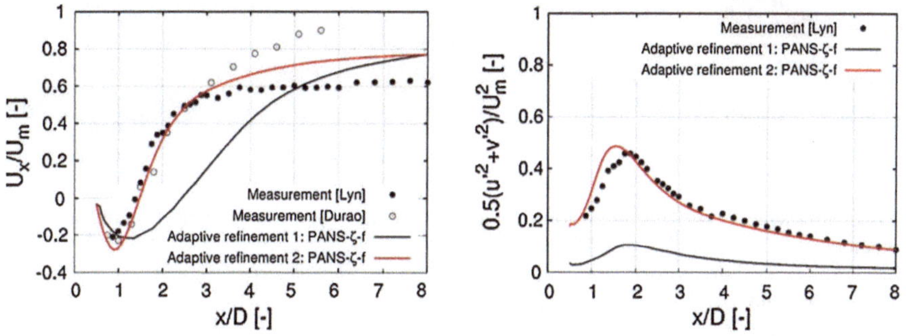

Fig. 5 Time-averaged velocity (*left*) and turbulent kinetic energy (*right*) along the *centre line* and behind the *square cylinder* computed on meshes which are refined based on f_k

meshes, the whole approach is even more practical especially for the industrial CFD users.

3.3 External Car Aerodynamics

There are number of publications dealing with the performance of RANS turbulence models for the external car aerodynamics. In general, the agreement with measurements for the drag coefficient is in the range of (+/−) 3 %. There are exceptional cases where the unsteady effects dominate and the error goes up to 10 %. The recent paper of Jakirlic et al. [13] presents improvements achieved with the PANS in comparisons with RANS and Unsteady RANS (URANS) approaches. It is generally accepted that the prediction of the lift coefficient brings more uncertainties when the RANS models are used. For the car aerodynamics test case, the simplified Volvo VRAK 1:1 experimental model is used. The complete description of the geometry is given in Krajnovic et al. [3]. The inlet velocity is given as 38.9 m/s. The numerical grid has 4.5 million computational cells. Even on so coarse mesh, the PANS model could predict small structures which cannot be captured with the RANS models, see Fig. 6. Predicted lift coefficient is shown in Table 1. A very good agreement with the measured data is achieved. Although the same numerical mesh was used, large improvements in the predicted lift coefficient have been observed.

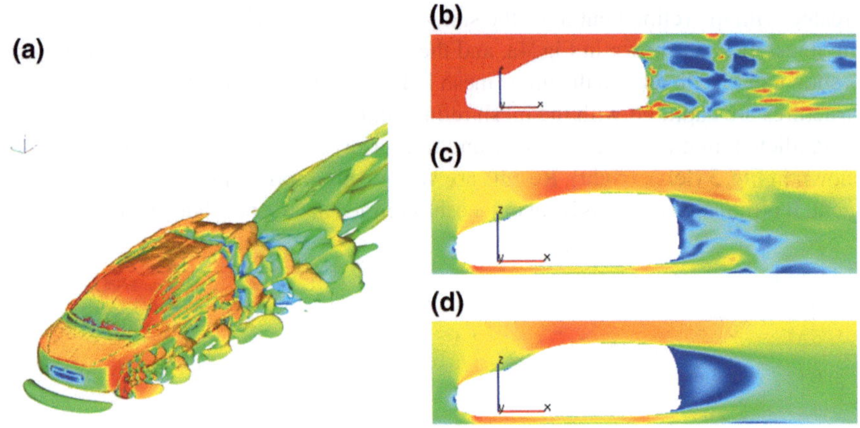

Fig. 6 Predicted flow structures around the car (**a**), f_k (**b**), instantaneous (**c**) and averaged velocities (**d**)

Table 1 Measured and calculated lift coefficient

Data	0.026
RANS k-ε-ζ-f	0.054
PANS k-ε-ζ-f	0.030

Figure 6b shows the resolution parameter f_k which varies from 0 (blue) to 1 (red). This is a typical distribution of f_k, the larger part of a domain in front and above the car is covered by the RANS model and in the wake with the full variable resolution PANS model (note that for $f_k = 1$ the PANS recovers into the RANS approach). Figure 6c shows the averaged velocity as predicted with the PANS. Calculation time needed for PANS is similar to LES calculations; the CPU needed for additional equations is small compared to time needed to average results.

3.4 Intake Port

Typical intake port configuration is shown in Fig. 7. The mesh consisted of 4 million cells have been used. The constant total pressure is specified at the inlet and the static pressure is specified at the outlet. The following parameters in the cylinder were compared:

(a) Torque: $M_t = \sum \rho_i(\vec{U}_i \times \vec{r}_i) \cdot \vec{n} \cdot U_{ax} \cdot A_i$ (r_i is the radius, $\vec{n}$ is the normal vector, U_{ax} axial velocity and A_i stands for the area)
(b) Swirl number: $SN = \frac{n_{Mt}}{n_{Eng}}$ (n_{Mt} is a torque speed and n_{Eng} is an engine speed).

Figure 7 (right) shows an instantaneous parameter f_k which determines the unresolved-to-total kinetic energy ratio which is in the range between 0 and 1. In the region where f_k is equal 1, the RANS k-ε-ζ-f model is used. Following this parameter, it can be seen where the mesh should be refined to further improve results.

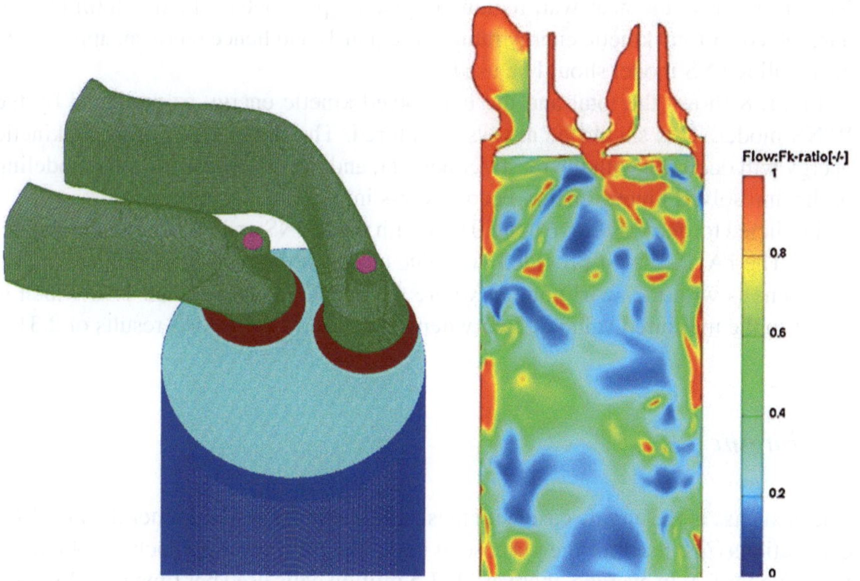

Fig. 7 The mesh for the intake port (*left*) and the predicted instantaneous resolution parameter f_k (*right*)

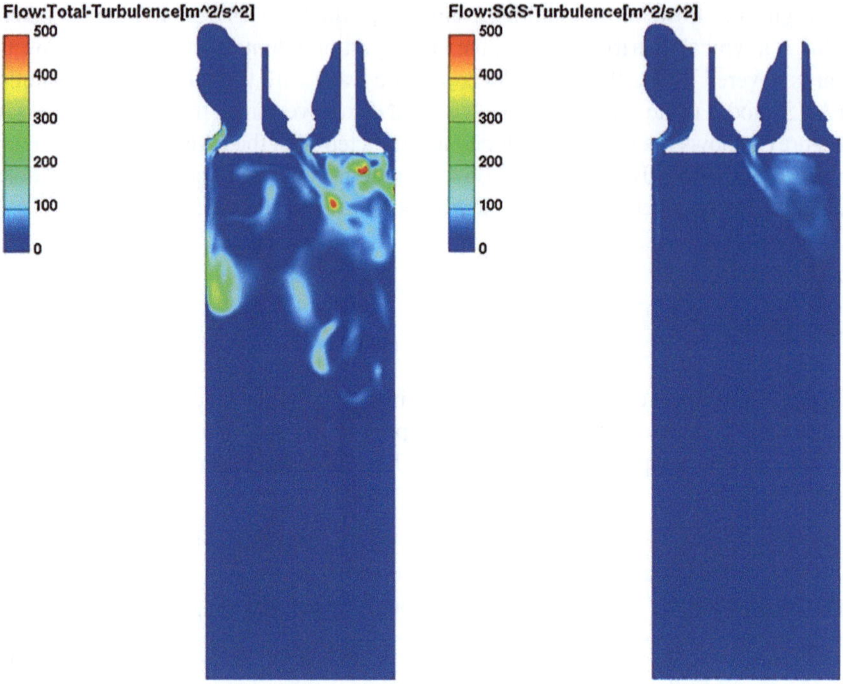

Fig. 8 Predicted total (*left*) and unresolved (*right*) turbulent kinetic energy

Note also that in the near wall region, a dynamic parameter which determines the unresolved-to-total kinetic energy ratio f_k is equal 1, and hence there, an appropriate near wall RANS model should be used.

Figure 8 shows the total and the unresolved kinetic energy as predicted by the PANS model. This should be always monitored. The unresolved (or SGS) kinetic energy will decrease with the mesh refinement, and the model used for the modeling of the unresolved kinetic energy becomes less important.

Predicted torque was 0.0161 and 0.014 with the RANS and PANS models respectively. The PANS k-ε-ζ-f provided the value closer to the measured value of 0.012. The same is with the swirl number where the PANS model reached 1.90 which is closer to the measured value of 1.65 when compared to the RANS results of 2.31.

3.5 Engine

The next case is the engine case with meshes suitable for RANS rather than for LES calculations (just used for the purpose of explaining the new approach for obtaining the value for f_k). In average, there are 1–1.5 million cells used per time step. Figure 9 shows the mesh just for the illustration of the computational domain and the turbulent

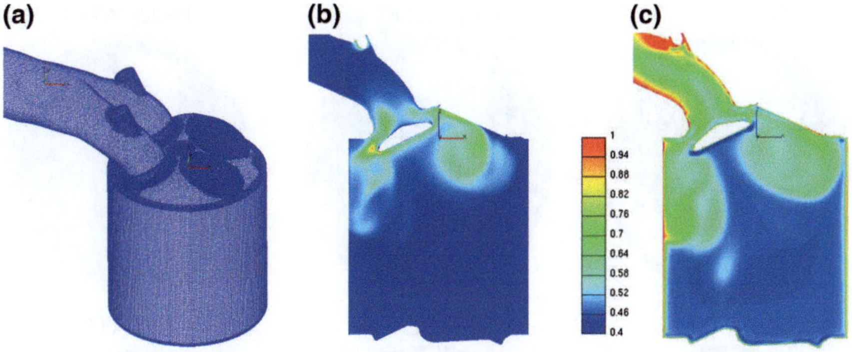

Fig. 9 Engine case: the standard mesh for RANS calculations (**a**), a turbulent kinetic energyas predicted by RANS (**b**) and the resolution parameter predicted by using Eq. (6) (**c**)

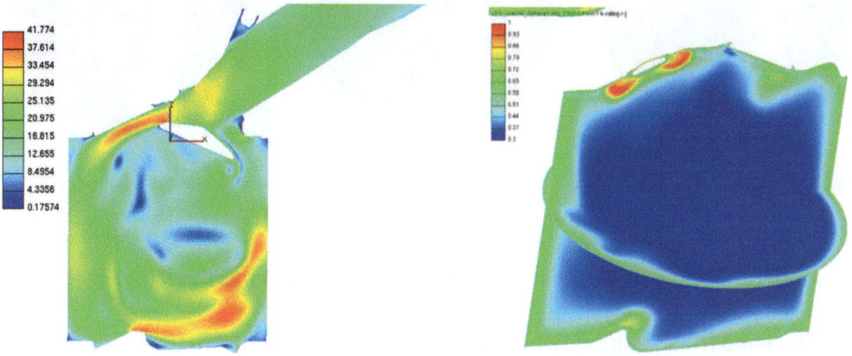

Fig. 10 Predicted instantaneous velocity magnitude (*left*) and the resolution parameter f_k (*right*)

kinetic energy as calculated by the RANS k-ε-ζ-f model (see Fig. 9a, b). This kinetic energy is the total energy represented by the RANS and it is used to estimate the suitability of this mesh for the LES or the hybrid RANS/LES calculations. Figure 9c shows the result of Eq. (6) with the predicted minimum value of 0.4 in the cylinder. It could be said that this is not an appropriate mesh for LES calculations but we have done it just for the purpose of the comparisons with PANS calculations.

In order to avoid averaging the velocity field and calculating the resolved kinetic energy, the formulation of Basara and Girimaji [4] given by Eqs. (16, 17) is used here. Now, this procedure is applied on the full engine case. Figure 10 (right) shows the instantaneous f_k which is different than one based on the RANS calculations shown in Fig. 9c, but overall has similarities. This is of course due to different turbulence level predicted by two approaches; see also the instantaneous velocity predicted by the PANS as shown in Fig. 10 (left). It should be also reported that for this particular mesh, cycle to cycle variations obtained by LES are double larger than obtained with PANS calculations (20 vs. 10 % which is closer to the measured value of 30 %).

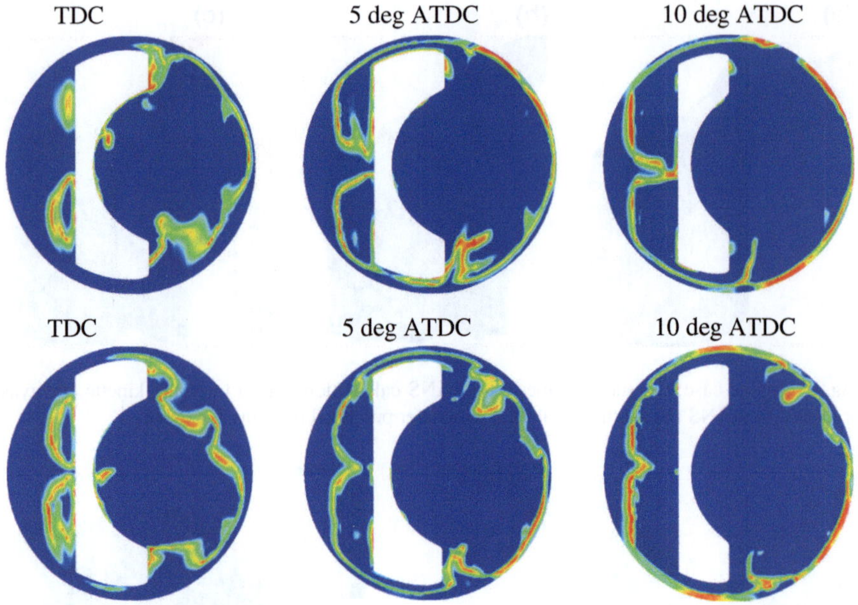

Fig. 11 The flame front position at different time steps as predicted with the PANS model

However, one could expect that these are spurious oscillations caused by inaccuracy of LES on the coarse mesh.

The flame front position as predicted with the PANS is shown in Fig. 11 for three different time steps in two neighbouring cycles. It is visible that the flame front position is different. Note that the combustion modelling was done the same way as for LES, the unresolved kinetic energy obtained by Eq. (8) just replaces the sub-grid scale kinetic energy provided by LES. The rest of modelling is the same.

4 Conclusions

The basis of the PANS method is briefly outlined and some representative results are shown. It should be pointed out that the code which employs the PANS model, in this case k-ε-ζ-f variant, can easily switch from the steady state RANS ($f_k = 1$ and no transient terms), to unsteady RANS ($f_k = 1$ and including transient terms) and to the variable resolution model, the full PANS approach ($f_k \neq 1$). Another advantage of this approach is that all calculations performed up to now have provided better results than those obtained by the RANS model as soon as the finer meshes are employed. The paper also shows the novel approach for calculations on moving meshes. Different variants of the PANS models as well as formulations of the resolution parameters will be tested and further improved.

Acknowledgments It is the author's pleasure to thank Dr. M. Bogensperger for setting up the engine test case.

References

1. AVL FIRE Manual, AVL List GmbH, Graz, Austria. CFD Solver Version 2013
2. Basara, B.: An eddy viscosity transport model based on elliptic relaxation approach. AIAA J. **44**(7), 1686–1690 (2006)
3. Basara, B., Krajnovic, S., Girimaji, S., Pavlovic, Z.: Near-wall formulation of the partially averaged Navier-Stokes turbulence model. AIAA J. **49**(12), 2627–2636 (2011)
4. Basara, B., Girimaji, S.: Modelling of the cut-off scale supplying variable in bridging methods for turbulence flow simulation. In: Proceedings of International Conference on Jets, Wakes, and Separated Flows, Nagoya, Japan, 17–21 Sept 2013
5. Foroutan, H., Yavuzkurt, S.: Partially-averaged Navier-Stokes modeling of turbulent swirling flow. In: American Physical Society, 66th Annual Meeting of the Division of Fluid Dynamics, Pittsburgh, Pennsylvania, 24–26 Nov 2013
6. Girimaji, S., Srinivasan, R., Jeong, E.: PANS turbulence models for seamless transition between RANS and LES; fixed point analysis and preliminary results. ASME Paper FEDSM45336, (2003)
7. Girimaji, S., Abdul-Hamid, K.S.: Partially-averaged Navier-Stokes model for turbulence: implementation and validation, AIAA Paper 2005–0502. Reno, NV (2005)
8. Girimaji, S.S.: Partially-averaged Navier-Stokes model for turbulence: a reynolds-averaged Navier-Stokes to direct numerical simulation bridging method. J. Appl. Mech. **73**, 413–421 (2006)
9. Girimaji, S.S., Wallin, S.: Closure modeling in bridging regions of variable-resolution (VR) turbulence computations. J. Turbul. **14**(1), 72–98 (2013)
10. Hussaini, M.Y., Thangam, S., Woodruff, S.L., Zhou, Y.: Development of a continuous model for simulation of turbulent flows. J. Appl. Mech. **73**, 441–448 (2006)
11. Hanjalic, K., Popovac, M., Hadziabdic, M.: A robust near-wall elliptic-relaxation eddy-viscosity turbulence model for CFD. Int. J. Heat Fluid Flow **25**(6), 1047–1051 (2004)
12. Iwamoto, K., Suzuki, Y., Kasagi, N.: Reynolds number effect on wall turbulence: toward effective feedback control. Int. J. Heat Fluid Flow **23**(5), 678–689 (2002)
13. Jakirlic, S., Kutej, L., Basara, B., Tropea, C.: Computational study of the aerodynamics of a realistic car model by means of RANS and hybrid RANS/LES approaches. SAE Int. J. Passeng. Cars Mech. Syst. **7**(2), (2014). doi:10.4271/2014-01-0594
14. Popovac, M., Hanjalic, K.: Compound wall treatment for RANS computation of complex turbulent flows and heat transfer. Flow Turbul. Combust. **78**, 177–202 (2007)
15. Speziale, C.G.: A combined large-eddy simulation and time-dependent RANS capability for high-speed compressible flows. J. Sci. Comput. **713**(3), 441–448 (1998)
16. Wallin, S., Girimaji, S.S.: Commutation error mitigation in variable-resolution PANS closure: proof of concept in decaying isotropic turbulence. In: 6th AIAA Theoretical Fluid Mechanics Conference, AIAA Paper 2011–3105. Honolulu, Hawaii, 27–30 June 2011
17. Wallin, S., Reyes, D.A., Girimaji, S.S.: Bridging between coarse and fine resolution in variable resolution turbulence computations. In: Proceeding of Turbulence, Heat and Mass Transfer, vol. 7, Palermo, Italy (2012)

Acknowledgments It is the author's pleasure to thank Dr. M. [illegible]

References

Hybrid LES/RANS of Internal Flows: A Case for More Advanced RANS

K. Hanjalić, D. Borello, G. Delibra and F. Rispoli

Abstract The Hybrid LES/RANS is emerging as the most viable modelling option for CFD of real-scale problems, at least in the aerospace design. Entrusting LES to resolve the intrinsic unsteadiness and three-dimensionality in the flow bulk reduces the modelling empiricism to a relatively small wall-adjacent RANS region, arguably justifying the use of very simple models. We argue, however, that for internal flows in complex passages, and involving heat and mass transfer, the role of the near-wall RANS should not be underestimated. The issue is discussed by two examples of flows in turbomachinery: a pinned internal-cooling passage in a turbine blade and tip leakage and wake in a compressor cascade with stagnant and moving casing. The examples illustrate the need for a topology-free wall-integration RANS model that accounts for versatile effects of multiple bounding walls. A HLR using an elliptic relaxation ($v^2/k - f$) RANS model coupled with a dynamic LES showed to perform well in the cases considered.

1 Introduction

"The use of CFD in the aerospace design process is severely limited by the inability to accurately and reliably predict turbulent flows with significant regions of separation. Advances in Reynolds-averaged Navier-Stokes (RANS) modeling alone are unlikely to overcome this deficiency, while the use of Large-eddy simulation (LES) methods will remain impractical for various important applications for the foreseeable future, barring any radical advances in algorithmic technology. Hybrid RANS-LES and

K. Hanjalić (✉)
Delft University of Technology, Delft, The Netherlands
e-mail: khanjalic@gmail.com

K. Hanjalić
Novosibirsk State University, Novosibirsk, Russia

D. Borello · G. Delibra · F. Rispoli
Sapienza University of Rome, Rome, Italy

S. Girimaji et al. (eds.), *Progress in Hybrid RANS-LES Modelling*,
Notes on Numerical Fluid Mechanics and Multidisciplinary Design 130,
DOI 10.1007/978-3-319-15141-0_2

wall-modeled LES offer the best prospects for overcoming this obstacle although significant modeling issues remain to be addressed here as well…" ([15], CFD Vision 2030).

The HLRM conference series has been without any doubt on the right track. The appeal of using a RANS model to overcome the formidable demands for near-wall resolution for LES of complex high-Re-number wall-bounded flows, initiated by the pioneering DES approach of Spalart et al. [18], has triggered a number of ideas and concepts. While it is practically impossible (and, in principle, unimportant) to classify all proposals into distinct categories, in order to assess their physical rationales, potentials and snares, it is useful to distinguish the key presumptions and implications underlining each concept.[1] Arguably, most methods can be grouped into three categories:

- *zonal* RANS/LES that employ separate models in different flow regions, a RANS model close to solid walls (or in the complete attached wall layer), usually modified to match the conditions at the predefined or dynamic interface, and the conventional LES remote from walls;
- *blended* models in which the two methods are blended by a set of continuous (empirical) functions applied to the local turbulent stress tensor or the effective turbulent viscosity in terms of the local characteristic grid size and turbulence length scale;
- *seamless* schemes in which a single model (usually a modified, "sensitized" RANS) is employed throughout the entire flow, which only asymptotically approaches the conventional RANS and LES (or DNS) in the limits of the wall-distance approaching zero or infinity, respectively; the sensitizing of the RANS model is usually accomplished by a "grid-detecting" parameter (computer needs to "feel" the grid and to adjust the model) in term of the typical grid size and the turbulence length scale (as in blended schemes).

Other schemes have also been proposed that could not fit precisely into the above classification. An insightful and a thorough comparison of the rationales of the various concepts, and especially the identification of the common denominators and major conceptual differences, is still lacking. However, extensive scrutiny of various methods in a variety of flows undertaken over the past decade showed varied success, but no indisputable "winner" has emerged yet. Instead, the testing have opened a number of questions, which, irrespective of the models and matching criteria used, to a large degree still remain as actual as ever.

Admittedly, the DES concept (and its variants DDES, IDDES), using the one-equation RANS model of Spalart-Allmaras (SA), is an exception: its robustness and economy have at present no serious competitor when it comes to external aerodynamics at real-scale Re numbers, e.g. a complete airplanes [9], where the only (verifiable) integral parameters (lift and drag) matter. However, its heuristic simplicity and non-transparency, and in particular the use of the wall distance to represent

[1] A comprehensive overview of various approaches and somewhat unorthodox classification can be found in Fröhlich and Terzi [10].

the turbulence length scale in the complete RANS area makes its applicability to geometrically complex internal flows questionable.

This paper attempts to make a case for more advanced RANS in the framework of HLR. It is argued that one of the key prerequisites, especially for internal flows of complex geometric configurations that may also include heat and mass transfer, particle deposition, phase change, near-wall chemical reactions, combustion and flames, as well as other surface-phenomena, is the choice of the adequate RANS model. The arguments are substantiated by discussing two cases relevant to turbomachinery: internal-cooling of gas- turbine blades and tip leakage in a compressor cascade.

2 Importance of RANS in HLR of Complex Internal Flows

Employing LES to resolve large eddies in the flow bulk, expected to reproduce well the intrinsic unsteadiness and the essential large-scale 3D turbulence dynamics, relaxes the burden on the RANS and reduces, in principle, its task to provide only the wall boundary conditions. The modelling empiricism is thus confined to the wall-adjacent, presumed to be relatively small, portion of the flow, arguably justifying the use of very simple one- or two-equation RANS models.

While in many external flows with identifiable wall-attached flow regions, such arguments can be acceptable, they may not hold for simulation of internal flows in complex passages encountered in many engineering flows and especially involving heat and mass transfer and other surface phenomena. One can think of turbomachinery (pinned, ribbed, grooved or dimpled internal blade cooling passages subjected to rotation, blade-tip—casing gaps, labyrinth seals), or IC engines (valves, cooling jackets), compact heat exchangers, electronic packages, and many other examples where complex geometric topology may impede the proper capturing of the near-wall phenomena.

The Re numbers are rarely excessive as in full-scale external aerodynamics, but often moderate and, one may argue, suitable for LES. However, because of multiple bounding walls and a lack of full flow periodicity (even in periodic configurations), wall-resolving LES is still very grid-demanding. On the other hand, because of the full three-dimensionality, intrinsic unsteadiness and complex vortical structures, treating the complete domain with a URANS may not return the essential physics. Thus, a hybridization of LES in the regions away from walls with a RANS in wall-adjacent regions seems for this kind of flows the most feasible option. But as stated above, the RANS model should be thoughtfully chosen bearing in mind a number of flow specificities:

- corrugated configurations with corners, tips, groves, protrusions, and accounting properly for the wall-blocking effect that may permeate over relatively large flow regions, require a model free from topological parameters;

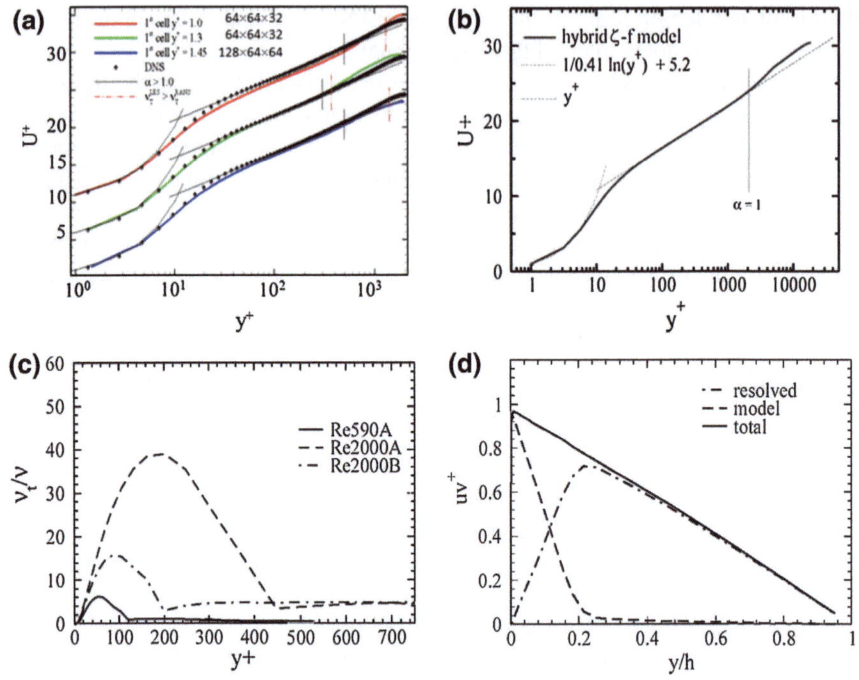

Fig. 1 A HLRM of channel flow: effect of grid clustering and location of RANS and LES interface. **a** Velocity profiles for $Re_\tau = 2{,}000$, **b** for $Re_\tau = 20{,}000$; **c** effective eddy viscosity for two Re numbers and grids; **d** turbulent shear stress

- multiple walls bounding the flows can exert strong influence beyond the thin boundary layers, which pose additional challenge especially if a relatively significant portion of the flow is entrusted to URANS (see Figs. 1 and 3);
- automatic gridding, common in handling complex geometries, may influence the location of the LES/RANS interface if it is based on the local cell size, changing/increasing the URANS region beyond that originally anticipated;
- complex topology will cause local separations, curvature, multiple vortex systems and their interactions, secondary flows, strong stress anisotropy, possible local laminarization (especially in rotating flows), all of which can, at least partially, occur within the URANS region;
- in most flows mentioned the confinement and topological periodicity does not allow long evolving time, the distortion scales are usually comparable with the intrinsic turbulence scales;
- the response of even a simple RANS to LES perturbation was found to be generally satisfactory [19] but it is reasonable to expect that more advanced anisotropy-capturing models will respond better than the more dissipative and diffusive liner one- or two-equation models.

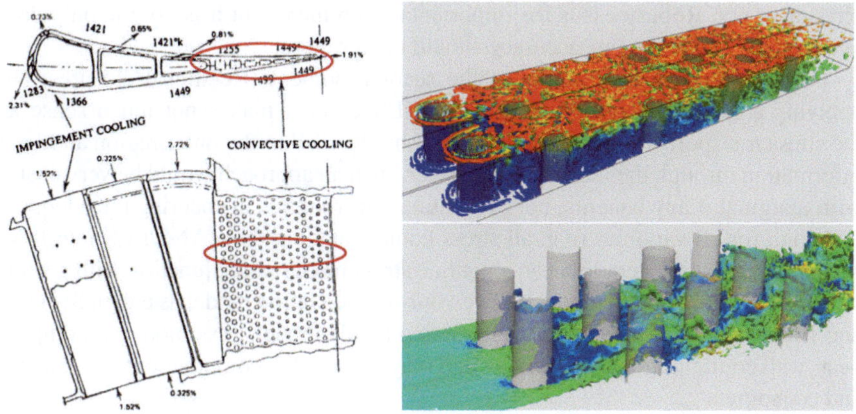

Fig. 2 *Left* a sketch of gas-turbine blade colling. *Right* LES of internal pinned passage (Re = 10,000); *top* vortical structures ($\nabla^2 p = 9$) coloured by temperature (*top* wall heated); *bottom* thermal plumes (T = 294 K) coloured by velocity magnitude (*bottom* wall heated)

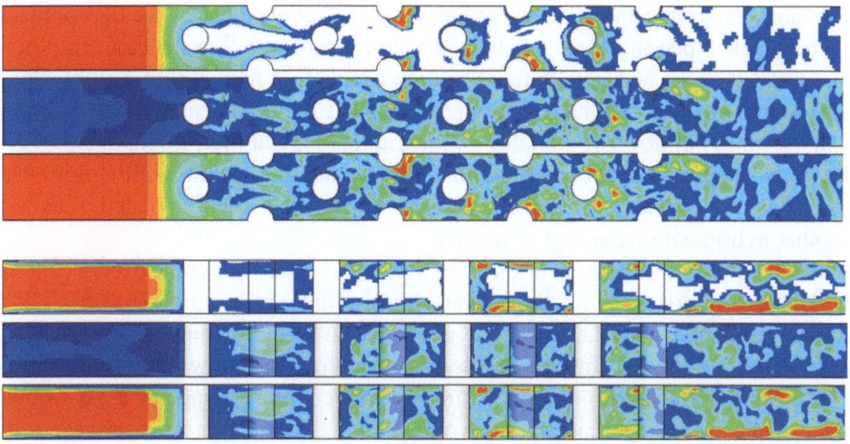

Fig. 3 Viscosity contours in the midplane (z = 1.0) and in the cross-plane ($y = 1.25$) cutting through pins 1, 3, 5 and 7. Pins 2, 4, 6 and 8 (laterally staggered) are also shown (*shaded*). *Top rows* ν_t^{RANS}; *center* ν_t^{LES}; *bottom* effective viscosity ν_t

All these (and other) issues and phenomena place a large responsibility on the RANS model, so that its choice in the HLR context matters. More physics is required and the model should be capable of reproducing versatile effects especially in response to the outer LES. A fringe benefit is that a more capable RANS model allows to extend the RANS region over a larger portion of the flow, thus reducing the demand on the grid size and the computing time. The models should also be more transparent to make it possible for the user to ascertain a priori if the model used contains sufficient physics or not. Robustness is the important requirement, but it is not the only one, and for many applications, accuracy should matter more.

The above requirements hint at the second-moment (Re-stress) closures, but applying a full differential model in the HLR context makes not much sense as the stress transport should be accounted for by the LES in the outer region and feed information through the interface. Moreover, such an approach would be very costly with marginal if any benefits, not to mention uncertainties in ensuring the adequate matching for all variables (e.g. all stress components) at the RANS/LES interface. A more viable option could be an algebraic stress model in conjunction with a wall-integration two-equation model (or even only one-equation model as e.g. in Schmidt and Breuer [16]). However, accommodating an algebraic stress model to complex near–wall configuration is not trivial, and may hamper the computational robustness and economy.

Another, perhaps still more attractive option is the elliptic-relaxation (ER) eddy-viscosity concept of Durbin [8], or one of several more robust model variants subsequently proposed. Although the ER-EVM involves solution of four transport equations, the experience shows that the computational penalty is modest since the equations are very simple and well coupled, ensuring fast convergence. The ER approach has been developed specifically to handle complex near wall flows without empirical damping functions and its elliptic relaxation accounting for inviscid wall blocking has a sound physical rationale. The approach seems thus well suited for HLR, especially for internal flows of complex geometric topology at high Re numbers where the RANS region may occupy more than just the attached portion of the flow. The arguments are considered valid irrespective of whether a zonal, blended, seamless or other hybridization concept is adopted.

In what follows we present briefly a HLR method that employs the ER concept for the URANS region, specifically, a robust linear elliptic relaxation $\zeta(= \upsilon^2/k) - f$ model hybridized with the LES. The performance of the method is illustrated by two examples from turbomachinery: flow and heat transfer in a wall-bounded matrix of staggered cylindrical pins (mimicking internal cooling passage of a gas-turbine blade), and tip leakage and trailing vortices in a linear compressor cascade with stationary and moving casing. The examples illustrate the need for a topology free wall-integration (WIN) RANS model, capable of accounting for versatile effects of the bounding walls. The same approach has earlier been used in the simulation of impinging jet flow and heat transfer [11], flow around short wall-bounded cylinder found in plate-fin-and-tube heat exchangers [2], and, of course, in a plane channel over a range of Re numbers.

3 A HLR with Elliptic Relaxation EVM

The ζ-f model, chosen for the URANS region, follows Durbin's υ^2-f model, but solves a transport equation for the υ^2/k ratio, which requires less stiff wall boundary conditions and proved generally to be more robust than the parent model. As a full

RANS model it proved to be successful and fast converging in a variety of flows (e.g. [12]). The rationale behind the HLR version [6, 11, 13] is to intervene in the sink term of the k-equation with a grid detecting parameter "α" by which the RANS eddy viscosity is damped to the subgrid-scale value of LES at the matching interface. A detailed description of the method can be found in Delibra et al. [6]. The full set of equations is provided in Appendix; here we list only the k-equation and the two limiters that control the switching from one to another model:

$$\frac{Dk}{Dt} = P_k - \alpha\varepsilon + \frac{\partial}{\partial x_j}\left[\left(\nu + \frac{\nu_t}{\sigma_k}\right)\frac{\partial k}{\partial x_j}\right] \tag{1}$$

$$\alpha = \max\left[1; L_{RANS}/L_{LES}\right] \quad \nu_t = \max(\nu_t^{RANS}, \nu_t^{LES}) \tag{2}$$

where k is the (RANS) modelled turbulent kinetic energy, ε its dissipation rate, $L_{RANS} = k^{3/2}/\varepsilon$ and $L_{LES} \equiv \Delta = C_\Delta(\Delta V)^{1/3}$. Close to a wall $\alpha = 1$ and the model acts as in the URANS formulation. Away from walls, where $L_{RANS} > \Delta$, $\alpha > 1$, k is damped, thus diminishing ν_t^{RANS}. Eventually, when $\nu_t^{RANS} < \nu_t^{LES}$, the second constraint is activated and the conventional LES is resumed.

The use of the above defined L_{RANS}, while more physical, may pose some stability problems and in the two examples here presented we used $L_{RANS} = \kappa x_n$, where x_n is the local distance from the nearest wall, in which case $C_\Delta = 1.3$. It is noted that L_{RANS} serves *only* to define the control parameter α, thus entering the turbulence model only in a narrow buffer region where $\alpha > 1$ (usually only few grid nodes) between the URANS and the full LES.

Figure 1 illustrates the performance of the method in a plane channel flow over a range of Re numbers and for different grid density and distribution. The model switching locations ($\alpha > 1$ and $\nu_t^{RANS} = \nu_t^{LES}$) denoted in Fig. 1a show that a relatively large portion of the flow is handled by the URANS, whereas the buffer zone between the two limiters varies depending on the grid clustering, but without noticeable effect on the mean velocity distribution. It is noted that no artificial forcing was used to smoothen the profiles in the buffer region.

4 Examples 1: Pinned Matrix Bounded by Heated Walls

Figure 2 shows a sketch of the typical gas-turbine blade cooling system in which the interior of the trailing part is fitted by a matrix of staggered cylindrical pins. The pins act primarily as promoters of vortex shedding and turbulence to enhance blade cooling by interior cold air flow. A simplified setup with parallel, differentially heated walls mimicking the experiment of [1] for two Re numbers (10^4 and 3×10^4) was simulated by URANS, LES and HLR aimed at testing the optimum computational strategy. Details have been described in Delibra et al. [5–7]. The complexity of the vortical and plume structures that govern heat transfer and its enhancement is illustrated by wall-resolved LES (available only for Re $= 10^4$) in Fig. 2 right. We discuss briefly some results obtained by HLR with moderate grids: 1.3×10^6 and 4.4×10^6 respectively for Re $= 10^4$ and 3×10^4, compared with 5×10^6 and 15×10^6 for LES (the latter grid proving insufficient). Regardless of the models applied, a much coarser grid used in the HLR cannot reproduce the wide spectrum of scales shown in Fig. 2 right, especially if URANS covers substantial portion of the flow, as illustrated in Fig. 3. But the key question is how important are the unresolved small scales for predicting the wall heat transfer.

Figures 4 and 5 show the time averaged distribution of the Nusselt number on the heated wall, compared with the experiments (Fig. 4) and also with URANS using the same ζ-f model in the whole domain. The HLR showed obviously superior results, and also (for Re $= 10^4$) in close agreement with the LES [6, 7].

Although in both the HLR and URANS the same (RANS) model is responsible for the wall heat transfer, the key difference is in the intensity of the outer forcing that influences the separation dynamics of vortex shedding especially behind the first pin row. As shown in Fig. 6, the velocity time records at a monitoring point just behind the pin (inset) show the HLR amplitude (well in agreement with the

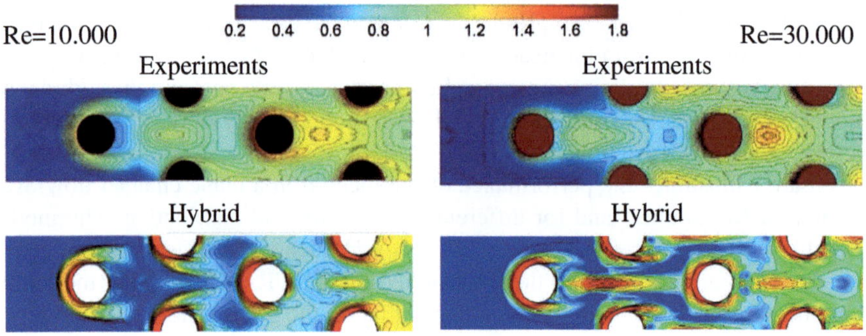

Fig. 4 Time averaged Nuseel number on the heated wall (the first 4 pins) (normalised with the area-averaged Nu_{av})

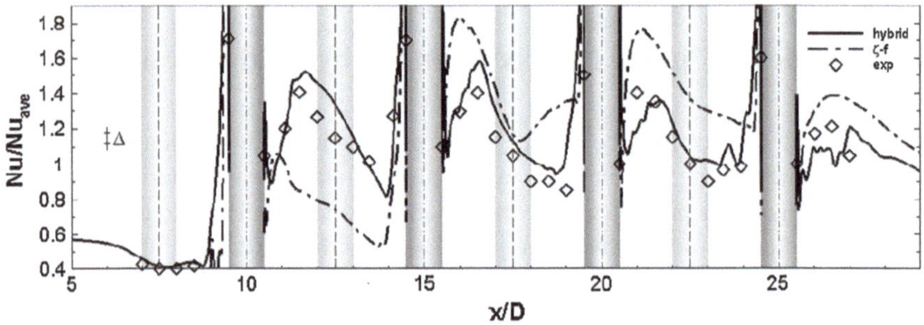

Fig. 5 Nusselt number Nu/Nu_{av} on the heated endwall along the line cutting pins 2, 4, 6 and 8 for $Re = 10^4$. From Delibra et al. [6]. *Symbols* experiments, Ames et al. [1]

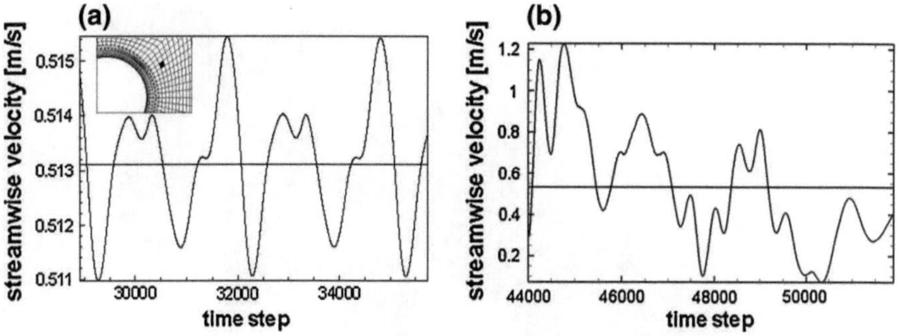

Fig. 6 Velocity fluctuations in the near-wake of the 1st pin (*inset*) in the midplane ($z = D$). **a** URANS, linear ζ-f, St = 0.24, mean velocity 0.513 m/s, maximum amplitude: 0.004 m/s. **b** Hybrid ζ-f-LES, St = 0.24, mean velocity 0.532 m/s, Maximum amplitude: 0.6 m/s

experiments, [6, 7]) to be two orders of magnitude larger than in the URANS, while both methods reproduce equally well the Strouhal number. Perhaps a more advanced RANS (nonlinear EVM or RSM) could do better but the here applied linear ζ-f EVM is too simple to capture the instabilities and complex 3D vortical structures around and behind the pins, as illustrated in Fig. 7.

5 Examples 2: Tip Leakage and Trailing Vortices in Compressor Cascade with Stagnant and Moving Casing

The efforts to reduce the unproductive, loss-bearing but unavoidable blade-tip leakage and secondary flows rely at present much on the CFD. A sketch of a simplified low-speed compressor cascade with a casing, Fig. 8, illustrates the critical phenomena that pose challenge to modelling [20]. We discuss briefly some salient flow features and challenges for CFD and show some results for stagnant and moving casing using the HLR and URANS. More details can be found in Borello et al. [3, 4].

Fast distortion dominated by pressure and inviscid effects in the tip leakage makes the results insensitive to turbulence modelling (Fig. 10), though the model plays a role in capturing the boundary layer properties on the moving casing. However, the formation and development of the tip vortex and its interaction with the blade wake (Fig. 9) puts a high demand on the model.

Figure 10 illustrates the insensitivity of computation within the tip clearance to the model used: the URANS and HLR give similar results, close to the experiments for the U and V component, though less satisfactory for W, presumably due to inadequate meshing or imperfect mimicking of the clearance configuration). The choice of the model affects, however, the formation and development of the tip vortex, the HLR capturing a broader spectrum of the scales even compared with the non-linear URANS, especially for moving casing, Fig. 11.

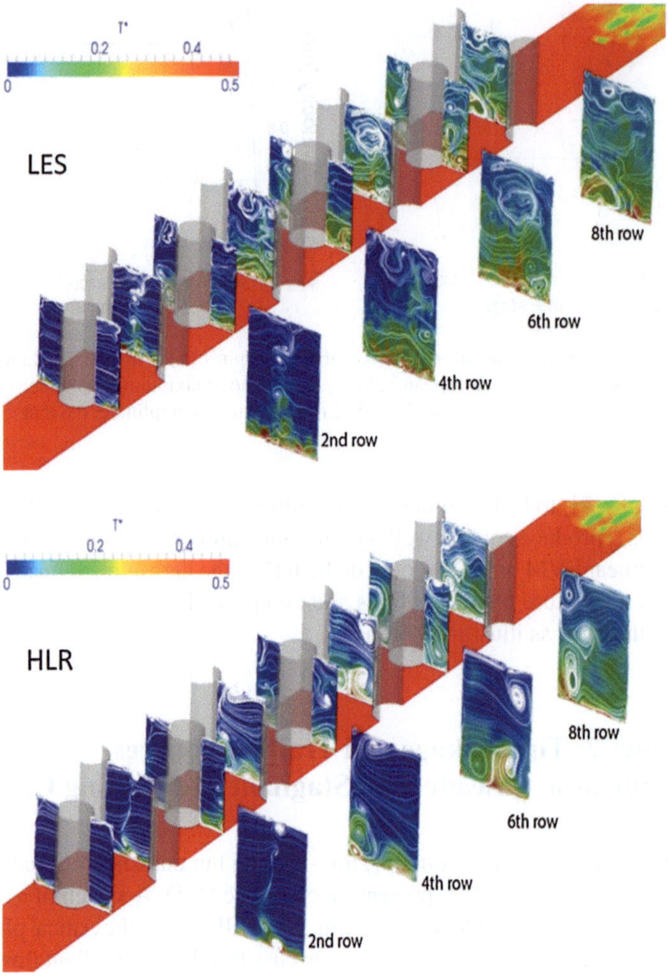

Fig. 7 Instantaneous streamlines and temperature field (*colours*) in the cut-planes show close structural similarity between LES and HLR

The superiority of the HLR over the URANS is further illustrated in Fig. 12 showing a view of vortical structures and streamlines for the case of moving casing. The hybrid simulation shows stronger, longer-lived, elongated structures in the trails of both the tip and hub vortices, expected to affect the wake. Both methods capture a horseshoe vortex at the hub, but surprisingly, the URANS returns a large and stronger recirculation immediately downstream the blade.

The results for the wake, however, showed to be more model-sensitive as illustrated in Figs. 13 and 14. The experimental data available in the planes indicated by red-quadrangles in Fig. 8, make it also possible to provide some quantitative test of the performances of the models considered. Figure 13, depicting the wall-normal

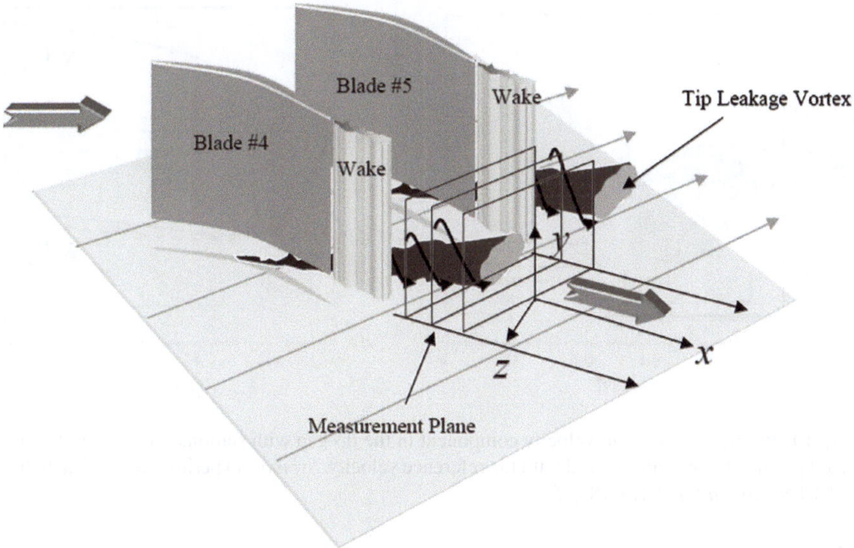

Fig. 8 A sketch of the cascade and tip lekage vortices; measuring planes indicated by *red* qudrangles behind the cascade. From Wang and Devenport [20]

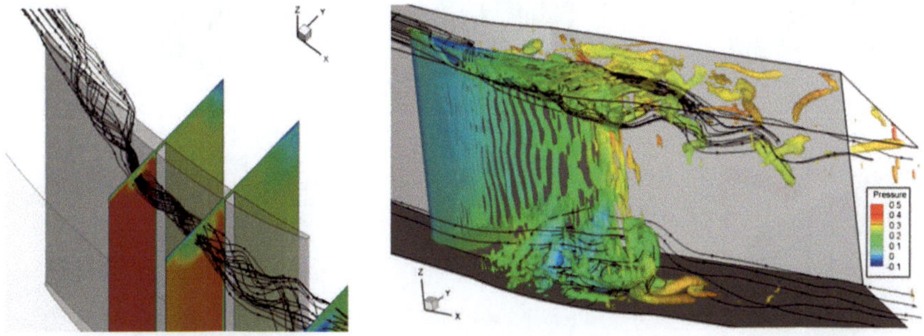

Fig. 9 Instantaneous HLR streamilnes over the blade tip (*left*) and vortical structures on the blade, over the tip and the hub identified by pressure-coloured isosurfaces of $\nabla^2 p = 60$ (*right*), over a linear compressor cascade with moving casing [4]

velocity field in the cross-plane at the first measuring plane ($x/c_a = 1.37$), shows that only the HLR captures properly the locations as well as the velocity intensity in the two counter-rotating tip-leakage vortices. The linear and non-linear ζ-f RANS models, when applied in the whole domain (URANS computations), show very similar patterns in between, as well as the vortices strengths and their distance as in the experiment, but shifted laterally towards the suction side, indicating at some wake and vortices deflection.

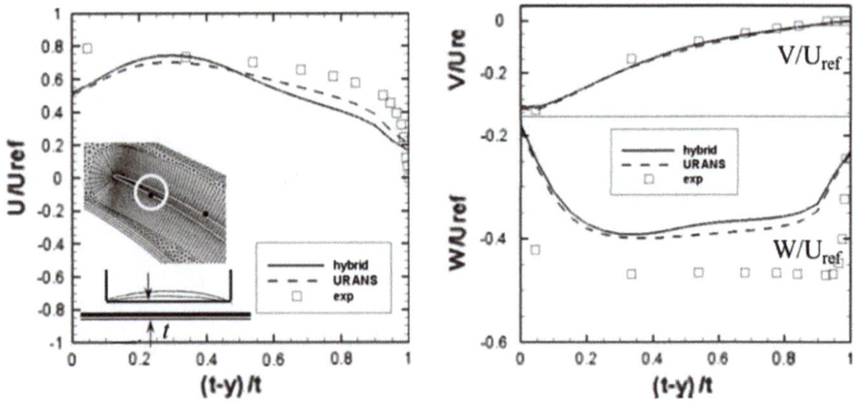

Fig. 10 Profiles of the mean velocity component in the tip-gap with stationary casing at the indicated position (*inset*), normalised with the reference velocity. *Symbols* experiments, *full line* hybrid ζ-f-LES; *dashed line* URANS ζ-f

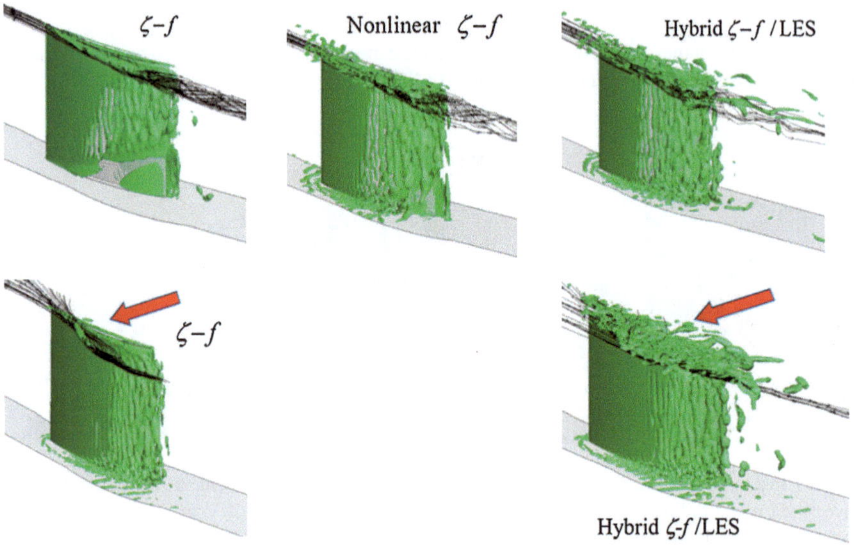

Fig. 11 Vortical structures in and behind the tip leakage, identified by Q = 35 isosurfaces for different models. *Top* stationary casing; *bottom* moving casing

The axial velocity fields, presented in Fig. 14 for two cross-planes (at $x/c_a = 1.37$ and 1.51), show similar agreement with the experiment, both in the intensity and distribution—though admittedly less satisfactory for $x/c_a = 1.51$. The velocity contours clearly disply the blade wakes (vortical patterns) and the tip vortices between the blades, close to the casing wall.

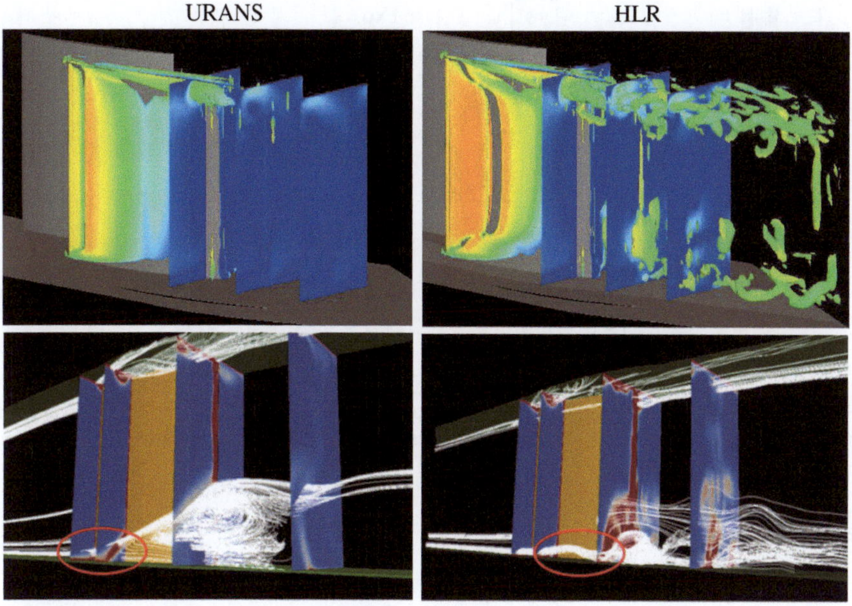

Fig. 12 URANS (*left*) and Hybrid LES/RANS (*right*) of flow in a linear compressor cascade with moving casing. *Top* Vortex structure (Q), coloured by velocity; *bottom* instantenous streamline pattern illustrating the horseshoe vortex (*encircled*), with focus on the hub vortex

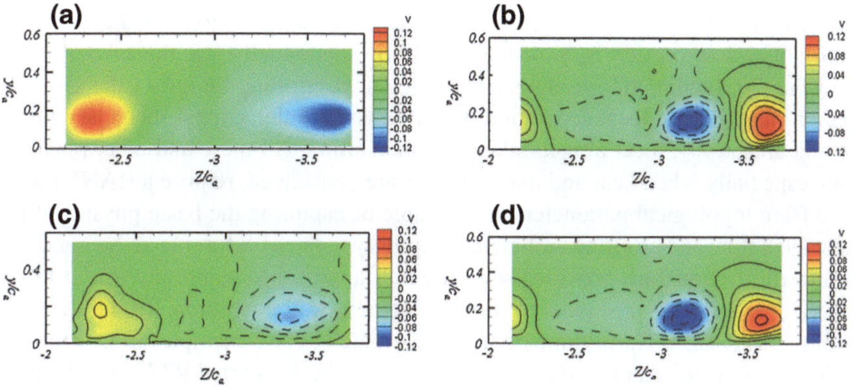

Fig. 13 Wall-normal velocity in the tip-vortex wake at $x/c_a = 1.37$. **a** Experiments Muthana and Devenport [14]. **b** Linear ζ-f RANS; **c** Hybrid linear ζ-f/LES; **d** Nonlinear ζ-f RANS

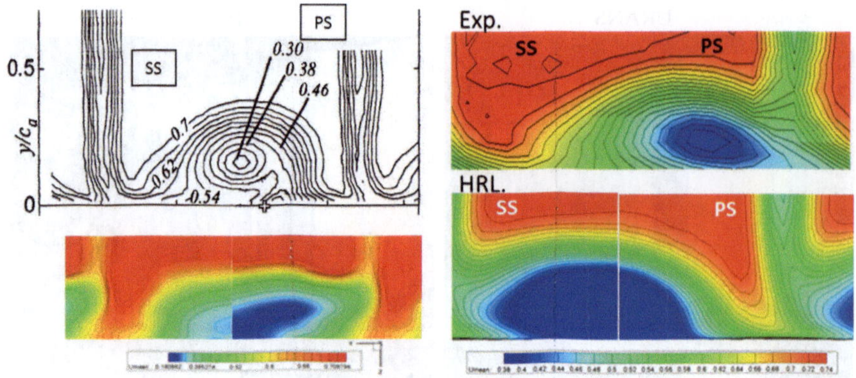

Fig. 14 Axial velocity in wake of the tip vortex for stationary casing at $x/c_a = 1.37$ (*left*) and for moving casing at $x/c_a = 1.51$ in a linear compressor cascade. *Top row* experiments; *bottom row* Hybrid ζ-f

6 Conclusions

The paper attempts to make a case for using more advanced and transparent RANS models in the Hybrid LES/RANS methods when targeting internal flows of industrial relevance. Most flows in questions are featured by complex geometrical configurations, full three-dimensionality and intrinsic unsteadiness even when the flow is steady in the bulk. The omnipresent bounding walls exert especially strong inviscid (blocking) effects that may permeate over large regions of the flow. If a substantial flow portion is entrusted to RANS, intentionally (for the sake of economy) or not (automatic gridding, difficulties in controlling the grid), the URANS region may include local separation, multiple vortex interactions, secondary flows, strong anisotropy, local transition and laminarization. All these and other phenomena, especially when heat and mass transfer are considered, require a RANS model free from topological parameters and capable of capturing the basic physics of the near-wall flow and ensuring sufficient receptivity to the LES forcing at the interface. Some illustrations are provided with a zonal approaches using an elliptic relaxation, $\zeta(= v^2/k) - f$, model to provide the near-wall conditions for the dynamic Smagorinsky LES. The arguments are substantiated by two examples from turbomachinery flows and heat transfer. It is noted that the here reported RANS can be used irrespective of whether a zonal, blended, seamless or other hybridization concept is adopted.

We further argue that reducing empirical contents and making the model more physically transparent and more understandable should be the general requirements, at least for treating internal flows of complex topology. Robustness and economy matter, but should not be decisive since, for credible prediction of unknown situations, physics matters more. Advanced and physically sounder models are inevitably numerically more challenging, require more computational time and effort, but often only by a margin. Admittedly, curing the robustness problem requires more skill, but serious design and optimization tasks should not be entrusted to non-specialist anyhow, except perhaps, for preliminary considerations and rough estimates.

As for an outlook, in the authors' opinion the CFD community has long been entrenched by too many constraints of the conventional rigid Navier-Stokes eddy-viscosity Ansatz and segregated solvers, which have in general hampered the use of more advanced models. The stability issues should be resolved by innovative algorithms tailored for specific modelled equation sets not necessarily coupled by eddy viscosity (e,g. second-moment closures) rather than insisting on simpler turbulence models. Arguing in favour of most simple heuristic turbulence models because they "work" and that all what matters is simplicity, robustness and economy, may be paralleled by an argument that one should stick to a simple first-order upwind convection scheme because it always "works". As noted by Spalart [17], "The more capable the RANS component is, the lower costs of the hybrid computations will be. Therefore, the switch to LES in some regions does not remove the incentive to further the RANS technology".

Acknowledgments This work was in part performed in the framework of the Lead Scientists Grant from the Government of Russian Federation (Grant No. 11.G34.31.0046, K. Hanjalić).

Appendix: The HLR ζ-f Model

$$\frac{Dk}{Dt} = \mathcal{D}_k + P_k - \alpha\varepsilon; \qquad \frac{D\varepsilon}{Dt} = \mathcal{D}_\varepsilon + \frac{C_{\varepsilon 1} P_k - C_{\varepsilon 2}\varepsilon}{\tau}$$

$$\frac{D\zeta}{Dt} = \mathcal{D}_\zeta + f - \frac{\zeta}{k} P_k; \qquad L^2 \nabla^2 f - f = \frac{1}{\tau}\left(C_1 + C_2 \frac{P_k}{\varepsilon}\right)\left(\zeta - \frac{2}{3}\right)$$

$$\alpha = \max\left(1, \frac{L_{RANS}}{\Delta}\right); \quad L_{RANS} = k^{3/2}/\varepsilon;$$

$$\Delta = C_\Delta (\Delta V)^{1/3}; \quad \mathcal{D}_\phi = \frac{\partial}{\partial x_j}\left[\left(\nu + \frac{\nu_t}{\sigma_\phi}\right)\frac{\partial \phi}{\partial x_j}\right];$$

$$\nu_t^{RANS} = c_\mu \tau \zeta k; \qquad \nu_t^{LES} = (c_s \Delta)^2 \overline{S}; \qquad \nu_t = \max(\nu_t^{RANS}, \nu_t^{LES})$$

$$\tau = \max\left[\min\left(\frac{k}{\varepsilon}, \frac{0.6}{\zeta c_\mu \sqrt{6 S^2}}\right), c_\tau \left(\frac{\nu}{\varepsilon}\right)^{1/2}\right];$$

$$L = c_L \max\left[\min\left(\frac{k^{3/2}}{\varepsilon}, \frac{k^{1/2}}{\zeta c_\mu \sqrt{6 S^2}}\right), c_\eta \left(\frac{\nu^3}{\varepsilon}\right)^{1/4}\right]$$

c_μ	$C_{\varepsilon 1}$	$C_{\varepsilon 2}$	C_1	C_2	σ_k	σ_ε	σ_ζ	c_τ	c_η	c_L	c_Δ
0.22	$1.4(1+0.012/\zeta)$	1.9	0.4	0.65	1.0	1.3	1.2	6	85	0.36	1.5

References

1. Ames, F.E., Nordquist, C.A., Dvorak, L.A.: Endwall heat transfer measurements in a staggered pin-fin array with an adiabatic pin. In: Proceedings of GT2007 ASME Turbo Expo, Montreal, Canada (2007)
2. Borello, D., Delibra, G., Hanjalić, K., Rispoli, F.: LES and hybrid LES/RANS study of flow and heat transfer around a wall-bounded short cylinder. In: Peinke, J., Oberlack, M., Talamelli, A. (eds.) Progress in Turbulence III, Springer Proceedings in Physics, vol. 131, pp. 147–150 (2008)
3. Borello, D., Delibra, G., Hanjalić, K., Rispoli, F.: Large-eddy simulations of tip leakage and secondary flows in an axial compressor cascade using a near-wall turbulence model. Proc. Inst. Mech. Engs, Pt A J. Power Energ. **223**(A6 SI), 645–655 (2009)
4. Borello, D., Delibra, G., Hanjalic, K., Rispoli, F.: Hybrid LES/RANS study of turbulent flow in a linear compressor cascade with moving casing. In: Paper GT2010-23755, Proceedings of ASME Turbo Expo 2010, Glasgow, UK (2010)
5. Delibra, G., Borello, D., Hanjalić, K., Rispoli, F.: URANS of flow and endwall heat transfer in a pinned passage relevant to gas-turbine blade cooling. Int. J. Heat Fluid Flow **30**, 545–560 (2009)
6. Delibra, G., Hanjalić, K., Borello, D., Rispoli, F.: Vortex structures and heat transfer in a wall-bounded pin matrix: LES with a RANS wall treatment. Int. J. Heat Fluid Flow **31**(5), 740–753 (2010)
7. Delibra, G., Borello, D., Hanjalić, K., Rispoli, F.: An LES insight into convective mechanism of heat transfer in a wall-bounded pin matrix. In: Paper IHTC14-23205, Proceedings of 14th International Heat Transfer Conference, Washington. D.C., USA, 8–13 Aug 2010
8. Durbin, P.: Near-wall turbulence closure modelling without 'damping functions'. Theor. Comput. Fluid Dyn. **3**, 1–13 (1991)
9. Forsythe, J.R., Squires, K.D., Wurtzler, K.E., Spalart, P.R.: DES of fighter aircraft at high alpha. In: AIAA Paper, 2002–0591 (2002)
10. Fröhlich, J., Von Terzi, D.: Hybrid LES/RANS methods for ten simulation of turbulent flows. Prog. Aerospace Sci. **44**, 349–377 (2008)
11. Hadžiabdić, M.: LES, RANS and combined simulations of impinging flows and heat transfer. Ph.D. Thesis, Delft University of Technology, The Netherlands (2006)
12. Hanjalić, K., Popovac, M., Hadžiabdić, M.: A robust near-wall elliptic relaxation eddy viscosity turbulence model for CFD. Int. J. Heat Fluid Flow **25**(6), 1047–1051 (2004)
13. Hanjalić, K.: Will RANS survive LES? A view of perspectives. ASME J. Fluids Eng. **127**, 831–839 (2005)
14. Muthanna, C., Devenport, W.J.: Wake of a compressor cascade with tip gap, Pt 1: mean flow and turbulence structure. AIAA J. **11**, 2320–2331 (2004)
15. Slotnick, J., Khodadoust, A., Alonso, J., Darmofal, D., Gropp, W., Lurie, E., Mavriplis, D.: CFD Vision 2030 Study: A Path to Revolutionary Computational Aerosciences, Contract NNL08AA16B, Task NNL12AD05T (2013)
16. Schmidt, S., Breuer, M.: Hybrid LES-URANS methodology for the prediction of non-equilibrium wall-bounded internal and external flows. Comp. Fluids **96**, 226–252 (2014)
17. Spalart, P.R.: Strategies for turbulence modelling and simulations. Int. J. Heat Fluid Flow **21**, 252–263 (2000)
18. Spalart, P.R., Jou, W.-H., Strelets, M., Allmaras, S.R.: Comments on the feasibility of LES for wings, and on a hybrid RANS/LES approach. In: Liu, C., Liu, Z. (eds.) Advances in DNS/LES. Greyden Press, OH, USA (1997)

19. Temmerman, L., Leschziner, M., Hadžiabdić, M., Hanjalić, K.: A hybrid two-layer URANS-LES approach for large-eddy simulation at high Reynolds numbers. Int. J. Heat Fluid Flow **26**, 173–190 (2005)
20. Wang, Y., Devenport, W.J.: Wake of a compressor cascade with tip gap. Part2: effects of endwall motion. AIAA J. **11**, 2332–2340 (2004)

Universal Reynolds Number of Transition and Derivation of Turbulent Models

V. Yakhot, C. Bartlett, H. Chen, R. Shock, I. Staroselsky and J. Wanderer

Abstract Renormalization or coarse-graining applied to basic equations governing multi -scale phenomena, leading to *effective* equations for large-scale properties is often called *model-building*. Unlike fluids in thermodynamic equilibrium, in case of high-Reynolds number turbulent flows the procedure leads to generation of an infinite number relevant high-order nonlinearities which are hard to deal with. In this paper, based on the recently discovered universality of transition to strongly non-Gaussian (anomalous) statistics of velocity derivatives, we show that in the infrared limit $k \to 2\pi/L$, where L is the integral scale corresponding to the top of inertial range, the lowest-order contributions to the renormalized perturbation expansion give asymptotically exact equations for the large-scale features of the flow. The quality of the derived models is demonstrated on a few examples of complex flows. At the small scales $\Delta < L$, an infinite number of $O(1)$ non-linear terms, generated by the procedure invalidate low-order models widely used for Large-Eddy-Simulations (LES) of turbulent flows.

1 Introduction

To describe all details of a typical atmospheric flow, one has to deal with at least $N \approx 10^{25}$ degrees of freedom which in foreseeable future is "mission impossible". In engineering applications, often, one is mainly interested in the large-scale flow-features and, therefore, iterative averaging over small-scale high-frequency modes may, in principle, lead to the coarse-grained equations. In the end of the sixties computers reached a critical power enabling direct numerical simulations (DNS) of strong turbulence. This was realized by Orszag in 1968–1969 who developed the first

V. Yakhot (✉) · C. Bartlett
Department of Mechanical Engineering, Boston University,
110 Cummington Street, Boston, MA 02215, USA
e-mail: vy@bu.edu

H. Chen · R. Shock · I. Staroselsky · J. Wanderer
EXA, 55 Network Drive, Burlington, MA01803, USA

S. Girimaji et al. (eds.), *Progress in Hybrid RANS-LES Modelling*,
Notes on Numerical Fluid Mechanics and Multidisciplinary Design 130,
DOI 10.1007/978-3-319-15141-0_3

numerical (spectral) method [14]. The first results of a "numerical experiment" on isotropic and homogeneous turbulence (HIT) were published by Orszag and Patterson in 1972. It became clear that even in the simplest case of HIT the spatial resolution requirements scaled at least as $N \propto Re^{\frac{9}{4}}$ and computational work $W \propto Re^3$. This means that even mere doubling of the Reynolds number required a new generation of hardware.

Almost simultaneously, Dearddorff introduced the concept of Large-Eddy-Simulations (LES) of turbulence [4]. On the first glance his prescription was strait-forward: since one is interested in the flow—features on the scales $l \geq \Delta$, where Δ is the size of computational mesh, write the model consisting of the Navier-Stokes equations with effective viscosity $\nu(\Delta) \gg \nu$. This way the unresolved ("sub-grid") small-scale ($l \leq \Delta$) fluctuations are overdamped while the large-scale dynamics can be accurately represented by the model. In case of success the method was supposed to give $W = O(Re^0)$. Based on dimensional reasoning Deardorff wrote $\nu(\Delta) \propto |S_{ij}|\Delta^2 \approx |u(x+\Delta) - u(x)|\ \Delta$ where the local rate-of-strain S_{ij} was calculated on the resolved (simulated) velocity field. Using this model, combined with the "mixing length" representation of turbulent viscosity, Deardorff conducted first LES of a channel flow and atmospheric boundary layer [4, 13].

In 1972 Launder and Spalding, generalizing Kolmogorov's ideas, came up with the two—equation model for description of the time-averaged flow-features and applied it to different flows of engineering interest [13]. Thus, by beginning of the seventies, the three methods for description of velocity fluctuations on different ranges of scales have been proposed.

The first DNS accurately describing a low-Reynolds number flow using $N \approx 64^3$ modes, demonstrated possibilities of the method. Modern computers, able to deal with $N \approx 4096^3$ modes, led to extremely accurate representation of fine structure of turbulence like intermittency, precise description of turbulence decay, dynamics of turbulence production in wall flows, structures in a convection cell etc. It is clear that DNS evolved into a powerful "experimental" tool.

The "two-equation" $\mathscr{K} - \mathscr{E}$ model and its time-dependent VLES variation, became an indispensable part of an engineering design cycle. The total 2012 annual sales in the rapidly growing market of commercial CFD codes (fluid-structure interaction excluded) was: $s \approx 5 \times 10^8 USD$. Only 3–4 % of designers use LES. The main problem with all existing LES schemes is illustrated on Fig. 1 where the results of the LES and DNS of HIT are compared. The DNS of HIT in a periodic box was conducted by Wanderer and Oberai [18] using $N = 256^3$ mesh points. Then, they run the variational dynamic LES model on a $N = (2\pi/\Delta)^3 = 32^3$ mesh. On Fig. 1 the ratios of structure functions computed by the two methods are presented:

$$s_p(N) = \frac{S(N\Delta)_{p,LES}}{S(N\Delta)_{p,DNS}} = \frac{\overline{((u^{LES}(N\Delta) - u^{LES}(0))^p)}^{\frac{1}{p}}}{\overline{((u^{DNS}(N\Delta) - u^{DNS}(0))^p)}^{\frac{1}{p}}}$$

If the LES scheme is accurate, the parameter $s_p(N\Delta)$ must be close to unity independently of the moment order p and displacement $l = N\Delta$. As one can see, at the

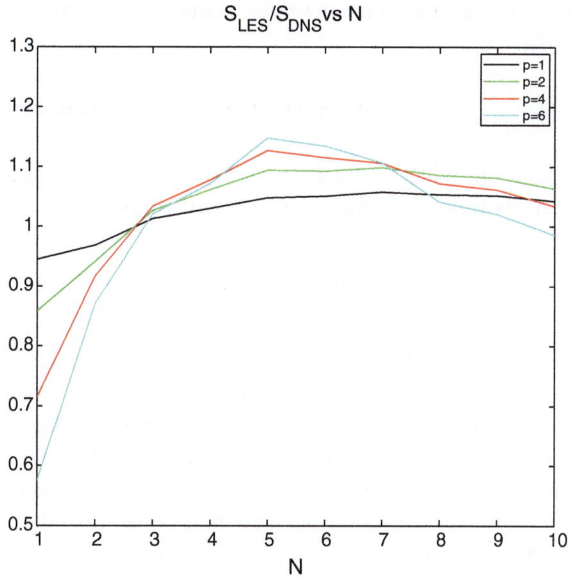

Fig. 1 Comparison of the structure functions $S_p(N\Delta)$ computed by two different methods: LES and DNS [18]

integral scale $N\Delta \approx 10$, the results are not bad. However, they rapidly deteriorate with decrease of the length scale $N\Delta$ and increase of the moment power p, meaning that the model is not capable of predicting powerful small-scale ($l \approx \Delta$) flow features. This makes LES very expensive, poorly scaling with the Reynolds number.

A typical LES is formulated in terms of a "small-scale-filtering procedure" removing unresolved or "sub-grid" modes. The resulting ad hoc equation is simply written down by **postulating** expression for effective viscosity in the Navier-Stokes-like equations for resolved velocity fluctuations, usually taken as $\nu_{SG} = a|S|\Delta^2$ with numerical factor a obtained from various local dynamic constraints. No systematic derivation of sub-grid models has ever been reported. The above relation is result of Kolmogorov's theory stating that effective viscosity, acting on the "slow" large-scale modes ($l \geq \Delta$), coming from interactions with small-scale ($l \leq \Delta$) fluctuations, is

$$\nu(\Delta) \approx (\overline{\mathscr{E}}\Delta^4)^{\frac{1}{3}}$$

The Smagorinsky model is obtained by *dropping the averaging procedure altogether* and instead, writing $\mathscr{E} = \nu(\Delta)S_{i,j}^2$ where the dissipation rate is not a mean but local fluctuating flow property. Due to intermittency this "simplification" leads to disastrous consequences discussed below in detail.

Disregarding intermittency, a turbulent flow is characterized by two length scales: integral $L \approx 1/\Lambda_f$ where energy is pumped into the "inertial range" and dissipation scale $\eta = 1/\Lambda_0$ at which viscous effects balance the non-linearity. *We would like to stress that even in the low Reynolds number flows, driven by the large-scale forcing, the small-scale dynamics of the viscous range are strongly non-linear.* This statement becomes clearer since the Fourier-transform of the Navier-Stokes equations can be written as:

$$(-i\omega + \nu k^2)\mathbf{u}(\mathbf{k}, \omega) = -NL(\mathbf{k}, \omega) + \mathbf{F}(\Lambda_f, \omega)$$

where, introducing the 4-vector $\hat{k} = (\mathbf{k}, \omega)$, the Fourier-transform of the nonlinear contribution to the Navier-Stokes equations reads:

$$NL(\mathbf{k}, \omega) = -\frac{i}{2}\mathscr{P}_{lmn}(\mathbf{k}) \int u_m(\hat{q})u_n(\hat{q} - \hat{k})d\hat{q}$$

In this equation $\mathscr{P}_{lmn} = k_m P_{ln}((k) + k_n P_{lm}(k))$ and projection operators $P_{ij} = \delta_{ij} - \frac{k_i k_j}{k^2}$. If turbulence is produced at the large scales, the forcing $\mathbf{F}(\mathbf{k}, \omega)$ is defined on the interval $k \approx \Lambda_f$ and in the "dissipation range" where $k \gg \Lambda_f$ and $F = 0$, the non-linearity is the only source of the small-scale excitations responsible for the energy balance. Thus, even in a flow with no developed inertial range, the small-scale $(k > \Lambda_f)$ dynamics are strongly non-linear. (Similar argument has been developed by Kuzmin and Patashinsky for the initial value problem in 1991 [11, 19]).

According to Kolmogorov's theory, neither L nor η can appear in the expression for the energy spectrum in the "inertial range". If this is so, the energy spectrum must be $E(k) \propto \mathscr{E}^{\frac{2}{3}} k^{-\frac{5}{3}}$. Due to the infra-red divergencies of renormalized perturbation expansions of turbulence theory, this qualitatively appealing result has never been derived directly from the Navier-Stokes equations. It is well-known that each term in Wyld's diagrammatic expansion is infra-red (i.r.) divergent, i.e. tends to infinity together with integral scale $L \approx 1/\Lambda_f$ [19]. Among the earliest attempts to "save" the theory was Kraichnan's Lagrangian History Direct Interaction Approximation (LHDIA) which was basically a one-loop closure written in Lagrangian coordinates eliminating the effects of a transport of small eddies by the large ones [10]. Still, the theory was unable to deal with subleading contributions appearing in higher orders. Moreover, according to modern experimental and numerical data, it is quite possible that it is the strong coupling of the integral- scale- fluctuations with the small-scale ones which is responsible for the observed anomalous scaling in both inertial and dissipation ranges.

1.1 Transition to Turbulence

We will loosely identify laminar flow as a pattern $\mathbf{u}_0$ *formed by a small set of excited modes supported in the range of wave-numbers* $k \approx \Lambda_f$. All modes with $k > \Lambda_f$ are strongly overdamped, i.e. $u(k) = 0$ for both $k \ll \Lambda_f$ and $k \gg \Lambda_f$.

Landau's theory. Here we mention just one work which is relevant for considerations presented below [12]. Assuming that in the vicinity of a transition point imaginary part of complex frequency is much smaller than the real one, Landau considered the linearized Navier-Stokes equations for incompressible fluid. Denoting the velocity field at a transition point $\mathbf{u}_0$ and introducing an infinitesimal perturbation $\mathbf{u}_1$ he wrote $\mathbf{u} = \mathbf{u}_0 + \mathbf{u}_1$ with $\mathbf{u}_1 = A(t)f(\mathbf{r})$. Based on general qualitative considerations, Landau proposed:

$$\frac{d|A|^2}{dt} = 2\gamma|A|^2 - \alpha|A|^4$$

where in the vicinity of transition point $\gamma = c(Re - Re_{tr})$ and $\alpha > 0$. In principle, $|A|^2$ must be considered as time- averaged. Landau noted, however, that $\mathbf{u}_1(\mathbf{k})$ is a slow mode and, since the averaging is taken over relatively short time -intervals, the averaging sign in the above equations is not necessary. At small times the solution exponentially grows, meaning that the basic flow is unstable.

1.2 Transition to Turbulence: Velocity Derivatives

A new way of looking at phenomenon of transition to turbulence was introduced in numerical simulations of a flow at a relatively low Reynolds number $R_\lambda = \sqrt{\frac{5}{3\mathscr{E}\nu}}u_{rms}^2 \geq 4.0$ [17]. In this approach transition to turbulence is identified with the first appearance of non-gaussian anomalous fluctuations of velocity derivatives including those of the dissipation rate $\mathscr{E}$. The flow in a periodic box was generated by a force in the right-side of the Navier-Stokes equation with driven by the force $\mathbf{F}(\mathbf{k},t) = \mathscr{P}\frac{\mathbf{u}(\mathbf{k},t)}{\sum' |\mathbf{u}(\mathbf{k},\mathbf{t})|^2}\delta_{\mathbf{k},\mathbf{k}'}$, where summation is carried over $\mathbf{k}_f = (1, 1, 2);\ (1, 2, 2)$. It is easy to see that the model with this forcing generates flows with constant energy flux $\mathscr{P} = \mathscr{E} = \nu\overline{(\frac{\partial u_i}{\partial x_j})^2} = const$ and the variation of the Reynolds number is achieved by variation of viscosity.

The results of Ref. [17] can be briefly summarized as follows: 1. Extremely well-resolved simulations of the low-Reynolds number flows at $R_\lambda \geq 9$–10 revealed a clear scaling range $\overline{(\frac{\partial u}{\partial x})^n} \propto Re^{\rho_n}$ with the anomalous scaling exponents ρ_n consistent with the inertial range exponents typically observed only in very high Reynolds number flows $Re \gg Re_{tr}$. Identical scaling exponents ρ_n were later obtained in some other flows [6] indicating possibility of a broad universality. 2. For $R_\lambda <$9–10 the flow was subgaussian indicating a dynamical system consisting of a small number of modes with the small-scale fluctuations strongly overdamped. This flow can be called "quasilaminar" or coherent. 3. At a transition point $R_{\lambda,tr} \approx 9$–10 the fluctuating velocity derivatives obey gaussian statistics and at $R_\lambda > 9$–10 a strongly anomalous scaling of the moments, typical of high-Reynolds number turbulence, is clearly seen. 4. It has also been noticed that transition is smooth, i.e. velocity field at $\mathbf{u}(R^-_{\lambda,tr}) - \mathbf{u}(R^+_{\lambda,tr}) \to 0$.

In Fig. 2 the moments of the dissipation rate computed in [17] are combined with the data obtained by Donzis et al. [6] in HIT generated by a completely different

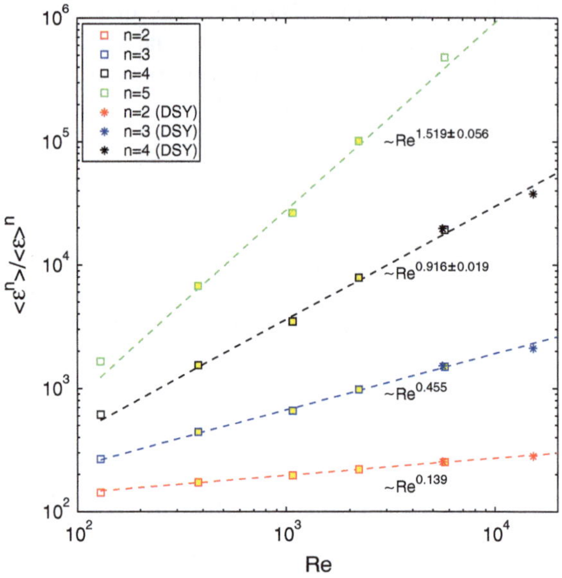

Fig. 2 Moments of the dissipation rate in homogeneous and isotropic turbulence. References [6, 17]

large-scale forcing. We can see that the scaling exponents, found in the range of very low Reynolds number in [17] hold in a much wider range of the Reynolds number variation. Moreover, these exponents have also been found in a channel flow [8] pointing to a possibility of a wide universality class. This means that in the range $R_\lambda \geq 10$ turbulence can be considered as fully developed.

In Fig. 3 the moments M_n of velocity derivatives are shown in the vicinity of a transition point. One can see that at $R_\lambda = 9$–10 a sharp transition from a sub-Gaussian at $R_\lambda < 9$–10 to anomalous scaling of the dissipation rate moments occurs independently on the driving force. This surprising result will be used below as a constraint on development of turbulence models.

2 The model

Based on these results, we consider a flow generated by the Navier-Stokes equations with a force $\mathbf{F}(\Lambda_f)$. Keeping the force $\mathbf{F} = const$ and the length-scale $L = 2\pi/\Lambda_f = const$, let us vary viscosity in the interval $0 \leq \nu \leq \infty$. In the range $\nu > \nu_{tr}$ or $Re < Re_{tr}$ the flow is laminar in a sense that it is described by a relatively small number of modes with $u(\mathbf{k})$ with $k \approx \Lambda_f$. At the transition point $Re_{\lambda,tr} \approx 9 - 10$ the transitional pattern $\mathbf{u}_0(\Lambda_f)$ is formed, so that:

$$L(\mathbf{u}_0, \nu_{tr}) = \frac{D\mathbf{u_0}}{Dt} + \nabla p - \nu_{tr}\nabla^2\mathbf{u_0} - \mathbf{F}(\Lambda_f) \equiv 0$$

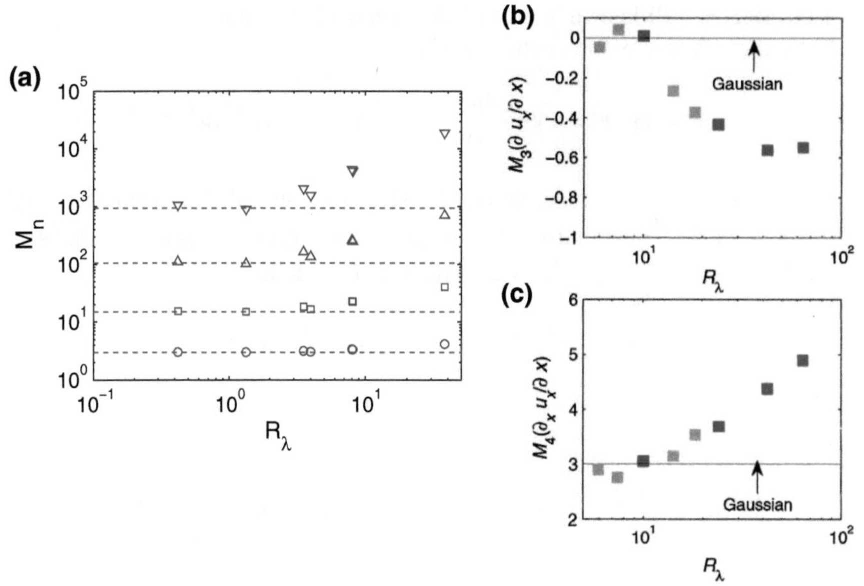

Fig. 3 Normalized moments of velocity derivatives in Isotropic and Homogeneous turbulence. Numerical simulations [6, 17]. *Right* Schumacher et al. [17]: homogeneous turbulence driven by a force $\mathbf{F}(\mathbf{k}, t) = \mathscr{P}\frac{\mathbf{u}(\mathbf{k},t)}{\sum' |\mathbf{u}(\mathbf{k},t)|^2}\delta_{\mathbf{k},\mathbf{k}'}$ (for more details, see above). *Left* Donzis et al. [6]: HIT drven by a gaussian large-scale noise. In the range $R_\lambda \leq 9.0$, we see a clear Gaussian behavior of a few moments with $S_3 = 0$; $M_4 \approx 3$; $M_6 \approx 15$; $M_8 \approx 105$, typical of Gaussian distribution. At $R_\lambda \geq 9 - 10$, all moments obey anomalous scaling of fully developed turbulence

Here $\frac{D}{Dt} \equiv \frac{\partial}{\partial t} + \mathbf{u_0} \cdot \nabla$. When $\nu \ll \nu_{tr}$ the Navier-Stokes equations read:

$$\frac{D\mathbf{u}}{Dt} = -\nabla p + \nu\nabla^2\mathbf{u} + \mathbf{F}(\Lambda_f) \equiv 0$$

If we write $\mathbf{u} = \mathbf{u_0} + \mathbf{v}$, the equation for the "turbulent" component $(k > \Lambda_f)$ of velocity field:

$$\frac{\partial \mathbf{v}}{\partial t} + \mathbf{v} \cdot \nabla\mathbf{v} = -\nabla p + \nu\nabla^2\mathbf{v} + \mathbf{f} \tag{1}$$

with

$$\mathbf{f} = \mathbf{f_1} + \mathbf{f_2} + \mathbf{f_3} = -\mathbf{u_0} \cdot \nabla\mathbf{v} - \mathbf{v} \cdot \nabla\mathbf{u_0} + (\nu - \nu_{tr})\nabla^2\mathbf{u}_0(\Lambda_f).$$

The first term $\mathbf{f}_1$ in this expression describes kinematic transfer of small "eddies" by the large ones. The second, $\mathbf{f}_2$, is responsible for turbulence production due to interaction of small-scale fluctuations with the large-scale, quasicoherent, flow $\mathbf{u}_0$.

This effect is well-known in the turbulence modeling literature.

Thus, the total energy production rate is:

$$\mathscr{P} = \overline{\mathbf{f_2} \cdot \mathbf{v}} = -\overline{v_i v_j} \frac{\partial u_{0,i}}{\partial x_j} \approx \nu_T \overline{(u^j_{0,i})^2} \approx \Lambda_f^{-\frac{4}{3}} \overline{(u^j_{0,i})^2}$$

The balance can be written for each scale $l = 2\pi/\Lambda(r)$ but with "turbulent viscosity" $\nu(r) \equiv \nu(l) \propto \mathscr{E}^{\frac{1}{3}} l^{\frac{4}{3}}$ and introducing the projection operator $\mathscr{P}_{lmn} = k_m P_{ln}(\mathbf{k}) + k_n P_{lm}(\mathbf{k})$ with $P_{i,j} = (\delta_{ij} - \frac{k_i k_j}{k^2})$ we have with $\hat{k} = (\mathbf{k}, \omega)$:

$$f_{2,l} = -\frac{i}{2} \mathscr{P}_{lmn}(\mathbf{k}) \int \nu(\hat{q}) u_{0,n}(\hat{k} - \hat{q}) d\hat{q}$$

so that $\overline{\mathbf{f}_2} = 0$ and

$$\mathscr{P}(\mathbf{k}) \propto k^2 \int \frac{q^{-\frac{13}{3}}}{\Omega^2 + q^{\frac{4}{3}}} (u^j_{0,i})^2 \delta(\hat{k} - \hat{q}) \delta(\omega - \Omega) d\hat{q} d\omega \propto k^{-3}$$

The experimental data of Refs. [8, 17] point to independence of small-scale features of turbulence on the nature of production mechanism. On the Fig. 4b the compensated energy spectrum in a flow past circular cylinder of diameter D is shown for the large—scale Reynolds number $Re = UD/\nu \approx 10^6$–10^7. The onset of Kolmogorov's inertial range can be clearly seen at the wave number $k \approx \Lambda_f = 2\pi/L$ separating inertial and non-universal, geometry-dependent energy—containing range of scales. This interval separating these two asymptotic limits is quite narrow, which points to the smallness of subleading contributions to the inertial range scaling of the energy spectrum. Therefore, for $\nu < \nu_{tr}$ we choose the well-known and well-studied model (1): where the random force, mimicking small-scale fluctuations is defined by the correlation function:

$$\overline{f_i(\mathbf{k}, \omega) f_j(\mathbf{k}', \omega')} = 2D_0 (2\pi)^{d+1} k^{-y} P_{ij}(\mathbf{k}) \delta(\mathbf{k} + \mathbf{k}') \delta(\omega + \omega'); \qquad (2)$$

Based on the above argument, the force (2) is *extrapolated* onto interval $\Lambda_f < k \leq \Lambda \equiv \Lambda_0$, so that $\overline{f_i F_j} = 0$ and, by construction, $f_i(k \leq \Lambda_f, t) = 0$.

2.1 The Renormalization or Coarse Graining

The renormalization group for fluid flows has been developed in Refs. [5, 7] and was generalized to enable computations of various dimensionless amplitudes in the low order in the ε-expansion in Refs. [15, 20]. Introducing velocity and length-and -time scales $U = \sqrt{D_0/(\nu_0 \Lambda_0^2)}$, $X = 1/\Lambda_0$ and $T = t\nu_0\Lambda_0^2$, respectively, the Eq. (1) can be written as (for simplicity we do not change notations for dimensionless variables):

$$\frac{\partial \mathbf{v}}{\partial T} + \hat{\lambda}_0 \mathbf{v} \cdot \nabla \mathbf{v} = -\hat{\lambda}_0 \nabla p + \nabla^2 \mathbf{v} + \frac{\mathbf{f}}{\sqrt{D_0 \nu_0 \Lambda^2}}$$

where the single dimensionless coupling constant ("bare" Reynolds number) is: $\hat{\lambda_0}^2 = \frac{D_0}{\nu_0^3 \Lambda_0^\varepsilon}$.

Projecting Navier-Stokes equation onto domain $k \le \Lambda_0 e^{-r}$ where $r \to 0$. **Technical details of all calculations presented below are best described in** [22]. Formally introducing modes $v^<(\mathbf{k}, t)$ and $v^>(\mathbf{k}, t)$ with k from the intervals $k \le \Lambda_0 e^{-r}$ and $\Lambda_0^{-r} \le k \le \Lambda_0$, respectively, and averaging over small-scale fluctuations $\mathbf{v}^>$, leads to equation for the large-scale modes:

$$\frac{\partial v_i^<}{\partial t} + v_j^< \cdot \nabla_j v_i^< = -\nabla_i p + \frac{\partial \sigma_{ij}^{(2)}}{\partial x_j} + \nu \nabla^2 v_i^< + f_i + \Delta f_i \tag{3}$$

where the second-order correction to the Reynolds stress $\sigma_{ij} = -\overline{v_i v_j}$ is [18]:

$$\begin{aligned}\sigma_{ij}^{(2)} &= \hat{\lambda}^2(r)\Delta\nu(r)S_{ij} - \hat{\lambda}_1^4(r)\nu(r)\frac{D}{Dt}[\tau(r)S_{ij}] \\ &- \hat{\lambda}_1^4(r)\nu(r)\tau(r)[\beta_1 \frac{\partial u_i}{\partial x_k}\frac{\partial u_j}{\partial x_k} + \beta_2(\frac{\partial u_i}{\partial x_k}\frac{\partial u_k}{\partial x_j} + \frac{\partial u_j}{\partial x_k}\frac{\partial u_k}{\partial x_i}) \\ &+ \beta_3 \frac{\partial u_k}{\partial x_i}\frac{\partial u_k}{\partial x_j}] + O(\hat{\lambda}_1^6) + \cdots \end{aligned} \tag{4}$$

where $\hat{\lambda}_1 = O(\hat{\lambda}_0(e^{-r} - 1))$ is a coupling constant generated by the scale-elimination and the time-constant $1/\tau(r) = \nu(r)\Lambda^2(r)$. In this limit the coefficients β_i can be explicitly calculated [15]. The "dressed" viscosity is denoted as $\nu(r) = \nu + \Delta\nu(r)$ with correction to "viscosity" written in the wave-number space as:

$$\Delta\nu = A_d \frac{D_0}{\nu_0^2}[\frac{e^{\varepsilon r} - 1}{\varepsilon \Lambda_0^\varepsilon} + O(\frac{k^2}{\Lambda_0^{\varepsilon+2}}\frac{e^{(\varepsilon+2)r} - 1}{\varepsilon + 2}) + O(\hat{\lambda}_0^4)] \tag{5}$$

$\varepsilon = 4 + y - d$ and $A_d = \hat{A}_d \frac{S_d}{(2\pi)^d}$; $\hat{A}_d = \frac{1}{2}\frac{d^2 - d}{d(d+2)}$. On the interval $k < \Lambda_0 e^{-r}$ the Eqs. (3)–(5) are equivalent to the original equations of motion defined on the interval $k \le \Lambda_0$.

Iterating scale-elimination procedure. The relations (3)–(5) are exact as long as the eliminated "slice" in the wave-vector space is very thin.

2.2 Recursion Relations

As a result of elimination of the first shell $\Lambda_0 e^{-r} \leq k \leq \Lambda_0$, the original uncorrected "bare" viscosity ν_0 disappears and instead the equations include only "dressed" viscosity $\nu(r)$. Starting with the Eqs. (3)–(5), we can eliminate the modes from the next shell of wave-numbers $\Lambda_0 e^{-(r+\delta r)} \leq k \leq \Lambda_0 e^{-r}$ and derive equations of motion with another set of corrected transport coefficients. The procedure can be iterated resulting in the cut-off-dependent viscosity, induced force etc. Therefore, with $\delta r \to 0$ the parameters in the coarse-grained "Navier-Stokes equations" for the "resolved" velocity field $\mathbf{u}^<$, defined at the scales $l \geq 2\pi/\Lambda(r)$, satisfy the differential relations:

$$\frac{\nu(r+\delta r)-\nu(r)}{\delta r} = A_d \frac{D_0}{\nu(r)^2}\frac{1}{\Lambda^{\varepsilon}(r)}[\sum_{n=0}^{\infty}\alpha_n \hat{\lambda}^n(r) + O(\frac{k^2}{\Lambda(r)^2})] \tag{6}$$

and, by Galileo invariance, all high-order non-linearities can be formally written as:

$$\begin{aligned} HOT = {} & [\sum_{n=2}^{\infty} \hat{\lambda_1(r)}^{2n} \tau(r)^{n-1}(\partial_t \mathbf{u}^< + \mathbf{u}^< \cdot \nabla)^{\mathbf{n}}]\mathbf{u}^< \\ & + O(\hat{\lambda}_2(r)^4 \nabla S_{ij}^2 \frac{1}{\Lambda^2(r)}) + \cdots \end{aligned} \tag{7}$$

with $\tau(r) \approx 1/(\nu(r)\Lambda^2(r))$. To asses the role of different contributions to (6) and (7), first we **assume** $\hat{\lambda}(r) \ll 1$ and $\hat{\lambda}_1(r) \ll 1$ and analyze the lowest-order terms only.

This leads to differential recursion equations: recalling that $\Lambda(r) = \Lambda_0 e^{-r}$, one obtains:

$$\frac{\nu(r+\delta r)-\nu(r)}{\delta r} = \frac{d\nu(r)}{dr} = A_d \nu(r)\hat{\lambda}^2(r)$$

$$\frac{d\hat{\lambda}^2}{dr} = \varepsilon\hat{\lambda}^2 - 3A_d\hat{\lambda}^4$$

where $\hat{\lambda}^2(r) = \frac{D_0}{\nu^3(r)\Lambda^{\varepsilon}(r)}$ and:

$$\nu(r) = \nu_0[1 + \frac{3A_d}{\varepsilon\nu_0^3}\frac{D_0 S_d}{(2\pi)^d}(\frac{1}{\Lambda^{\varepsilon}(r)} - \frac{1}{\Lambda_0^{\varepsilon}})]^{\frac{1}{3}}$$

$$\hat{\lambda} = \hat{\lambda}_0 e^{\frac{\varepsilon r}{2}}[1 + \frac{3A_d}{\varepsilon\nu_0^3}\frac{D_0 S_d}{(2\pi)^d}(\frac{1}{\Lambda^{\varepsilon}(r)} - \frac{1}{\Lambda_0^{\varepsilon}})]^{-\frac{1}{2}}$$

The solution for the "induced" coupling constant $\hat{\lambda}_1$ is :

$$\hat{\lambda}_1(r) = \frac{\sqrt{\varepsilon} e^{\frac{\varepsilon r}{2}}}{\sqrt{\frac{\varepsilon}{\hat{\lambda}_1^2(0)} + 3A_d(e^{\frac{\varepsilon r}{2}} - 1)}} \tag{8}$$

For $\varepsilon = 4$, corresponding to Kolmogorov's energy spectrum, in the limit $\varepsilon r \gg 1$ the coupling constants tend to the fixed point

$$\hat{\lambda}_* \to (\frac{\varepsilon}{3A_d})^{\frac{1}{2}} \approx 1.29\sqrt{\varepsilon} \approx 2.58. \tag{9}$$

It is also clear that $\hat{\lambda}_{1,*} \approx \hat{\lambda}_*$. *This result means that for $k > \Lambda_f$, the above truncation of the expansion is, in general,* **incorrect** *and high-order non-linearities generated by procedure are* **not small**. *Now we consider a special case of the transport approximation $\Lambda(r) \to \Lambda_f = 2\pi/L$.*

Parameters. All low-order calculations leading to dimensionless amplitudes presented below, are best described in great detail in Ref. [22], where it was shown:

$$\nu(k) = (\frac{3}{8}A_d 2D_0)^{\frac{1}{3}} k^{-\frac{4}{3}} \approx 0.42(\frac{2D_0 S_d}{(2\pi)^d})^{\frac{1}{3}} k^{-\frac{4}{3}}$$

and from the linearized equation at the fixed point we derive Kolmogorov's spectrum valid in the range $k \geq \Lambda_f$:

$$\begin{aligned} E(k) &= \frac{1}{2}\frac{S_d k^2}{(2\pi)^{d+1}} \int_{-\infty}^{\infty} Tr V_{ij}(\mathbf{k}\omega) d\omega \\ &= \frac{1}{2(\frac{3}{8}\hat{A}_d)^{\frac{1}{3}}} (2D\frac{S_d}{(2\pi)^d})^{\frac{2}{3}} k^{-\frac{5}{3}} \\ &= 1.186(2D_0\frac{S_d}{(2\pi)^d})^{\frac{2}{3}} k^{-\frac{5}{3}} \end{aligned} \tag{10}$$

where $(2\pi)^{d+1} V_{ij}(\mathbf{k}, \omega) = \frac{u_i^<(\mathbf{k},\omega) u_j^<(\mathbf{k}',\omega')}{\delta(\mathbf{k}+\mathbf{k}')\delta(\omega+\omega')}$. In the so called EDQNM approximation, which is exact at the Gaussian fixed point (see below), the force amplitude D_0 can be related to the mean dissipation rate [3, 22]:

$$2D_0 S_d/(2\pi)^d \approx 1.59\mathscr{E};\ E(k) = C_K \mathscr{E}^{\frac{2}{3}} k^{-\frac{5}{3}};\ C_K = 1.61 \tag{11}$$

Let us identify the infra-red cut-off $\Lambda_f = \Lambda(r) \approx 2\pi/L$ with the wave-number corresponding to the top of the inertial range. In the large Re-limit $\Lambda_0/\Lambda_f \gg 1$, the total energy of the inertial range turbulent fluctuations is evaluated readily:

$$\mathcal{K} = \int_{\Lambda_f}^{\infty} E(k)dk = \frac{3}{2} C_K (\frac{\mathcal{E}}{\Lambda_f})^{\frac{2}{3}}$$

$$= \frac{3}{2} 1.61 (\frac{3}{8} \hat{A}_d 1.59)^{\frac{1}{3}} \frac{\mathcal{E}}{\nu(\Lambda_f)\Lambda_f^2} \approx 1.19 \frac{\mathcal{E}}{\nu(\Lambda_f)\Lambda_f^2} \tag{12}$$

and, setting $k = \Lambda_f$ gives the expression for effective viscosity in equation for the large-scale dynamics in the interval of scales $k < \Lambda_f$:

$$\nu_T \equiv \nu(\Lambda_f) \approx 0.084 \frac{\mathcal{K}^2}{\mathcal{E}}; \qquad 10.0 \times \nu(\Lambda_f)^2 \Lambda_f^2 = \mathcal{K} \tag{13}$$

3 Fixed-Point Reynolds Number and Irrelevant Variables

The expression (13) gives effective viscosity accounting for all turbulent fluctuations from the interval $1/\Lambda_0 \le r < L = 1/\Lambda_f$ acting on the almost-coherent-large scale flow on the scales $r \approx L = 1/\Lambda_f$. Using (13)–(15) we can calculate the effective $R_{\lambda,f} = 2\mathcal{K}\sqrt{5/(3\mathcal{E}\nu(\Lambda_f))} = \sqrt{20/(3 \times 0.084)} = 9.0$. The same parameter can be expressed in terms of the fixed-point coupling constant:

$$\hat{\lambda}^* = \sqrt{\frac{D_0 S_d/(2\pi)^d}{\nu_T^3 \Lambda_f^4}} = \sqrt{\frac{0.8\mathcal{E}}{\nu_T^3 \Lambda_f^4}} = \frac{\sqrt{0.8 \times 400 \mathcal{E} \nu_T}}{u_{rms}^2}$$

$$= \frac{\sqrt{0.8 \times 400 \times \frac{5}{3}}}{R_\lambda^{fp}} = \sqrt{\frac{4}{3\hat{A}_d}} = 2.58$$

and $R_{\lambda,f} \approx 9.0$ very close to Reynolds number of transition $R_\lambda \approx 9.$, obtained from direct numerical simulations of Refs. [6, 17]. This result agrees with the observation that in the flows past various bluff bodies, the Reynolds number based on the measured "turbulent viscosity" and large-scale velocity field is $R_{\lambda,T} = O(10)$, independent on the "bare" (classic) Reynolds number calculated with molecular viscosity.

As $\Lambda(r) \to L = 2\pi/\Lambda_f$, the effective viscosity $\nu(r) \to \nu_{tr}$ and $Re(r) \to Re_{tr}$, which is the most important and surprising outcomes of the theory. If, as was found numerically, transition to turbulence is "smooth", the velocity field $\mathbf{u}_0$ must come out from equations of motion obtained by the scale-elimination and Navier-Stokes equations for quasi-laminar flow at a transition point. Therefore,

$$\hat{L}(\mathbf{u}_0, \nu_{\Lambda_f}) - HOT = \hat{L}(\mathbf{u_0}, \nu_{tr}) \equiv 0 \tag{14}$$

If, in addition, the transition is universal, i.e. is independent on initial conditions we conclude that as $\Lambda(r) \to \Lambda_f$, and, according to the above derivation $\nu(\Lambda_f) \to \nu_{tr}$, the nonlinearities generated by the scale elimination procedure

$$HOT \to 0$$

All we can definitely say is: if indeed transition is continuous in the limit $\Lambda(r) \to \Lambda_f$, the nonlinearities $HOT(\Lambda(r)) \to 0$. To understand the way it tends to zero, let us consider the linearized equation of motion in the vicinity of the fixed point where $\mathbf{u} = \mathbf{u}_0 + \mathbf{u}_1$:

$$\frac{\partial \mathbf{u}_1}{\partial t} + \mathbf{u}_0 \cdot \nabla \mathbf{u}_1 + \mathbf{u} \cdot \nabla \mathbf{u}_0 = -\nabla p_1 + \nu \nabla^2 \mathbf{u}_1 + HOT \qquad (15)$$

If $\mathbf{u}_1 \propto Ae^{i\omega t}$, then according to phenomenological Landau's theory of transition: $u_1 \propto A_{max} \propto \sqrt{Re - Re_{tr}}$ and

$$HOT \approx \mathbf{u}_0 \cdot \nabla \mathbf{u}_1 \approx u_0^2 \Lambda_f \sqrt{Re - Re_{tr}}$$

4 Turbulence Modeling

The above results mean that in the inertial range all high-order terms are relevant and no existing Smagorinsky-like LES models, based on the low-order truncation of expansion is valid. However, the turbulence dynamics on the scales $l \approx L = \frac{2\pi}{\Lambda_f}$ are governed by the following equation ($\rho = 1$):

$$\partial_t \mathbf{V} + V_i \nabla_i \mathbf{V} = 2\nu_T S_{ij}^2 - \nabla p + \nabla(\nu_0 + \nu_T)\nabla \mathbf{V} \qquad (16)$$

where $\nu_T = 0.084\frac{\mathscr{K}^2}{\mathscr{E}}$. To close the problem we need equations for kinetic energy $\mathscr{K}$ and mean dissipation rate $\mathscr{E}$ obtained directly from the Navier-Stokes equations:

$$\partial_t \mathscr{K} + V_i \nabla_i \mathscr{K} = \tau_{ij} \frac{\partial V_i}{\partial x_j} - \mathscr{E} - \nabla_i(\overline{v_i' K}) - \overline{\nabla_i v_i' p} + \nu \nabla^2 \mathscr{K} \qquad (17)$$

$$\begin{aligned} \partial_t \mathscr{E} + V_i \nabla_i \mathscr{E} = &- 2\nu\overline{(\nabla_j v_i')(\nabla_j v_l')(\nabla_l v_i')} - 2\nu^2\overline{(\nabla_j \nabla_l v_i')^2} \\ &- 2\nu\overline{(\nabla_j v_i)(\nabla_i \nabla_j p)} + \nu \nabla^2 \mathscr{E} \end{aligned} \qquad (18)$$

To obtain equations for the mean or better to say "resolved" component of velocity field $\mathbf{V}(\mathbf{r}, t)$ on the scales $r \geq \mathscr{L}$, one has to derive expression for the Reynolds stress τ_{ij} which is equivalent to solving "the turbulence problem". The difficulty with analytic representation of this term has been realized long before introduction of renormalized perturbation expansions and analytic theories of turbulence.

4.1 LB-BGK-$\mathcal{K} - \mathcal{E}$ Model

This model, briefly described in Ref. [18], is **not** based on equations of hydrodynamics but on a more basic "quasymicroscopic" kinetic equations for a "gas" of imaginary interacting particles described by a kinetic equation:

$$\partial_t f + \mathbf{v} \cdot \nabla f = -\mathscr{C}(f) = -\frac{f - f^{eq}}{\tau} \tag{19}$$

where the collision integral $\mathscr{C}(f) = 1/\tau$ in the right-side is written in the relaxation time approximation. The kinetic equation (20) is a generator of various hydrodynamic approximations from the Euler to Navier-Stokes and Burnett equations of different complexities. The relaxation time τ has to be chosen to reflect basic features of inter-particle interactions and various dynamic constraints, like conservation laws, equations of state etc. In homogeneous and isotropic turbulence (HIT) the only available time-scale is $\tau_{ih} \propto \mathcal{K}/\mathcal{E}$ and with $\tau_{ih} \propto \frac{\mathcal{K}}{\mathcal{E}} \times \frac{\mathcal{K}}{k_B T}$, the Eq. (20) generates the Navier-Stokes-like equations but with viscosity $\nu_T \propto \mathcal{K}^2/\mathcal{E}$ which, combined with the equations for the scalars, is the 'standard' two-equation model.

In a strongly sheared flow, the smaller and, therefore, dominating scale is $\tau_s \propto 1/|\overline{S_{ij}}|$ and the dimensionless parameters reflecting relative importance of both effects is: $\eta = \tau_{ih}/\tau_s$. Thus, in general $\tau = \tau_0 + \Psi(\mathcal{K}, \mathcal{E}, 1/|S_{ij}|, G)$ where G stands for contributions from various external forces coming, for example, from electro- magnetic, gravitational, centrifugal and other fields. In addition, it has to include effects of stratification, local swirl, acceleration etc. The choice of the relaxation time is discussed in Refs. [1, 2]. Combined with the equations for kinetic energy and mean dissipation rate [22], the model consists of kinetic equation (20) and two-equations for two scalars and a general expression for the relaxation time: $(\tau_0 + \Psi(\mathcal{K}, \mathcal{E}, 1/|S_{ij}, G))k_B T = \nu_T$.

The equations for two scalars $\mathcal{K}$ and $\mathcal{E}$, used in all calculations presented below are:

$$(\partial_t \mathcal{K} + V_i \nabla_i \mathcal{K}) = \nu_T \left(\frac{\partial V_i}{\partial x_j} \frac{\partial V_i}{\partial x_j} \right) - \mathcal{E} + \nabla_i (\nu + \alpha_K \nu_T \nabla_i \mathcal{K}) \tag{20}$$

$$\begin{aligned} \partial_t \mathcal{E} + V_i \nabla_i \mathcal{E} =& C_{\varepsilon 1} \frac{\mathcal{E}}{K} \nu_T \left(\frac{\partial V_i}{\partial x_j} \frac{\partial V_i}{\partial x_j} \right) - [C_{\varepsilon 2} + C_\mu \frac{\eta^3 (1 - \frac{\eta}{\eta_0})}{1 + \beta \eta^3}] \frac{\mathcal{E}^2}{\mathcal{K}} \\ &+ \nabla_i (\nu + \alpha_{\mathcal{E}} \nu_T \nabla_i \mathcal{E}); \end{aligned}$$

The constants $\alpha_K = \alpha_\varepsilon = 1.37$, $C_\mu = 0.084$, $C_{\varepsilon,1} = 1.42$ and $C_{\varepsilon,2} = 1.68$ have been derived in the low order in [22] and justified in this paper. The expression for the second term in the $\mathcal{E}$-equation, coming from: $\mathcal{R} = 2\nu \overline{S_{ij} \frac{\partial u_m}{\partial x_i} \frac{\partial u_m}{\partial x_j}} \approx C_\mu \frac{\eta^3(1-\eta/\eta_0)}{1+\beta\eta^3}$ was derived in Ref. [21] on the basis of dimensional considerations with the single "underived" constant $\eta_0 \approx 4.38$ fixed to produce von Karman constant $\kappa \approx 0.4$. The boundary conditions (wall functions), necessary for numerical solution are described in detail in Ref. [1].

4.2 Examples

All examples presented below were computed with Eqs. (19) and (20) including the single set of theoretically derived numerical constants. Numerical simulations were conducted with PowerFlow^{TM} commercial package developed by EXA Corporation.

4.2.1 Flow Past 3D Cylinder at $Re = UD/\nu = 1.2 \times 10^6$

This is a classic example [1] of a flow past bluff body used for studies of transition to and physics of developed of turbulence. Also, it is often applied for calibration and comparison of numerical codes. The resolution study was conducted on three cases $N = \frac{D}{\Delta} = 256;\ 128;\ 64$. The computed pressure coefficient agreed well with experimental data. It has to be mentioned that not many studies exist in this range of high Reynolds numbers. In Fig. 4 statistical properties of a computed field are presented. The signal was collected at four different probes shown in Fig. 4a, b shows the compensated energy spectrum $E(k)/(1.61\mathscr{E}^{\frac{2}{3}}k^{-\frac{5}{3}}) \approx 1.$ in a wide range of the wave-number variation. The structure functions $S_2 \propto (\delta x)^{\frac{2}{3}}$ and the Kolmogorov's relation $S_3 \propto \delta x$ are also shown. We can also notice that, while S_2 was accurately computed on both 128^3 and 256^3 meshes, the high-order moment S_3 was much more resolution—sensitive: the results of the 128^3-run are less impressive than the high-resolution computation (256^3). We believe, it is the first example of two-equation turbulence model giving almost two decades of Kolmogorov's inertial range. We will see below that this feature is crucial for prediction of pressure fluctuations and turbulent acoustics.

4.2.2 Flow Past 3D Inclined Back Steps

This is the case of the European Researh Community on Flow, Turbulence and Combustion (ERCOFTAG) database [9]. Experimental data are from Ruck and Makiola [16]. Reynolds number $Re = UH/\nu = 64{,}000$ and expansion ratio $ER = h_2/h_I = 1.48$. Inclination angles $10^0;\ 15^0;\ 20^0;\ 25^0;\ 30^0;\ 90^0$ were investigated. Below, on Fig. 5, the results for two inclination angles $\alpha = 10^0$ and $\alpha = 30^0$ are presented in a good agreement with the data.

4.2.3 Full Simulations of the BMW530i

This is a part of EADE study. The speed of the car is $U = 45$ m/s ($\approx$162 km/h), corresponding to $Re = UH/\nu \approx 4.5 \times 10^6$. Here H is the hight of the vehicle. A few cases with yaw varying in the interval between 0 and 10° were considered. In Fig. 6 the center-line lift coefficient C_p versus x for yaw of 0^0 and 10^0 are shown.

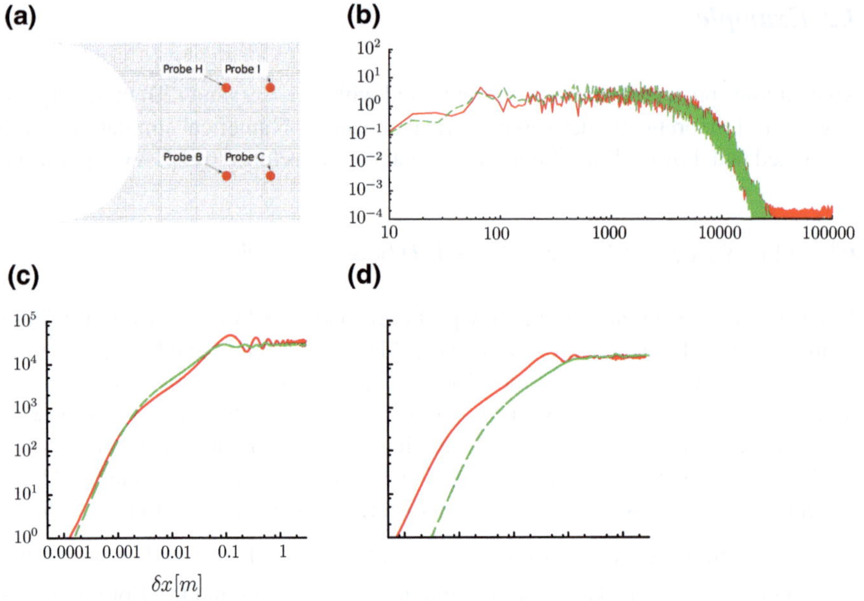

Fig. 4 Statistical characteristics of a flow past 3-dimensional cylinder. $Re = \frac{UD}{\nu} = 1.2 \times 10^6$. **a** Probe layout in the wake. **b** Computed compensated energy spectrum $E(k)/(1.6\mathscr{E}^{\frac{2}{3}}k^{-\frac{5}{3}})$ **c** Structure function $S_2 = \overline{((u(x+\delta x) - u(x))^2} \propto (\delta x)^{\frac{2}{3}}$. **d** Structure function $S_3 = \overline{((u(x+\delta x) - u(x))^3} \propto \delta x$ versus δx. The axis are identical to the ones on Fig. 4c. Resolution $N_{red} = \frac{D}{\Delta} = 256^3$ and $N_{green} = 128^3$. The runs with $N = 64^3$, not shown, were not that accurate

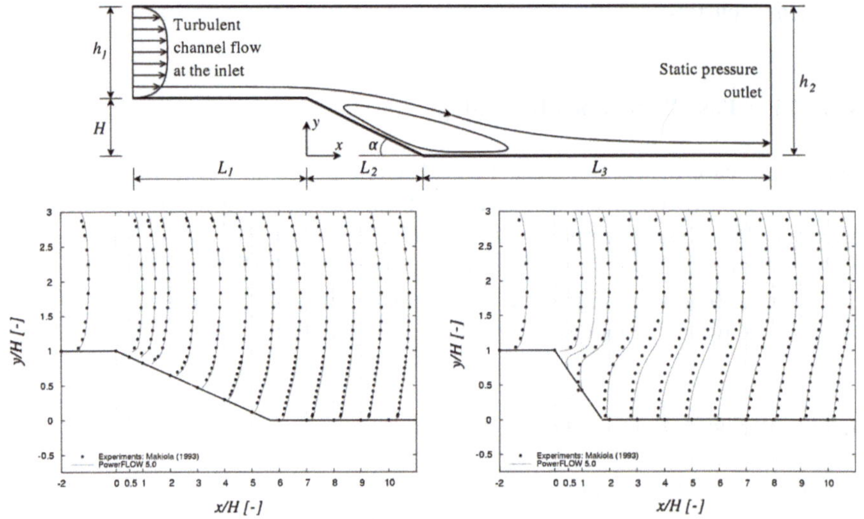

Fig. 5 Flow past 3D inclined back steps. *Top* flow setup. *Left* schematic representation of the flow with separation bubble. *Right* compared computed and measured velocity profiles for inclination angles $\alpha = 10^0$ and $\alpha = 30^0$ shown on the *left* and *right* panels, respectively [1, 9]

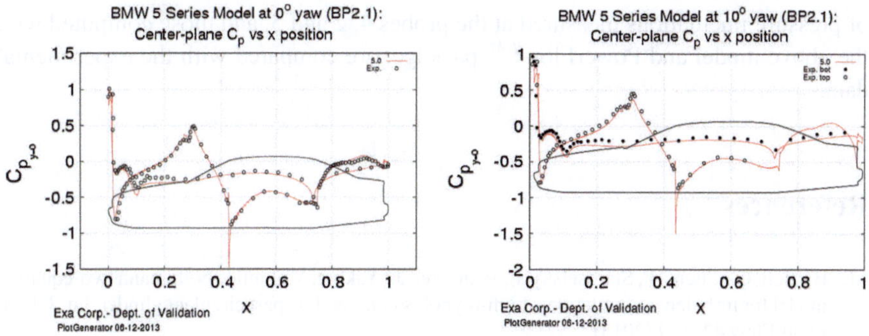

Fig. 6 BMW 5-series. Comparison of the pressure coefficient obtained from numerical simulations with experimental results. Pressure coefficient $C_p(x)$ at the *centerline*. *Left* yaw 0^0; *right* 10^0. Here x is the distance to stagnation point

4.2.4 Pressure Fluctuations on Daihatsu Wedge Box

Three-dimensional "wedge box" designed for testing various numerical approaches for simulating drag, lift and acoustics shown in Fig. 7. Sensitive microphones, installed at the five positions on a sidewall are also shown in Fig. 7. The spectra

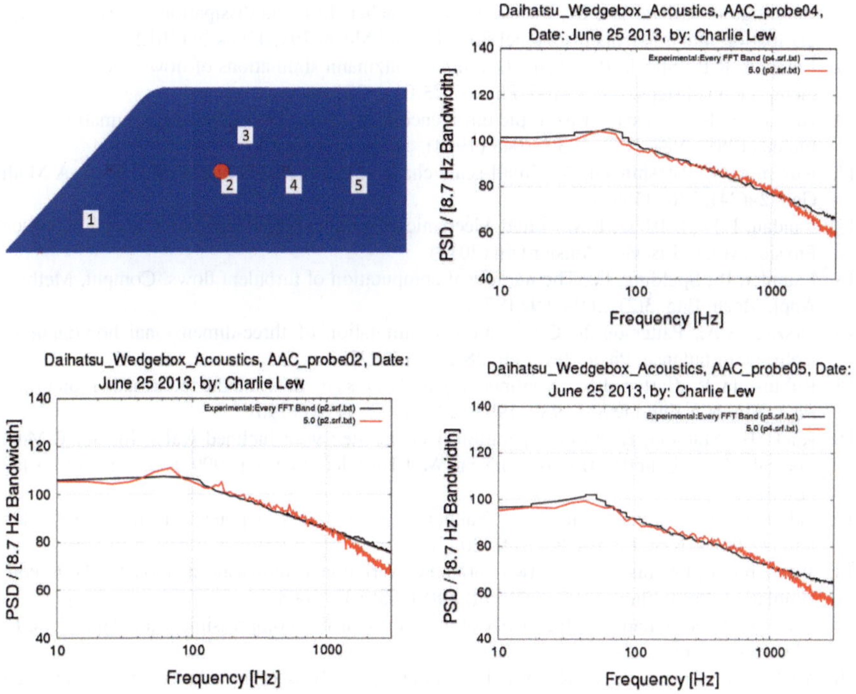

Fig. 7 Pressure fluctuation spectra computed for a Wedgebox. Comparison of the outcome of computations with experimental data on the probes 4, 2, and 5

of pressure fluctuations measured at the probes 4, 2 and 5, and those computed with the above model and PowerFlow^{TM} package, are compared with the experimental data.

References

1. Bartlett, C., Chen, H., Staroselsky, I., Wanderer, J., Yakhot, V.: Lattice boltzmann two-equation model for turbulence simulations: high-reynolds number flow past circular cylinder. Int. J. Heat Fluid Flow **42**, 1–9 (2013)
2. Chen, H., Orszag, S.A., Staroselsky, I., Succi, S.: Expanded analogy between boltzmann kinetic theory of fluids and turbulence. J. Fluid Mech. **519**, 301–314 (2004)
3. Dannevik, W.P., Yakhot, V., Orszag, S.A.: Analytical theories of turbulence and the ε expansion. Phys. Fluids (1958–1988) **30**(7), 2021–2029 (1987)
4. Deardorff, J.W.: A numerical study of three-dimensional turbulent channel flow at large reynolds numbers. J. Fluid Mech. **41**(02), 453–480 (1970)
5. DeDominicis, C., Martin, P.: Energy spectra of certain randomly-stirred fluids. Phys. Rev. A **19**(1), 419 (1979)
6. Donzis, D., Yeung, P., Sreenivasan, K.: Dissipation and enstrophy in isotropic turbulence: resolution effects and scaling in direct numerical simulations. Phys. Fluids (1994–present) **20**(4), 045108 (2008)
7. Forster, D., Nelson, D.R., Stephen, M.J.: Large-distance and long-time properties of a randomly stirred fluid. Phys. Rev. A **16**(2), 732 (1977)
8. Hamlington, P., Krasnov, D., Boeck, T., Schumacher, J.: Local dissipation scales and energy dissipation-rate moments in channel flow. J. Fluid Mech. **701**, 419–429 (2012)
9. Kotapati, R.B., Shock, R., Chen, H.: Lattice-boltzmann simulations of flows over backward-facing inclined steps. Int. J. Mod. Phys. C **25**(01), (2014)
10. Kraichnan, R.H.: Decay of isotropic turbulence in the direct-interaction approximation. Phys. Fluids (1958–1988) **7**(7), 1030–1048 (1964)
11. Kuz'min, G., Patashinskii, A.: Small-scale chaos at low reynolds numbers. J. Phys. A Math. Gen. **24**(24), 5763 (1991)
12. Landau, L.D., Lifshitz, E.M.: Fluid Mechanics: Landau and Lifshitz: Course of Theoretical Physics, vol. 6. Elsevier, Amsterdam (2013)
13. Launder, B., Spalding, D.: The numerical computation of turbulent flows. Comput. Methods Appl. Mech. Eng. **3**(2), 269–289 (1974)
14. Orszag, S.A., Patterson Jr, G.: Numerical simulation of three-dimensional homogeneous isotropic turbulence. Phys. Rev. Lett. **28**(2), 76 (1972)
15. Rubinstein, R., Barton, J.M.: Nonlinear reynolds stress models and the renormalization group. Phys. Fluids A Fluid Dyn. (1989–1993) **2**(8), 1472–1476 (1990)
16. Ruck, B., Makiola, B.: Flow separation over the step with inclined walls. In: So, R.M.C., Speziale, C.G., Launder, B.E. (eds.) Near-Wall Turbulent Flows, p. 999. Elsevier, Amsterdam (1993)
17. Schumacher, J., Sreenivasan, K.R., Yakhot, V.: Asymptotic exponents from low-reynolds-number flows. New J. Phys. **9**(4), 89 (2007)
18. Wanderer, J., Oberai, A.A.: A two-parameter variational multiscale method for large eddy simulation. Phys. Fluids (1994–present) **20**(8), 085107 (2008)
19. Wyld Jr, H.: Formulation of the theory of turbulence in an incompressible fluid. Ann. Phys. **14**, 143–165 (1961)
20. Yakhot, V., Orszag, S.A.: Renormalization-group analysis of turbulence. Phys. Rev. Lett. **57**(14), 1722 (1986)

21. Yakhot, V., Orszag, S., Thangam, S., Gatski, T., Speziale, C.: Development of turbulence models for shear flows by a double expansion technique. Phys. Fluids A Fluid Dyn. (1989–1993) **4**(7), 1510–1520 (1992)
22. Yakhot, V., Smith, L.M.: The renormalization group, the εexpansion and derivation of turbulence models. J. Sci. Comput. **7**(1), 35–61 (1992)

Part II
LES and Embedded LES

Overset DNS with Application to Sound Source Prediction

R.A.D. Akkermans, R. Ewert, S.M.A. Moghadam, J. Dierke and N. Buchmann

Abstract In this contribution, we present an application of a computational aeroacoustics code as a hybrid Zonal DNS tool. The extension of the Non-Linear Perturbation Equations (NLPE) with viscous terms is presented as well as information related to the numerical method. The applicability of the simulation tool is illustrated with two testcases, i.e., a 2D circular cylinder in a uniform flow at moderate Reynolds numbers and a 3D decaying flow initialised with Taylor-Green vortices. Both testcases provide results which match well with data reported in literature. The cylinder testcase verifies that the viscous terms are indeed correctly implemented (at least in 2D) and the Taylor-Green vortex case illustrates that the numerical scheme introduced minimal numerical dissipation.

1 Introduction

The PIANO code of DLR is a block-structured, high-order, low dispersive and dissipative Computational AeroAcoustics (CAA) code used for the simulation of aeroacoustic noise and its propagation [1]. As a CAA-code, PIANO can solve perturbation equations (i.e., Linearised Euler Equations, Acoustic Perturbation Equations, or non-linear Euler Equations in primitive disturbance form) over a time-averaged background flow. This steady meanflow is usually, but not necessarily, obtained from a Reynolds Averaged Navier-Stokes (RANS) simulation.

All these implemented disturbance equations neglect the direct influence of fluctuating viscous terms on sound generation and propagation, which are deemed essential for the direct simulation of sound generation and propagation problems. The non-linear Euler equations have been augmented with fluctuating viscosity terms and

R.A.D. Akkermans (✉) · N. Buchmann
Institute of Fluid Mechanics, TU Braunschweig, Braunschweig, Germany
e-mail: r.akkermans@tu-braunschweig.de

R. Ewert · S.M.A. Moghadam · J. Dierke
Department of Technical Acoustics, German Aerospace Center (DLR), Braunschweig, Germany

S. Girimaji et al. (eds.), *Progress in Hybrid RANS-LES Modelling*,
Notes on Numerical Fluid Mechanics and Multidisciplinary Design 130,
DOI 10.1007/978-3-319-15141-0_4

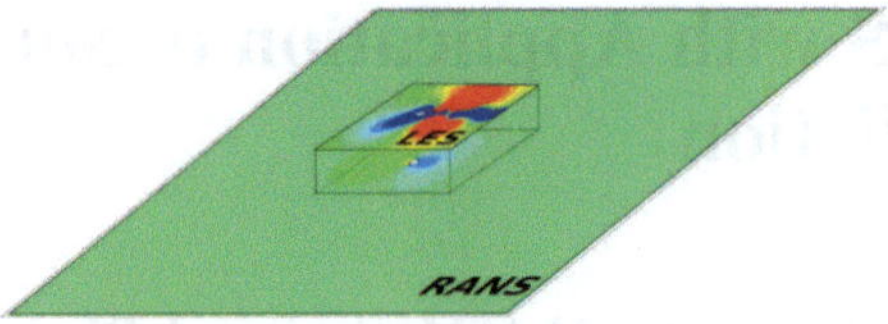

Fig. 1 Schematic of overset-DNS illustrating that the perturbation DNS is performed over the background flow. Here, the latter is provided by a RANS simulation

therefore represent the fully non-linear Navier-Stokes equations formulated in perturbation form over a given base flow (see, e.g., [2, 3]). As such, the CAA-code PIANO can be applied to directly simulate laminar or turbulent sound generation in the framework of Direct Numerical Simulations (DNS) or even Direct Noise Computation (DNC). See [4] for an overview of DNC.

In the above described simulation approach, PIANO is applicable as a hybrid Zonal tool. Contrary to an embedded approach, the here proposed procedure is denoted as Overset DNS to emphasise that the DNS is performed over the mean background flow. As such, it constitutes a hybrid RANS/DNS approach similar to what Terracol [5] proposed, albeit formulated in primitive variables. This Overset DNS principle is illustrated in Fig. 1. Furthermore, the computed perturbations provide information about the correctness of the underlying RANS, i.e., the non-zero steady part of the perturbation gives the correction for the RANS Background flow. For a recent review concerning Hybrid LES/RANS methods the reader is referred to [6].

In this contribution, we present shortly the implementation and the progress made with PIANO towards a CFD DNS-tool for turbulent flow problems. Validation of the implementation has been performed for a circular cylinder in uniform flow at moderate Reynolds numbers, and a 3D decaying flow initialised with Taylor-Green Vortices. From these validation cases it is seen that indeed we are able to accurately capture the relevant physics.

This proceeding is organised as follows. First the governing equations are presented, followed by some details concerning the numerical method. In the fourth Sect. the results of the test cases are presented, i.e., the 2D cylinder in a uniform flow and the Taylor-Green Vortex flow. Finally, the conclusion is presented and an outlook is given.

2 Governing Equations: Viscous NLPE

The basis of the here used equations are the extended viscous Non-Linear Perturbation Equations (NLPE). The interested reader is referred to Ewert et al. [2] for a detailed derivation. The use of primitive variables (i.e., density ρ, velocity $\boldsymbol{v}$, and pressure p) has the advantage that it delivers directly the desired variable of interest in aeroacoustics, the (fluctuating) pressure (see, e.g., [7]). Shock capturing is a potential disadvantage of a numerical code based on a non-conservative formulation [8]. The latter might not be too problematic as a perturbation simulations is performed on top of a background flow, which is often obtained from the strongly conservative RANS-code TAU (see, e.g., [9]).

$$\begin{aligned}
&\frac{\partial \rho'}{\partial t} + \boldsymbol{v}\cdot\nabla\rho' + \boldsymbol{v}'\cdot\nabla\rho^0 + \rho\nabla\cdot\boldsymbol{v}' + \rho'\nabla\cdot\boldsymbol{v}^0 = 0 \\
&\frac{\partial \boldsymbol{v}'}{\partial t} + (\boldsymbol{v}\cdot\nabla)\,\boldsymbol{v}' + (\boldsymbol{v}'\cdot\nabla)\,\boldsymbol{v}^0 + \frac{\rho'}{\rho}\left(\boldsymbol{v}^0\cdot\nabla\right)\boldsymbol{v}^0 + \frac{\nabla p'}{\rho} = \frac{\nabla\cdot\tau'}{\rho} + \frac{\rho^0}{\rho}\boldsymbol{r}_2^0 \qquad (1)\\
&\frac{\partial p'}{\partial t} + \boldsymbol{v}\cdot\nabla p' + \boldsymbol{v}'\cdot\nabla p^0 + \gamma p\nabla\cdot\boldsymbol{v}' + \gamma p'\nabla\cdot\boldsymbol{v}^0 = \\
&\qquad (\gamma-1)\left[(\tau'\cdot\nabla)\cdot\boldsymbol{v} + \left(\tau^0\cdot\nabla\right)\cdot\boldsymbol{v}' - \nabla\cdot\boldsymbol{q}'\right] + r_3^0.
\end{aligned}$$

In these equations, the viscous stress tensor and heat fluxes are denoted by τ and $\boldsymbol{q}$, respectively. Furthermore, γ represents the specific heat ratio. The base flow variables are indicated by the superscript 0 and perturbations by a prime. Variables that do not have a superscript or subscript are total quantities. Note that τ' denotes the fluctuating part of the stress tensor, the viscous fluctuations. Turbulent stresses from the base flow on the other hand will be indicated by τ^{t0} (in a similar way also for the turbulent heat fluxes from the base flow). The residual right-hand source terms, $\boldsymbol{r}_2^0$ and r_3^0 in Eq. (1), represent the residual turbulent viscous stresses and heat fluxes, whose exact definition depends on the definition of the base flow. These residual source terms are give by

$$\boldsymbol{r}_2^0 = -\frac{\nabla\cdot\tau^{t0}}{\rho^0} \qquad (2)$$

$$r_3^0 = -(\gamma-1)\left[\left(\tau^{t0}\cdot\nabla\right)\cdot\boldsymbol{v}^0 - \nabla\cdot q^{t0}\right]. \qquad (3)$$

Here it has been used that the base flow is obtained from a (steady) compressible RANS solution yielding constant residual terms. In general the only assumption made is that the background flow is also a solution of the Navier-Stokes equations. Note that the here presented extended viscous NLPE are preferred above the original viscous NLPE as they require less memory storage (see [2]). By using the momentum equation applied to the base flow, the divergence of the mean flow tensor can be eliminated and thus the storage requirement of the mean flow variables is relaxed a bit (10 mean flow variable permanently stored, i.e., 6 for the viscous mean-stress tensor and 4 for the residual right-hand source terms).

3 Numerical Methods

The CAA-Code PIANO [1] applies curvilinear, structured grids. Besides the above described extended viscous NLPE equations, it also provides the computation of the Linearized Euler Equations (LEE), Acoustic Perturbation Equations (APE) and non-linear Euler equations in primitive disturbance form as governing equations. Spatial gradients are approximated by using the Dispersion Relation Preserving (DRP) scheme, as proposed by Tam and Webb [10]. The temporal discretization is achieved with the 4th-order low-dispersion Runge-Kutta (LDDRK) algorithm proposed by Hu et al. [11]. Furthermore the possibility of filtering of contaminating short wave components of the wave spectrum is provided by artificial selective damping. The additional terms in the equation as compared to PIANO as a pure aeroacoustics simulation code are the viscous and heatflux terms. These additional terms involve second derivatives, which could be tackled by a successive application of the first derivative operator. This would encompass the computation of 27 derivatives, as compared to 12 for the implemented version. The interested reader is referred to Moghadam [3] for a detailed description of the developed divergence computation.

4 Circular Cylinder in Uniform Flow Testcase

4.1 Background Flow, Computation Setup, and Initialisation

The DLR TAU-code is utilised to obtain a background flow as a RANS solution (as the flow is laminar, this constitutes a Navier-Stokes solution). For this purpose, a half cylinder is chosen so as to prevent (unsteady) vortex shedding besides the smaller computational domain and time. The boundary condition applied to the half cylinder wall is a viscous no-slip condition and a slip-wall one is applied to the symmetry plane. The laminar simulations have been performed for one Reynolds number (Re $= 150$) and for three Mach numbers, i.e., 0.1, 0.2, and 0.3. We will mainly present results of the latter Mach number. With a speed of sound a equal to 340.26 m/s, this corresonds to $U_\infty = 102.078$ m/s. The background flow for the Overset simulation is obtained upon mirroring the RANS solution of the half cylinder about the streamwise symmetry plane. For more detailed concerning the TAU simulations, see [3]. The background flow is subsequently interpolated on the Overset grid, whose details are elucidated below.

The Overset DNS grid consists of 16 blocks with a total of approximately 400,000 points, ranging from $x \in [-100D; 100D]$ and $y \in [-100D; 100D]$. On the surface of the cylinder, having unit diameter D (equivalent to 1 m), 169 point are distributed. Based on the Reynolds number, the initial grid spacing to the wall is chosen as $\delta = 0.05$. A zoom-in of the grid topology around the cylinder is depicted in Fig. 2, whose center is located at $x = y = 0$. In order to resolve the acoustic wave a minimum of 10 points per wavelength (PPW) is chosen (above the minimum of the

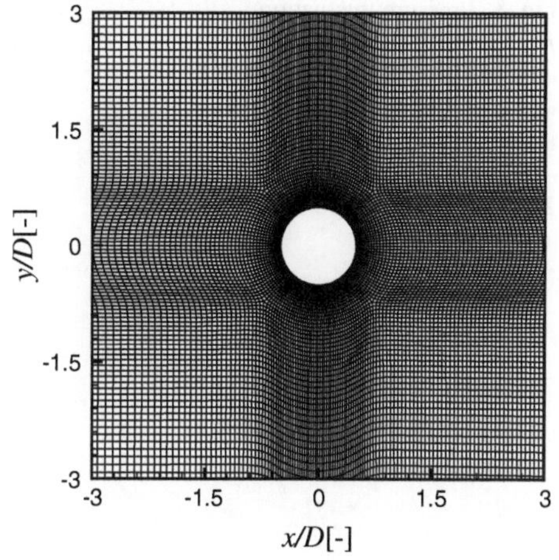

Fig. 2 Zoom-in of grid around the cylinder showing the topology. Note that every one out of three points is shown out of clarity of presentation

DRP scheme). As we are in fact conducting a multi-scale simulation here, we need to resolve the acoustic as well as the vorticity scales. The latter is the restricting length scale, and amounts to $\lambda/D \approx 5.46$. In the remainder a grid spacing of $0.5D$ is taken (resulting from 10 PPW).

On the cylinder surface adiabatic and no-slip wall boundary conditions are enforced, while radiation conditions are set to the outer domain walls. Furthermore, a sponge layer is set to the outflow boundary to damp out the strong hydrodynamic fluctuations before they reach the downstream outflow border. Based on test simulations, the thickness of this sponge layer was set to $30D$, equating to 150 grid points .The integration time step Δt was set to 4.3×10^{-6} s, based on the inviscid CFL criteria. To expedite the transition to unsteady vortex shedding of the cylinder, the simulation is initialised with a Taylor-like vortex at $t = 0$, initially located at $(x, y) = (1, 0)$. With time, this initialisation vortex convects out of the computational domain, where after data acquisition is commenced.

4.2 Results of the Cylinder in Uniform Flow

In Fig. 3 the instantaneous pressure fluctuations $\tilde{p}'$ are presented in isolines at $t = 1,750$. Note that for aeroacoustics the correct temporal evolution of the pressure fluctuations is important and the (residual) mean pressure should be subtracted as this masks the acoustic content, i.e., $\tilde{p}' = p'_{\text{PIANO}} - \overline{p'}$ (where p'_{PIANO} is the computed perturbation pressure from PIANO). In Fig. 3 it is seen that the hydrodynamic fluctuations in the cylinder wake are much stronger (2 orders of magnitude) then the acoustic ones. The pressure fluctuations strongly radiate in the flow-normal direction (dipole radiation character) and the convective amplification can be appreciated in the upstream direction. Note that this is a Direct Noise Computation (DNC) result, on top of being a DNS.

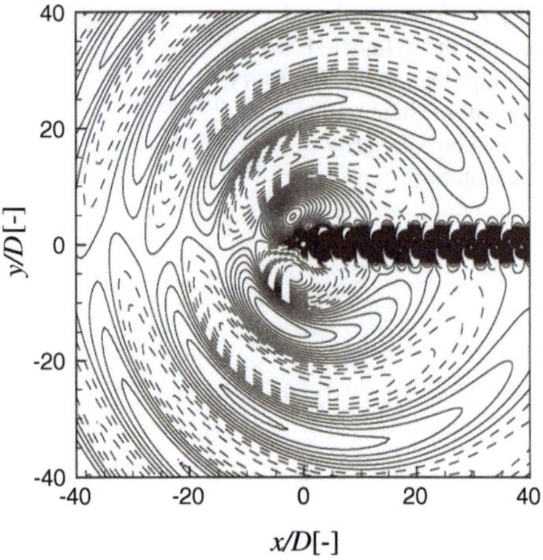

Fig. 3 Pressure fluctuation $\tilde{p}'$ around the 2D-cylinder in uniform flow (M = 0.3) and Re = 150 plotted with isolines (levels from $-1.0\,M^{2.5}$ to $1.0\,M^{2.5}$ with $2.5 \times 10^{-3}\,M^{2.5}$ increments)

For the considered Reynolds number, the Strouhal number St can be estimated from the empirical relationship $\mathrm{St} = 0.2684 - 1.0356\mathrm{Re}^{-0.5}$ as 0.183 (see [12]). This yields a vortex shedding frequency $f = (\mathrm{St} \cdot \mathrm{M} \cdot a)/D \approx 18.67$, where a denotes the speed of sound. Two snapshots of the vorticity are depicted in Fig. 4, separated by half the vortex shedding frequency. Indeed it can be seen that the snapshots are each others mirror about the $y = 0$ axes. The shedding frequency is further illustrated in Fig. 5. In this figure, sound pressure level of the pressure fluctuations L_p and acoustic particle velocity level L_v are presented. The latter is defined as follow

$$L_v = 20\log(\frac{|\mathbf{v}|}{v_0}), \tag{4}$$

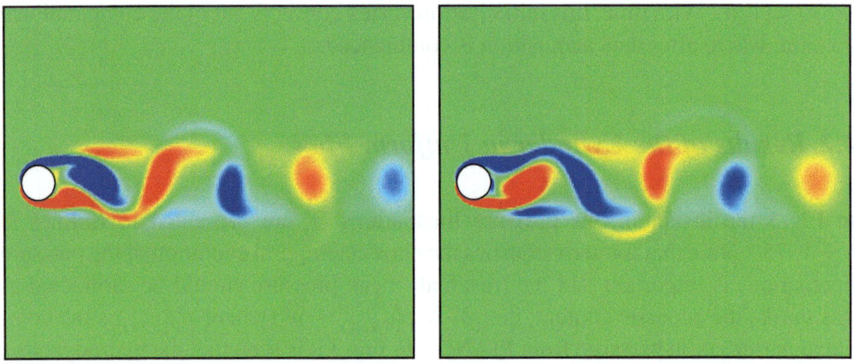

Fig. 4 Snapshots of vorticity, separated by half the vortex shedding period for M = 0.3 and Re = 150

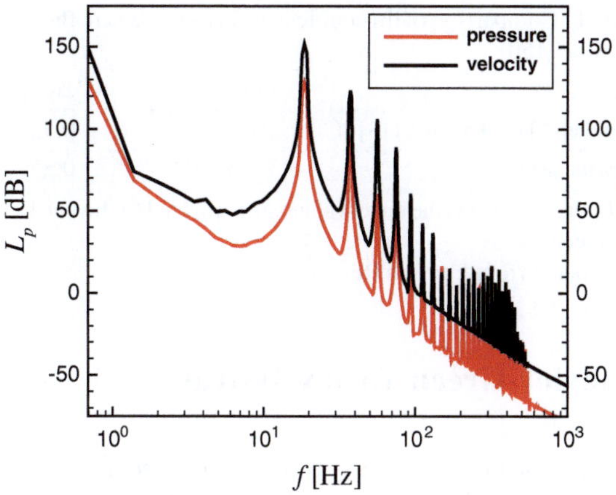

Fig. 5 Spectrum of the pressure fluctuation L_p and the velocity L_v for the case M = 0.3 and Re = 150

with v_0 the reference particle velocity 5.0×10^{-8} m/s. It is observed in Fig. 5 that, as expected, the acoustics and hydrodynamic fluctuations exhibit the same frequency (see, e.g., [13]). The numerical value of this shedding frequency matches the empirical prediction exactly.

Snapshots of the pressure fluctuations are presented as function of distance r in Fig. 6 (left), where different lines correspond to different time instances. The peak values decrease with increasing distance. This tendency is further illustrated in the right picture of Fig. 6 where the decay of the peak pressure fluctuation levels are presented. Comparison with the theoretical expected decay of $r^{-0.5}$ reveals an excellent agreement for the different considered Mach numbers.

Finally, in Table 1 values for the lift and drag coefficients are presented. From this table it can be clearly seen that the mean as well as the amplitude are in good agreement with the values reported in Inoue and Hatakeyama [13].

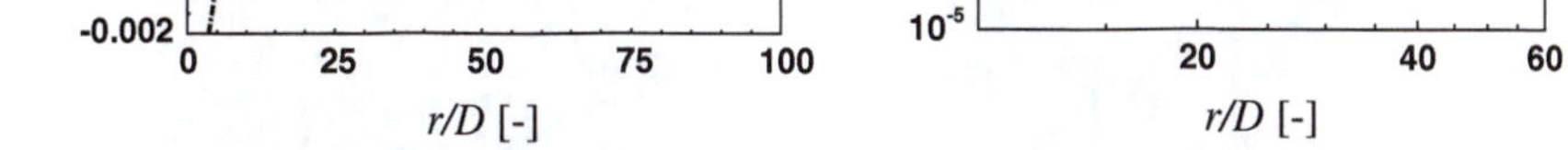

Fig. 6 *Left* propagation of pressure waves perpendicular to mean flow of 2D-Cylinder for M = 0.3 and *right* decay of pressure peaks compared to the theoretical decay of $r^{-1/2}$ for two Mach numbers. For both pictures Re = 150

Table 1 Comparison of lift coefficients (c_l) and drag coefficients (c_d) of the cylinder case (M = 0.3 and Re = 150)

	$\overline{c_l}$ [–]	Δc_l [–]	$\overline{c_d}$ [–]	Δc_d [–]
Inoue and Hatakeyama [13]	0.0	0.52	1.38[a]	2.6×10^{-2}
Present study	8.5×10^{-4}	0.53	1.35	2.8×10^{-2}

For both coefficients the mean values are presented (indicated by an overbar), as well as the amplitude (indicated by Δ)
[a] Estimated from Fig. 3 in Ref. [13]

5 Taylor-Green Vortex Testcase

The Taylor-Green Vortex is an often used test case for the determination of the accuracy of scale resolving, higher-order codes. The initial condition consists of a 3D periodic cube on which a relatively simple analytical vortex structure is set. Initially, this vortex structure has only one length scale L (the vortex length scale, equal to 1 m) and will decay with time into increasingly smaller length scales so that finally homogeneous isotropic turbulence develops. The initial condition is given by

$$u = \mathrm{M} \cdot \sin\left(\frac{x}{L}\right) \cdot \cos\left(\frac{y}{L}\right) \cdot \cos\left(\frac{z}{L}\right)$$
$$v = -\mathrm{M} \cdot \cos\left(\frac{x}{L}\right) \cdot \sin\left(\frac{y}{L}\right) \cdot \cos\left(\frac{z}{L}\right)$$
$$w = 0$$
$$p = \frac{\mathrm{M}^2}{16} \cdot \left[\cos\left(\frac{2x}{L}\right) + \cos\left(\frac{2y}{L}\right)\right] \cdot \left[\cos\left(\frac{2z}{L}\right) + 2\right].$$

Figure 7 (left) shows the initial condition, i.e., the regular (periodic) array of vortices in the cube with 3 periodic direction ($x, y, z \in [-\pi L; \pi L]$). The resolution

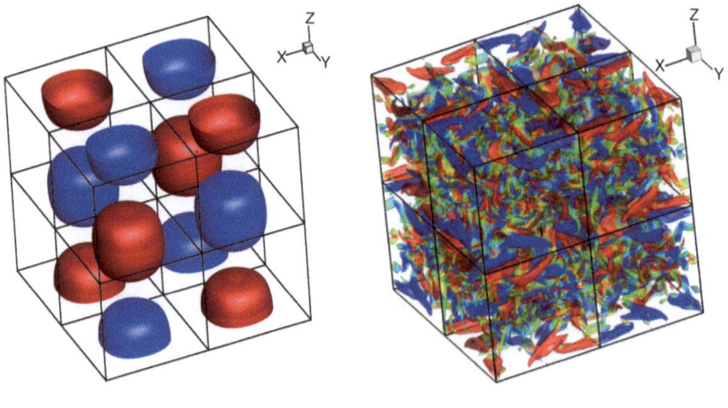

Fig. 7 *Left* Initialisation with Taylor-Green Vortex showing magnitude of vorticity as iso-surfaces and contours the z-component of the vorticity. *Right* Snapshot at $t = 20t_c$ with iso-surface indicating the magnitude of vorticity and contours the Helicity

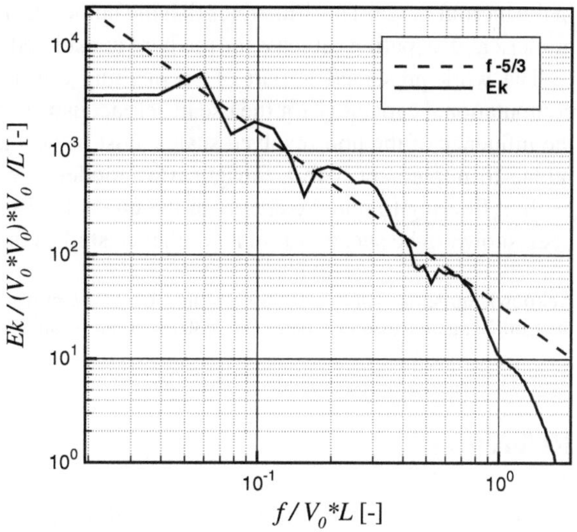

Fig. 8 One dimensional kinetic energy spectrum over frequency

of the box (uniform Cartesian grid), is chosen such that length scales are resolved up to the Kolmogorov length, yielding 3.37×10^6 points. Based on this resolution and a convective Velocity (M $= 0.1$, $a = 340.26$ m/s), a timestep Δt of 1×10^{-6} s is chosen. In Fig. 7 (right), a snapshot is displayed of the fluctuating vorticity magnitude illustrating the small scales that develop with time. The energy spectrum is presented in Fig. 8 together with the expected $-5/3$ scaling of the inertial range. The late departure from this scaling is an indication that the applied 4th-order central difference scheme is indeed low dissipative.

6 Conclusion and Outlook

Computational Aeroacoustic codes exhibit certain characteristics which make them attractive candidates for extension to a scale-resolving simulation tool, based on perturbation equations and applicable as a zonal tool. The governing equations have been shortly presented as well as some numerical aspects. The quality of the results have been demonstrated with the aid of two test cases, i.e., the cylinder in uniform flow and the Taylor-Green vortex. The former testcase illustrates the ability to simultaneously act as a DNS as well as a Direct Numerical Noise (DNC) tool. A very good agreement with respect to Strouhal number, pressure decay behaviour, and lift and drag coefficients was shown between the here presented results and Inoue and Hatakeyama [13]. From this 2Dtestcase it was concluded that the viscous terms are correctly implemented. The second testcase constitutes of a 3D periodic box, which was initialised with Taylor-Green Vortices and then left to decay. Good qualitative agreement was shown, as well as the expected $-5/3$ scaling of the energy spectrum in the inertial range. This illustrates that the numerical method introduces only minor numerical dissipation (at least for the here considered case).

The work presented is part of an ongoing effort to build up an Overset LES computational capability for the use in aeroacoustic source studies. For this purpose, the influence of the non-resolved scales must be accounted for by some means. We intend to implement here a Smagorinsky model (see, e.g., [14]) as well as to use explicit filtering to remove energy at the unresolved scales. The latter approach has been shown to be successful in a number of studies (see, e.g., Marsden et al. [15]).

Acknowledgments This work has been funded by the German Science Foundation (DFG) within the Collaborative Research Centre CRC880 (Sonderforschungsbereich 880, SFB880).

References

1. Delfs, J.W., Bauer, M., Ewert, R., Grogger, H.A., Lummer, M., Lauke, T.G.W.: Numerical Simulations of Aerodynamic Noise with DLRs aeroacoustic code PIANO, vol. Handbook Version 5.2, (2008)
2. Ewert, R., Bauer, M., Lummer, M.: A review of state-of-the-art aeroacoustic prediction approaches. In: Dénos, R., Lecomte, E., Kors, E., Schram, C. (eds.) Aircraft Noise, Monograph VKI Lecture Series 2012–02, Aircraft Noise, von Kármán Institute for Fluid Dynamics, Brussels (2012)
3. Moghadam, M.S.A.: Implementation of viscous terms into the computational aeroacoustics code PIANO. MSc.-thesis TU Braunschweig (2012)
4. Bailly, C., Bogey, C., Marsden, O.: Progress in direct noise. Int. J. Aeroacoust. **9** (2010)
5. Terracol, M.: A zonal RANS/LES approach for noise sources prediction. Flow Turbul. Combust. **77** (2006)
6. Fröhlich, J., von Terzi, D.: Hybrid LES/RANS methods for the simulation of turbulent flows. Prog. Aerosp. Sci. **44** (2008)
7. Sesterhenn, J.: A characteristic-type formulation of the equations for high order upwind schemes. Comput. Fluids **30** (2001)
8. Anderson, J.D.: Computational Fluid Dynamics: Basics with Applications. McGraw-Hill, New York (1995)
9. Schwamborn, D., Gerhold, T., Kessler, R.: DLR-TAU code—an overview. In: 1st ONERA/DLR Aerospace Symposium (1999)
10. Tam, C.K., Webb, J.C.: Dispersion-relation-preserving flnite difference schemes for computational aeroacoustics. J. Comp. Phys. **107** (1993)
11. Hu, F.Q., Hussaini, M.Y., Manthey, J.: Low-dissipation and low-dispersion Runge-Kutta schemes for computational acoustics. J. Comput. Phys. **124** (1996)
12. Fey, U., König, M., Eckelmann, H.: A new Strouhal-Reynolds-number relationship for the circular cylinder in the range $47 \leq Re \leq 2 \times 10^5$. Phys. Fluids **10** (1998)
13. Inoue, O., Hatakeyama, N.: Sound generation by a two-dimensional circular cylinder in a uniform flow. J. Fluid Mech. **471** (2002)
14. Sagaut, P.: Large Eddy Simulation for Incompressible Flows: An Introduction, 3rd edn. Springer, Berlin (2006)
15. Marsden, O., Bogey, C., Bailly, C.: Direct noise computation of the turbulent flow around a zero-incidence airfoil. AIAA J. **46** (2008)

High-Order Flux Reconstruction Schemes for LES on Tetrahedral Meshes

Jonathan R. Bull and Antony Jameson

Abstract The use of the high-order Flux Reconstruction (FR) spatial discretization scheme for LES on unstructured meshes is investigated. Simulations of the compressible Taylor-Green vortex at $Re = 1{,}600$ demonstrate that the FR scheme has low numerical dissipation and accurately reproduces the turbulent energy cascade at low resolution, making it ideal for high-order LES. To permit the use of subgrid-scale models incorporating explicit filtering on tetrahedral meshes, a high-order filter acting on the modal form of the solution (i.e. the Dubiner basis functions) is developed. The WALE-Similarity mixed (WSM) model using this filter is employed for LES of the flow over a square cylinder at $Re = 21{,}400$, obtaining reasonable agreement with experiments. Future research will be directed at improved SGS models and filters and at developing high-order hybrid RANS/LES methods.

1 Introduction

Advances in computing hardware are bringing large eddy simulation (LES) of real applications at high Reynolds numbers into the realms of possibility. However, most LES solvers are based on dissipative second-order accurate ($p = 2$) numerical schemes which have poor wave propagation properties. This limits the usefulness of LES in many applications where the turbulent motions must be accurately resolved far downstream from their source. These severe problems can be overcome by using high-order accurate ($p > 2$) spatial discretizations such as Discontinuous Galerkin (DG) [17], Spectral Difference (SD) [23] and Flux Reconstruction (FR) [18] methods. A family of energy-stable FR schemes have been developed by the Aerospace

J.R. Bull (✉) · A. Jameson
Department of Aeronautics and Astronautics, Stanford University,
Stanford, CA 94305, USA
e-mail: jbull@stanford.edu

S. Girimaji et al. (eds.), *Progress in Hybrid RANS-LES Modelling*,
Notes on Numerical Fluid Mechanics and Multidisciplinary Design 130,
DOI 10.1007/978-3-319-15141-0_5

Computing Lab (ACL) at Stanford University [34] based on the work of Huynh [18]. By varying a stabilization parameter, a range of existing schemes can be recovered including nodal DG and SD. The FR method has several advantages over DG: it is conceptually simpler, easier to implement (including on unstructured meshes) and cheaper to compute due to casting the equations in differential form (avoiding costly integration procedures). The FR schemes are proposed as an ideal basis for turbulence-resolving simulations due to their flexibility, stability and low numerical dissipation. As a demonstration of this, the compressible Taylor-Green vortex benchmark problem at $Re = 1{,}600$ was simulated. Results show that the FR schemes are highly accurate compared to DNS and stable at low resolution.

Various stabilized high-order schemes have been successfully used for implicit LES (ILES) of turbulent flows (i.e. exploiting numerical dissipation as an ersatz turbulence model) [5, 6, 21]. Turbulent simulations using the FR method to recover the SD scheme have also been successful, including channel flow at $Re_\tau = 400$ [22] and transitional flow over an airfoil at $Re = 60{,}000$ [9]. However, in the case of separating flow over a square cylinder at $Re = 21{,}400$ it was found that a subgrid-scale (SGS) model was needed to improve accuracy in the turbulent wake [24]. Other studies have also shown that high-order schemes benefit from the inclusion of an SGS model [11, 13, 30]. To permit the use of advanced SGS models based on explicit filtering, including the similarity and dynamic models, discrete high-order filters were developed for hexahedral elements by Lodato et al. as a tensor product of 1D filter operators [24]. The filters were designed to have consistent cutoff length scales at different polynomial orders and at all solution points within the element. Support for the filters is limited to the element interior to maintain the efficient parallelization of the FR method. In this paper, a filter for tetrahedral elements is developed, enabling SGS models based on explicit filtering to be applied to unstructured meshes. The filter acts on the modal form of the solution and is based on the Spectral Vanishing Viscosity (SVV) stabilization technique for DG methods [16, 20, 28, 31]. The flow over a square cylinder at $Re = 21{,}400$ was simulated using the FR method with the WALE-Similarity Mixed (WSM) model, with reasonably accurate results obtained on a relatively coarse unstructured mesh. Future research will develop better SGS models and near-wall models including high-order hybrid RANS-LES techniques.

2 High-Order Flux Reconstruction

The compressible Navier-Stokes equations are discretized using the high-order Flux Reconstruction scheme. We write the equations in conservative form in a 3D domain Ω with spatial coordinates $\mathbf{x} = \{x_1, x_2, x_3\}$ and time t:

$$\frac{\partial \mathbf{u}}{\partial t} + \frac{\partial \mathbf{f}}{\partial \mathbf{x}} = 0\,, \tag{1}$$

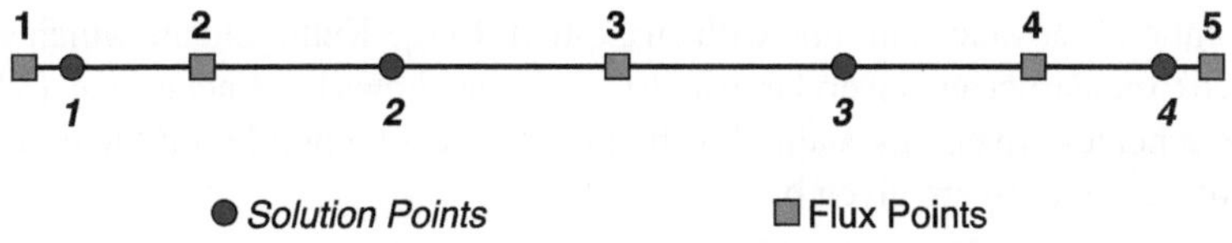

Fig. 1 Location of solution and flux points in 1D

where $\mathbf{u} = (\rho\ \rho u\ \rho v\ \rho w\ \rho e)^T$ are the conservative variables and $\mathbf{f}$ is the flux. The domain is split into non-overlapping elements Ω_i. Elements are transformed to a reference element described by local coordinates $\xi = \{\xi_1, \xi_2, \xi_3\}$. For simplicity, we consider the reference element in one dimension ($\xi = [-1 : 1]$) and transform the 1D analog of (1) to the reference coordinates. Inside the reference element a degree $p = N - 1$ polynomial is defined on a set of N solution points, resulting in an Nth-order accurate scheme. The Gauss-Legendre quadrature points (Fig. 1) are chosen as the solution points as they were found to minimize aliasing errors for nonlinear problems [19]. The Gauss-Lobatto points are used as the flux points (Fig. 1). The piecewise-continuous pth-order solution polynomial u is defined

$$u(\xi) = \sum_{i=1}^{N} u_i l_i(\xi)\,, \tag{2}$$

where u_i are the nodal solution values and l_i is a set of basis functions, in this case the Lagrange polynomials. A similar expression is used to obtain the $(p+1)$th-order flux polynomial f. Note that the flux is discontinuous across element boundaries; common interface fluxes are found using a two-point upwind-biased flux formula such as Roe's method [32]. The next step is to construct a globally continuous flux polynomial. In the FR method this is achieved by adding a flux correction polynomial Δf to the discontinuous flux f. The correction satisfies: (a) $f + \Delta f$ equals the common interface fluxes, and (b) the corrected flux optimally represents the discontinuous flux in the element interior. Δf is given by

$$\Delta f(\xi) = [f_L^* - f(-1)]g_L(\xi) + [f_R^* - f(1)]g_R(\xi)\,, \tag{3}$$

where f_L^*, f_R^* are the common interface fluxes at left and right interfaces and $g_L(\xi), g_R(\xi)$ are order $p+1$ polynomial *correction functions* satisfying $g_L(-1) = g_R(1) = 1,\ g_L(1) = g_R(-1) = 0,\ g_L(\xi) = g_R(-\xi)$. Now the corrected, globally C_0-continuous flux f^C is given by $f^C = f + \Delta f$. The final stage of the FR process is to calculate the divergence of f^C at each solution point ξ_i using the expression

$$\frac{\partial f^C}{\partial \xi}(\xi_i) = \sum_{j=1}^{N} f_j \frac{\partial l_j}{\partial \xi}(\xi) + [f_L^* - f(-1)]\frac{\partial g_L}{\partial \xi}(\xi_i) + [f_R^* - f(1)]\frac{\partial g_R}{\partial \xi}(\xi_i). \tag{4}$$

The solution is advanced in time with an explicit Runge-Kutta scheme which avoids the need to construct and invert large matrices. It was shown by Vincent et al. [34] that the FR schemes are energy-stable for the linear advection problem if the correction functions g_L and g_R are given by

$$g_L = \frac{(-1)^p}{2}\left[L_p - \left(\frac{\eta_p(c)L_{p-1} + L_{p+1}}{1 + \eta_p(c)}\right)\right], \tag{5}$$

$$g_R = \frac{(-1)^p}{2}\left[L_p - \left(\frac{\eta_p(c)L_{p-1} + L_{p+1}}{1 + \eta_p(c)}\right)\right], \tag{6}$$

where L_p is the degree p Legendre polynomial, $\eta_p(c) = (c(2p+1)(a_p p!))^2)/2$ and $0 \leq c \leq \infty$ is a user-defined stability parameter. The proof was extended to linear advection-diffusion problems by Castonguay et al. [10]. The baseline case, $c = 0$, corresponds to a nodal DG scheme which is susceptible to aliasing-driven instabilities. The particular choice of $c > 0$ determines the amount of additional stabilizing dissipation and allows one to recover other schemes such as SD [34]. Allaneau and Jameson [3] showed that $c > 0$ corresponds to damping of the highest-order solution mode. Stabilization techniques are commonly used with DG, either by direct filtering [15] or by including an equivalent high-order dissipative term [7, 31]. As a demonstration of the stability and accuracy of FR for turbulent flow computations, results of the compressible Taylor-Green vortex are presented next.

3 Taylor-Green Vortex

The Taylor-Green vortex is a simple yet powerful tool for studying the ability of a numerical method to represent the turbulent energy cascade. Starting from an analytical solution defining a single length scale, nonlinear interactions between eddies generate a broadband turbulent spectrum which decays in a well-defined manner. Here, we run the compressible benchmark problem at $Re = 1{,}600$ (part of the 1st and 2nd International Workshops on High-Order CFD Methods [1, 2]) and compare results to high-order DG computations [5] and DNS [12]. Figure 2a–d show the volume-averaged kinetic energy $\langle k \rangle$ and dissipation rate $-d\langle k \rangle/dt$ using FR to recover the fourth-order-accurate SD scheme on regular tetrahedral meshes formed by splitting hexahedral meshes of 16^3, 32^3 and 64^3 elements. On the 64^3 mesh the kinetic energy (Fig. 2a) and dissipation rate (Fig. 2b) predictions match the DNS data [12] and the high-order results at equal resolution [5], showing that numerical dissipation dissipates the turbulent energy at the correct rate. At lower mesh resolution the kinetic energy decays more rapidly, implying that the numerical dissipation is too strong. Figure 2c shows the results of the third to sixth order accurate

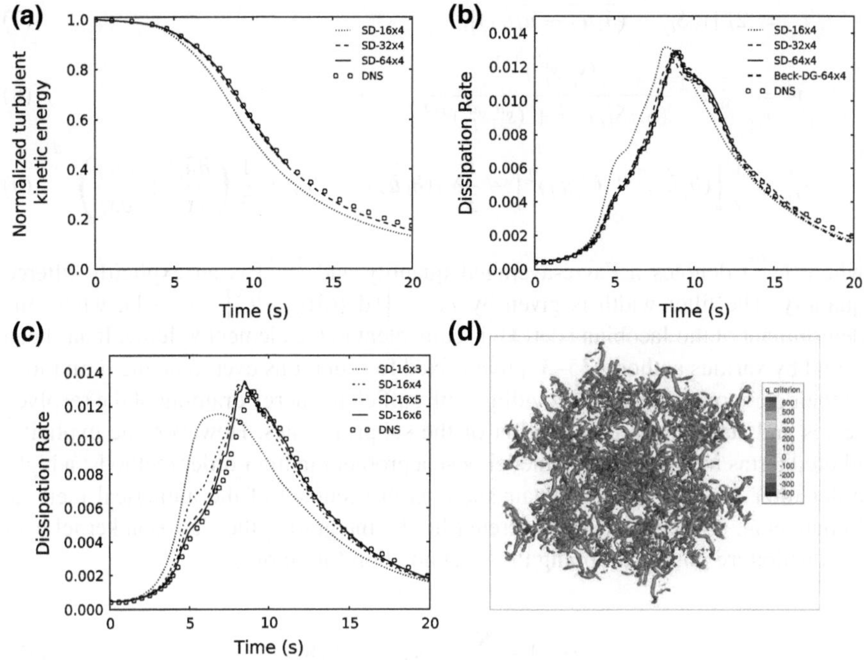

Fig. 2 Taylor-Green vortex results on tetrahedral meshes **a** Evolution of average kinetic energy $\langle k \rangle$. Dissipation rate $-d\langle k \rangle/dt$ with different **b** meshes and **c** orders **d** Isosurface of Q-criterion at 10.75 s on 64^3 mesh at 4th order, showing vortex filaments. 'SD-$M \times N$' refers to M^3 mesh, Nth-order accurate SD scheme. (- - -) 4th-order DG on 64^3 mesh [5]; (○) DNS [12]

SD schemes on the 16^3 tetrahedral mesh. At third order the dissipation peak is shifted to the left, indicating that numerical dissipation is too strong. As the order increases, the accuracy improves, with the biggest change being from third to fourth order. Instabilities tend to occur for orders higher than six, particularly on coarse meshes, implying that fourth order may be optimal for high-order LES. Finally Fig. 2d shows an isosurface of the Q criterion at 10.75 s, demonstrating the ability of the high-order method to maintain a finely detailed turbulent spectrum after a long time by virtue of the low dissipation.

4 LES Modeling

Previous work by Lodato et al. [24] found that the addition of an SGS model to the SD method improved the accuracy of first and second moments in a separated turbulent flow. A mixed SGS model was employed, comprising the WALE eddy-viscosity model [29] and the similarity model [4]:

$$\tau_{ij} = 2\overline{\rho}[\nu_t \tilde{S}_{ij} - (\overline{\tilde{u}_i \tilde{u}_j} - \overline{\tilde{u}}_i \overline{\tilde{u}}_j)], \tag{7}$$

$$\nu_t = C_w^2 \, \Delta^2 \, \frac{(s_{ij}^d s_{ij}^d)^{3/2}}{(S_{ij} S_{ij})^{5/2} + (s_{ij}^d s_{ij}^d)^{5/4}}, \tag{8}$$

$$s_{ij}^d = \frac{1}{2}\left[(\partial_j \tilde{u}_i)^2 + (\partial_i \tilde{u}_j)^2\right] - \frac{1}{3}\delta_{ij}(\partial_k \tilde{u}_k)^2, \quad S_{ij} = \frac{1}{2}\left(\frac{\partial \tilde{u}_i}{\partial x_j} + \frac{\partial \tilde{u}_j}{\partial x_i}\right), \tag{9}$$

where $(\tilde{\ldots})$ denotes a Favre-averaged quantity and $\overline{(\ldots)}$ is an explicitly filtered quantity. The filter width is given by $\Delta = \left[|\det(\mathbf{J}(\xi))|^{1/3}\right]/(p+1)$, where the determinant of the Jacobian $|\det(\mathbf{J})|$ is equivalent to the element volume. It has been found by various authors [35–37] that mixed formulations overcome the limitations of their component models, providing both sufficient energy draining of the resolved scales and structural representation of the subgrid scales. However, the explicitly filtered terms in the similarity model pose a problem in high-order methods: a high-order filter is required to maintain the overall accuracy of the numerical method. Lodato et al. [24] developed a discrete filter by integrating the Gaussian kernel over the quadrature points (dropping the $(\tilde{\ldots})$ notation for clarity):

$$\overline{u}(\xi_s) = \sum_{i=1}^{N} G(\xi_s - \xi_i) u_i w_i, \tag{10}$$

where u_i are the nodal values of the solution, w_i are the quadrature weights and $G(\xi_s - \xi_i)$ is the Gaussian filter kernel centered at ξ_s evaluated at a point ξ_i. A filter for hexahedral elements is then simply the tensor product of three 1D filters.

An alternative approach in high-order methods is to filter the *modal* form of the solution (in 1D):

$$u = \sum_{i=0}^{p} \hat{u}_i L_i(\xi), \tag{11}$$

where $\hat{u}_i$ are the *modal coefficients* and $L_i(\xi)$ are the normalized Legendre polynomials. Since the Legendre polynomials form an orthogonal basis, the modal form (11) can be viewed as a pseudo-spectrum and the action of damping the high-order modes is analogous to low-pass filtering. Modal filtering is commonly applied as a stabilization technique in high-order methods on structured grids by taking tensor products of 1D modal filters [16, 20, 31]. A natural basis for high-order discretizations on simplex elements (triangles and tetrahedra) is the *Dubiner basis* [14, 33], given by (in 3D):

$$\phi_{ijk}(r,s,t) = P_i^{(0,0)}(a)P_j^{(2i+1,0)}(b)P_k^{(2i+2j+2,0)}(c), \tag{12}$$

$$a = \left(\frac{2(1+r)}{-s-t} - 1\right)\left(\frac{-s-t}{2}\right), \; b = \left(2\frac{1+s}{1-t} - 1\right)\left(\frac{1-t}{2}\right), \; c = t\,, \tag{13}$$

where $P_j^{(A,B)}$ is the 1D jth-order Jacobi polynomial and r, s, t are local coordinates on the reference tetrahedron $\mathbf{T}^3 = \{(r,s,t) \in \mathbf{R}^3 | r, s, t \geq -1, r+s+t \leq 0\}$. The Dubiner polynomials (12) form an orthogonal basis $\mathbf{P} = \text{span}\,\{r^i s^j t^k \mid 0 \leq i, j, k \leq p;\; 0 \leq i+j+k \leq p\}$. The tetrahedral modal solution is then expressed as

$$u = \sum_{i+j+k \leq p} \hat{u}_{ijk}\phi_{ijk}(r,s,t). \tag{14}$$

The Dubiner basis can be directly filtered, avoiding the issue of not being able to define a tensor product of 1D operators in simplex elements. Following the work of Meister et al. [28] in 2D, a 3D pth-order exponential filter operator is defined by

$$\overline{u} = \sum_{i+j+k \leq p} \sigma_{ijk}\,\hat{u}_{ijk}\,\phi_{ijk}(r,s,t)\,, \tag{15}$$

$$\sigma_{ijk}(\eta) = \exp(-\alpha\eta^{2p}), \quad \eta = \frac{i+j+k}{p+1}, \quad \alpha = C_\sigma p, \tag{16}$$

where C_σ is a filter strength coefficient. It was found that $C_\sigma \approx 0.1$ resulted in good behavior of the similarity model. In future, this parameter could be estimated using a dynamic procedure.

5 LES of Flow Over a Square Cylinder

Using the FR method to recover the fourth-order accurate SD scheme, the flow over a square cylinder of side D at $Re = 21{,}400$ and Mach 0.3 was simulated, for which LDV experimental data is available [26, 27]. The domain measured $21D \times 12D \times 3.2D$ and a tetrahedral mesh of 87,178 elements was generated, which with 20 solution points per element resulted in 1.74 M degrees of freedom. A fourth-order five-stage explicit RK scheme was used and a total time of 250 s was simulated, with time-averaged quantities calculated over the last 100 s ($\approx$5 flow-through periods). The WSM model (7) based on the tetrahedral modal filter (15) was used, with the three-layer wall model [8] used within 0.2D of the wall. The computation took around 60 h on 7 GPUs in the lab's own cluster. Figure 3a, b show the normalized mean streamwise and vertical velocity components $\langle u\rangle/u_B$ and $\langle v\rangle/u_B$ respectively along several vertical lines in the wake. Figure 4a, b show the normalized mean Reynolds stress components $\langle u'u'\rangle/u_B^2$ and $\langle u'v'\rangle/u_B^2$. The high-order

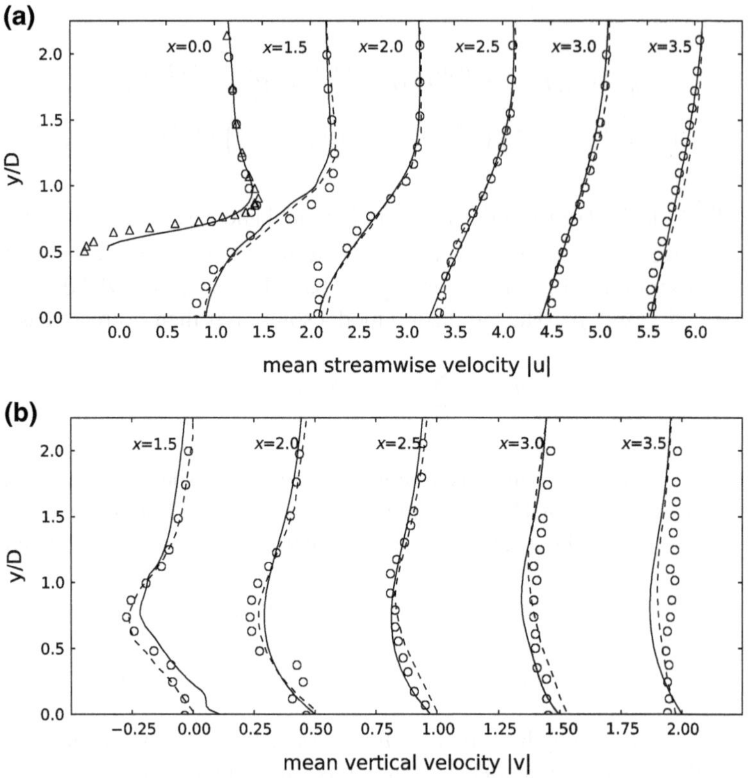

Fig. 3 **a** Mean streamwise and vertical velocity along *vertical lines* in the wake. (—) current results, (- - -) 4th order SD+WSM on hexahedral mesh by Lodato and Jameson [24], (△) LES by Lubcke et al. [25], (○) LDV experiments by Lyn et al. [26, 27]

LES results on a hexahedral mesh by Lodato and Jameson [24] are also plotted for comparison. Mean velocities are reasonably approximated, but near the cylinder the accuracy deteriorates. The Reynolds stresses are less well predicted, particularly near the cylinder. One reason for the inaccurate results is that the near-wall mesh is quite coarse. Another reason is that the three-layer wall model [8] is not well suited to separated flows, as seen in the first subplot of Fig. 3a which shows the mean streamwise velocity profile through the boundary layer on top of the cylinder ($x = 0.0$) compared to LES by Lubcke et al. [25]. It is clear that the recirculation bubble is not predicted by the wall model. The use of more suitable wall modeling approaches is left as a matter for future research.

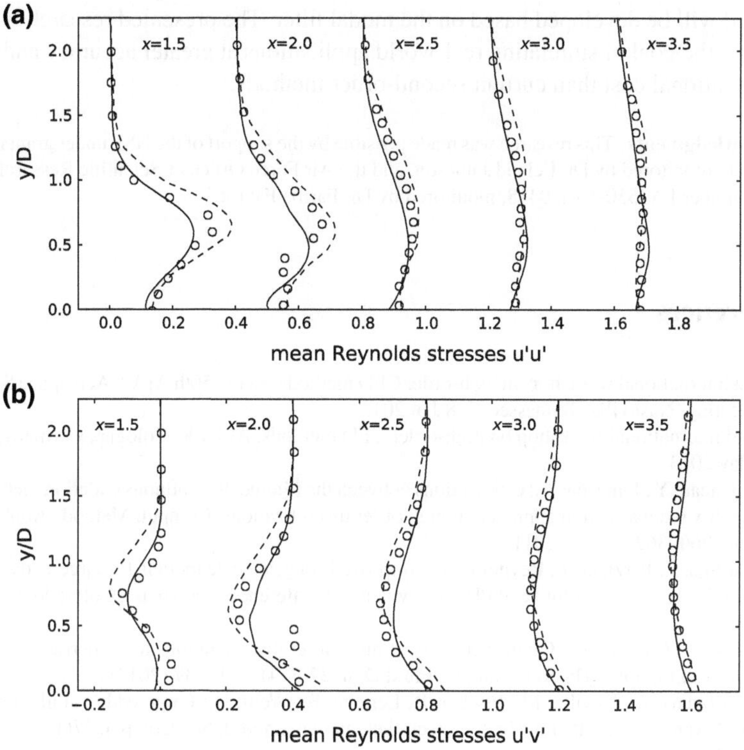

Fig. 4 **a** Mean Reynolds stresses along vertical lines in the wake. (—) current results, (- - -) 4th order SD+WSM on hexahedral mesh by Lodato and Jameson [24], (○) LDV experiments by Lyn et al. [26, 27]

6 Conclusions

The high-order Flux Reconstruction (FR) method was applied to turbulent flows on tetrahedral meshes. Simulated energy dissipation rates for the Taylor-Green vortex were in excellent agreement with DNS, suggesting that the FR method's low numerical dissipation and good stability make it a strong candidate for high-order LES. A high-order modal filter was developed, enabling the use of SGS models based on explicit filtering including the mixed WALE-similarity model tested here. Results of the flow over a square cylinder were reasonably close to experimental data at low mesh resolution. More work is needed to improve the accuracy of high-order LES with the FR method, including studying the sensitivity to the mesh, choice of high-order scheme, order of accuracy, SGS model and wall model. In future, work will focus on modeling of the near-wall region in complex turbulent flows, in particular with hybrid RANS-LES methods. In addition, a high-order dynamic LES method will be developed based on the modal filter. The presented research is a step towards the goal of simulating real-world applications at greater accuracy and lower computational cost than current second-order methods.

Acknowledgments This research was made possible by the support of the NSF under grant number 1114816, monitored by Dr. Leland Jameson, and the Air Force Office of Scientific Research under grant number FA9550-10-1-0418, monitored by Dr. Fariba Fahroo.

References

1. 1st international workshop on high-order CFD methods. At the 50th AIAA Aerospace Sciences Meeting, Nashville, Tennessee, 7–8 Jan 2012
2. 2nd international workshop on high-order CFD methods. At DLR, Cologne, Germany, 27–28 May 2013
3. Allaneau, Y., Jameson, A.: Connections between the filtered discontinous Galerkin method and the flux reconstruction approach to high order discretizations. Comput. Methods Appl. Mech. Eng. **200**, 3628–3636 (2011)
4. Bardina, J., Ferziger, J., Reynolds, W.: Improved subgrid-scale models for large-eddy simulation. In: 13th AIAA Fluid and Plasma Dynamics Conference. Snowmass, Colorado, 4–16 July 1980
5. Beck, A., Gassner, G.: On the accuracy of high-order discretizations for underresolved turbulence simulations. Theoret. Comput. Fluid Dyn. **27**(3–4), 221–237 (2012)
6. Bouffanais, R., Deville, M., Fischer, P., Leriche, E., Weill, D.: Large-eddy simulation of the lid-driven cubic cavity flow by the spectral element method. J. Sci. Comput. **27**(1–3), 151–162 (2006)
7. Boyd, J.: The Legendre-Burgers equation: when artificial dissipation fails. Appl. Math. Comput. **217**, 1949–1964 (2010)
8. Breuer, M., Rodi, W.: Large-eddy simulation of turbulent flow through a straight square duct and a 180 bend. Direct and Large-Eddy Simulation I, pp. 273–285. Springer, Berlin (1994)
9. Castonguay, P., Liang, C., Jameson, A.: Simulation of transitional flow over airfoils using the spectral difference method. In: AIAA, pp. 2010–4626 (2010)
10. Castonguay, P., Williams, D., Vincent, P.E., Jameson, A.: Energy stable flux reconstruction schemes for advection-diffusion problems. Comput. Methods Appl. Mech. Eng. **267**, 400–417 (2013)
11. Chapelier, J.B., Plata, M., Renac, F.: Inviscid and viscous simulations of the Taylor-Green vortex flow using a modal Discontinuous Galerkin approach. In: AIAA paper, pp. 2012–3073 (2012)
12. Debonis, J.: Solutions of the Taylor-Green vortex problem using high-resolution explicit finite difference methods. In: AIAA paper, pp. 2013–0382 (2013)
13. Diosady, L., Murman, S.: Design of a variational multiscale method for turbulent compressible flows. In: AIAA paper, pp. 2013–2870 (2013)
14. Dubiner, M.: Spectral methods on triangles and other domains. J. Sci. Comput. **6**(4), 345–390 (1991)
15. Fischer, P., Mullen, J.: Filter-based stabilization of spectral element methods. C. R. Acad. Sci. I-Math. **332**(3), 265–270 (2001)
16. Gottlieb, D., Hesthaven, J.: Spectral methods for hyperbolic problems. J. Comput. Appl. Math. **128**, 83–131 (2001)
17. Hesthaven, J., Warburton, T.: Nodal discontinuous Galerkin Methods: algorithms, analysis, and applications. Springer, New York (2007)

18. Huynh, H.: A flux reconstruction approach to high-order schemes including discontinuous Galerkin methods. In: AIAA paper (2007). 18th AIAA Computational Fluid Dynamics Conference, Miami, FL, 25–28 Jun 2007
19. Jameson, A., Vincent, P.E., Castonguay, P.: On the non-linear stability of flux reconstruction schemes. J. Sci. Comput. **50**, 434 (2011)
20. Karamanos, G., Karniadakis, G.: A spectral vanishing viscosity method for large-eddy simulations. J. Comput. Phys. **163**(1), 22–50 (2000)
21. Lee, E., Gunzburger, M.: A finite element, filtered eddy-viscosity method for the Navier-Stokes equations with large Reynolds number. J. Math. Anal. Appl. **385**, 384–398 (2012)
22. Liang, C., Premasuthan, S., Jameson, A., Wang, Z.: Large eddy simulation of compressible turbulent channel flow with spectral difference method. In: AIAA paper **402** (2009). 47th AIAA Aerospace Sciences Meeting Including the New Horizons Forum and Aerospace Exposition, p. 15. Orlando, FL, 5–8 Jan 2009
23. Liu, Y., Vinokur, M., Wang, Z.J.: Spectral difference method for unstructured grids I: basic formulation. J. Comput. Phys. **216**, 780–801 (2006)
24. Lodato, G., Castonguay, P., Jameson, A.: Discrete filter operators for large-eddy simulation using high-order spectral difference methods. Int. J. Numer. Meth. Fluids **72**(2), 231–258 (2013)
25. Lubcke, H., Schmidt, S., Rung, T., Thiele, F.: Comparison of LES and RANS in bluff-body flows. J. Wind Eng. Ind. Aerodyn. **89**, 1471–1485 (2001)
26. Lyn, D., Einav, S., Rodi, W., Park, J.: A laser-Doppler velocimetry study of ensemble-averaged characteristics of the turbulent near wake of a square cylinder. J. Fluid Mech. **304**(1), 285–319 (1995)
27. Lyn, D., Rodi, W.: The flapping shear layer formed by flow separation from the forward corner of a square cylinder. J. Fluid Mech. **267**, 353–376 (1994)
28. Meister, A., Ortleb, S., Sonar, T.: On spectral filtering for discontinuous Galerkin methods on unstructured triangular grids. Math. Schriften Kassel (2009)
29. Nicoud, F., Ducros, F.: Subgrid-scale stress modelling based on the square of the velocity gradient tensor. Flow Turbul. Combust. **62**(3), 183–200 (1999)
30. Parsani, M., Ghorbaniasl, G., Lacor, C., Turkel, E.: An implicit high-order spectral difference approach for large eddy simulation. J. Comput. Phys. **229**, 5373–5393 (2010)
31. Pasquetti, R.: High-order LES modeling of turbulent incompressible flow. C.R. Méc. **333**(1), 39–49 (2005)
32. Roe, P.L.: Approximate Riemann solvers, parameter vectors and difference schemes. J. Comput. Phys. **43**, 357–372 (1981)
33. Taylor, M., Wingate, B.: The natural function space for triangular and tetrahedral spectral elements. SIAM J. Numer. Anal. (1998)
34. Vincent, P.E., Castonguay, P., Jameson, A.: A new class of high-order energy stable flux reconstruction schemes. J. Sci. Comput. **47**(1), 50–72 (2011)
35. Wang, B., Bergstrom, D.J.: A dynamic nonlinear subgrid-scale stress model. Phys. Fluids **17**, 1 (2005)
36. Winckelmans, G.S., Wray, A.A., Vasilyev, O.V., Jeanmart, H.: Explicit-filtering large-eddy simulation using the tensor-diffusivity model supplemented by a dynamic Smagorinsky term. Phys. Fluids **13**(5), 1385–1404 (2001)
37. Zang, Y., Street, R.L., Koseff, J.R.: A dynamic mixed subgrid-scale model and its application to turbulent recirculating flows. Phys. Fluids A-Fluid **5**(12), 3186–3196 (1993)

Forced Synthetic Turbulence Approach to Stimulate Resolved Turbulence Generation in Embedded LES

Daniela G. Francois, Rolf Radespiel and Axel Probst

Abstract A synthetic turbulence generator is implemented in the DLR TAU-Code, to reduce the grey area that is typical for embedded LES approaches. The selected synthetic turbulence generator builds up a velocity field of fluctuations that reproduces the anisotropy and non-homogeneity of the boundary layer taken as reference from the RANS solution. In addition, the generator is significantly improved in order to reduce the numerical divergence of the generated velocity field and also to provide convective motion of the synthetic eddies. The implementation was tested in a flat plate boundary layer and verified against experimental data for the HGR-01 airfoil. The results show that the implementation shortens the transition distance to fully developed turbulence. The obtained flow profiles for the HGR-01 airfoil compare nicely to available PIV data.

1 Introduction

Advanced versions of RANS models can provide accurate solutions of attached boundary layer and mildly separated flows. But, they become unreliable when massively separated flows need to be computed. Generally, proper approaches to perform this task are DNS and LES. Unfortunately these are often too expensive in terms of computational cost. Therefore, Detached Eddy Simulation (DES) has emerged as a potential alternative to LES. Up to now there are several versions of DES. One of them is the Delayed DES (DDES) [1], which is prone to suffer from poor shielding of the attached BL, switching prematurely to LES within the attached BL, and also from poor detection of the separation onset [2]. A new version of DDES called Algebraic DDES (ADDES) was recently implemented in the DLR TAU-Code which, utilizing

D.G. Francois (✉) · R. Radespiel
Institute of Fluid Mechanics, Technische Universität Braunschweig, 38108 Braunschweig, Germany
e-mail: d.g.francois@tu-braunschweig.de

A. Probst
Institute of Aerodynamics and Flow Technology, DLR, 37073 Göttingen, Germany

S. Girimaji et al. (eds.), *Progress in Hybrid RANS-LES Modelling*,
Notes on Numerical Fluid Mechanics and Multidisciplinary Design 130,
DOI 10.1007/978-3-319-15141-0_6

algebraic sensors provides accurate shielding of the attached boundary layer and precise switch to LES at the separation onset [3]. Nevertheless, due to the difference between RANS and LES in their approaches to solving turbulence, computations performed with ADDES often exhibit a significant grey area until the turbulence of the flow is regenerated downstream of the RANS-LES interface. The present work aims to solve this problem of the embedded LES approaches, forcing the formation of resolved turbulence by means of synthetic turbulence which is added as a volumetric source term into the governing equations at the RANS-LES interface.

2 The Synthetic Turbulence Generator

2.1 Original Model

In the present work, the synthetic turbulence (ST) generator presented in [4] was adopted. This is an extension of the method described in [5] for the generation of turbulent fluctuations for noise modelling. It defines the fluctuating velocity at each point in space and in time as a superposition of Fourier modes (Eq. 2.1).

$$\boldsymbol{v}'(r,t) = \sqrt{6} \cdot \sum_{n=1}^{N} \sqrt{q^n} \cdot \left[\boldsymbol{\sigma^n} \cdot cos\left(k^n \cdot \boldsymbol{d^n} \cdot + \varphi^n + S^n \frac{t}{\tau} \right) \right] \tag{2.1}$$

The amplitude of each mode q^n is obtained from a modified von Karman spectrum (Eqs. 2.2 and 2.3).

$$q^n = \frac{E(k^n)\Delta k^n}{\sum_{n=1}^{N} E(k^n)\Delta k^n} \tag{2.2}$$

$$E(k^n) = \left(\frac{k^n}{k_e}\right)^4 \left[1 + 2.4\left(\frac{k^n}{k_e}\right)^2\right]^{-17/6} f_\eta f_{cut} \tag{2.3}$$

Here, k_e is the wave number of the most energetic mode which is determined through the local length scale of the turbulence. The local length scale is calculated based on field turbulence properties provided by the RANS solution taken as reference.

$$l_e = \min(2 \cdot \mathrm{d_w}, \mathrm{C_l l_t}) \tag{2.4}$$

The normalized amplitude, q^n, as a function of the characteristic length scale of the turbulence, l_e, together with the parameter τ, provide the generated velocity field of fluctuations with realistic space and time correlation features. Further information can be found in [5].

Equation 2.1 provides an isotropic and homogeneous velocity field of fluctuations. To reproduce the anisotropy and non-homogeneity of the turbulent boundary layer, the auxiliary velocity field $\boldsymbol{v}'(\boldsymbol{r},t)$ is multiplied by the matrix $\boldsymbol{A}$ which is obtained by performing Cholesky decomposition to the Reynolds stress tensor, $\boldsymbol{R} = \boldsymbol{A}^T\boldsymbol{A}$.

$$\boldsymbol{u}'(\boldsymbol{r},t) = \boldsymbol{A}\cdot\boldsymbol{v}'(\boldsymbol{r},t) \tag{2.5}$$

The velocity field of fluctuation given by Eq. 2.5 returns a second moment tensor equal to the Reynolds stress tensor used to obtain the matrix $\boldsymbol{A}$.

2.2 Changes Implemented in the Original Approach

2.2.1 Correction to the Numerical Divergence

In the original formulation for the auxiliary velocity field (Eq. 2.1), $\boldsymbol{\sigma^n}$ has to be perpendicular to $\boldsymbol{d^n}$. This condition comes from [5] in order to satisfy analytically the zero divergence condition $\partial\boldsymbol{v}'(\boldsymbol{r},t)/\partial\boldsymbol{r} = 0$ for the isotropic and homogeneous velocity field generated by that equation.

When $\boldsymbol{\sigma^n}$ is perpendicular to $\boldsymbol{d^n}$, $\boldsymbol{\sigma^n}\cdot\boldsymbol{d^n}$ is equal to zero, hence the right side of Eq. 2.6 becomes zero.

$$\frac{\partial\boldsymbol{v}'(\boldsymbol{r},t)}{\partial\boldsymbol{r}} = \sqrt{6}\sum_{n=1}^{N}\sqrt{q^n}\left[\boldsymbol{\sigma^n}k^n\boldsymbol{d^n}sin\left(k^n\boldsymbol{d^n r} + \Phi^n + S^n\frac{t}{\tau}\right)\right] \tag{2.6}$$

This constraint loses its effect when the auxiliary velocity field $\boldsymbol{v}'(\boldsymbol{r},t)$ is multiplied by tensor $\mathbf{A}$ (Eq. 2.5). In addition, in a non-homogeneous boundary layer, q^n and A_{ij} vary with the wall distance z, increasing the velocity field divergence. In this case, the divergence for the anisotropic and non-homogeneous velocity field (Eq. 2.5) is given by Eq. 2.7.

$$\begin{aligned}\nabla\cdot u'(r,t) = &-\sqrt{6}\sum_{n=1}^{N}\sqrt{q^n}[(A_{11}\sigma_x^n)\cdot d_x^m + (A_{21}\sigma_x^n + A_{22}\sigma_y^n)\cdot d_y^n \\ &+ (A_{31}\sigma_x^n + A_{32}\sigma_y^n + A_{33}\sigma_z^n)\cdot d_z^n]\cdot sin(...) \\ &+ \sqrt{6}\sum_{n=1}^{N}\{\frac{1}{2}\frac{1}{\sqrt{q^n}}\frac{\partial(q^n)}{\partial z}\cdot(A_{31}\sigma_x^n + A_{32}\sigma_y^n + A_{33}\sigma_z^n))\cdot cos(...) \\ &+ \sqrt{q^n}\cdot\left(\frac{\partial A_{31}}{\partial z}\sigma_x^n + \frac{\partial A_{32}}{\partial z}\sigma_y^n + \frac{\partial A_{33}}{\partial z}\sigma_z^n\right)\cdot cos(...)\}\end{aligned} \tag{2.7}$$

After a meticulous study, it was found that the first term of Eq. 2.7 is the major contributor to the divergence. However, its contribution can be reduced following a similar criterion to [5], by defining $\boldsymbol{d}^n$ according to Eq. 2.8.

$$\boldsymbol{d}^n \perp \left\{ \begin{array}{c} A_{11,ref}\,\sigma_x^n \\ A_{21,ref}\,\sigma_x^n + A_{22,ref}\sigma_y^n \\ A_{31,ref}\,\sigma_x^n + A_{32,ref}\sigma_y^n + A_{33,ref}\sigma_z^n \end{array} \right\} \tag{2.8}$$

Here, $A_{ij,ref}$ is a suitable average along the boundary layer of A_{ij} which is kept constant in order to preserve the spatial correlation of the generated velocity field.

The second term of Eq. 2.7 also produces a high increase in the divergence of the velocity field at the outer part of the boundary layer. This effect comes from an abrupt change of the most energetic scale of turbulence at the edge of the boundary layer. Although the turbulence intensity outside of the boundary layer is negligible, the maximum of the normalized amplitude q^n (Eq. 2.2) is suddenly shifted to the smallest wave numbers and consequently it produces a large gradient of q^n in the wall normal direction $(\partial q^n/.\partial z)$.

In order to compensate this effect, the spectral distribution was modified by keeping it constant at wall distances larger than 0.8 of the boundary layer thickness, $z/\delta_0 = 0.8$. This will not affect considerably the generated velocity field of fluctuations as the amplitudes of the fluctuations at these locations are very small.

The third term of the Eq. 2.7 is the minor contributor to the divergence, having a negligible effect when compared to the first term of Eq. 2.7. No compensation was found for this term.

Figure 1 displays the total divergence of the fluctuation velocity field at different locations of the boundary layer depending on the maximum assumed wave number considered for the summation of Fourier modes that generates the velocity

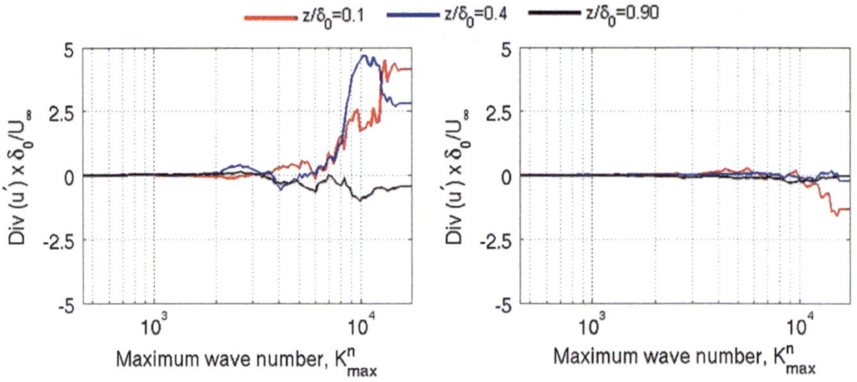

Fig. 1 Divergence of the generated velocity field of fluctuations based on the maximum wave number considered k^n_{max}. *Left* original approach [4], *Right* new approach with divergence correction

field (Eq. 2.5). The divergence is considerably reduced with the application of the previously mentioned corrections. Note, that for high maximum wave numbers, the divergence rapidly increases despite the corrections. This is because the smaller length scales of turbulence result in a high discretization error due to grid resolution.

2.2.2 Synthetic Turbulence Convection

Most of the synthetic turbulence approaches introduce the synthetic turbulence in a plane at the inlet of the LES domain. In doing so, the added fluctuations are carried through the LES domain by the mean flow. Whereas, in the present work, the synthetic turbulence is added into a volume domain behind the RANS-LES interface. As a result, the fluctuations are created within this volume changing their amplitude locally with certain time correction which is defined by the parameter τ of Eq. 2.1. But, they do not move with the stream, thus violating Taylor's frozen hypothesis. In order to correct this behavior, Eq. 2.1 was modified as shown in Eq. 2.9.

$$v'(\boldsymbol{r}, t) = \sqrt{6} \sum_{n=1}^{N} \sqrt{q^n} \left[\boldsymbol{\sigma}^n cos \left(k^n \boldsymbol{d}^n \boldsymbol{r}' + \Phi^n + S^n \frac{t'}{\tau} \right) \right]$$

$$\text{with } \boldsymbol{r}' = (x_0 \underset{ei}{\rightharpoonup} + y_0 \underset{ej}{\rightharpoonup} + z_0 \underset{ek}{\rightharpoonup}) \text{ and } t' = (t - \frac{\Delta r}{U_{conv}}) \tag{2.9}$$

In Eq. 2.9, $\boldsymbol{r}'$ is a suited reference location of the synthetic turbulence domain where the actual fluctuation was located at a previous time t' equal to the current time minus the convective time required for the fluctuation to move from the reference location to the current location. Figure 2 illustrates this concept.

Note that, in the present work, the convective velocity, U_{conv}, is chosen as 60 % of the boundary layer edge velocity.

Fig. 2 Sketch of the convection's formulation

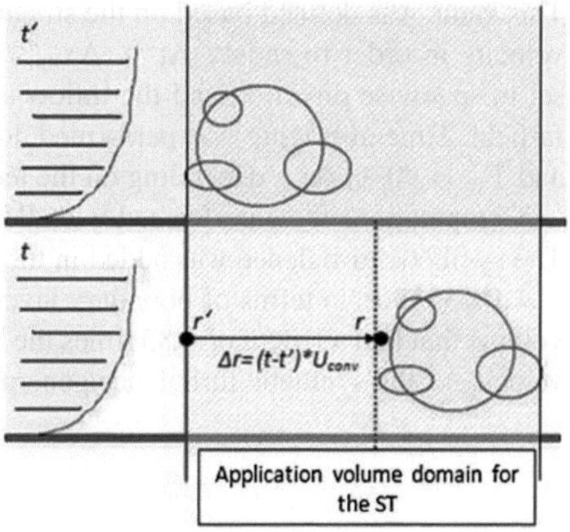

3 Implementation

The new synthetic turbulence generator was implemented in a development version of the DLR TAU-Code [6] as a volume source term of the Navier-Stokes equations. All the computations presented here were performed with JHh-v1 RSM [7] or with JHh-v2 RSM [8] as background RANS models. The non-zonal ADDES (Algebric Delayed DES) was applied as DES approach which applies algebraic boundary layer sensors in order to switch between the RANS domain and the LES domain [3]. In order to avoid excessive damping of the modelled subgrid-scale-stresses [1], the ADDES implementation of the DLR TAU-Code uses a low-Re correction that switches off the damping functions of the RSM in the LES domain [3]. Here, switching off these damping functions of the underlying RANS model in the LES domain is referred to as High-Re configuration and as Low-Re configuration when they are kept on.

4 Test Cases

4.1 Zero Pressure Gradient Flat Plate

The synthetic turbulence implementation was extensively tested for a 3D flat plate test case. The length of the flat plate was $L = 1$ m and the transition between RANS and LES was fixed at $x = 0.65\,L$. The LES domain extended from there up to the end of the flat plate, while the test section was restricted to the range between $x = 0.65$ L and $x = 0.75$ L, in order to save computational cost. Two different grids were used which only differ in their grid resolution in spanwise direction (y). The resolutions of these grids are $\Delta x^+ = 19$ (streamwise), $\Delta y^+ = 19$ (spanwise), $z^+ = 0.5$ (wall-normal) and, for the second finer mesh, $\Delta x^+ = 19$, $\Delta y^+ = 11$, $z^+ = 0.5$ [9]. All simulations were performed at a Reynolds number based on the flat plate length equal to 2.3×10^6 and a free stream velocity of 35 m/s. The physical time step was 5×10^{-6}. This value was defined based on the streamwise grid resolution and the free stream velocity in order to satisfy $\Delta t \leq \Delta x_{min}/U_\infty$. Periodic boundary conditions were set in spanwise direction and the inflow and outflow boundaries were modeled as farfield. Time averaging was performed during a period between $T_{av} = 60\ \delta_0 / U_\infty$ and $T_{av} = 90\ \delta_0 / U_\infty$ depending on the test cases.

Computations were performed with JHh-v2 RSM as RANS background model. The synthetic turbulence was added in the volume defined between $x = 0.65\,L$ and $x = 0.655\,L$ or, in terms of boundary layer thickness at the transition location, in a volume that had a length of 0.85 times the boundary layer thickness. The convective velocity of the synthetic turbulence generator was set to $U_{conv} = 0.6\,U_{edge}$.

4.2 HGR-01 Airfoil

In order to assess the methodology for local flow separation, the HGR-01 airfoil was computed and validated against PIV data and pressure measurements. The wind tunnel data used for validation were acquired in the MUB facility of the Technische Universität Braunschweig [10]. The present results with ST are compared with previous data using JHh-v1 RSM ADDES without ST [3]. To get comparable results, the same grid and settings of the computations without synthetic turbulence were applied for the present computation.

The 3D grid was constructed from a hybrid 2D mesh with 460 × 92 grid points in the structured part and 75,000 grid points in the unstructured part, which was uniformly extruded in the spanwise direction up to a distance of 0.15 chord lengths by using 64 layers.

The computations were performed at $Re = 0.65 \times 10^6$, $Ma = 0.07$ and $\alpha = 12°$.

The chosen angle of attack represents a flow condition close to airfoil stall, with a significant amount of trailing edge separation. The physical time step was $\Delta t = 2 \times 10^{-5}$, as recommended in [11].

For this application, synthetic turbulence was added on the upper airfoil surface in a volume between 0.77 and 0.80 chord lengths. The airfoil chord length is $c = 1$ m.

Periodic boundary conditions were set in the spanwise direction, and the farfield condition was used for the in- and outflow boundaries located at a distance of 100 chord lengths away from the airfoil.

5 Results

5.1 Flat Plate

In order to evaluate the influence of ST features on the resolved turbulence, the length scale of the turbulence (Eq. 2.4) in the von Karman spectrum (Eq. 2.3) was increased by a factor of two ($l_e \times 2$) and compared with the results of the original length scale ($l_e \times 1$). In addition, the influence of the underlying RANS model on the resolved solution of the LES domain was assessed by performing computations with JHh-v2 RMS in both Low- and High-Re configuration.

It is observed that by the end of the range where the synthetic turbulence was added ($x/\delta_0 = 0.85$) (Fig. 3a), the two computations performed with the same design turbulent length scale ($l_e \times 2$) converge to quite similar Reynolds shear stress profiles whose maximum amplitudes are in the order of the reference RANS solution. On the other hand, computations performed with the original turbulence scale $l_e \times 1$ largely overpredict the turbulence content (Fig. 3a) at the end of the region with turbulence injection, but this quickly decays downstream.

On the other hand, at the end of the LES domain (Fig. 3b), the two solutions obtained with the ST length scale $l_e \times 2$ are quite different while the solutions

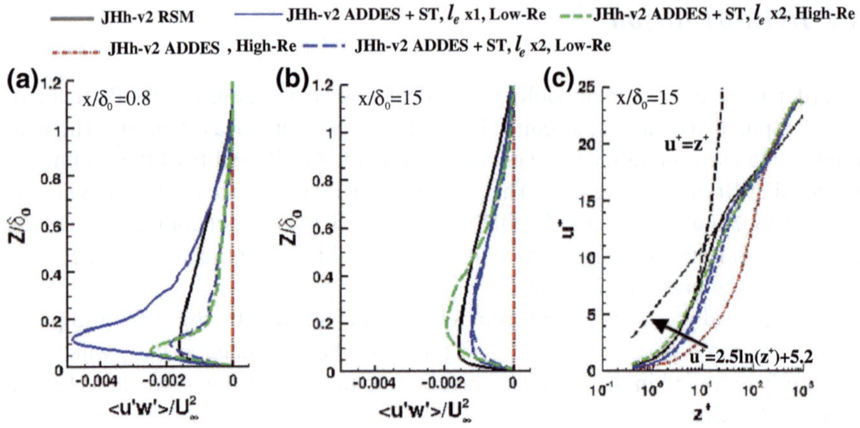

Fig. 3 Reynolds shear stress and mean velocity profiles at different location of the LES domain for different settings of the synthetic turbulence generator and of the JHh-v2 RSM

obtained with the same near-wall treatment (Low Re) converge towards the same resolved shear stress profile. Hence, the fully developed solution downstream of the grey area depends mainly on the characteristics of the background turbulence model used for modelling subgrid stresses and the quality of the grid.

It can also be seen that without applying synthetic turbulence the JHh-v2 ADDES does not generate resolved turbulence in any part of the LES domain and the mean velocity profile does not satisfy the log-law of the turbulent boundary layer (Fig.3a–c).

None of the computed cases reproduces the RANS skin friction coefficient taken as reference, Fig. 4. The case computed with Low-Re configuration and $l_e \times 2$ is the one that converges to its developed solution most rapidly (in about $x/\delta_0 = 6$). It is observed that the Reynolds shear stress profile displays only small variations

Fig. 4 Mean skin friction coefficient distribution along the test section

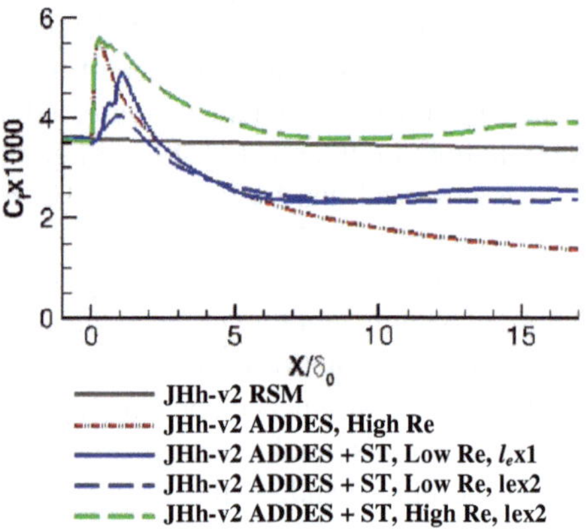

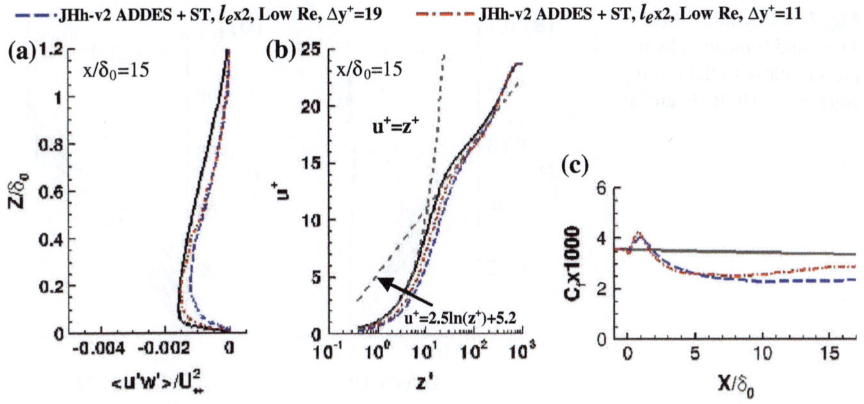

Fig. 5 Comparison of the coarse and fine grid for the flat plate. **a** Reynolds shear stress profile, **b** Streamwise mean velocity profile, **c** Mean skin friction coefficient distribution

between its solution at the end of the ST application range ($x/\delta_0 = 0.85$) and its fully developed profile ($x/\delta_0 = 15$).

The computation with Low-Re configuration and $l_e \times 2$ was also performed for the fine grid with $\Delta y^+ = 11$. As shown in Fig. 5, increasing the grid resolution in the spanwise direction notably improves the development of resolved turbulence in the near wall area (Fig. 5a), thus enhancing slightly the streamwise mean velocity behavior in the viscous sublayer (Fig. 5b). It also improves the estimation of skin friction (Fig. 5c), but the required distance in order to reach the fully developed solution appears somewhat longer than the one for the coarse grid ($x/\delta_0 > 15$).

Fig. 6 displays the skin friction coefficient development along the test section computed with the Low-Re configuration of the JHh-v2 RSM and $l_e \times 2$ by applying the generator with and without the correction for synthetic turbulence convection. It is observed that both cases converge to the same developed solution. However, when non-convective ST is applied, the development of the fully developed solution requires a larger distance ($x/\delta_0 > 15$).

Fig. 6 Mean skin friction coefficient

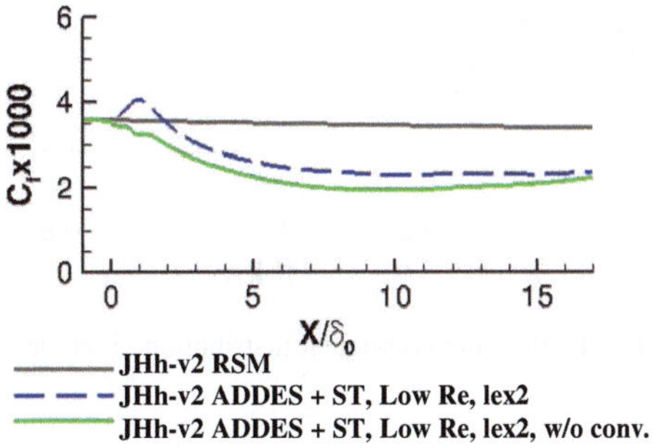

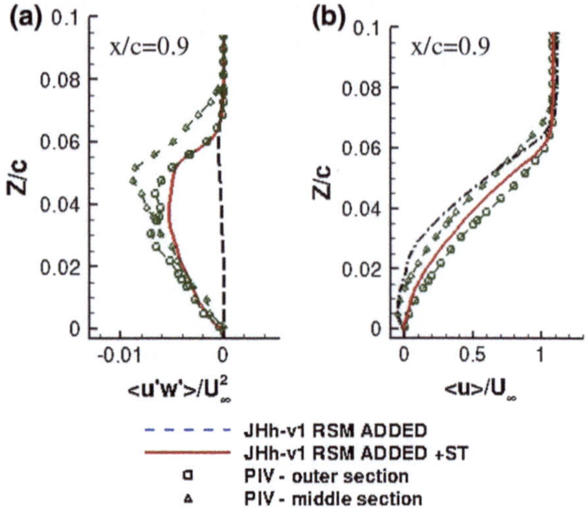

Fig. 7 **a** Reynolds shear stress and **b** mean velocity profiles close to the trailing edge of the HGR-01 airfoil

5.2 HGR-01 Airfoil

As for the flat plate test case, computations performed on the HGR-01 airfoil with JHh-v1 ADDES without synthetic turbulence were unable to create resolved turbulence along the whole LES domain above the airfoil surface (Fig. 7). Computations performed with JHh-v1 ADDES and the synthetic turbulence generator yield very good agreement with the measured Reynolds shear stress (Fig. 7a) and mean velocity (Fig. 7b) profiles.

As for the time-averaged pressure distribution around the airfoil, the computation performed with this new implementation also fits very well with the experimental data (Fig. 8).

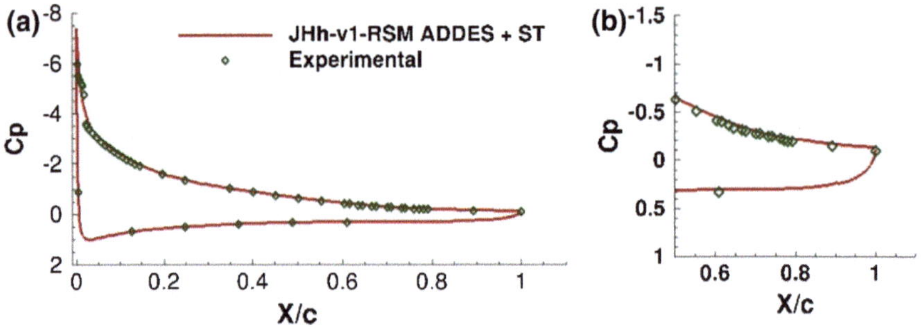

Fig. 8 Pressure coefficient distribution. **a** whole airfoil surface, **b** trailing edge

6 Conclusions

A synthetic turbulence generator based on a superposition of Fourier modes was adopted and adapted in order to improve its divergence properties and to allow for synthetic turbulence convection when using it in a source-term approach applied to a certain volume of the computational domain. From computations of a flat plate, it is concluded that the speed and smoothness of the transition from modeled to resolved turbulence depend on the quality of the generated synthetic turbulence, the background model used to represent the subgrid scales of turbulence and the grid density. We find that generating locally resolved LES embedded into a RANS solution at high Reynolds numbers is feasible. The results of the HGR-01 airfoil show the effectiveness of the implementation with an onset of turbulent flow separation. The implementation of the synthetic turbulence generator was very successful in speeding up the transition from modeled to resolved turbulence, providing physically plausible turbulence content in areas where, without ST, no resolved turbulence was built up.

Acknowledgments The authors gratefully thank the "Deutsche Forschungsgemeinschaft" (DFG) for funding this research within the DFG FOR1066 Research unit and the "NorddeutscheVerbundfür Hoch- und Höchstleistungsrechnen" (HLRN) for providing the computer resources that enabled the simulations presented here.

References

1. Spalart, P., Deck, S., Shur, M., Squires, K., Strelets, M., Travin, A.: A new version of detached-eddy simulation, resistant to ambiguous grid densities. Theor. Comput. Fluid Dyn. **20**(3), 181–195 (2006)
2. Probst, A., Radespiel, R.: A comparison of detached-eddy simulation and reynolds-stress modelling applied to the flow over a backward-facing step and an airfoil at stall. AAIA-2010-920 (2010)
3. Probst, A., Radespiel, R., Knopp, T.: Detached-eddy simulation of aerodinamic flows using a reynolds-stress background model and algebraic RANS/LES sensors. AIAA-2011-3206 (2011)
4. Adamian, D., Andrey, T.: An efficient generator of synthetic turbulence at RANS-LES interface in embedded LES of wall-bounded and free shear flows. In: Computational Fluid Dynamics 2010, pp. 739–744, Russia, Springer (2011)
5. Bechara, W., Bailly, C., Lafon, P., Candel, S.M.: Stochastic approach to noise modeling for free turbulent flows. AIAA J. **32**, 455–463 (1994)
6. Schwamborn, D., Gerhold, T., Heinrich, R.: The DLR TAU-code: recent applications in research and industry. In: Wesseling, P., Oñate, E., Périaux, J. (eds.) European Conference on Computational Fluid Dynamics (ECCOMAS CFD). TU Delft, The Netherlands (2006)
7. Probst, A., Radespiel, R.: Implementation and extension of a near-wall reynolds-stress model for application to aerodinamic flows on unstructured mesh. AIAA-2008-770 (2008)
8. Cécora, R.D., Eisfeld, B., Probst, P., Crippa, S., Radespiel, R.: Differential reynolds stress modeling for aeronautics. AIAA 2012–0465 (2012)
9. Roidl, B.: Development of a zonal method to efficiently simulate viscous flows. Dissertation, der Rheinisch Westfälische Technische Hochschule Aachen (2012)

10. Wokoeck, R., Krimmelbein, N., Ortmanns, J., Ciobaca, V., Radespiel, R.: RANS simulation and experiments on stall behaviour of an airfoil with laminar separation bubbles. AIAA-2006-0244, Reno, Nevada (2006)
11. Spalart, P.R.: Young person's guide to detached-eddy simulation grids. NASA CR-2001-211032 (2001)

Assessment of Local LES-Resolution Sensors for Hybrid RANS/LES Simulations

S. Reuß, T. Knopp, A. Probst and M. Orlt

Abstract Different sensors that provide a measure for the resolution of the LES content in hybrid RANS/LES computations are proposed and investigated. In an a-priori test on isotropic turbulence a suitable sensor is identified. Based on that sensor an automatic local mesh refinement is performed for an IDDES of the flow over a backward facing step. The results obtained on locally adapted grids are compared to results on globally refined grids. It is shown, that the proposed sensors can detect underresolved LES regions and that the local mesh refinement can help to reduce resolution errors caused by a too coarse grid spacing.

1 Introduction

The assessment of the grid resolution in hybrid RANS/LES methods is crucial to ensure the reliability of the results [1]. Therefore, only a very careful mesh design can provide sufficient resolution in the LES regions while keeping the total number of points as low as possible. In complex cases, where the location of the LES regions is not known in advance, this may even be impossible.

Thus, robust sensors are proposed in this work to assess the LES resolution. They are applied in combination with the automatic grid adaptation module, that is provided by the DLR-TAU code [12].

The paper is organized as follows: In Sect. 2 the numerical method that is used throughout all simulations and the adaptation algorithm are described. Specific details about the resolution sensors are given in Sect. 3 and the proposed sensors are assessed in a-priori tests on isotropic turbulence. Actual mesh refinements based on one of the proposed sensors are performed for the flow over a backward facing step in Sect. 4. In Sect. 5 conclusions are drawn and an outlook is given.

S. Reuß (✉) · T. Knopp · A. Probst · M. Orlt
Institute of Aerodynamics and Flow Technology DLR (German Aerospace Center), Bunsenstrasse 10, 37073 Göttingen, Germany
e-mail: silvia.reuss@dlr.de

S. Girimaji et al. (eds.), *Progress in Hybrid RANS-LES Modelling*,
Notes on Numerical Fluid Mechanics and Multidisciplinary Design 130,
DOI 10.1007/978-3-319-15141-0_7

2 Numerical Method

The simulations are performed using the DLR TAU code, an unstructured finite volume solver for the compressible Navier-Stokes equations, that provides hybrid RANS/LES methods based on different RANS models. The influence of the numerical settings on the quality of scale resolving simulations is subject to an in-depth study in [11]. The reference settings that are described there are used in this work.

2.1 The Hybrid RANS/LES Method

All simulations that are shown here use the Improved Delayed DES (IDDES) [13] based on the Menter-SST RANS model. In [4] it was shown, that a modification of the filter width in the hybrid model can help to reduce the grey area in Zonal DES. Following the same route here, the filter width in the IDDES is modified, using $\Delta_\omega = \sqrt{N_x^2 \Delta_y \Delta_z + N_y^2 \Delta_z \Delta_x + N_z^2 \Delta_x \Delta_y}$, with $\mathbf{N}$ the normalized vorticity vector. In the present unstructured dual grid metric Δ_x is the maximal extent in x-direction of all connected dual-edges. In the IDDES the original length scale comprises three different parts: the maximum length h_{max} and the maximum wall normal extent h_{wn} of all connected dual edges and the wall distance d_w: $\Delta = \min(\max[C_w d_w; C_w h_{max}; h_{wn}]; h_{max})$. The maximum edge length h_{max} also occurs in $\alpha = 0.25 - d_w / h_{max}$. In the modified filter width h_{max} is replaced by Δ_ω, however, only the second occurrence in the length scale equation is replaced, in order to preserve the wall-modelled LES capabilities. In [11] this procedure is validated for the fully-developed channel flow.

2.2 The Adaptation Algorithm in the DLR-TAU Code

The TAU-adaptation module [2] was originally designed for the equidistribution of differences or gradients of a preliminary flow solution by local hierarchical grid re- and derefinement. The main steps of the adaptation algorithm include a preparing loop over all elements. Considering the geometry and the implemented subdivision cases of the element type, some of the edges are excluded from subdivision. This check is repeated until all edges that would lead to invalid refinement cases are forbidden.

In the next step all edges allowed for subdivision that are marked by the edge indicator are subdivided. A following loop over all elements checks and ensures the validity of all element refinements by subdivision of additional edges and faces if needed. In this step all the bridging subdivisions between the refined and unrefined grid areas are created.

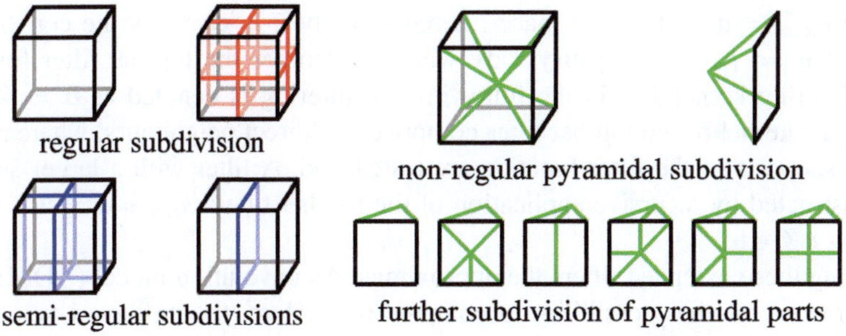

Fig. 1 Regular (*left*) and irregular bridging (*right*) hexahedral subdivision cases

The new point coordinates at the surface are determined by a quadratic spline interpolation to reconstruct the curvature. In the inner grid parts a linear interpolation is used. Finally, the new elements are constructed and the current solution is interpolated to the new points. The result is a locally refined grid with bridging elements introduced into the original grid structure. In Fig. 1 the implemented subdivisions for hexahedral elements are shown.

3 Resolution-Sensors for Local Mesh Adaptation

In this work the original adaptation indicator is replaced by a suitable sensor that attempts to measure the quality of the LES resolution. Several sensors have been proposed in the literature and some of them are presented here.

Since a grid refinement in hybrid RANS/LES simulations not only reduces the numerical error but also changes the model, in [3, 7] systematic grid and model variations have been performed to distinguish the numerical from the modelling error. For complex test cases these exhaustive variations are not affordable. Instead, a resolution estimation that can be obtained from a single simulation is preferable. One concept for such a sensor, proposed by Pope [10], is the ratio of the resolved turbulent kinetic energy (TKE) k_{res} to the total TKE k_{tot}. In practice the total TKE is split into the resolved grid scale and the modelled subgrid-scale (sgs) TKE:

$$S = \frac{k_{res}}{k_{total}} = \frac{k_{res}}{k_{res} + k_{sgs}} \ . \tag{1}$$

The resolved TKE can be computed directly from the resolved quantities, $k_{res} = 1/2\langle(\mathbf{u} - \langle\mathbf{u}\rangle)^2\rangle$, where $\mathbf{u}$ is the velocity and $\langle\mathbf{u}\rangle$ denotes the mean. But for the subgrid-scale TKE an appropriate model is required. In [8] it is proposed to estimate the sgs TKE contribution by explicitly filtering the solution:

$$k_{sgs} = \frac{1}{2}\langle\mathbf{u}_{sgs}^2\rangle \ , \tag{2}$$

where $\mathbf{u}_{sgs} = \mathbf{u} - \langle\overline{\mathbf{u}}\rangle$. The space-filtered velocity $\overline{\mathbf{u}}$ is given by the convolution integral $\overline{\mathbf{u}} = \int_{\Omega_1} \mathcal{G}(\mathbf{x}-\mathbf{y})\mathbf{u}(\mathbf{y},t)d\mathbf{y}$ where $\mathcal{G}$ denotes the top hat filter function with filtering kernel Ω_1. In the following the filtering is denoted as $\overline{\mathbf{u}} = \mathcal{G} * \mathbf{u}$. In [8] the kernel of the top hat filter comprises all direct neighbours, whereas here larger supports of the filter function are tested, too. A filter with a larger support is constructed by recursive application of the top hat filter: $\mathcal{G}_{\Omega n} * \mathbf{u} = \mathcal{G}^n * \mathbf{u} = \mathcal{G} * \cdots * \mathcal{G} * \mathbf{u}$.

Using these extended filters the approximate deconvolution model (ADM) [14], [5] provides another possibility to compute $\mathbf{u}_{sgs}$. Here, the unfiltered velocity is constructed using the approximated inverse of a filter function. The idea is to use a filter $\mathcal{G}$ with a filter width larger than the local mesh spacing. The filtered solution $\mathbf{u}_2 = \mathcal{G} * \mathbf{u}$ contains only scales larger than the smallest scales that can be resolved on the grid. The solution $\mathbf{u}_1$ that would be obtained using the local mesh spacing as filter width is reconstructed by the deconvolution of the coarser solution: $\mathbf{u}_1 = \mathcal{G}^{-1} * \mathbf{u}_2$. This approach is used in order to get an estimation of the exact solution as follows: Let $\mathbf{u}_{DNS}$ be the solution of a DNS and $\mathcal{G}$ a filter operator, then the solution of an LES is $\mathbf{u}_{LES} = \mathcal{G} * \mathbf{u}_{DNS}$ or $\mathbf{u}_{DNS} = \mathcal{G}^{-1} * \mathbf{u}_{LES}$. If $H = Id - \mathcal{G}$ is replaced into the geometric series

$$\sum_{j=0}^{\infty} H^j = 1/(1+H) = (1+H)^{-1} \text{ then follows } \mathcal{G}^{-1} = \sum_{j=0}^{\infty} (1-\mathcal{G})^j \ .$$

Now the series representation is truncated at a finite N. In [14] is stated that $N = 3$ already yields acceptable results, and for $N > 5$ no further improvement can be expected, so $N = 4$ is chosen here. This gives

$$\mathcal{G}^{-1} = 5 - 10\mathcal{G} + 10\mathcal{G}^2 - 5\mathcal{G}^3 + \mathcal{G}^4 \ .$$

With the decomposition $\mathbf{u}_{DNS} = \mathbf{u}_{LES} + \mathbf{u}_{sgs}$ the sgs velocity is given by

$$\mathbf{u}_{sgs} = (Id - \mathcal{G}^{-1}) * \mathbf{u}_{LES} \ . \tag{3}$$

Another model for the subgrid-scale TKE is proposed by Lilly [9]:

$$k_{sgs} = \langle \nu_t \rangle^2 / (c\Delta)^2, \text{ with } c = 0.094 \ . \tag{4}$$

If the basic RANS model provides a model-equation for the TKE (e.g. $k-\omega$, $k-\varepsilon$), it is also possible to use the mean of that quantity:

$$k_{sgs} = \langle k \rangle \ . \tag{5}$$

3.1 Validation of the Sensors Based on Isotropic Homogeneous Turbulence

For an a-priori assessment of the different sensor formulations, they are evaluated for the initial field of the well known decaying isotropic turbulence (DIT) test case. The computational domain is a cubic volume with an edge length $L = 2\pi$. The number of points in the fine, medium and coarse grid $G1$, $G2$ and $G3$ are $n_1 = 128^3$, $n_2 = 64^3$ and $n_3 = 32^3$ respectively. For the DIT measured spectra are available, from which experimental values for the sensors can be estimated. The cut-off wave-number is $\kappa_c = n_{pnt}/2$. The total TKE is the area under the power spectrum $k_{tot} = \int_0^\infty E(\kappa)\mathrm{d}\kappa$ while the resolved TKE is $k_{res} = \int_0^{\kappa_c} E(\kappa)\mathrm{d}\kappa$.

The initial solution is constructed as a superposition of all frequencies from the measured spectrum. From this initial solution the different sensor formulations are evaluated. In this test the homogeneous turbulence is regarded as frozen and the temporal average is replaced by a spacial average. Thus an a-priori evaluation is possible. In Fig. 2 the resulting sensor values are listed. On the x-axis the relative cell size w.r.t. to the finest grid cell size is plotted.

The most complex sensor S_2, which is based on the approximate deconvolution model, reproduces the slope of the experimental TKE ratio the best. However, it gives an overall far too pessimistic estimation of the grid resolution, therefore it is not chosen for the sample application to the backward facing step in Sect. 4.

Apart from S_2, the slope of all sensors is similar except for a constant offset. They all estimate the resolution of the coarsest grid too optimistic and underestimate the resolution of the fine grid. The sensor S_1 with the smaller explicit filter size as well as the sensor $S4$ also overestimate the medium grid.

Since an underpredicted grid resolution is considered the conservative (i.e. "safer") approach the sensor S_1 with the larger support of $n = 4$ is chosen for the sample application in Sect. 4.

	Eq. for k_{sgs}	$G1$	$G2$	$G3$	
Exp.	-	0.92	0.75	0.55	——
$S_1, n=2$	(2)	0.83	0.78	0.71	- - -
$S_1, n=4$	(2)	0.77	0.71	0.64	——
S_2	(3)	0.58	0.47	0.37	——
S_3	(4)	0.80	0.74	0.68	——
S_4	(5)	0.85	0.81	0.76	——

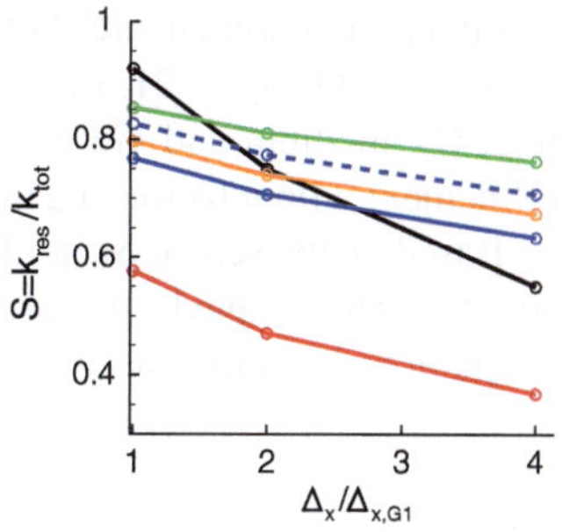

Fig. 2 A-priori sensor values for different grids for the DIT test case

3.2 Modified Adaptation Indicator

The above sensor definitions provide a local value for each grid point. However, since the sensors are only applicable to LES, a certain amount of local resolved turbulence is necessary, to draw conclusions in hybrid RANS/LES computations. Therefore, the evaluation of the sensor is limited to regions where a specified threshold of the resolved turbulence intensity $T = 1/u\sqrt{2/3(\langle u'^2\rangle + \langle v'^2\rangle + \langle w'^2\rangle)}$ is exceeded. In this work $T_0 = 0.1$ is chosen, as this value is typically reached in a turbulent boundary layer.

For the adaptation algorithm an edge-wise indicator is constructed: The edge is marked for subdivision if the sensor indicates an insufficient grid resolution at one of the adjacent points, i.e. $S < S_0$. Since in [10] it is stated that an LES should resolve at least 80 % of the turbulence by the grid, the threshold is chosen as $S_0 = 0.8$. Additionally, the mode of the hybrid RANS/LES model is taken into account: If the model operates in RANS mode, i.e. the IDDES hybrid switching-function $\tilde{f}_d = 1$, the edge is not refined.

A smoothing of the refined region is performed by spreading the indicator over further elements. Elements touched by the edge indicator and their neighbours up to a specified number of element layers around them are subdivided isotropically to make sure that the marked grid parts reside completely in the isotropically refined grid area. In the present work, marking four additional layers of elements leads to acceptably smooth regions.

4 Application to a Backward Facing Step Flow

The test case of the flow over a backward facing step (BFS) is determined in terms of the step hight $h = 0.0127$ m. Experimental data for a Reynolds-number based on h of $Re_h = 37500$ are given by Driver and Seegmiller [6]. The dimensions of the computational domain are shown in Fig. 3. The point-numbers in the coarse grid $G3$ are $n_{x,1} = 11$, $n_z = 21$, $n_{x,2} = 50$, $n_h = 24$ and $n_y = 15$. A family of grids—$G1$ fine, $G2$ medium and $G3$ coarse—has been constructed based on $G3$ by a global refinement with a factor of 2 in each direction.

Based on the sensor S_1 (cf. Fig. 2) the medium and coarse grids have been locally adapted. The adapted grids are referred to as $G2A$ and $G3A$ hereafter. Since in the adaptation algorithm the edges are bisected, the resolution in the refined regions is the

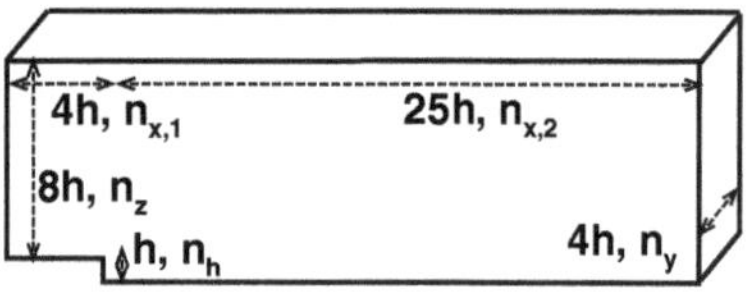

Fig. 3 Dimensions and denotation of the computational domain of the flow over a backward facing step

Table 1 Number of grid points in the globally refined and the locally adapted grids

	$G1$	$G2A$	$G2$	$G3A$	$G3$
n_{pnt}	2,127,000	810,000	274,000	132,000	36,000

same as in the next globally refined grid, i.e. $G1$ and $G2A$ have the same resolution in the adapted LES region, $G2$ and $G3A$ respectively. A comparison of the total point-numbers in the different grids is shown in Table 1.

First, the sensor values resulting from the different mesh resolutions are compared in Fig. 4. Regions with a local turbulence intensity lower than $T_0 = 0.1$ are blanked. In the left column the globally refined grids are shown, the respective adapted coarser grid is shown in the same line in the right column, the adapted regions are outlined by the black lines.

Here it can be seen how the sensor reacts to the mesh refinement: The values in the adapted regions are at the same level as those on the globally refined meshes.

In Fig. 5 the spanwise averaged sensor values at a position of $x/h = 4$ behind the step at half the step height are compared. On the x-axis the relative cell size w.r.t. to the finest grid cell size is plotted. The sensor values on the globally refined grids and those on the locally adapted grids almost match exactly. All sensors indicate that in the finest grid more than 80 % of the turbulence is resolved. In contrast to the a-priori test for the DIT, the sensors now show a clearer distinction. The sensors S_3 and S_4, that are directly based on ν_t and k_{RANS} from the background RANS model, do not

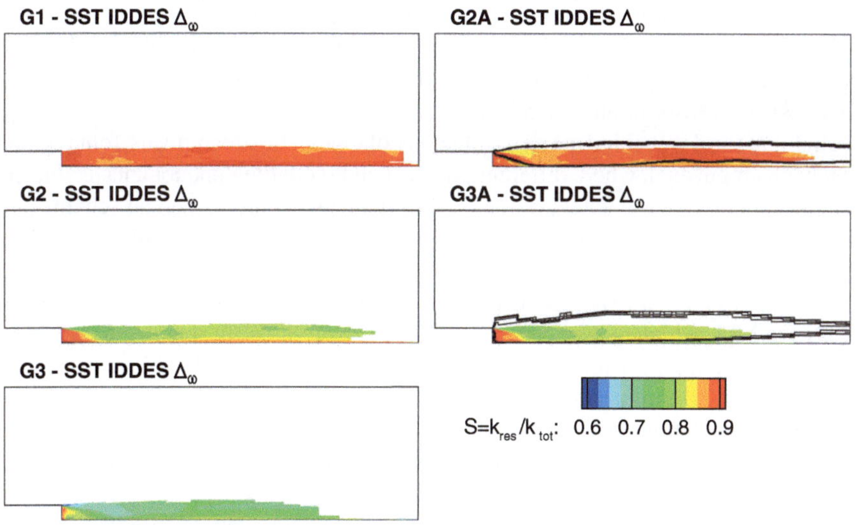

Fig. 4 Comparison of the grid sensor values on different grids. The values are blanked if the turbulence intensity is smaller than $T_0 = 0.1$. On the *left* the globally refined grids are shown, on the *right* the locally adapted grids are shown

	$G1$	$G2A$	$G2$	$G3A$	$G3$
$S_1, n = 2$	0.93	0.93	0.87	0.87	0.79
$S_1, n = 4$	0.87	0.87	0.78	0.78	0.68
S_2	0.86	0.87	0.71	0.72	0.48
S_3	0.98	0.98	0.98	0.98	0.99
S_4	0.95	0.95	0.94	0.93	0.94

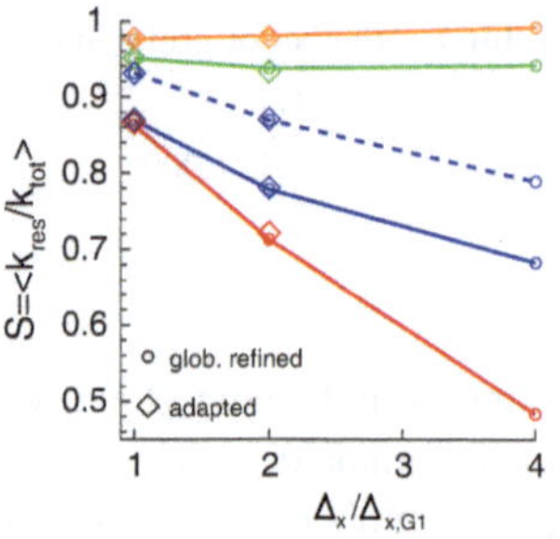

Fig. 5 Sensor values for different grids for the BFS test case

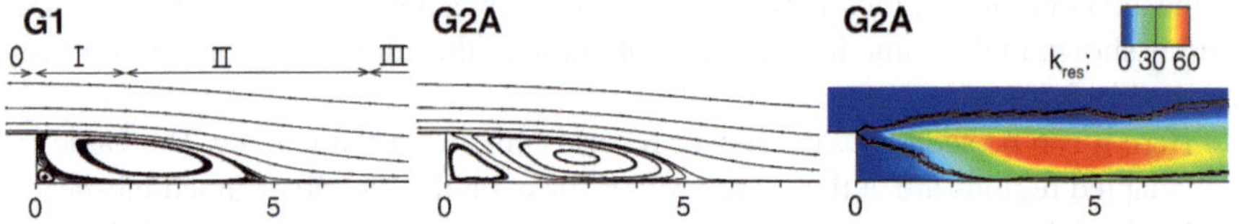

Fig. 6 Streamlines in the recirculation area in the fine (*left*) and adapted medium (*middle*) grid. Resolved turbulence on the adapted medium grid $G2A$ (*right*), the adapted region is outlined

react to the grid refinement, so that they are not suited as an indicator for the grid resolution in this test case. Again the values from S_1 with the two different explicit filter widths share the same slope but exhibit an offset.

Remarkably, the sensor S_2 that still shows the steepest slope now yields a positive estimation for the finest grid. One explanation might be, that the a-priori test is not suited for this sensor, because it does not only depend on the subgrid velocity but also on the mean velocity itself which is zero for the DIT test case. It seems worthwhile to assess this sensor in other applications.

In Fig. 6 the flow behind the step is shown. Following the streamtraces, four regions can be distinguished, where different physical and modelling phenomena characterize the simulation. The extent of the regions depends on the numerical scheme and on the grid design. Upstream of the step, for $x/h < 0$ an attached turbulent boundary layer is modelled in RANS mode, *region 0*. In the first region of interest, *region I* ($x/h = 0$ to $x/h \approx 2$) immediately behind the step, the hybrid model switches from RANS to LES, which may be delayed due to the "grey-area" problem. An important feature in region I is the topology of the secondary recirculation bubble in the corner of the step. Its size influences the primary recirculation, by pushing the centre of the primary recirculation further downstream. The second region, *region II* ($x/h \approx 2$ to $x/h \approx 7$) around the reattachment point, is dominated by the LES and characterized by high values of resolved TKE. The third region, *region III* ($x/h > 7$) downstream of the reattachment, is essentially a resolved attached boundary layer with turbulent inflow. The different characteristics that dominate the flow make this a good testcase to examine the validity as well as the limits of the proposed adaptation strategy.

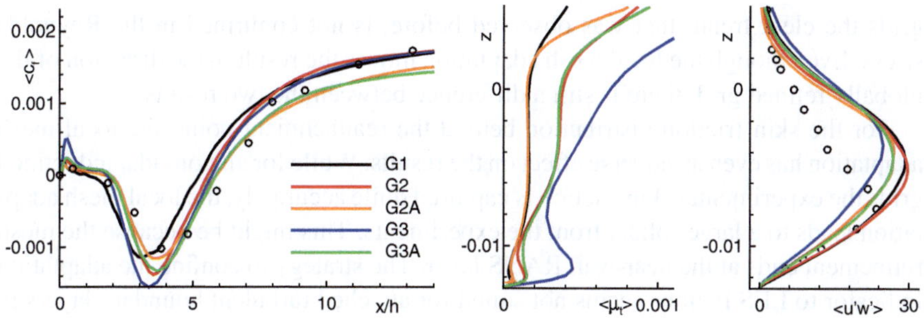

Fig. 7 Skin friction distribution (*left*), resolved Reynolds-stresses (*middle*) and eddy-viscosity in a section at $x/h = 4$ (*right*) for different grids

It was found that using the filter width Δ_ω reduces the grey area in region I significantly compared to the Δ_{edge}-filter (not shown here) and that resolved turbulence develops quickly behind the step. However, in the grids $G2$ and $G3$ the resolution in the centre region is rather coarse, and due to a low level of resolved turbulence intensity this region is excluded from refinement. The coarse resolution of the upper part of the shear layer can have a negative influence on its development.

As depicted in Fig. 4 the region of the secondary flow is considered well resolved according to the sensor and thus not refined in the medium grid $G2A$. Nevertheless, the size of this secondary recirculation is much larger than in grid $G1$. Both, this and the resolution, influence the downstream behaviour of the flow. In Fig. 7 (left) the skin friction distributions for all grids are compared. The plateau at $x/h = 0$ to $x/h \approx 2$ is the footprint of the secondary corner flow which has the smallest extent on the finest grid $G1$. All other simulations are comparable in this regard.

The highest values of resolved turbulence are reached in region II. Here, the minimum of the skin friction, i.e. the strongest backflow, is obtained. The location of the minimum is influenced by the upstream flow, but in terms of the absolute minimum value the effect of the grid refinement can be compared. Through the global grid refinement the absolute values decrease. As expected, the locally adapted meshes yield almost the same minimum values.

In Fig. 7 (middle) the mean values of the eddy-viscosity $\langle\mu_t\rangle$ are shown for a cut at $x/h = 4$. The global mesh refinement leads to lower levels of modelled turbulence. The local mesh-adaptation has the same effect. The solutions from the adapted meshes, $G3A$ (green) and $G2A$ (orange), coincide with the corresponding globally refined grids, $G2$ (red) and $G1$ (black).

In Fig. 7 (right) the resolved Reynolds stresses in the x-z plane $\langle u'w'\rangle$ are shown. Compared to the experimental data all of the simulations exhibit large deviations. But also when comparing the simulations among each other, no clear trend is found. The profiles on the medium grid $G2$ agrees well with the fine grid $G1$. However, due to the upstream flow the cut corresponds to different positions in the recirculation region. The solution on the locally adapted grid $G2A$ has a larger offset from the globally refined solution than from the coarse-grid solution. Also for the coarser grids the clear trend, that was observed before, is not confirmed in the Reynolds stress. Even though the local mesh adaptation moves the result in the direction of the globally refined grid, there is still a difference between the two results.

For the skin friction distribution behind the reattachment point, the local mesh adaptation has even an adverse effect on the results. While for all non-adapted refined grids the experimental skin friction is captured quite accurately, the local mesh adaptation leads to a larger offset from the experiments. This might be because the mesh refinement ends at the near-wall RANS layer. The strategy to confine the adaptation indicator to LES regions seems not suited for attached turbulent boundary layers in wall-modelled LES mode. Instead, it may be necessary to extend the refined region down to the wall in this case.

5 Conclusion

Different formulations of a mesh sensor that assesses the resolution in the LES region have been investigated. In the case of isotropic turbulence the proposed sensors have been compared to experimental values. One of the sensors has been evaluated for a family of grids for a backward facing step and was used as indicator for a local mesh adaptation. The results of the globally and locally refined grids have been compared, and the range of validity as well as the limits of the sensor have been examined.

It has been shown, that the proposed sensors are able to detect underresolved regions in an LES. Moreover, the sensors can be used as input for a local mesh refinement. This is an important step to automatically reduce the uncertainty w.r.t. mesh design in hybrid RANS/LES modelling.

In the future the sensors will be applied to more complex testcases. In the case of wall-modelled LES the possibility to improve the results by extending the refined region down to the wall will be assessed.

Acknowledgments The authors gratefully acknowledge the "Deutsche Forschungsgemeinschaft" (German Research Foundation) who funded this work in the scope of the Research Group DFG FOR 1066.

References

1. Addad, Y., Gaitonde, U., Laurence, Dominique Rolfo, S.: Optimal unstructured meshing for large eddy simulations. In: Meyers, J., Geurts, B.J., Sagaut, P. (eds.) Quality and Reliability of Large-Eddy Simulations, pp. 93–103. Ercoftac Series. Springer, Netherlands (2008)
2. Alrutz, T., Orlt, M.: Parallel dynamic grid refinement for industrial applications. In: Wesseling, P., Oñate, E., Périaux, J. (eds.) Proceedings of ECCOMAS CFD 2006. The Netherlands, TU Delft (2006)
3. Celik, I.B., Cehreli, Z.N., Yavuz, I.: Index of resolution quality for large eddy simulations. J. Fluids Eng. **127**(5), 949 (2005)

4. Chauvet, N., Deck, S., Jacquin, L.: Zonal detached eddy simulation of a controlled propulsive jet. AIAA J. **45**(10), 2458–2473 (2007)
5. Chow, F.K., Street, R.L., Xue, M., Ferziger, J.H.: Explicit filtering and reconstruction turbulence modeling for large-eddy simulation of neutral boundary layer flow. J. Atmos. Sci. **62**, 2058–2077 (2005)
6. Driver, D.M., Seegmiller, H.L., Marvin, J.G.: Time-dependent behavior of a reattaching shear layer. AIAA J. **25**(7), 914–919 (1987)
7. Klein, M.: An attempt to assess the quality of large eddy simulations in the context of implicit filtering. Flow Turbul. Combust. **75**(1–4), 131–147 (2005)
8. Knopp, T., Zhang, X., Kessler, R., Lube, G.: Enhancement of an industrial finite-volume code for large-eddy-type simulation of incompressible high Reynolds number flow using near-wall modelling. Comput. Methods Appl. Mech. Eng. **199**(13–16), 890–902 (2010)
9. Lilly, D.K.: The representation of small-scale turbulence in numerical simulation experiments. NCAR Manuskript No, 281 (1966)
10. Pope, S.B.: Turbulent Flows. Cambridge University Press, Cambridge (2000)
11. Probst, A. Reuß, S.: Scale-resolving simulations of wall-bounded flows with an unstructured compressible flow solver. In: 5th Symposium on Hybrid RANS-LES Methods TEXAS A&M University, College Station, Houston, USA, 19–21 March 2014
12. Schwamborn, D., Gerhold, T., Heinrich, R.: The DLR TAU-Code: recent applications in reaserach and industry. In: Wesseling, P., Oñate, E., Périaux, J. (eds.) Proceedings of ECCOMAS CFD 2006. TU Delft, The Netherlands (2006)
13. Shur, M.L., Spalart, P.R., Strelets, M.K., Travin, A.K.: A hybrid RANS-LES approach with delayed-DES and wall-modelled LES capabilities. Int. J. Heat Fluid Flow **29**(6), 1638–1649 (2008)
14. Stolz, S., Adams, N.A., Kleiser, L.: An approximate deconvolution model for large-eddy simulation with application to incompressible wall-bounded flows. Phys. Fluids **13**(4), 997 (2001)

Constrained Large-Eddy Simulation for Aerodynamics

Zhenhua Xia, Zuoli Xiao, Yipeng Shi and Shiyi Chen

Abstract A constrained large-eddy simulation (CLES) method for wall-bounded compressible flow is introduced and validated via simulations of several typical flow configurations, including compressible turbulent channel flow, flow past a NACA0021 airfoil at 60° angle of attack, and compressible flow past a Delta wing. In the wall-bounded compressible CLES method, the whole flow domain is solved using large-eddy simulation (LES) technique, but the subgrid-scale (SGS) stress and heat flux are constrained by given models for the Reynolds stress and heat flux in the near-wall region. For attached flows, CLES method can eliminate the non-physical Log-Layer Mismatch phenomenon appearing in hybrid RANS/LES methods, and can predict the mean velocity and temperature profiles more accurately as compared with traditional LES and detached-eddy simulation (DES) approaches. For detached flows, CLES method can calculate the skin friction force more precisely than traditional LES method, and is comparable to DES technique in prediction of the aerodynamic statistics. For both cases, CLES method can capture fruitful multi-scale turbulent structures, which are lacking in DES, and can successfully overcome the coarse-grid effect observed in traditional LES method. Therefore, it is suggested that the present CLES method could be a promising numerical simulation tool for wall-bounded compressible turbulent flows in the realm of aerodynamics.

Z. Xia (✉)
SKLTCS, College of Engineering, Peking University,
Beijing 100871, People's Republic of China
e-mail: xiazh1006@gmail.com

Z. Xiao · Y. Shi · S. Chen
SKLTCS & CAPT, College of Engineering, Peking University,
Beijing 100871, People's Republic of China
e-mail: zxiao@coe.pku.edu.cn

Y. Shi
e-mail: syp@mech.pku.edu.cn

S. Chen
e-mail: syc@pku.edu.cn

S. Girimaji et al. (eds.), *Progress in Hybrid RANS-LES Modelling*,
Notes on Numerical Fluid Mechanics and Multidisciplinary Design 130,
DOI 10.1007/978-3-319-15141-0_8

1 Introduction

In recent years, computational fluid dynamics (CFD) techniques have played more and more important roles in aerodynamic and aeroacoustic engineering applications and have received increasing attention. Among various CFD methods, the simplest one is direct numerical simulation (DNS), which is free of turbulence closures. However, DNS method suffers from the high computational consumption and is only limited to simple flow configurations at low or moderate Reynolds numbers. Reynolds Averaged Navier-Stokes (RANS) approach has been widely used in simulations of real engineering flows because of its low requirement for computational meshes. However, the application of RANS method to unsteady non-equilibrium separated flows remains questionable and unsatisfactory, since the turbulence models used in RANS are lack of generality. Large-eddy simulation (LES) method, whose computational cost is between DNS and RANS approaches, possesses the capability to predict three-dimensional (3D) unsteady flow fields by solving the filtered Navier-Stokes equations with the subgrid-scale (SGS) models, which are inherently more "universal" than the turbulence models used in RANS. However, the application of LES method to wall-bounded turbulent flows of engineering interest is still far from practical due to its fairly fine grids requirement.

Thus, the numerical simulations of wall-bounded turbulent flows with massive separations have sparked continuous interest in the development of hybrid RANS/LES methods [2], among which the most commonly employed one is detached-eddy simulation (DES) technique [13]. In hybrid RANS/LES method, the computational domain is usually divided into two regions, i.e., the near-wall region and the outer region. The RANS equations are solved in the near-wall region, while the LES equations are integrated in the outer region. The hybrid RANS/LES method takes advantages both from RANS solution in the near-wall region and from LES solution in the outer region. Although the hybrid RANS/LES approaches, e.g. DES, have achieved satisfactory success and have been incorporated into the commercial CFD software for Engineering application, this family may suffer from problematic drawbacks. For instance, a super buffer layer forms in the mean velocity profile when DES technique is used to simulate simple turbulent channel flow [11]. Such log-layer mismatch (LLM) phenomenon may cast doubts on the applicability of DES approach for simulation of wall-bounded flows when the near-wall flow structures play a significant role. For more details of the descriptions and discussions on hybrid RANS/LES methods, readers are referred to the review articles by Fröhlich and von Terzi [2] and Spalart [13].

It has been suggested that specific constraints shall be satisfied in turbulence modeling for CFD [7, 8, 10]. In consideration of these pioneer works, Shi et al. developed a dynamic mixed SGS model with an SGS energy dissipation constraint [12]. LES of 3D incompressible isotropic turbulence shows that the constrained SGS (C-SGS) model not only predicts the mean SGS dissipation more accurately, but calculates the high-order velocity structure functions more precisely in inertial range as compared with traditional mixed SGS models. Chen et al. [1] developed a Reynolds stress constrained large-eddy simulation (RS-CLES) method for incompressible wall-bounded turbulent flows to overcome the barrier posed by the massive computational grids requirement in traditional LES. In the RS-CLES method, the mean SGS stress in the near-wall region is determined by the difference between a given Reynolds stress model and the resolved Reynolds stress, while traditional SGS model is employed in the remaining region of the flow domain. It is found that the RS-CLES approach successfully combines the positive points both from traditional LES and DES, and can accurately predict the mean quantities and flow structures of both attached and detached flows. Recently, Jiang et al. [6] extended the RS-CLES method and introduced a Reynolds heat flux constrained model for the SGS heat flux required in LES of wall-bounded compressible turbulent flows. The performance of the CLES method for wall-bounded compressible flows is superior to those of traditional LES and DES methods when they are compared in the prediction of compressible turbulent channel flow [6]. In this paper, the compressible CLES technique is tested and validated *a posteriori* in the simulations of both attached and detached flows in order to ascertain its capability to predict complex and wall-bounded compressible turbulent flows on relatively coarse grids.

The rest contents of this paper are organized as follows. First a short review of the CLES method for wall-bounded compressible flows is given. Then, the compressible CLES approach is applied to the simulations of turbulent flow in several different geometries, including compressible turbulent channel flow, compressible flow past a NACA0021 airfoil with 60° angle of attack (AOA), and compressible flow past a Delta Wing. The results are compared with previous experimental and numerical data. Finally, conclusions and discussions are made.

2 Constrained Large Eddy Simulation

In LES of compressible turbulent flows, we solve the following filtered Navier-Stokes equations

$$\frac{\partial \bar{\rho}}{\partial t} + \frac{\partial (\bar{\rho}\widetilde{u}_i)}{\partial x_i} = 0, \tag{1}$$

$$\frac{\partial (\bar{\rho}\widetilde{u}_i)}{\partial t} + \frac{\partial (\bar{\rho}\widetilde{u}_i\widetilde{u}_j + \bar{p}\delta_{ij})}{\partial x_j} = \frac{1}{Re}\frac{\partial \widetilde{\sigma}_{ij}}{\partial x_j} + \frac{\partial \tau_{ij}^{LES}}{\partial x_j}, \tag{2}$$

$$\frac{\partial (\bar{\rho}\widetilde{e})}{\partial t} + \frac{\partial [\widetilde{u}_i(\bar{\rho}\widetilde{e} + \bar{p})]}{\partial x_i} = \frac{\partial \widetilde{q}_i}{\partial x_i} + \frac{1}{Re}\frac{\partial (\widetilde{\sigma}_{ij}\widetilde{u}_j)}{\partial x_i} + \frac{\partial q_i^{LES}}{\partial x_i} + \frac{\partial \tau_{ij}^{LES}\widetilde{u}_j}{\partial x_i}, \tag{3}$$

and the thermodynamic equation of state

$$\bar{p} = \frac{\bar{\rho}\widetilde{T}}{\gamma M_\infty^2}. \tag{4}$$

In the above equations, all the physical quantities are nondimensionalized by given characteristic parameters, i.e., a characteristic length scale D, the free-stream velocity U_∞, density ρ_∞, temperature T_∞, dynamic viscosity μ_∞ and thermal conductivity κ_∞. Here, $\overline{f}$ denotes a spatially-filtered field, and $\widetilde{f} = \frac{\overline{\rho f}}{\overline{\rho}}$ is the Favre-filtered field. $\overline{\rho}$ is the density, $\widetilde{u}_i$ is the velocity vector, $\overline{p}$ is the pressure, $\widetilde{T}$ is the temperature, $\widetilde{e} = C_v\widetilde{T} + \frac{1}{2}\widetilde{u}_i\widetilde{u}_i$ is the total energy per unit mass, $C_v = 1/[\gamma(\gamma - 1)M_\infty^2]$ is the specific heat at constant volume, γ is the ratio of specific heats, $C_p = \gamma C_v$ is the specific heat at constant pressure, $M_\infty = U_\infty/\sqrt{\gamma R T_\infty}$ is the Mach number, R is the specific gas constant, $Re = \rho_\infty U_\infty D/\mu_\infty$ is the Reynolds number,

$$\widetilde{\sigma}_{ij} = \mu(\widetilde{T})\left(\frac{\partial \widetilde{u}_i}{\partial x_j} + \frac{\partial \widetilde{u}_j}{\partial x_i} - \frac{2}{3}\frac{\partial \widetilde{u}_k}{\partial x_k}\delta_{ij}\right) \tag{5}$$

and

$$\widetilde{q}_i = -\frac{C_p\mu(\widetilde{T})}{RePr}\frac{\partial \widetilde{T}}{\partial x_i} \tag{6}$$

are the viscous stress tensor and the heat flux vector, respectively. $P_r = C_p\mu_\infty U_\infty^2/\kappa_\infty T_\infty$ is the molecular Prandtl number and μ is the dimensionless viscosity determined via Sutherland's law

$$\mu = \frac{\tilde{T}^{3/2}(1 + S)}{\tilde{T} + S}, \tag{7}$$

with $S = 110.3K/T_\infty$. The SGS stress tensor

$$\tau_{ij}^{LES} = -\bar{\rho}(\widetilde{u_i u_j} - \widetilde{u}_i\widetilde{u}_j), \tag{8}$$

and the SGS heat flux vector

$$q_i^{LES} = -\bar{\rho}C_p(\widetilde{Tu_i} - \widetilde{T}\widetilde{u}_i), \tag{9}$$

must be modeled based on the resolved fields before we can numerically solve Eqs. (1)–(4).

Considering that both the Reynolds stress and the Reynolds heat flux are poorly estimated by traditional LES approaches in the near-wall region of wall-bounded compressible flows, Jiang et al. introduced a dual-constraint SGS (DC-SGS) model for LES of such flows [6]. The formulations of the DC-SGS models are reviewed briefly as follows. In the near-wall region, the SGS stress τ_{ij}^{LES} and the SGS heat flux q_i^{LES} are split into a mean part and a fluctuation part, namely,

$$\tau_{ij}^{LES} = \langle\tau_{ij}^{LES}\rangle + \tau_{ij}^{LES\prime} \tag{10}$$

and

$$q_i^{LES} = \langle q_i^{LES} \rangle + q_i^{LES\prime}. \tag{11}$$

The mean SGS stress and heat flux are determined by the following constraint conditions

$$\langle \tau_{ij}^{LES} \rangle = \tau_{ij}^{RANS} + \langle \overline{\rho} \rangle (|\tilde{u}_i \tilde{u}_j| - |\tilde{u}_i||\tilde{u}_j|), \tag{12}$$

and

$$\langle q_i^{LES} \rangle = q_i^{RANS} + \langle \bar{\rho} \rangle C_p (|\tilde{u}_i \tilde{T}| - |\tilde{u}_i||\tilde{T}|). \tag{13}$$

Here, $\langle \cdot \rangle$ denotes an ensemble average, and $| \cdot |$ represents a Favre average. The total Reynolds stress τ_{ij}^{RANS} and the total Reynolds heat flux q_i^{RANS} are prescribed by selected RANS models based on the resolved LES fields. Recognizing the lack of generality of RANS models, one is suggested to choose appropriate models for the total Reynolds stress and Reynolds heat flux when simulating flow in different geometries. In the present paper, the one equation Spalart-Allmaras (S-A) model [14] is employed. The SGS stress and SGS heat flux fluctuations can be modeled according to traditional compressible Smagorinsky models [9]. Finally, the SGS stress and SGS heat flux models in the near-wall region are written as

$$\begin{aligned} \tau_{ij}^{LES} =& \tau_{ij}^{RANS} + \langle \overline{\rho} \rangle (|\tilde{u}_i \tilde{u}_j| - |\tilde{u}_i||\tilde{u}_j|) \\ &+ 2C_s \left[\bar{\rho} \Delta^2 |\widetilde{S}| \left(\widetilde{S}_{ij} - \frac{1}{3} \widetilde{S}_{kk} \delta_{ij} \right) - \left\langle \bar{\rho} \Delta^2 |\widetilde{S}| \left(\widetilde{S}_{ij} - \frac{1}{3} \widetilde{S}_{kk} \delta_{ij} \right) \right\rangle \right] \\ &- \frac{2}{3} C_I \left[\bar{\rho} \Delta^2 |\widetilde{S}|^2 - \langle \bar{\rho} \Delta^2 |\widetilde{S}|^2 \rangle \right] \delta_{ij}, \end{aligned} \tag{14}$$

$$\begin{aligned} q_i^{LES} =& q_i^{RANS} + \langle \bar{\rho} \rangle C_p (|\tilde{u}_i \tilde{T}| - |\tilde{u}_i||\tilde{T}|) \\ &+ \frac{C_s}{Pr_T} \left(\Delta^2 \bar{\rho} C_p |\widetilde{S}| \frac{\partial \widetilde{T}}{\partial x_i} - \left\langle \Delta^2 \bar{\rho} C_p |\widetilde{S}| \frac{\partial \widetilde{T}}{\partial x_i} \right\rangle \right). \end{aligned} \tag{15}$$

Here, C_s and C_I are the Smagorinsky constants, and Pr_T is the SGS Prandtl number. These model coefficients can be prescribed empirically or calculated instantaneously through a dynamic procedure. For details of the derivation of the DC-SGS models, readers are referred to the article by Chen et al. [1] for incompressible flow and the one by Jiang et al. [6] for compressible flow. In the rest region of the flow domain, traditional Smagorinsky models for the SGS stress and SGS heat flux can be utilized [9].

3 Applications of CLES

To test and validate the performance of the compressible CLES method summarized in Sect. 2, it was applied to the simulations of several flow problems, including fully-developed compressible turbulent channel flow [6], flow past a NACA0021 airfoil at 60° AOA, compressible flow past a circular cylinder [5], compressible flow over periodic hills [18], compressible flow past a Delta Wing, and flow past a commercial plane at 14° AOA. In what follows, we shall briefly present some of the results obtained in the above simulation works.

3.1 *Compressible Turbulent Channel Flow*

The compressible CLES approach was first validated in the simulation of fully developed compressible turbulent channel flow at various Reynolds and Mach numbers [6]. The simulations were carried out in a box of size $4\pi \times 2 \times 4/3\pi$ with $64 \times 65 \times 64$ grid resolution. The width of the first wall-normal grid in wall units is $y^{+}(1) \approx 1$, which satisfies the requirement for LES of wall-bounded flows. The isothermal-wall boundary condition is used.

Figure 1 shows the profiles of mean streamwise velocity with van Driest transform (Fig. 1a) and mean temperature (Fig. 1b) in wall units obtained from CLES, DES, and traditional LES at bulk Reynolds number $Re = \rho_b U_b \delta/\mu = 15{,}000$ and $Ma = 1.5$. Here, the friction velocity and friction temperature are defined as $u_\tau = \sqrt{\langle\tau_w\rangle/(Re\langle\rho_w\rangle)}$ and $T_\tau = -\langle q_w\rangle/(\langle\rho_w\rangle c_p u_\tau)$, respectively. The van Driest

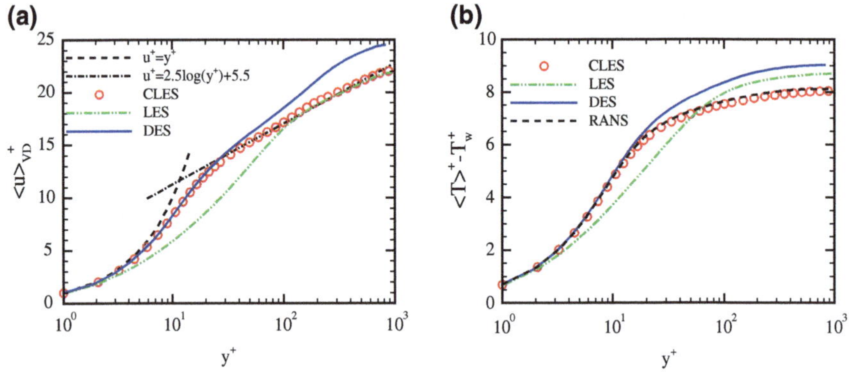

Fig. 1 Profiles of **a** mean streamwise velocity and **b** mean temperature in wall units calculated in CLES (*circles*), DES (*solid line*), and traditional LES (*dashed-double-dotted line*) at bulk Reynolds number $Re = \rho_b U_b \delta/\mu = 15{,}000$ and Mach number $Ma = 1.5$. The linear law (*dashed line*) and the log law (*dashed-dotted line*) in panel (**a**) and the temperature profile obtained using RANS simulation (*dashed line*) in panel (**b**) are also shown for references

transformed mean velocity is defined as $\langle u\rangle^+_{VD} = \int_0^{\langle u\rangle^+} (\langle\rho\rangle/\langle\rho_w\rangle)^{1/2}\, d\langle u\rangle^+$. It is clearly seen in Fig. 1a that the mean velocity distribution calculated by CLES agrees pretty well with the linear relation in viscous sublayer and the log-law in the logarithmic layer. There is an obvious super-buffer layer followed by a shifted log-law region in the mean velocity profile predicted by DES as observed in incompressible case (the so-called LLM phenomena) [1, 11]. Meanwhile, the velocity profile given by traditional LES deviates strongly from that by CLES in a wide range across the buffer layer. Since the mean velocity in van Driest transform is suggested to satisfy the log law of the wall, i.e., $\langle u\rangle^+_{VD} = 2.5\ln y^+ + 5.5$, the performance of CLES method is much better than those of traditional LES and DES methods. As can be seen in Fig. 1b, the mean temperature profile given by CLES nearly coincides with that by RANS simulation, while the temperature profile calculated by DES deviates strongly from that by RANS in the logarithmic region. Traditional LES, however, fails to predict the mean temperature distribution nearly in the whole flow field, i.e., overestimating the mean temperature in the logarithmic region while underestimating it in the buffer-layer region. If one takes the mean temperature predicted by RANS simulation as the benchmark, these results manifest that CLES technique is superior to traditional LES and DES approaches.

3.2 Flow Past a NACA0021 Airfoil at 60° AOA

One aspect of a successful simulation tool is the capability in predicting flows with massive separations. A frequently invoked example of such flows is the flow over an airfoil in a deep stall state. The performance of the compressible CLES method was also tested in simulation of flow past a 3D NACA0021 airfoil with a chord length c and a spanwise length $L_z = c$ at an AOA $\alpha = 60°$. The free-stream Reynolds number $Re_\infty = U_\infty c/\nu$ and Mach number $M_\infty = U_\infty/a$ are, respectively, 2.7×10^5 and 0.1. O-type computational mesh was generated with about 2.9×10^6 grid cells.

Shown in Fig. 2 are the pressure distributions on the upper/lower airfoil surfaces predicted by different simulation tools (present RANS: dashed-dotted line, present DES: dashed-double-dotted line, and present CLES: solid line). The previous experimental data (circles) and numerical data using DES (dotted line) reported by TUB group are also plotted in the same figure for comparison. It is found that the present RANS simulation fails to predict the pressure distributions on the upper surface, where massive separations occur, resulting in large deviations in both lift and drag coefficients. The pressure calculated by the present DES based on the S-A model is more accurate than that by the present RANS, and is in a reasonable agreement with that by previous DES, although obvious departures of pressure on the upper surface from the experimental data are observed. The pressure distributions predicted by the present CLES method show pretty good consistency with the experimental data. The lift and drag forces are also calculated for different runs, and the lift and drag

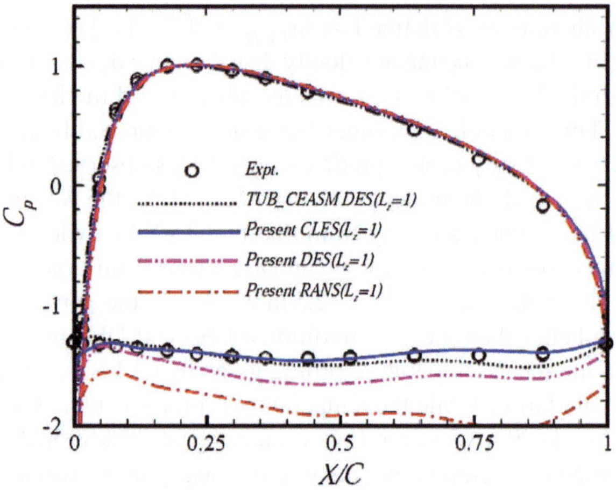

Fig. 2 Pressure distributions on the airfoil surface from different simulations: present CLES (*solid line*), present DES (*dashed-double-dotted line*), and present RANS (*dashed-dotted line*). The corresponding results from the experimental measurement conducted by Swalwell (*circles*) and DES by TUB_CEASM (*dotted line*) are also plotted for comparison

Table 1 Summaries of the normalized forces calculated by different methods

Approach	Lift coefficient (C_L)	Drag coefficient (C_D)
Experiment	0.931	1.517
S-A DES (NUMECA)	1.026	1.688
S-A DDES (DLR)	1.001	1.548
X-LES (NLR)	1.082	1.567
$k-\omega$ OEM DES (IMFT)	1.093	1.796
SALSA DES(TUB)	0.984	1.592
LLR DES (TUB)	0.985	1.620
Present CLES	0.987	1.555
Present DES	1.060	1.699
Present RANS	1.189	1.924

The cited experimental and numerical data are from Ref. [3]

coefficients are listed in Table 1. The corresponding values obtained in the numerical simulations from DESider group [3] and the experimental measurements from Swalwell [15] are also listed for comparison. It is apparent that the CLES can predict the aerodynamic forces very well, with errors of 6.1 and 2.5 % for lift and drag forces, respectively, which are among the best results listed in Table 1.

3.3 Flow Past a Delta Wing

Flow over a nonslender Delta wing at large AOA has recently become a topic of increased interest and is another flow model characterized by massive flow separations. Therefore, the compressible CLES method was further used to simulate the flow past a $\Lambda = 50°$ sweep Delta wing introduced in literatures [4, 16, 17], with the root chord length being $c = 0.31$ m, and the relative thickness $t/c = 1.6\,\%$. The AOA α is 15°, the free stream Mach number M_∞ is 0.1, and the Reynolds number Re is 2×10^5. Multiblock grids were generated with a total of 2.6×10^7 grid cells.

The computational results from CLES are compared with those from the present DES on the same grids and experimental measurement conducted by Gursul et al. [4]. The averaged lift coefficients calculated in DES and CLES are 0.812 and 0.809, respectively. As compared with the experimental value of 0.795, the errors are about 2.1 and 1.8 %, both of which are in an acceptable range.

We display in Fig. 3 the instantaneous snapshots of vortex structures over a half wing obtained in DES (panel a) and CLES (panel b). The vortex structures are identified by the isosurface of the second invariant of the strain rate tensor (denoted by Q) with $Q = 2{,}000$. The colors represent various values of the streamwise vorticity. A clear and large core of the primary vortex is figured out in

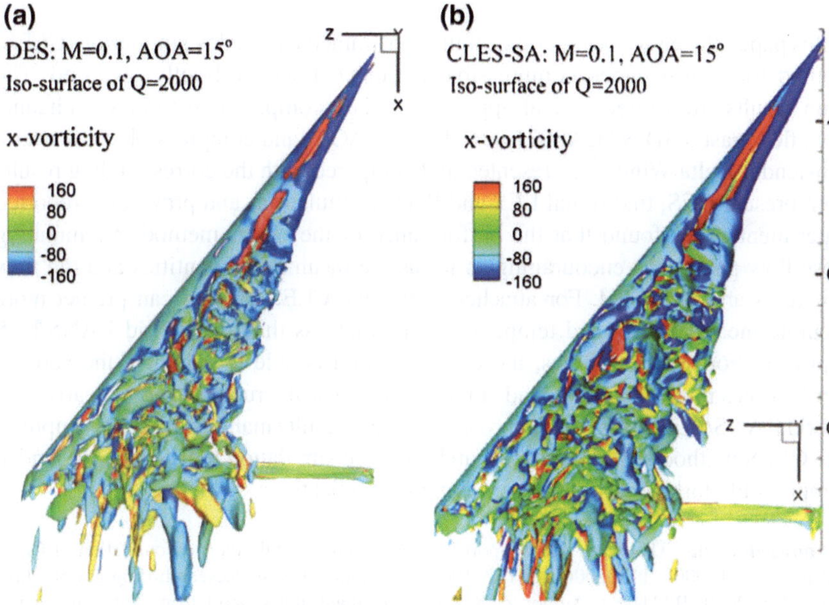

Fig. 3 Instantaneous vortex structures over a $\Lambda = 50°$ sweep Delta wing at $\alpha = 15°$ AOA obtained in different simulations: **a** DES, and **b** CLES. The vortex structures are identified by the isosurface of the second invariant of the strain rate tensor (Q) with $Q = 2{,}000$. The *colors* are visualized by the values of streamwise component of vorticity

CLES, while the primary vortex core given by DES is more diffuse and much smaller in size. Phenomenologically, there are two types of vortex breakdown processes, i.e., the spiral breakdown and the bubble-type breakdown. The bubble-type breakdown is characterised by the appearance of small-scale separation bubble or the thickening of the vortex core, which is terminated at its downstream end by a pinch off region. The onset of the bubble-type vortex breakdown is clearly observed in CLES, but is blurry in DES. A second primary vortex is also clearly seen both in CLES and DES, in agreement with the experimental observation. Another feature observed in Fig. 3 is that the vortex structures downstream of the bubble-type breakdown given by CLES are more abundant and chaotic that those by DES. This is because the CLES solves the LES equations in the whole simulation domain, while DES gives smooth RANS solution in the near-wall region, which shall influence the fluctuating properties of the flow beyond the interface. The pictures observed in Fig. 3 are consistent with a previous conclusion drawn for incompressible turbulent flows [1]. Therefore, we conclude that the compressible CLES technique may yield more reasonable results than hybrid RANS/LES methods when simulating flow over a Delta wing at high incidence.

4 Conclusions and Discussions

In this paper, the basic formulations of the constrained large-eddy simulation (CLES) method for wall-bounded compressible turbulent flows are briefly reviewed. The main results from three typical applications, i.e., compressible turbulent channel flow, flow past a 3D NACA0021 airfoil at 60° AOA, and compressible flow over a nonslender Delta Wing, are presented and compared with the corresponding results from present DES, traditional LES and RANS simulations and previously reported experiments. It is found that the performance of the CLES method in simulating these flow patterns is encouraging as far as the mean flow quantities and the flow structures are concerned. For attached flows, the CLES method can predict more accurate mean velocity and temperature distributions than the hybrid RANS/LES approach. For detached flows, the CLES method is able to calculate the aerodynamic forces more accurately and simulate the vortical structures more clearly than hybrid RANS/LES and RANS approaches. These results manifest that the compressible CLES method shall be a useful and effective simulation tool for wall-bounded compressible turbulent flows in aerodynamic applications.

Acknowledgments This work was support from National Natural Science Foundation of China (Grant No. 91130001, 11302006 and 11221061) and National Basic Research Program of China (Grant No. 2009CB724101). Author Z. Xia wants to thank the support from National Science Foundation for Postdoctoral Scientists of China (Grant No. 2012M520109).

References

1. Chen, S.Y., Xia, Z.H., Pei, S.Y., Wang, J.C., Yang, Y.T., Xiao, Z.L., Shi, Y.P.: Reynolds-stress-constrained large eddy simulation of wall bounded turbulent flows. J. Fluid. Mech. **703**, 1–28 (2012)
2. Fröhlich, J., von Terzi, D.: Hybrid LES/RANS methods for the simulation of turbulent flows. Prog. Aerosp. Sci. **44**, 349–377 (2008)
3. Garbaruk, A., Shur, M., Strelets, M., Travin, A.: DESider-A European Effort on Hybrid RANS-LES Modelling, Chap. NACA0021 at 60° Incidence, pp. 127–139. Springer, Berlin (2009)
4. Gursul, I., Gordnier, R., Visbal, M.: Unsteady aerodynamics of nonslender delta wings. Prog. Aerosp. Sci. **41**, 515–557 (2005)
5. Hong, R.K., Xia, Z.H., Shi, Y.P., Xiao, Z.L., Chen, S.Y.: Constrained large-eddy simulation of compressible flow past a circular cylinder. Commun. Comput. Phys. **15**(2), 388–421 (2014)
6. Jiang, Z., Xiao, Z.L., Shi, Y.P., Chen, S.Y.: Constrained large eddy simulation of wall-bounded compressible turbulent flows. Phys. Fluids **25**, 106102 (2013)
7. Kraichnan, R.H.: Theoretical approaches to turbulence. In: Dwoyer, D.L., Hussaini, M.Y., Voigt, R.G. (eds.) Applied Mathematical Sciences Series, vol. 58, p. 91. Springer, New York (1985)
8. Kraichnan, R.H., Chen, S.: Is there a statistical mechanics of turbulence? Physica D **37**, 160–172 (1989)
9. Martín, M.P., Piomelli, U., Candler, G.V.: Subgrid-scale models for compressible large-eddy simulations. Theor. Comput. Fluid Dyn. **13**, 361–376 (2000)
10. Meneveau, C.: Statistics of turbulence subgrid-scale stresses: necessary conditions and experimental tests. Phys. Fluids **6**, 815–833 (1994)
11. Nikitin, N.V., Nicoud, F., Wasistho, B., Squires, K.D., Spalart, P.R.: An approach to wall modeling in large-eddy simulations. Phys. Fluids **12**, 1629–1632 (2000)
12. Shi, Y., Xiao, Z., Chen, S.: Constrained subgrid-scale stress model for large eddy simulation. Phys. Fluids **20**, 011701 (2008)
13. Spalart, P.R.: Detached-eddy simulation. Annu. Rev. Fluid Mech. **41**, 181–202 (2009)
14. Spalart, P.R., Allmaras, S.R.: A one-equation turbulence model for aerodynamic flows. Rech. Aerospatiale **1**, 5–21 (1994)
15. Swalwell, K.E.: The effect of turbulence on stall of horizontal axis wind turbines. Ph.D. thesis, Monash University (2005)
16. Taylor, G., Gursul, I.: Unsteady vortex flows and buffeting of a low sweep delta wing. AIAA Paper AIAA-2004-1066 (2004)
17. Taylor, G., Kroker, A., Gursul, I.: Passive flow control over flexible non-slender delta wings. AIAA Paper, AIAA-2005-0865 (2005)
18. Xia, Z.H., Shi, Y.P., Hong, R.K., Xiao, Z.L., Chen, S.Y.: Constrained large-eddy simulation of separated flow in a channel with streamwise-periodic constrictions. J. Turbul. **14**(1), 1–21 (2013)

Part III
Detached Eddy Simulations

Grey-Area Mitigation for the Ahmed Car Body Using Embedded DDES

N. Ashton, A. Revell and R. Poletto

Abstract The Ahmed car body represents a generic car geometry which exhibits many of the flow features found in real-life cars despite its simplified geometry. It is a challenging test case for the turbulence modelling community as it combines both 3D separation and the formation of counter-rotating vortices, which interact together to produce a recirculation region behind the car body. It is shown that none of the RANS models tested are able to correctly predict the size of the recirculation region, regardless of modelling level, mesh resolution or the choice of the length scale (i.e. ω or ε). All of these models under-predict the turbulence levels over the slanted back and as a consequence over-predict the separation region. The DDES simulations (regardless of the underlying URANS model) offer an improved predictive capability compared to the RANS models when the mesh resolution is sufficient. When the mesh resolution is insufficient the DDES models produces worse results than either of the URANS models. In both cases, the grey area problem is demonstrated, wherein a lack of both modelled and resolved turbulence in the initial separated shear layer results in an over-prediction of the separation region. A one-way embedded DDES approach is shown to give the best compromise between accuracy and simulation cost. It accurately predicts the level of resolved turbulence in the initial separated shear layer and thus compared to non-embedded DDES and URANS, the injection of synthetic turbulence upstream of the separation point allows for the correct level of turbulence at the onset of separation. The resulting separation zone is correctly predicted and the grey-area problem is reduced.

N. Ashton (✉) · A. Revell · R. Poletto
School of Mechanical, Aerospace & Civil Engineering, University of Manchester, Manchester, UK
e-mail: neil.ashton@manchester.ac.uk

A. Revell
e-mail: alistair.revell@manchester.ac.uk

R. Poletto
e-mail: ruggero.poletto@postgrad.manchester.ac.uk

S. Girimaji et al. (eds.), *Progress in Hybrid RANS-LES Modelling*,
Notes on Numerical Fluid Mechanics and Multidisciplinary Design 130,
DOI 10.1007/978-3-319-15141-0_9

1 Introduction

Computational Fluid Dynamics has increasingly provided the methodology behind an important design tool for the automotive industry, used to supplement experimental studies. With a desire to reduce noise levels and improve fuel efficiency, reliable CFD simulations of the complex separated turbulent flow around vehicles is becoming an ever more crucial goal. Validating high-order turbulence simulation methods on an actual car geometry remains computationally expensive and conclusions are often case specific. The Ahmed car body [1] represents a generic car geometry with a slanted back and a flat front and has been extensively tested in the literature. While it is a much simplified version of a real car, it nevertheless provides many of the flow features found in real-life cars such as the complex vortex interactions that occur in its wake and the large 3D separation region behind the car body itself.

This geometry was first proposed and studied by Ahmed et al. [1] (Fig. 1a). The wake behind the car body is a result of the interaction between the counter-rotating vortices produced by the slant side edges and the separated flow over the rear of the body. The angle of the slant back section was found to be influential in the structure of the wake and the reattachment point. At 35° the counter-rotating vortices are weaker, which results in the flow being completely separated over the entire slant back of the vehicle. At 25° the counter-rotating vortices are strong enough to help to bring enough momentum into the separation region to reattach the flow half way down the slant back.

More recently this geometry has also been investigated by Lienhart et al. [12], who performed a more detailed experimental study (at a lower Reynolds number), which included LDA measurements of the mean and fluctuating velocities as well as on-surface oil flows. This was performed for slant back angles of both 25° and 35°, and the results confirmed many of the observations that were made by Ahmed in their initial investigation.

This case has been the focus of several CFD investigations, most notably the 9th and 10th ERCOFTAC workshops on refined turbulence modelling [8, 13], where a range of RANS models were investigated, and also the DFG-CNRS program: LES of Complex Flows [17], where several LES and DES formulations were evaluated.

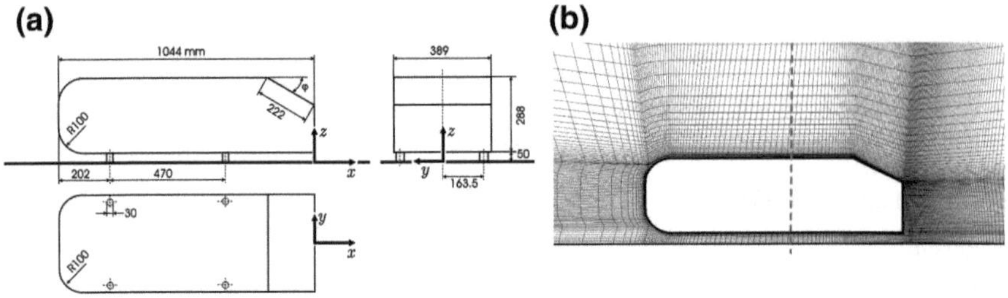

Fig. 1 Dimensions of the Ahmed car body [17] and fine mesh for the Ahmed car body

It has also been studied within the FLOMANIA [6] and DESider [7] EU projects, where a range of RANS, LES and hybrid RANS-LES models were investigated.

The general conclusions from the studies involving RANS models [6, 8, 13] were dependant on the slant angle. At 35° (where the separation occurred along the entire slant back) most of the RANS approaches (both simple linear $k - \varepsilon$ and more complex Reynolds stress models) captured the correct level of the turbulent stresses and, as a result, showed good agreement with the experimental results for the separation and reattachment points. However at 25° (where the complex interaction between the counter-rotating vortices and the separated flow results in a shorter separation region) the majority of RANS models,regardless of mesh refinement or wall treatment, failed to predict the flow correctly. In general, they either failed to predict separation completely, or even when they did, they did not predict the correct separation point and thus were unable to capture the correct size of the recirculation region due to an under-prediction of the turbulent stresses [17].

More recently, a number of LES studies have been performed on this case, mainly at the more challenging 25° slant angle [7, 10, 15, 17]. These studies were performed with a range of sub-grid scale models and wall-treatments and, while some were more successful than others, many failed to capture fully the correct recirculation region. The high-Reynolds number ($Re = 7.68 \times 10^5$) meant that even with meshes of up to 48 million cells, the resolution requirements were still not ideal for a wall-resolved LES. One conclusion from Serre et al. [17] was that due to the high cost of a wall-resolved LES, hybrid RANS-LES methods represent an attractive alternative, even if the mesh generation requires greater care with regards to the RANS-LES zones.

2 Turbulence Models

One of the more common hybrid RANS-LES methods in usage is the Delayed Detached-Eddy Simulation (DDES) approach [18] which is an improved version of the original Detached-Eddy Simulation (DES) method of [19]. DDES can be seen as a three-dimensional, unsteady model based on an underlying 'off-the-shelf' RANS model. It seamlessly joins a sub-grid scale model in regions where the grid is fine and outside of the attached boundary layer to a RANS model in all other regions.

The first version of DES was based on the Spalart-Allmaras (SA) model, as this was seen as 'the most convenient length scale to inject Δ and turn a RANS model into a SGS model' [20]. Since then, many RANS models have been applied to DES and DDES, however the most popular models for commercial codes are still the SA and $k - \omega$ Shear-Stress Transport (SST) DDES models.

In this study an alternative version of DDES is also used; the $\varphi - f$ DDES model [4] which uses the $\varphi - f$ underlying RANS model [11]. This has previously shown promising performance on several canonical flows and is used to assess the influence (if any) of the underlying RANS model.

Furthermore, both the SST and $\varphi - f$ models in a URANS framework is also used to assess the advantage of hybrid RANS-LES methods over RANS models.

In addition, a low-Reynolds number Reynolds-Stress model; the Elliptic-Blending Reynolds-Stress model (EB-RSM) [14] is used to ascertain whether a second-moment closure model offers any improvement over linear eddy-viscosity models.

2.1 Synthetic Turbulence Generation: Synthetic Eddy Method

For the embedded DDES simulations, a synthetic turbulence generator is required at the RANS-LES interface. In this study the recently developed Divergence-Free Synthetic Eddy Method (DFSEM) [16] is used. This present DFSEM is based on the methodology described in [9] where synthetic eddies, each of which represents a set of velocity fluctuations, are convected through a box that entirely surrounds the inlet plane upon which a turbulent velocity field is required. These eddies, which are defined by their centre and a formulation for the velocity fluctuation distribution around it, are convected at each time step by the locally imposed mean velocity. Once they have traversed and exited the box they are regenerated at a random location on the box inlet plane. In general, the steps of the DFSEM algorithm are summarized as:

1. User selection of inlet surface Ω.
2. User definition of average velocity $\mathbf{u}(\mathbf{x})$, Reynolds stresses and turbulence length-scales $\sigma(\mathbf{x})$, for $\mathbf{x} \subset \Omega$.
3. Eddy Bounding Box taken as: $\max\{\mathbf{x} + \sigma\}$, $\min\{\mathbf{x} - \sigma\}$ for $\mathbf{x} \subset \Omega$.
4. Definition of the number of eddies.
5. Assigning random positions $\mathbf{x}^k$ and intensities α^k to all the eddies.
6. Eddies being convected through the eddy box, by $\mathbf{x}^k = \mathbf{x}^k + \mathbf{U}_b * \Delta t$, where $\mathbf{U}_b = \int_\Omega \mathbf{u} ds / \int_\Omega ds$ is the bulk velocity calculated from the user imposed average velocity. Eddies that leave the Bounding Box are re-generated at the opposite surface.
7. $\mathbf{u}'(\mathbf{x})$ calculated and superimposed to $\mathbf{u}$ to generate the inlet condition.
8. Repeat steps 6–7 for all the subsequent time steps.

The reader is referred to [9] for more details about the overall procedure.

3 Computational Grid and Boundary Conditions

The car body was mounted on four stilts $y/H = 0.174$ (where $H = 0.288$ m is the height of the body) above the ground in the experiment to model the effect of the car's height with wheels. However to aid the mesh generation in the CFD simulations these stilts were not used and thus the body is fixed at $y/H = 0.174$ above the floor. Only the 25° slant back angle has been investigated in this study, as it is this angle that produces the most challenging flow features; the strong influence of the counter-rotating vortices on the reattachment point. The body has a length of $L/H = 3.625$, a height of $H/H = 1$ and a width of $W/H = 1.35$. Two meshes were used; firstly, a structured mesh of 2.7 million cells and secondly, a much finer mesh of 16 million cells with greater refinement in the wake and separation locations (Fig. 1b) courtesy of Professor Krajnovic of Chalmers University of Technology.

The flow is at a Reynolds number of $Re = 768{,}000$ based on the body height H and the free-stream velocity $U_\infty = 40\,\text{ms}^{-1}$. An inlet condition is imposed $x/H = -7.3$ upstream of the body and an outlet condition is imposed $x/H = 20.3$ downstream. A no-slip wall condition is imposed on the ground floor and car body, with slip walls applied to the wind tunnel walls. In the case of the simulations with the synthetic eddy method then the whole geometry was cut at $x/H = -1.73$ and fluctuating velocity components were imposed based upon the time averaged velocity and turbulent quantities from a pre-courser RANS simulation. The time step is set to $\Delta t U_\infty/L = 2 \times 10^{-4}$ and $\Delta t U_\infty/L = 1 \times 10^{-4}$ for the URANS and DDES simulations respectively, which ensures a maximum CFL number of less than one. Each simulation was run for a total of 30 convective transit times ($=TU_\infty/L$); time-averaging began after the initial 10 transit times. All calculations were performed using the open-source software *Code_Saturne* [2, 5] developed by EDF R&D. The temporal discretization is second order, while a hybrid numerical scheme based on a blend of central differencing (for the LES zones) and upwinding (for the RANS zones) is used to discretize spatially the convective terms [3].

4 Results

4.1 URANS Results

Figure 2b, e show the mean streamlines for the SST and $\varphi - f$ URANS models at the centreline ($y = 0$) position. Compared to the experimental data (Fig. 2a), both models show a much larger separation region over the slanted back and behind the car body. This is more clearly observed for the mean stream wise velocity in Fig. 3a. Both the URANS models on both meshes under predict the turbulent kinetic energy in the initial separated shear layer (Fig. 3b) and as a consequence over predict the separation zone. It is also the case that both URANS models under predict the strength of the counter-rotating vortices, which is a key mechanism for the small separated region compared to the 35° case where the lack of these vortices causes the flow to completely separate over the slanted back.

Figure 3a, b also show that even the use of a low-Reynolds number Reynolds-Stress model is unable to capture the correct level of turbulence. This is in agreement with previous studies that have shown that no RANS model regardless of its modelling level can capture the flow correctly, This suggests either that these RANS models are not capturing some fundamental physical mechanism or the large scale unsteadiness is only captured by higher-order methods [13].

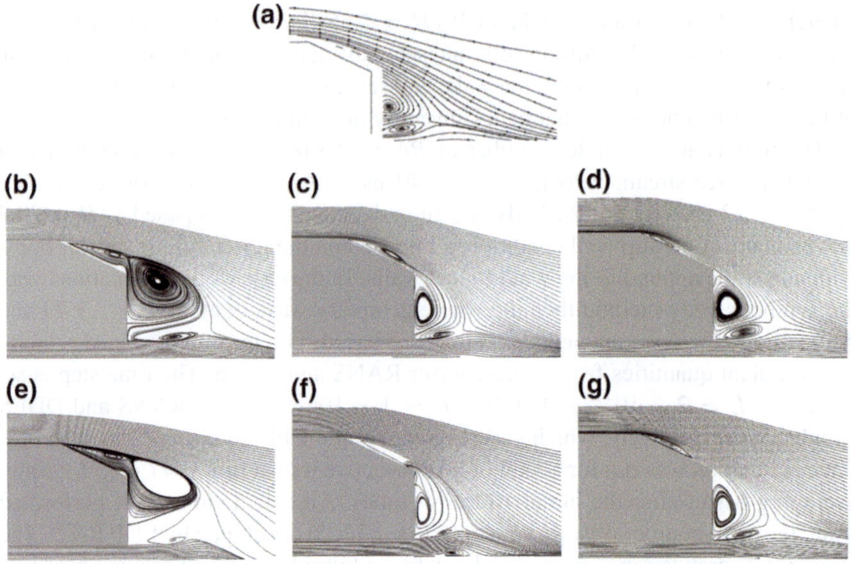

Fig. 2 Mean streamlines over the Ahmed car body for the **a** experiment **b** SST-URANS **c** SST-DDES **d** SST E-DDES **e** $\varphi - f$ URANS **f** $\varphi - f$ DDES **g** $\varphi - f$ E-DDES

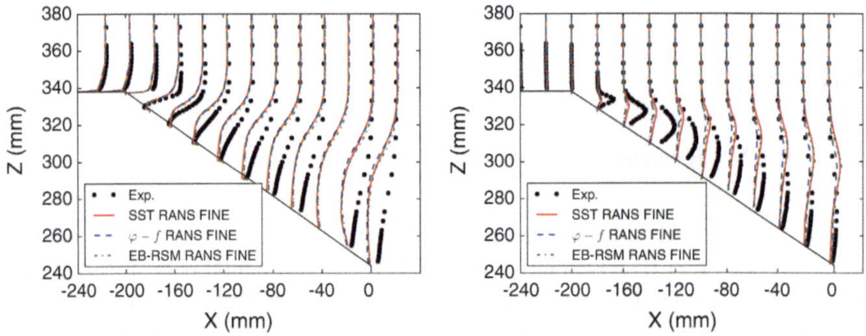

Fig. 3 Mean stream-wise velocity and turbulent kinetic energy profiles for the Ahmed car body using the SST RANS, $\varphi - f$ RANS and EB-RSM model for the fine mesh. Experimental data from [12]

4.2 DDES Results

Compared to the URANS results the mean streamlines (Fig. 2c, f) and the mean stream wise velocity (Figs. 5 and 7) for the SST and $\varphi - f$ DDES models on the fine mesh show a reduced separation region, which is more in-line with the experimental results. This is directly related to increased turbulence levels compared to the URANS (Figs. 6 and 8) which allow for increased turbulent mixing and less recirculation. This shows that the transition from RANS to LES mode occurs in the separated region although the grey-area problem is observed; the initial separated shear layer turbulence is under-predicted due to a lack of resolved turbulence. The modelled turbulence is almost gone but the resolved turbulence does not develop quickly enough to produce the correct level of turbulence at the very beginning of the separated region.

By following the flow as it traverses the car body it is clear that the counter-rotating vortices are correctly predicted so the failure to predict the correct reattachment point is due to an under prediction of the turbulence in the initial separated shear layer.

4.2.1 Effect of Mesh Resolution

A very important observation from this work for other DDES users is the poor performance of DDES compared to URANS when the mesh resolution is too coarse. Figure 4 shows the clear difference in the prediction of the flow between the 2.7 million and 16 million cell mesh, with the coarse mesh actually providing a worse solution than either of the URANS models on the same mesh. There is a complete lack of turbulence in the initial separated region, with neither modelled nor resolved turbulence, a very clear example of a long grey-area. It is therefore important to note that a DDES simulation may only deliver improvements over standard URANS models when the grid resolution is sufficient.

4.3 Embedded DDES Results

For the embedded results there is a clear improvement in the prediction of the separation zone which is most clearly observed for the mean stream wise velocity on the centreline (Figs. 5 and 7). This is due to the turbulent kinetic energy being more accu-

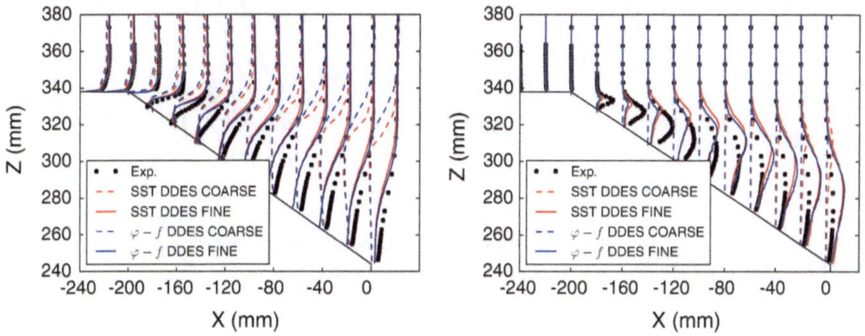

Fig. 4 Mean stream-wise velocity profiles and turbulent kinetic energy for the Ahmed car body using the SST DDES and $\varphi - f$ DDES models for the coarse and fine mesh. Experimental data from [12]

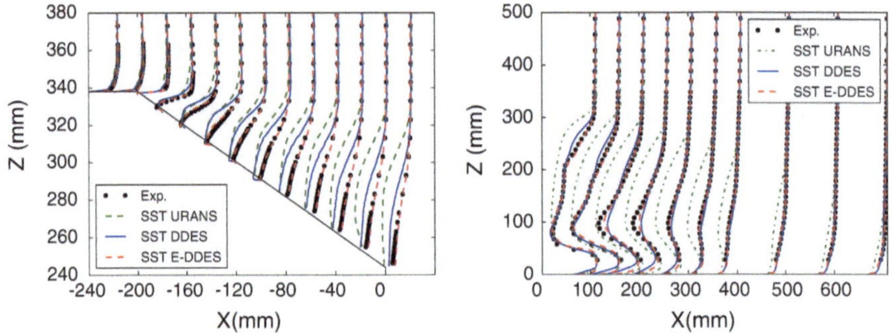

Fig. 5 Mean stream-wise velocity profiles for the Ahmed car body using the SST RANS, SST DDES and SST E-DDES models. Experimental data from [12]

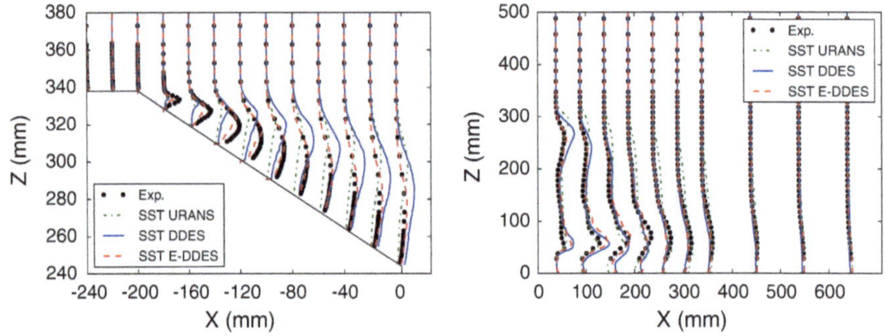

Fig. 6 Mean turbulent kinetic energy profiles for the Ahmed car body using the SST RANS, SST DDES and SST E-DDES models. Experimental data from [12]

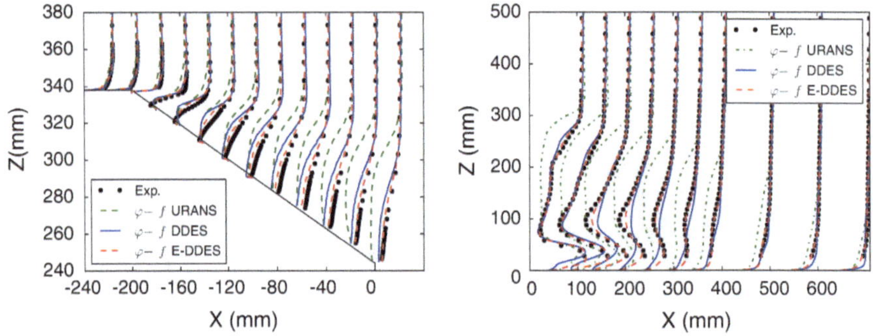

Fig. 7 Mean stream-wise velocity profiles for the Ahmed car body using the $\varphi - f$ RANS, $\varphi - f$ DDES and $\varphi - f$ E-DDES models. Experimental data from [12]

rately predicted by a quicker transition to resolved turbulence compared to DDES (Figs. 6 and 8). Compared to the DDES results, there is almost no modelled turbulence even in the initial separated shear layer, which is why the grey-area problem is

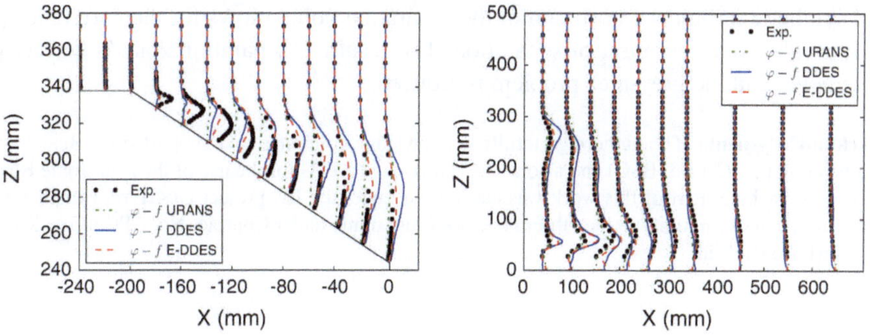

Fig. 8 Mean turbulent kinetic energy profiles for the Ahmed car body using the $\varphi - f$ RANS, $\varphi - f$ DDES and $\varphi - f$ E-DDES models. Experimental data from [12]

mitigated when synthetic turbulence is imposed $x/H = 1.73$ upstream of the slanted back of the body. This means the DDES model does not need the same development length compared to a non-embedded approach to move from modelled to resolved turbulence. It can be argued that this is an unintended consequence of using an embedded approach, but nevertheless it shows that synthetic turbulence can be used to address the grey-area problem associated with DDES models on non-massively separated flows.

5 Conclusions

The results point to three main conclusions. Firstly RANS models, eddy-viscosity or second moment closure (utilizing either ω or ε for the turbulent length scale), with sufficient near-wall resolution ($y^+ < 1$) and which have reached mesh independence, cannot capture the correct level of turbulence in the initial separated shear layer. The consequence of this under-prediction of the turbulence is reduced turbulent mixing and an over-prediction of the separated region over the slant of the car body.

Secondly a DDES simulation (regardless of the underlying URANS model) with insufficient grid resolution produces worse results than either URANS model. The grey area problem is clearly observed, with a lack of both modelled and resolved turbulence in the initial separated shear layer, i.e. the transition to resolved turbulence is delayed such that an over-prediction of the separation region is observed. A DDES simulation on a finer mesh (16 million cells) improves the prediction of the overall turbulence in the initial separated shear layer (due to a faster production of resolved turbulence than the coarse mesh) compared to both the URANS and coarse DDES results.

Finally it is shown that an embedded one-way coupling method based upon the framework of DDES, accurately predicts the level of resolved turbulence in the initial separated shear layer. Compared to non-embedded DDES and URANS, the injection of synthetic turbulence upstream of the separation point allows for the correct level of turbulence at the onset of separation. The resulting separation zone is correctly predicted and the grey-area problem is reduced.

Acknowledgments The authors gratefully acknowledge computational support from Barcelona Supercomputer Centre (BSC) and also to the Hartree and STFC for the use of the Blue Joule Blue Gene Q machine. Part of this work was carried out under the EU project Go4Hybrid funded by the European Community in the 7th Framework Programme under Contract No. APC3-GA-2013-605361-Go4Hybrid.

References

1. Ahmed, S.R., Ramm, G., Faltin, G.: Some salient features of the time averaged ground vehicle wake. SAE-Paper 840300 (1984)
2. Archambeau, F., Mechitoua, N., Sakiz, M.: A finite volume method for the computation of turbulent incompressible flows—industrial applications. Int. J. Finite **1**, 1–62 (2004)
3. Ashton, N., Prosser, R., Revell, A.: A hybrid numerical scheme for a new formulation of delayed detached-eddy simulation (DDES) based on elliptic relaxation. J. Phys. Conf. Ser. **318**, 042043 (2011). doi:10.1088/1742-6596/318/4/042043
4. Ashton, N., Revell, A., Prosser, R., Uribe, J.: Development of an alternative delayed detached-eddy simulation formulation based on elliptic relaxation. AIAA J. **51**(2), 513–519 (2013)
5. Fournier, Y., Bonelle, J., Moulinec, C., Shang, Z., Sunderland, A., Uribe, J.: Optimizing code saturne computations on Petascale systems. Comput. Fluids **45**(1), 103–108 (2011). doi:10.1016/j.compfluid.2011.01.028
6. Haase, W., Aupoix, B., Bunge, U., Schwamborn, D. (eds.): FLOMANIA—A European Initiative on Flow Physics Modelling. Notes on Numerical Fluid Mechanics and Multidisciplinary Design, vol. 94. Springer, New York (2006)
7. Haase, W., Braza, M., Revell, A.: DESider—A European Effort on Hybrid RANS-LES Modelling. Notes on Numerical Fluid Mechanics and Multidisciplinary Design, vol. 103. Springer, New York (2007)
8. Jakirlić, S., Jester-Zurker, R., Tropea, C.: Report on 9th ERCOFTAC/IAHR/COST workshop on refined turbulence modelling. In: ERCOFTAC Bulletin, Darmstadt University of Technology (2002), pp. 36–43
9. Jarrin, N.: Synthetic inflow boundary conditions for the numerical simulation of turbulence. Ph.D. thesis, Manchester University (2008)
10. Krajnovic, S., Davidson, L.: Large-eddy simulation of the flow around simplified car model. In: SAE World Congress, Detroit (2004)
11. Laurence, D.L., Uribe, J.C., Utyuzhinkov, S.V.: A robust formulation of the $\overline{v^2} - f$ model. Flow Turbul. Combust. **73**, 169–185 (2004)
12. Lienhart, H., Becker, S.: Flow and turbulent structure in the wake of a simplified car model. SAE **01**(1), 0656 (2003)
13. Manceau, R., Bonnet, J.P., Leschziner, M., Menter, F.R.: 10th Joint ERCOFTAC(SIG-15)/IAHR/QNET-CFD Workshop on Refined Flow Modelling. Universite de Poitiers (2002)
14. Manceau, R., Hanjalic, K.: Elliptic blending model: a new near-wall Reynolds-stress turbulence closure. Phys. Fluids **14**, 744–754 (2001)
15. Minguez, M., Pasquetti, R., Serre, E.: High-order large-eddy simulation of flow over the Ahmed body car model. Phys. Fluids **20**(9), 095,101 (2008). doi:10.1063/1.2952595
16. Poletto, R., Craft, T., Revell, A.: A new divergence free synthetic eddy method for the reproduction of inlet flow conditions for LES. Flow Turbul. Combust. 519–539 (2013). doi:10.1007/s10494-013-9488-2

17. Serre, E., Minguez, M., Pasquetti, R., Guilmineau, E., Deng, G.B., Kornhaas, M., Schäfer, M., Fröhlich, J., Hinterberger, C., Rodi, W.: On simulating the turbulent flow around the Ahmed body: a FrenchGerman collaborative evaluation of LES and DES. Comput. Fluids **78**, 10–23 (2013). doi:10.1016/j.compfluid.2011.05.017
18. Spalart, P.R., Deck, S., Shur, M.L., Squires, K.D., Strelets, M.K., Travin, A.: A new version of detached-eddy simulation, resistant to ambiguous grid densities. Theor. Comput. Fluid Dynam. **20**(3), 181–195 (2006). doi:10.1007/s00162-006-0015-0
19. Spalart, P.R., Jou, W.H., Strelets, M., Allmaras, S.R.: Comments on the feasibility of LES for wings and on a hybrid, RANS/ES approach. In: Advances in DNS/LES, Proceedings of 1st AFOSR International Conference on DNS/LES **1**, 137–147 (1997)
20. Travin, A., Shur, M., Strelets, M., Spalart, P.R.: Detached-eddy simulations past a circular cylinder. Flow Turbul. Combust. **63**, 293–313 (2000)

Detached-Eddy Simulation of Separated Wake Flow Around Complex Helicopter Fuselage Configuration

M. Fuchs, F. Le Chuiton, C. Mockett, J. Sesterhenn and F. Thiele

Abstract The contribution presents a detailed CFD study of a massively separated flow around a detailed EC135 helicopter fuselage in forward flight. Delayed detached-eddy simulation (DDES) is compared to conventional steady and unsteady RANS. The performance of all approaches is evaluated for two different geometrical configurations at three angles of attack. Excellent agreement of DES to different sets of experimental benchmark data is achieved particularly for the separated wake behind the cabin backdoor. An improvement of predictive accuracy is seen over URANS for both mean statistical quantities as well as unsteady solution content. Limitations to the comparability between CFD and experiment are furthermore discussed.

1 Introduction

The optimisation of drag and hence fuel consumption is one of the central goals of helicopter design and part of the strategic long-term goals set by ACARE 2020 for the aircraft industry. In forward flight, a major part of total drag originates from the massively separated vortical wake behind the cabin backdoor. The wake flow furthermore interacts with the empennage downstream, which significantly influences unsteady flight dynamics of the helicopter.

The current state-of-the-art industrial method, i.e. unsteady RANS, is known to exhibit strong modelling deficits for massively separated flows, where as hybrid RANS-LES methods have been shown to improve predictive accuracy significantly.

M. Fuchs (✉) · J. Sesterhenn
Technische Universität Berlin, Berlin, Germany
e-mail: marian.fuchs@cfd.tu-berlin.de

F. Le Chuiton
Airbus Helicopters Deutschland GmbH, Donauwörth, Germany
e-mail: frederic.lechuiton@airbus.com

C. Mockett · F. Thiele
CFD Software-Entwicklungs and Forschungsgesellschaft mbH, Berlin, Germany
e-mail: charles.mockett@cfd-berlin.com

S. Girimaji et al. (eds.), *Progress in Hybrid RANS-LES Modelling*,
Notes on Numerical Fluid Mechanics and Multidisciplinary Design 130,
DOI 10.1007/978-3-319-15141-0_10

Nevertheless, hybrid RANS-LES methods are not yet routinely applied in industry, which is in part due to the elevated CPU costs compared to URANS and in part due to low software and process maturity.

The work presented here aims to contribute to the analysis of an existing fuselage geometry in terms of aerodynamic loads and interactional effects using state-of-the-art hybrid RANS/LES methods. Furthermore, the adopted DES process chain is assessed for complex helicopter applications.

2 Numerical Methodology and Setup

The presented study is for a full scale EC135 helicopter configuration in free forward flight, i.e. without wind-tunnel support struts. Figure 1 gives a visual impression of the geometry and an instantaneous flow field taken from SST-DDES. The configuration features a high level of geometric detail, where exhausts, landing skids and small-scale "roughness elements" (e.g. window frames, door handles) are accurately captured. All rotating parts of the helicopter were excluded from the simulation, i.e. the main rotor and the fenestron located at the empennage. Two different geometrical configurations with and without landing skids (denoted as CR and CLR respectively) were simulated for three different angles of attack and zero side-slip. The flight conditions correspond to cruise flight at 140 kt and a height of 500 m in standard atmosphere, resulting in a Reynolds number per unit length of $Re = 4.7 \times 10^6/\text{m}$.

The open source software package OpenFOAM® was used, which is an unstructured finite-volume based code of second order accuracy in space and time. Key features to perform high-fidelity turbulence-resolving simulations were implemented and carefully verified [1], including standardised DES model variants, an implicit pressure-based solver for incompressible flows and the hybrid convection scheme of Travin et al. [6]. The convection scheme applies a localised blending between

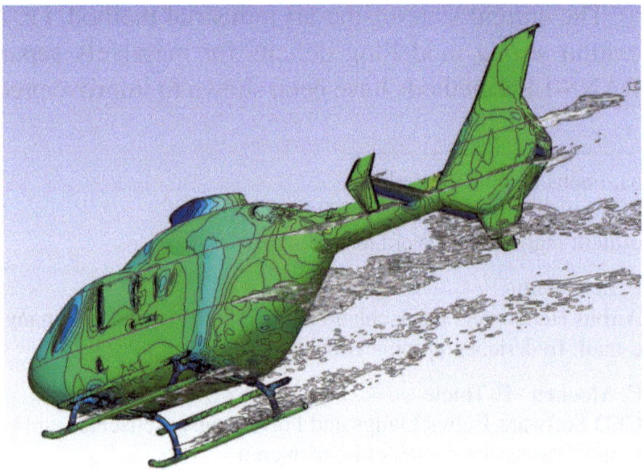

Fig. 1 Geometry and instantaneous flow field from SST-DDES for configuration with landing skids (CR)

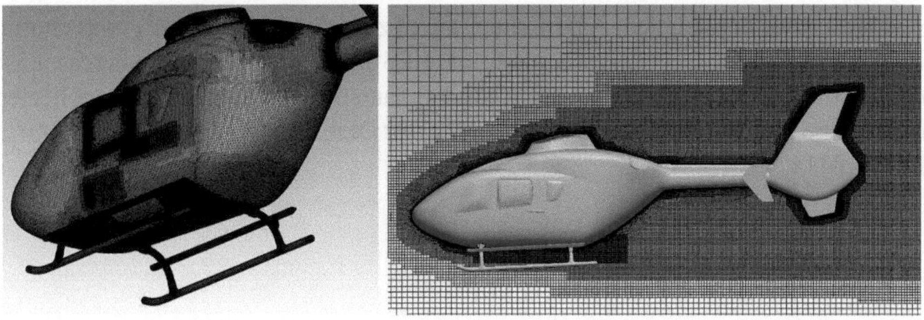

Fig. 2 Surface grid and mid-section slice through the grid for test case configuration CR

low-dissipation second order central differences in well-resolved turbulent regions and a robust second order accurate upwind scheme in coarse or irrotational flow regions.

Due to the high complexity of the geometry, an unstructured grid methodology is essential. The tool HEXPRESS™ of Numeca International was applied to generate high quality hex-dominant unstructured meshes with near-wall prism layers, where the surface grid and a mid-section slice through the grid for the configuration with landing skids are shown in Fig. 2. The grid was designed to achieve a non-dimensional wall distance of $y^+_{max} \approx 50$ on large parts of the fuselage, where a wall function valid for all y^+-values was used. The productive grids for both configurations consist of 24 M (CLR) and 32 M (CR) purely hexahedral cells, where conjunctions between different grid levels were realised using 2:1 jumps in cell size. Clustered rectangular refinement boxes were used to achieve adequate LES resolution (except for skids region where sufficient LES resolution is unfeasible) based on a precursor simulation, where the target grid spacing size in the wake focus region behind the cabin is $\Delta_0 \approx 0.012 L_y$ (where L_y is the fuselage width). To minimise the influence of the 2:1 jumps in cell size on the resolved scales in LES-mode, the grid resolution behind the fuselage was kept constant until behind the empennage.

With regard to the turbulence modelling approach, LES is known to improve prediction of massively separated flows, such as the main wake behind the fuselage. As large areas of the fuselage surface are covered with thin boundary layers, wall-resolved and wall-modelled LES are computationally too prohibitive for this application. On the contrary, zonal methods are not practicable here due to the high level of geometrical detail and the complex interaction of different separation zones. The delayed DES [3] approach was therefore chosen in this study, where a shield function ensures RANS treatment of all attached boundary layers. The complex shape of the EC135 and the topology of the flow field featuring pressure-induced separation from gently-curved surfaces (e.g. from the cabin backdoor) furthermore implies a high sensitivity of DES to the underlying RANS model, as was identified in a previous study of a more simplified EC145 fuselage [5]. As an outcome of that study, the k-ω SST model of Menter was chosen as background model, for which steady and unsteady RANS were additionally performed. All simulations were conducted in "fully turbulent" mode, meaning that no boundary layer transition location was prescribed or predicted. A DES time step of $\Delta t = 0.0027 \times L_y/U_\infty$ was chosen to guarantee $\text{CFL}_{max} \leq 1.0$ throughout the LES-mode focus region, whereas a 5× coarser time step was applied for URANS. For statistical analysis, data was collected over a time sample of $t_{avg} \approx 12-38\ L_y/U_\infty$ for DDES and $t_{avg} \approx 12-17\ L_y/U_\infty$ for URANS.

3 Experimental Benchmark Data

Two experimental data sets were provided by Airbus Helicopters, which stem from different measurement campaigns conducted at the TU Munich [2]. The corresponding geometries and measurement locations are shown in Fig. 3. Both experiments were conducted at considerably lower Reynolds numbers than in the CFD study, ranging between $0.28 \times 10^6/\text{m} \leq Re \leq 0.54 \times 10^6/\text{m}$. As no boundary layer tripping was prescribed in the experiments, laminar flow regions on the front part of the cabin as well as on the skids, the wings and end plates of the fuselage are expected in the experiments, which are not accounted for in CFD.

Mean force and moment coefficients for different angles of attack along with steady and unsteady surface pressure information was provided in the ECO-HC data set. Mean velocity components were extracted via PIV for several locations within the separated wake in the ADHeRo campaign.

A significant geometric discrepancy occurs in the underbelly region, which is significantly smoother in the ECO-HC configuration than is the case for the ADHeRo

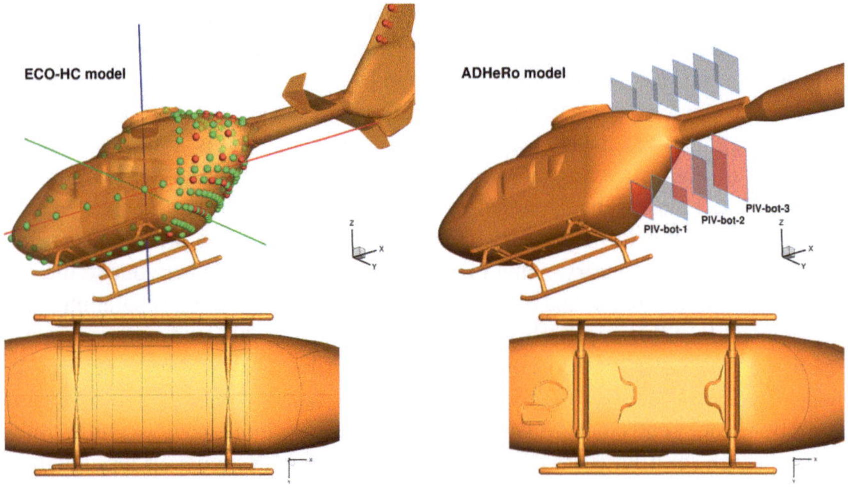

Fig. 3 Wind tunnel geometries of EC135 measurement campaigns. Moment reference axis (*solid lines*), steady pressure tapings (*green spheres*), unsteady pressure transducers (*red spheres*) and PIV windows (*red and grey frames*). Magnified views of underbelly geometry below

and CFD geometries (which are identical here). Both the underfloor equipment bays and the landing skid mounting cavities are smoothed over in the ECO-HC configuration. Furthermore, for the configurations without skids, the mounting cavities are left open in ADHeRO and CFD whereas the holes through which the landing skid struts were threaded in the ECO-HC were sealed with tape. No empennage is present for the ADHeRo geometry, where the wind tunnel support strut is directly connected to the tail boom.

4 Results and Discussion

4.1 Prediction of Mean Flow Field

As one of the central quantities of interest, the predicted global mean force and moment coefficients for the configuration with landing skids are shown in Fig. 4. A fairly good agreement between CFD and experiment is seen for the drag coefficient (C_{FX}), which is slightly higher in all CFD. A clear improvement of prediction is seen here for DES over RANS and URANS, both in absolute value and in trend.

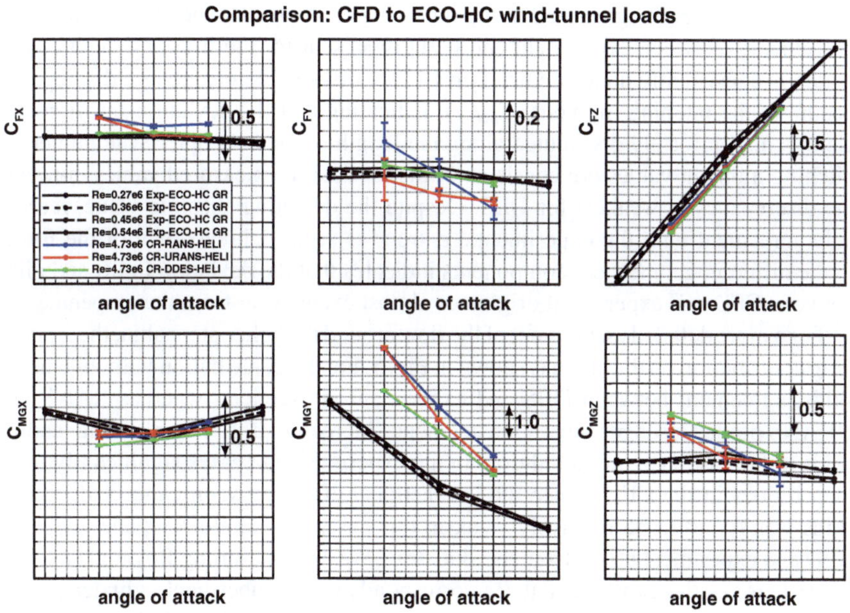

Fig. 4 Comparison of global mean force and moment coefficients for all CFD and ECO-HC data, CR configuration. Error bars on the CFD data denote statistical 95 % confidence intervals on the means [4]. Note that plots exhibit different scaling to account for the different ranges of the coefficients

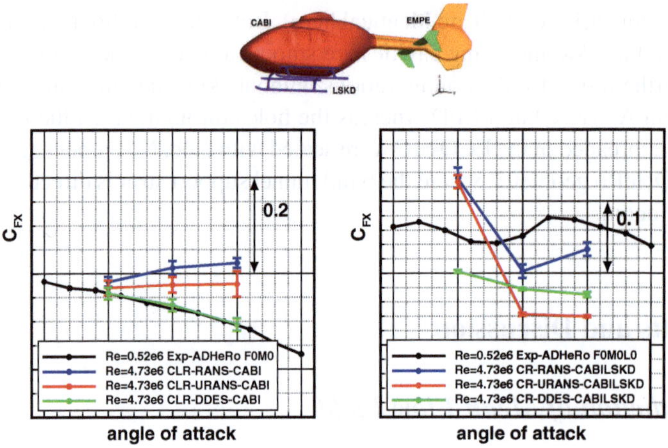

Fig. 5 Drag coefficient on isolated cabin (*left*) and cabin + landing skids (*right*), comparison of CFD configurations to ADHeRo measurements

The global side force (C_{FY}) shows good agreement to the measurement, although the wrong trend with angle of attack is present for the lowest simulated angle. The disagreement in side force transfers also to the roll moment (C_{MGX}) and yaw moment (C_{MGZ}), which both agree well at the other angles of attack. The most pronounced discrepancy between experiment and CFD is present for the lift coefficient (C_{FZ}), which is consistently underpredicted in CFD. Correspondingly, a large discrepancy occurs in the pitching moment (C_{MGY}), indicating that the deviation in lift coefficient is not uniformly distributed along the length of the helicopter. The numerical pitching moment values are consistently higher than the measured values, however the discrepancy is weakest for DES. Despite the deviating absolute value, reasonable agreement with the experimental α trend is seen. A breakdown of the forces to the different geometrical components indicates that the deviation in lift and pitch between CFD and experiment originates almost exclusively from the empennage.

For a more detailed assessment of the flow around the helicopter cabin, the isolated drag contribution on the cabin is shown in Fig. 5. For the contribution without landing skids, agreement between DES and experiment is excellent, whereas RANS and URANS predict stronger separation and hence increased drag. When adding landing skids to the fuselage, measured drag in the experiment increases significantly more than it does in the CFD. Since the interactional effect between cabin and landing skids was found to be relatively weak in this case, the main discrepancy between CFD and experiment arises from the skid drag. We expect a strong Reynolds number effect here, as the experimental Re corresponds to the subcritical cylinder regime based on the strut diameter. Furthermore, sufficient LES grid resolution around the high aspect ratio skids is unfeasible, so that the DES is clearly under-resolved here.

Mean surface pressure contours on the cabin backdoor are shown in Fig. 6 for RANS, URANS and DDES, respectively. Confirming the statements made for drag

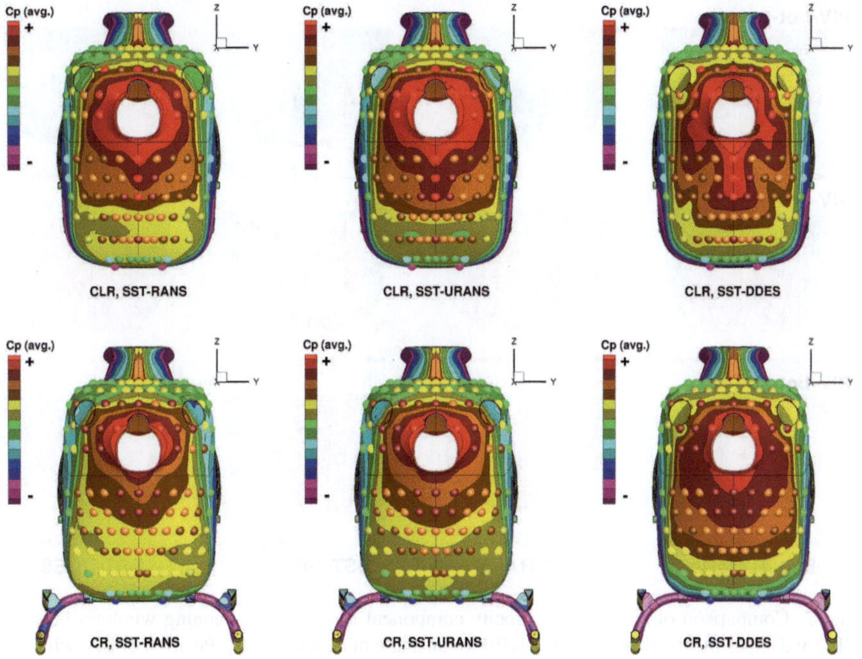

Fig. 6 Mean surface pressure coefficient from all simulations (contours) compared with experimental data from ECO-HC (scatter) for an angle of attack $\alpha = -5°$

prediction on the isolated cabin, DDES provides a more accurate representation of the pressure field topology, where RANS and URANS exhibit lower values especially on the lower part of the backdoor. The predicted wake topology is shown in Fig. 7 for 3 scanning windows downstream of the cabin backdoor. DES shows excellent agreement to the PIV measurements, whereas stronger separation and lower flow velocities in the wake are predicted by (U)RANS.

4.2 Prediction of Unsteady Solution Content

Surface pressure fluctuations (standard deviation of C_p) are plotted in Fig. 8 at a mid-section slice through the cabin backdoor region, where all CFD shows exaggerated pressure fluctuations compared to the experiment. As unsteady pressure data has been measured in the ECO-HC experiment with the smoother underbelly geometry, this is seen as a potential cause for this discrepancy. Nevertheless, DES gives much lower fluctuations compared to URANS as well as greater symmetry about the centre plane.

Power spectral densities of C_p at selected locations are depicted in Fig. 9. For the location inside the wake closely below the tail boom, which is not much affected

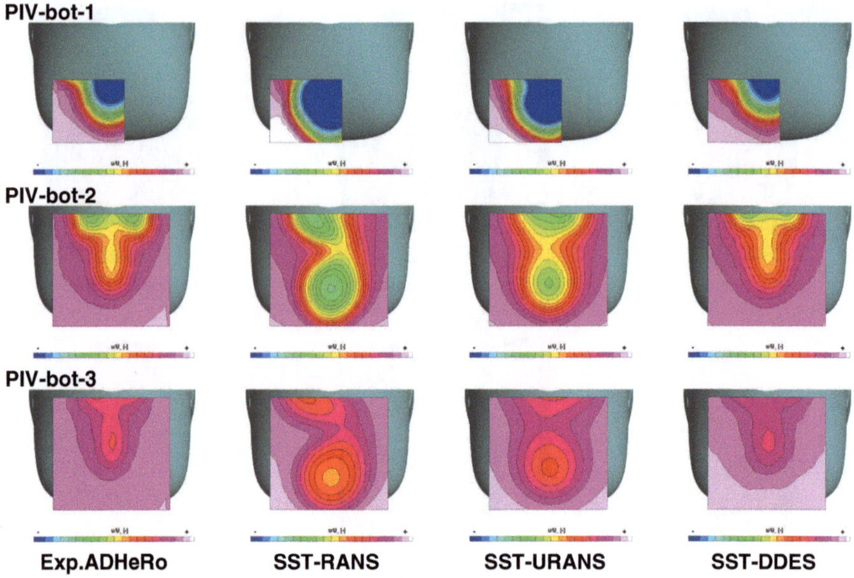

Fig. 7 Comparison of streamwise velocity component for different scanning windows between CFD and experimental data from ADHeRo for an angle of attack $\alpha = 0°$. Position of PIV windows is shown in Fig. 3

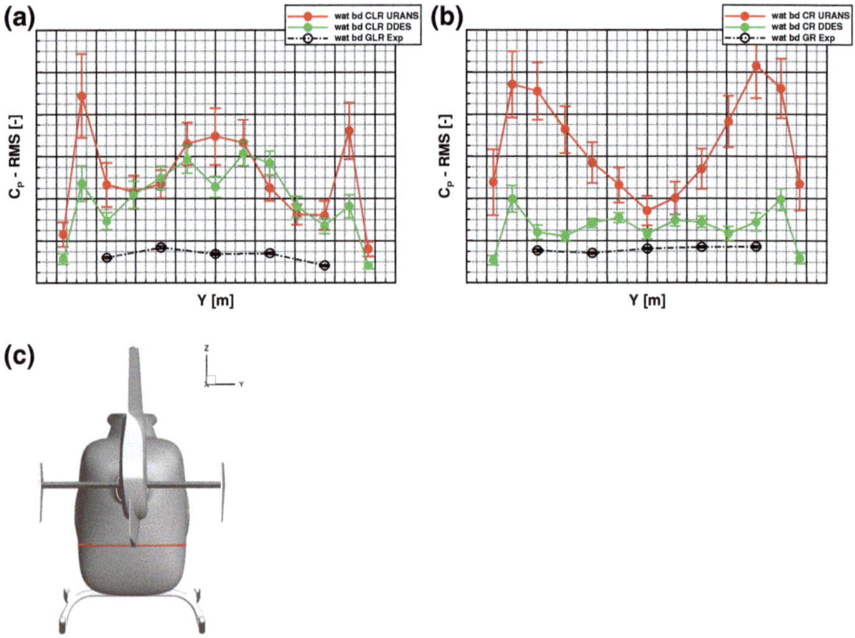

Fig. 8 Comparison of surface pressure fluctuations at mid-section slice, $\alpha = -5°$. The error bars denote statistical 95 % confidence intervals [4]

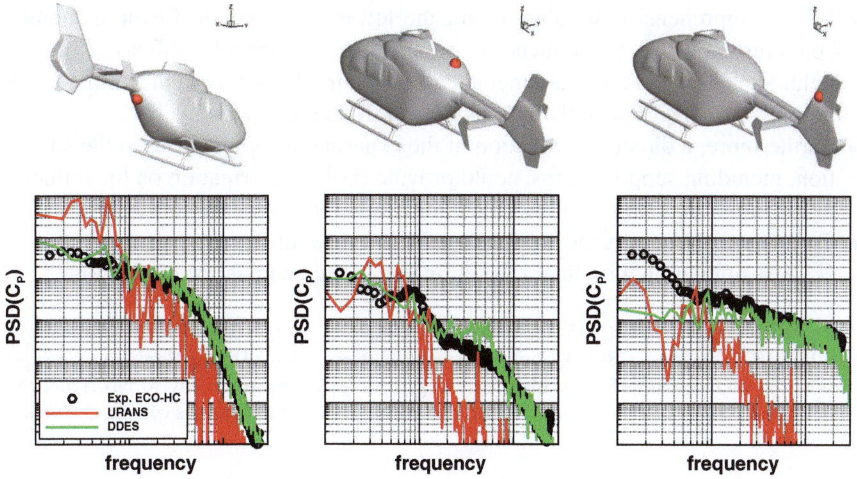

Fig. 9 Power spectral densities of surface pressure coefficient for CFD and ECO-HC experimental data for selected locations, $\alpha = 0°$

by the underbelly discrepancy, spectra from DES and experiment are in very good agreement, whereas strong damping of high frequencies is seen for URANS. This also holds for the location downstream of the rotor-head fairing on top of the cabin. The vortical wake structure that separates from the fairing then impinges on the tail fin. Pressure spectra there also correspond well to the experiment, thus indicating that DES is able to capture important unsteady wake/empennage interactional effects (e.g. tail shake).

5 Conclusions

A detailed study comparing state-of-the-art DDES with conventional (U)RANS was carried out for a highly-complex helicopter geometry. For situations with direct comparability (i.e. identical underbelly geometry and weak Reynolds number effects), excellent agreement between DDES and experiment was achieved. DDES significantly improves prediction of the massively separated wake behind the cabin back-door compared to both steady RANS and URANS. No case could be identified where DDES shows inferior performance to RANS or URANS for this flow region.

Discrepancies between CFD and experiment were observed in relation to the landing skid drag and flow prediction around the empennage. Geometric discrepancies between CFD and experiment significantly compromises comparability of results for the underbelly region as well as at the empennage due to modified upwash.

The following suggestions for future work are given:

- More comprehensive simulations for the lower experimental Reynolds number could further clarify the influence of possible Reynolds number effects on the flow field. Such simulations could include transition modelling to predict laminar flow regions, which significantly influences overall drag prediction.
- Furthermore, a direct reproduction of the experimental geometries in the simulation, including support stings, could provide further information on the influence of the mounting sting for flow prediction around the empennage.
- To investigate the influence of suspected under-resolution of the near wakes of the skids, a simulation with finer resolution in this region could be carried out.

Acknowledgments The EC135 simulations were funded by the EU FP7 CleanSky project HELIDES (CSGA-2010-278415). Computing resources were provided by the North German Supercomputing Alliance (HLRN, www.hlrn.de). The geometries were provided by Airbus Helicopters Deutschland GmbH. The authors also wish to acknowledge fruitful discussions with M. Grawunder, TU Munich.

References

1. Fuchs, M.: Assessment, validation and enhancement of detached-eddy simulation in OpenFOAM. Master's thesis, Technische Universität Berlin (2013)
2. Grawunder, M., Reß, R., Breitsamter, C., Adams, N.A.: Flow characteristics of a helicopter fuselage configuration including a rotating rotor head. In: Proceedings of the 28th Congress of the International Council of the Aeronautical Sciences, Brisbane, Australia, 23–28 Sept 2012
3. Gritskevich, M., Garbaruk, A., Schütze, J., Menter, F.: Development of DDES and IDDES formulations for the k-ω shear stress transport model. Flow Turbul. Combust. **88**(3), 431–439 (2012)
4. Mockett, C., Knacke, T., Thiele, F.: Detection of initial transient and estimation of statistical error in time-resolved turbulent flow data. In: Proceedings of the 8th International Symposium on Engineering Turbulence Modelling and Measurements (ETMM8), Marseille, France, 9–11 June 2010
5. Mockett, C., Le Chuiton, F., Fuchs, M., Thiele, F.: Helicopter fuselage wake prediction using Detached-Eddy Simulation. In: Progress in Hyrid RANS-LES Modelling, NNFM, vol. 117, pp. 413–425. Springer (2012)
6. Travin, A., Shur, M., Strelets, M., Spalart, P.: Physical and numerical upgrades in the detached-eddy simulation of complex turbulent flows. In: Proceedings of the 412th Euromech Colloquium on LES and Complex Transitional and Turbulent Flows, Munich, Germany (2000)

Zonal Detached Eddy Simulation (ZDES) Using Turbulent Inflow and High Order Schemes: Application to Jet Flows

F. Gand, V. Brunet and G. Mancel

Abstract This paper presents a numerical investigation of a round jet using a synthetic method to generate some free-stream turbulence into the jet core. Besides, this work also aims at evaluating the added advantage of the use of high-order numerical schemes—namely the AUSM+P spatial scheme with MUSCL extrapolation of the 3rd and 5th order—for this type of simulations. Academic test cases are presented to illustrate the main properties of the numerical methods used, then the Zonal Detached Eddy Simulation (ZDES) of a round jet is scrutinized with an emphasis on the influence of the injected synthetic turbulence.

1 Introduction

In the framework of sustainable aviation, it is now commonly acknowledged that jet flow simulations using eddy-resolving approaches could trigger major improvements for the prediction of airframe noise, acoustic sources locations, pylon unsteady loading, thermal fatigue, etc. While RANS/LES and LES methods were proven to be mature enough to reproduce quantitatively jet flows, these methods need to reproduce accurately the flight conditions to be of use in an industrial context.

However, the turbulence rate at the engine primary and secondary exhausts is usually neglected in simulations while this type of feature has a strong impact on the physics of the jet [1, 2]. Apart from the turbulent content generation, an increase in simulations accuracy can also be achieved using high order numerics, or at least by reducing the amount of numerical dissipation in the simulations.

This paper presents numerical investigations of turbulent content generation for eddy-resolving simulations of simple test cases of core jet and wall turbulence. The potential advantage of the use of high-order numerical schemes for this type of simulations is evaluated. The main findings are used to simulate a low speed round jet and the results are compared to an experimental database.

F. Gand (✉) · V. Brunet · G. Mancel
The French Aerospace Lab, 92190 Meudon, France
e-mail: fabien.gand@onera.fr

S. Girimaji et al. (eds.), *Progress in Hybrid RANS-LES Modelling*,
Notes on Numerical Fluid Mechanics and Multidisciplinary Design 130,
DOI 10.1007/978-3-319-15141-0_11

2 Numerical Methods

2.1 Flow Solver

The simulations presented in this paper have been performed with the ***elsA*** software [3] developed at Onera which solves the compressible Navier-Stokes equations on structured multiblock meshes.

The time integration is performed using an implicit LU-SSOR algorithm and a second-order accurate backward Gear scheme. The number of sub-iterations is adjusted to reach a convergence of one order of magnitude of the inner iteration residuals to achieve second-order time accuracy.

For the spatial integration, the diffusive fluxes are discretized using a second-order-accurate centered scheme. The convective terms are treated with the hybrid centered/upwind second-order-accurate AUSM+P scheme [4] using MUSCL extrapolation of the third or fifth order. This version of the AUSM scheme involves a "wiggle" sensor to minimize numerical viscosity by applying some upwinding only in areas where the solution displays numerical oscillations, while the scheme is actually centered everywhere else [4]. The AUSM+P scheme is well suited to the low-speed applications presented in this paper since its dissipation is proportional to the fluid velocity.

2.2 Physical Modelling

Several physical approaches are considered according to the test-case investigated. In Sect. 3, no turbulence model is used for the convection test case and Large Eddy Simulations (LES) are performed using the Mixed Scale Model (MSM) subgrid scale model for boundary layer simulations.

In Sect. 4, hybrid RANS/LES simulations are performed using the Zonal Detached Eddy Simulation (ZDES) [5]. In the present study, modes 1 and 2 of the ZDES are used. The ZDES mode 1, introduced to treat separations triggered by the geometry, relies on a user-defined zonal decomposition of the computational domain in RANS and DES areas. The hybrid length scale $\tilde{d}^I_{ZDES}$ entering the Spalart-Allmaras (SA) model is equal to:

$$\tilde{d}^I_{ZDES} = \begin{cases} d_{wall} \text{ in RANS areas} \\ \min(d_{wall}, C_{DES}\tilde{\Delta}^I_{ZDES}) \text{ in DES areas} \end{cases} \quad (1)$$

where d_{wall} is the wall distance; C_{DES} is the original DES97 constant and $\tilde{\Delta}^I_{ZDES} = \Delta_{vol} = (\Delta x \Delta y \Delta z)^{1/3}$ or $\tilde{\Delta}^I_{ZDES} = \Delta_\omega = \sqrt{\bar{S}_\omega}$, $\bar{S}_\omega$, is the average cross section of the cell normal to the vorticity vector ω. In the present study, the subgrid length scale based on the local vorticity vector direction Δ_ω was chosen since there is a preferred mean vorticity direction. Furthermore, the near-wall functions of the SA model are removed in LES areas for mode 1: $f_{v1}=1, f_{v2}=0, f_w=0$.

The ZDES mode 2 [5] was developed to deal with separations over smooth surfaces. Therefore it relies on a protection function f_d—similar to the one employed in DDES [6]—to shield the attached boundary layers. The hybrid length scale reads:

$$\tilde{d}^{II}_{ZDES} = d_{wall} - f_d \max(0, d_{wall} - C_{DES}\tilde{\Delta}^{II}_{ZDES}) \quad (2)$$

The mode 2 of ZDES provides a specific definition of the subgrid length scale according to the flow resolution. Thus, in attached regions $\tilde{\Delta}^{II}_{ZDES} = \Delta_{\max}$ (Δ_{max} being the characteristic mesh length necessary to shield the attached boundary layers) but in detached areas, the subgrid length scale revolves to $\Delta^{II}_{ZDES} = \Delta_{\omega}$ or Δ_{vol}. The switch between the two subgrid length scales is done automatically using a threshold value f_{d0}= *0.8* for the f_d function (see [5]).

2.3 Synthetic Eddy Method

To generate realistic turbulent inflow content, the Synthetic Eddy Method (SEM) [7] is used. The SEM is based on the generation of velocity fluctuations carried by synthetic eddies which are superimposed to a target mean flow to obtain a synthetic unsteady turbulent field used to feed an inlet boundary condition. The formulation relies on the prescription by the user of a target Reynolds stress tensor. As suggested by Jarrin, in the present study an isotropic Reynolds stress tensor with 2/3 k for the normal stresses is used as target (the turbulent kinetic energy profile is either extracted from a RANS boundary layer profile or computed from a target turbulent rate for jets cases).

The length scale of turbulence σ has also to be prescribed. For wall turbulence, Jarrin has proposed to compute σ according to RANS provided turbulent kinetic energy and dissipation rate profiles, or it can be explicitly given by the user.

The SEM formulation of Jarrin and its modifications presented above are implemented in a python module which is coupled with the ***elsA*** software [3] using the external coupling feature for boundary conditions.

3 Preliminary Verifications on Academic Cases

3.1 Turbulence Convection

3.1.1 Test Case and Simulations Parameters

The same test case of turbulence convection introduced in Ref. [1] is used in the present study. The computational domain is a channel without any wall, side faces being periodic (there are no walls in this test case), of dimensions in the streamwise (x), normal (y) and transverse (z) directions 5LxLxL (L = 0.2 m). The mesh is cartesian in the region of interest (up to x = 0.8 L) of dimension $(L/100)^3$. The cells are stretched in the streamwise direction from x = 0.8 L to create a sponge layer at the exit of the computational domain. In total, the grid contains 4×10^6 cells. The inlet Mach number is set at M = 0.2 (U_0 = 70 m/s) with atmospheric pressure and temperature conditions.

The inlet turbulence level is set to $Tu_0 = \sqrt{(2/3k)}/U_0 = 1\,\%$, which is used within the SEM formulation to build an isotropic target Reynolds stress tensor as described in Sect. 2. Two values of turbulence length scale have been tested: σ = L/20 and σ = L/5 to assess the effect of the mesh resolution of the eddies generated. For the present test case, no turbulence modelling is used. MUSCL extrapolation of the 3rd and 5th order is used along with the AUSM+P scheme. The time step is 10^{-5} s (leading to a CFL number based on the maximal acoustic velocity of 2), and the number of sub-iterations of the Gear scheme has been set to 8 to ensure the decrease of one order of magnitude of the inner iteration residuals.

3.1.2 Results and Discussion

Flow visualizations are shown in Fig. 1. The turbulent structures generated by the SEM are convected through the computational domain, then are dissipated when entering the sponge layer as expected.

The streamwise evolution of the turbulence level is plotted in Fig. 2a. First, with the standard 3rd order MUSCL extrapolation, it appears that the initial turbulence level is not sustained with only 5 grid points per eddy (case σ = L/20) whereas it is maintained with 20 grid points when σ = L/5. However, this situation seems to be improved when using a 5th order reconstruction with σ = L/20. As a matter of fact, Fig. 2a shows that there seems to be a significant additional amount of turbulent content resolved when the 5th order MUSCL extrapolation is used, compared to the 3rd one which is confirmed by the velocity spectra plotted in Fig. 2b. Even more, in the case σ = L/20, it seems that the solution recovers the −5/3 slope of the energy cascade, which could mean that the lack of grid points covering one eddy of length scale σ = L/20 is somewhat compensated by the increase of numerical accuracy.

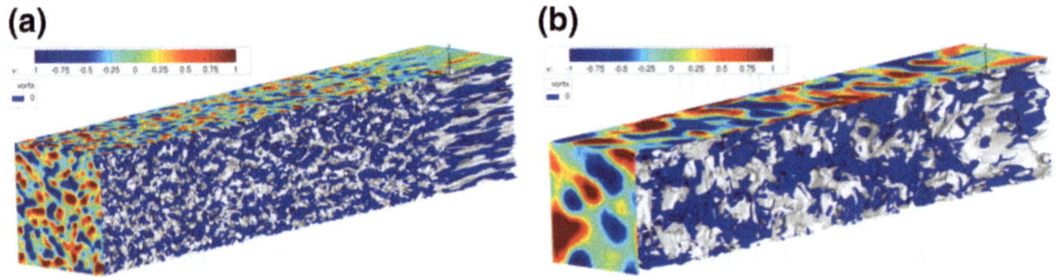

Fig. 1 Flow visualizations for the turbulence convection test case. Isosurface of Q criterion colored by the streamwise vorticity and contours of transverse velocity. **a** $\sigma = L/20$, **b** $\sigma = L/5$

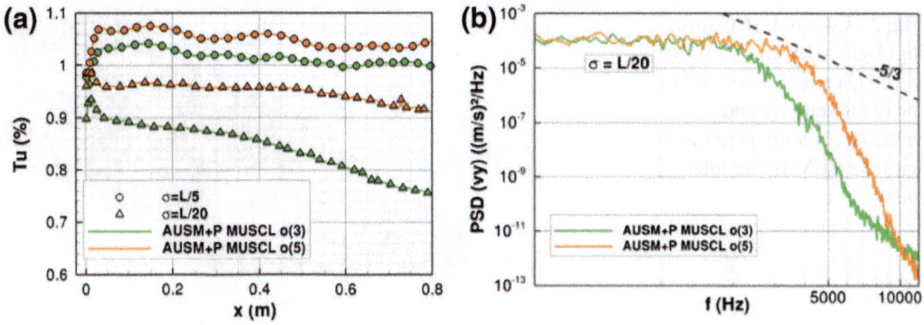

Fig. 2 Simple convection test case results: streamwise evolution of the turbulence rate (**a**) and PSD at x = 0.6L (**b**). **a** Turbulence rate, **b** power spectral density

These results indicate that the SEM needs to be carefully set-up according to the inlet grid density (the reverse is also true, especially if inflow data is available from experiments) to achieve targeted inflow turbulence levels. Nevertheless, it appears that some information on the spectral content is also required to accurately describe the physics of the initial turbulence to be generated by the SEM. In the next section, the effect of the numerical accuracy is further evaluated to extend the application of the above conclusions to a more realistic case with a non-cartesian grid.

3.2 Wall Turbulence Simulations

In this section, the effect of the spatial scheme order in a Wall Resolved LES (WRLES) context is assessed.

3.2.1 Test Case and Simulations Parameters

A spatially developing turbulent boundary layer at a moderate Reynolds number ($Re = 4 \times 10^6$/m) is investigated in the present section. The freestream Mach number is equal to M = 0.3 and the freestream total pressure is equal to $Pi_\infty = 21{,}063$ Pa.

The inlet boundary layer thickness is $\delta_0 = 6 \times 10^{-3}$ m. The computational domain has been extended to 40 δ_0 in the streamwise direction (x), 4 δ_0 in the transverse direction (z) and 10 δ_0 in the wall normal direction (y). The cell sizes are $\Delta x^+ = 50$, $\Delta z^+ = 12$ and $\Delta y^+ = 1$. The cell stretching ratio in the wall normal direction is less than 15 %. The mesh comprises 3.3×10^6 cells. Reference DNS data for the station $Re_\theta = 1{,}410$ are available in Ref. [8].

The simulations parameters are presented in Sect. 2. The time step was set to 5×10^{-7} s so that the CFL number based on the maximum acoustic velocity is lower than 15. The effect of the order of the MUSCL extrapolation (3rd and 5th order reconstruction) is assessed. Eight sub-iterations of the Gear scheme were necessary to achieve a decrease of one order of magnitude of the inner iteration residuals.

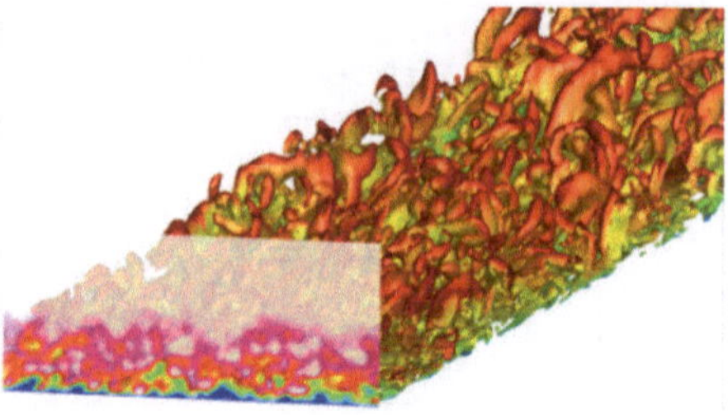

Fig. 3 Coherent structures highlighted by an isosurface of the Q criterion colored by the axial velocity and velocity contours generated with the SEM in the inlet *plane*

3.2.2 Results and Discussion

A visualisation of the coherent structures resolved in a LES simulation performed in the present study is shown in Fig. 3. One can see that the typical boundary layer structures, including hairpin vortices, seem to be properly captured by the simulation.

Figure 4a provides a more quantitative insight into the LES results and more specifically the effect of the increase of the accuracy of the spatial discretization. While the error on the friction coefficient at the reference location $x_{Re\theta(x)} = 1{,}410$ is of 13 % (compared to the RANS-SA solution) when using a standard 3rd order MUSCL reconstruction—which is a classical result for this type of simulation using this type of numerics—the error decreases to 7 % with a 5th order reconstruction.

In order to better understand the reasons of the increased accuracy of the simulations with high order MUSCL reconstruction, the Power Spectral Density (PSD) of the streamwise velocity at the location $x_{Re\theta(x)} = 1{,}410$ at y+ $\approx$ 100 is provided in Fig. 4b. The typical broadband shape of the boundary layer spectra is found, the energy cascade characterized by a slope of −5/3 of the velocity spectra is reproduced, although this phenomena occurs on a rather limited frequency bandwidth at this rather low Reynolds number. The focus is put on the frequency range

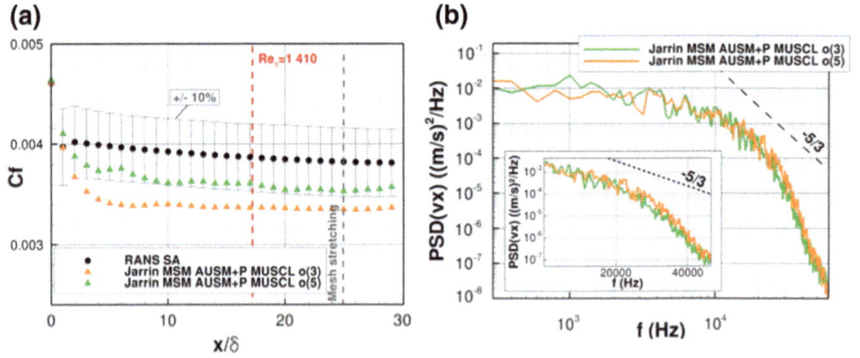

Fig. 4 Boundary layer test case results: streamwise evolution of the friction coefficient (*left*) and PSD of the streamwise velocity *velocity at* Re_θ = *1,410*, $y^+ \approx 100$ (*right*). **a** Friction coefficient, **b** PSD

[10; 50] kHz to highlight some increase of the turbulent content resolved by the simulations when the 5th order flux reconstruction is used.

As an intermediate conclusion, it can be noted that the use of high order MUSCL reconstruction with the AUSM+P scheme seems to be relevant in the framework of LES of wall turbulence. This is consistent with the results of Sect. 3.1. However, the use of such schemes for more technical configurations (namely the round jet presented thereafter) leads to robustness issues which are still under examination.

4 Application to a Low Speed Round Jet Flow

Axisymmetric jets have been widely investigated in the literature, both experimentally (e.g. [9]) and numerically (e.g. [2]). In particular, the influence of the initial level of perturbations in initially laminar boundary layers on the vortex pairing mechanisms and the associated noise generation is thoroughly documented. In the present study, initial turbulent boundary layers are considered because it seems closer to industrial configurations and the focus is put on the effect of the core-jet perturbations, which is less commonly found in the literature.

4.1 Test Case and Simulations Parameters

The round jet investigated is similar to the one of Ref. [10]. The Reynolds number based on the nozzle exit velocity $U_0 = 20$ m/s and diameter $D = 0.15$ m is equal to $Re_D = 2.1 \times 10^5$.

A sketch of the computational domain is presented in Fig. 5 (the actual grid is tri-dimensional, x is the axial coordinate). 40 grid points are used to discretize the nozzle boundary layer and the mesh contains 40×10^6 cells in total. Three simulations were performed:

- Simulation 1: reference without any inflow turbulence, hence using mode 0 in the nozzle;

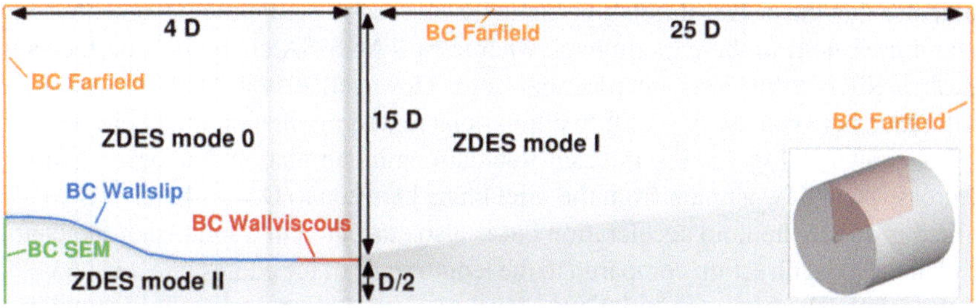

Fig. 5 Round jet test case. Slice in the 3D computational domain, ZDES settings and location of the boundary conditions

- Simulation 2: same as simulation 1, using the DDES approach [6] to assess the influence of the subgrid length scale used in the LES area;
- Simulation 3: simulation 1 with turbulence added at the nozzle inlet using SEM ($Tu_0 = 2\,\%$ and $\sigma = 0.02\,\text{m}$, wild guess parameters) as shown in Fig. 5. In this case, mode 2 is retained inside the nozzle to allow both the transport of the injected turbulence and the RANS simulation of the attached boundary layer.

Only the 3rd order MUSCL reconstruction is assessed in the present paper since a simulation with the 5th order has shown some robustness issues, the solving of which is underway. The time step is 2×10^{-6} s so that the convective CFL number based on the maximal acoustic velocity is lower than 15 in the mixing layer, 8 sub-iterations are used for the Gear scheme. The unsteady results are time averaged over 200 ms after a transient stage estimated of around 120 ms.

4.2 Results and Discussion

Visualizations of simulations 1, 2 and 3 are provided in Fig. 6. The length scale used in DDES creates unjustified high values of eddy viscosity due to anisotropic cells downstream of the nozzle, hence the undesired damping of the instabilities. Conversely, the quick growth of instabilities in the ZDES simulations is attributed to the use of $\Delta\omega$ (see Eq. 1), thanks to which the eddy viscosity scales on the grid density normal to the main vorticity direction. In this case, the grid density actually allows the resolution of fine turbulent scales in the round mixing layer. See Ref. [5] for a complete analysis on this topic. Simulation 3 depicted in Fig. 6c exhibits little visual difference with simulation 1 in the mixing layer area.

The nozzle exit conditions are shown in Fig. 7a. The experimental mean boundary layer profile is correctly reproduced by the simulations operating in RANS mode in this area and one can see that, in simulation 3, some of the core jet turbulence actually penetrates the RANS modelled boundary layer, and is somewhat amplified. It is important to remind that the addition of perturbations into the nozzle boundary layer is not the purpose of this study so it is not considered a drawback of the simulations to exhibit such low resolved turbulence levels in the attached boundary layer.

The streamwise evolution of the velocity along the jet axis shown in Fig. 7b indicates that the ZDES simulations actually succeed in predicting the potential core length measured in the experiments, whereas a RANS-SA simulation on the same mesh predicts a too long jet potential core. However, it is difficult to assess the difference between the two ZDES simulations regarding the potential core length.

As shown in Fig. 7b, the injected fluctuations in simulation 3 undergo a strong decrease right downstream from the inlet plane between $\text{x}/\text{D} = -4$ and $\text{x}/\text{D} = -3$, which is due to the fluid acceleration but is also attributed to a somewhat too coarse grid in the axial direction compared to the length scale of the eddies injected ($\Delta\text{x} \sim \sigma$ at the inlet). Therefore the turbulence level reached at the nozzle exit in simulation 2 is $0.5\,\%$ only, which makes a small difference with simulation 1 but is actually consistent with the residual turbulence level measured experimentally.

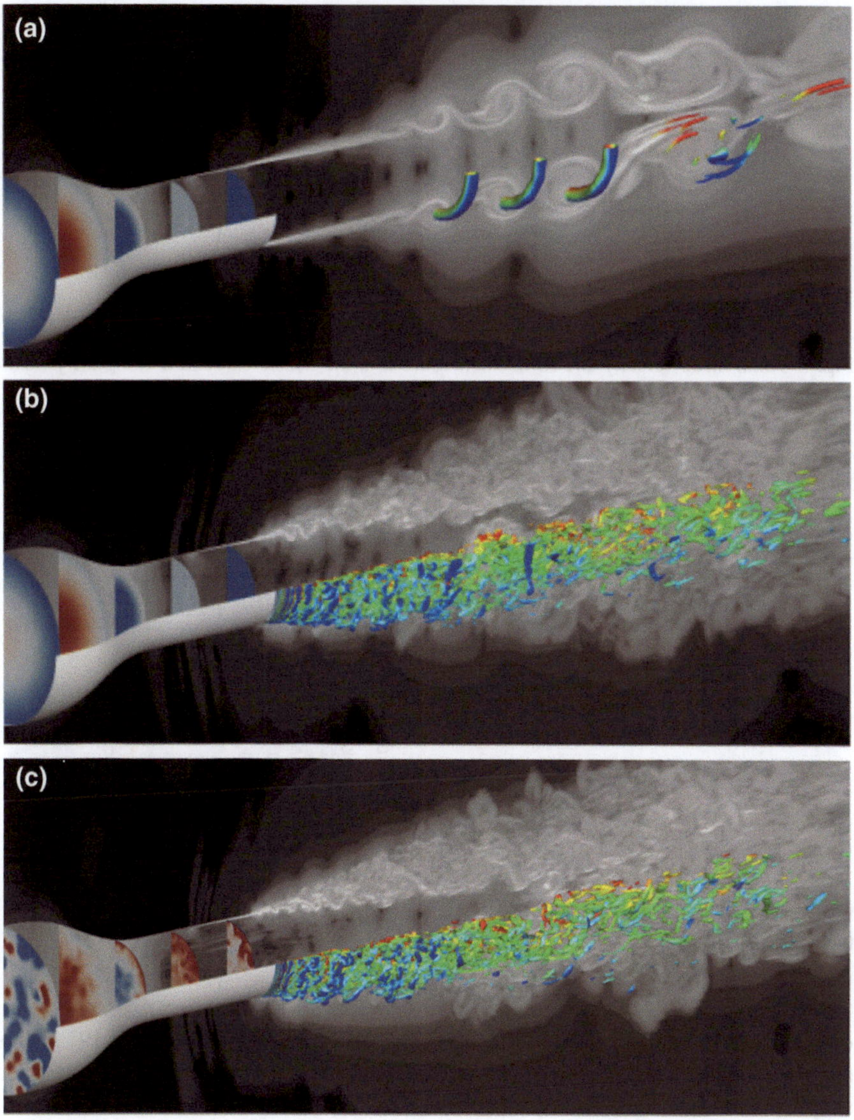

Fig. 6 Flow visualizations. Isosurface of Q criterion, contours of density gradient (*grayscale*) and streamwise velocity inside the pipe (*arbitrary scale*). **a** DDES, no inlet turbulence (simulation 2), **b** ZEDS, no inlet turbulence (simulation 1), **c** ZEDS, $Tu_0 = 2\,\%$

The very start of the formation of instabilities in the mixing layer is illustrated in Fig. 8. The quick development of turbulence is confirmed for simulation 1, which indicates that no added fluctuations are required to develop the turbulent content in the mixing layer with ZDES mode 1. However, the double peaked shaped evolution

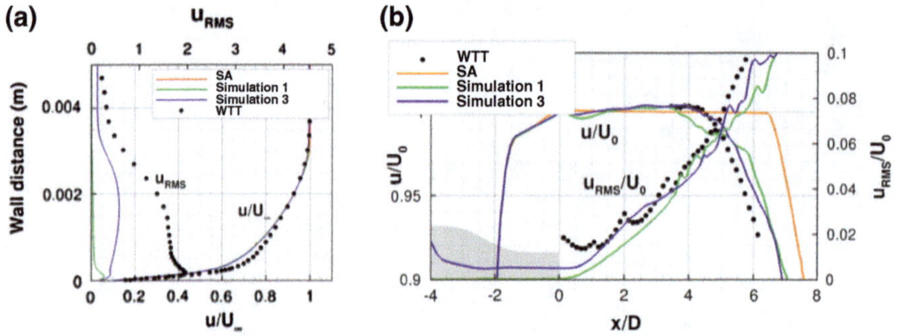

Fig. 7 Axial mean velocity and RMS fluctuations **a** at the nozzle exit and **b** along the jet axis. **a** At the nozzle exit, **b** Along the jet axis

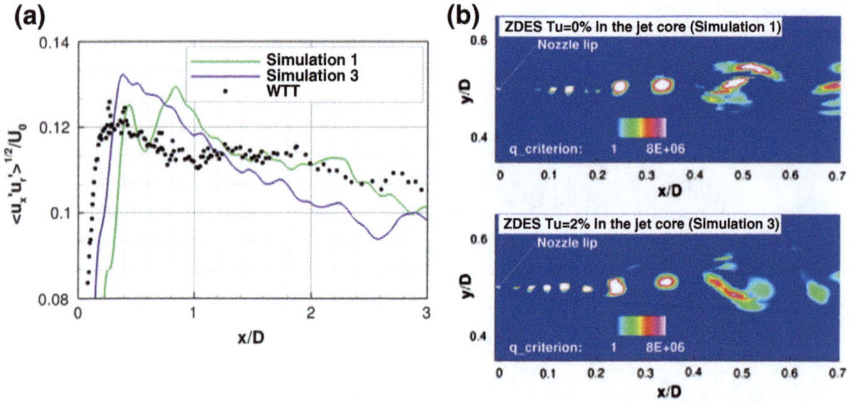

Fig. 8 Evolution of $\sqrt{\overline{u'_x u'_x}}/U_0$ in the shear layer (**a**) and Q criterion contours in the vicinity of the nozzle lip (**b**)

of $\sqrt{\overline{u'_x u'_x}}/U_0$ could be related the first stage of strong vortex pairings [9] in this simulation which are not observed in the experimental curve in Fig. 8a. No such shape is found in simulation 3 which, in this respect, is in better agreement with experimental data. Figure 8b illustrates the quick development of the mixing layer in simulations 1 and 3.

The evolution of the frequency distribution of the turbulent energy along the jet axis and the mixing layer is plotted in Fig. 9. No differences are identified within the mixing layer. On the other hand, the flow clearly exhibits the signature of the inflow turbulence along the jet axis up to x = 2D. Increased low frequencies in simulation 2 are attributed to the eddies injected, but some spurious accumulation of energy at high frequencies is also observed. This issue is still under analysis but it could be a major limitation to the use of SEM methods in the framework of acoustic studies.

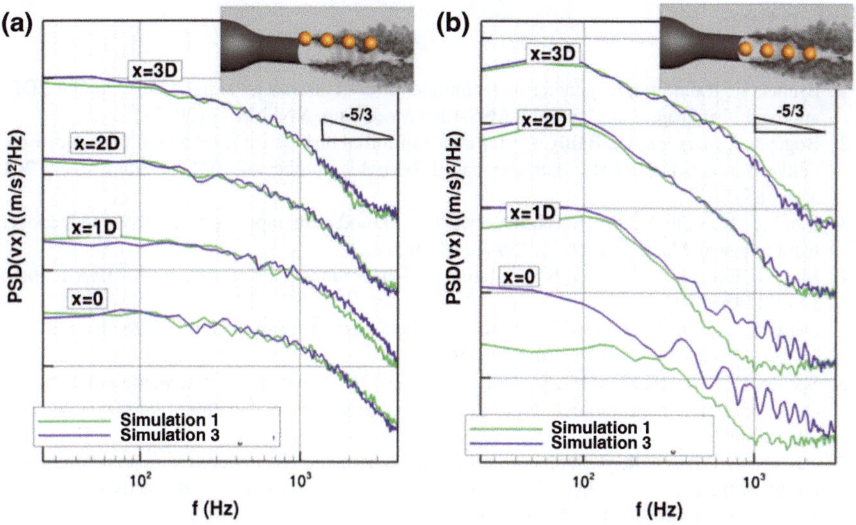

Fig. 9 Power spectral density of the axial velocity fluctuations. **a** In the mixing layer, **b** Along the jet axis

5 Summary, Conclusions and Future Work

Some preliminary work on turbulence injection and high order schemes in the context of RANS/LES simulations has been presented.

The use of the SEM allows a straightforward turbulence injection for jet flows, however this imposes new constraints on the mesh in order to correctly advect the injected fluctuations.

The advantage of using high order MUSCL reconstruction has been demonstrated on very basic test cases and remains to be confirmed on a more technical configuration for which robustness issues have been encountered.

The round jet simulation has highlighted the capability of the ZDES approach to simulate such flow without any delay in the formation of the instabilities even without the need for inflow turbulence. In particular, the use of a flow dependent subgrid length scale is advocated and shows that the removal of the eddy viscosity in the LES areas is not necessary to avoid delays in the formation of instabilities. The possibility to take into account the turbulent content of the jet core has been illustrated using ZDES mode 2 to transport the inflow turbulence.

In the end, it is important to note that the possibility to add free stream turbulence for propulsive jet simulations brings into focus the need for detailed experimental data for these parameters to increase the experimental-numerical complementarity.

Acknowledgments The simulations presented in Sect. 4 were performed using HPC resources from GENCI-TGCC (Grant 2014-t20142a7215).

References

1. Brunet, V.: Random flow generation technique for civil aircraft jet simulations with the ZDES approach. In: Progress in Hybrid RANS-LES Modelling, NNFM 117 (2012)
2. Bogey, C., Marsden, O., Bailly, C.: Influence of initial turbulence level on the flow and sound fields of a subsonic jet at a diameter-based Reynolds number of 10^5. J. Fluid Mech. **701**, 352–385 (2012)
3. Cambier, L., Heib, S., Plot, S.: The Onera elsA CFD software: input from research and feedback from industry. Mech. Ind. **14**(3), 159–174 (2013)
4. Mary, I., Sagaut, P.: Large eddy simulation of flow around an airfoil near stall. AIAA J. **40**(6), 1139–1145 (2002)
5. Deck, S.: Recent improvements of the zonal detached eddy simulation (ZDES) formulation. Theoret. Comput. Fluid Dyn. **26**, 523–550 (2012)
6. Spalart, P., Deck, S., Shur, M., Squires, K., Strelets, M., Travin, A.: A new version of detached-eddy simulation, resistant to ambiguous grid densities. Theoret. Comput. Fluid Dyn. **20**, 181–195 (2006)
7. Jarrin, N., Prosser, R., Uribe, J.-C., Benhamadouche, S., Laurence, D.: Reconstruction of turbulent fluctuations for hybrid RANS/LES simulations using a synthetic-eddy method. Int. J. Heat Fluid Flow **30**(3), 435–442 (2009)
8. Spalart, P.: Direct simulation of a boundary layer up to Rt = 1,410. J. Fluid Mech. **187**, 61–98 (1988)
9. Zaman, K., Hussain, A.: Vortex pairing in a circular jet under controlled excitation. Part 1 general jet response. J. Fluids Mech. **101**, 449–491 (1980)
10. Davoust, S., Jacquin, L., Leclaire, B.: Dynamics of m = 0 and m = 1 modes and of streamwise vortices in a turbulent axisymmetric mixing layer. J. Fluids Mech. **709**, 408–444 (2012)

A Renormalized Detached Eddy Simulation Method Without Log-Layer Mismatch

Ning Hu, Han-Dong Ma and Wei-Min Zhang

Abstract In order to correct the log-layer mismatch (LLM) commonly encountered in detached eddy simulation (DES) of attached or weakly-detached flows, a multi-layered order function $\ell_{\mathrm{M,ML}}$ is defined. One of the layers of $\ell_{\mathrm{M,ML}}$ is a plateau and is crucial to the log-law of the mean velocity profile. The plateau height can be determined theoretically from the usual log-law formulation. A target function $\ell_{\mathrm{M,target}}$ is defined which inherits near wall scaling of SA model and extends it to reach the correct plateau. The function is used to renormalize (stretching/compressing) turbulent fluctuations during DES calculation. This renormalized DES (RNDES) is tested in both an incompressible channel flow and a Ma $= 2.25$ compressible boundary layer (CBL) flow. RNDES predicts correct mean velocity profiles free of LLM and correct turbulent intensities as well, indicating that RNDES develops correct turbulent structural ensemble under the constraint of statistical quantities.

1 Introduction

Large eddy simulations (LES) of high Reynolds number wall-bounded flows require prohibitively high resolution to resolve the near-wall structures since the dynamically important, energy-containing eddies in the near-wall region are scaled by the viscous scale. Hybrid RANS/LES methods are proposed to relax the near-wall resolution requirement. Detached Eddy Simulation (DES [1]) is a typical one of such methods, which uses RANS in the near-wall region and switches into LES mode far enough from the wall. DES has been successfully applied in the simulation of massively separated flows. However, in flows with moderate or no separation, this method has a problem which is known as log layer mismatch (LLM), namely, an unphysical overestimated log layer of mean velocity profile. This problem, first found in an incompressible channel flow [2] and also in a compressible boundary layer afterwards [3], leads to a severe underestimation of the skin friction and even artificial separation.

N. Hu (✉) · H-D. Ma · W-M. Zhang
China Academy of Aerospace Aerodynamics, Mailbox 7201-56, Beijing 100074, China
e-mail: h_ning@pku.org.cn

S. Girimaji et al. (eds.), *Progress in Hybrid RANS-LES Modelling*,
Notes on Numerical Fluid Mechanics and Multidisciplinary Design 130,
DOI 10.1007/978-3-319-15141-0_12

Piomelli et al. [4] found that the time scales and spatial scales are mismatched for DES at the RANS/LES interface. This leads to a "gray area" with slow development of turbulent eddies beyond the interface, and hence insufficient total Reynolds stress. They applied stochastic forcing to excite turbulent fluctuations at the interface and obtained improved results. The delayed DES (DDES) [5] reduces the extent of the "gray area" by introducing a delay function, and fixes the artificial separation caused by LLM. The improved DDES (IDDES) [6] can further fix the LLM problem itself, but with the expense of more additional empirical functions. The zonal Detached Eddy Simulation (ZDES) [7] also reduces the gray area by artificially setting the position of RANS/LES interface. However, its performance is strongly affected by the position of interface.

In this paper, we propose a method that corrects the log-layer mismatch problem by re-normalizing the resolved turbulent fluctuations according to an order function. Within the frame of structural ensemble dynamics (SED) theory [8, 9], order functions, which are defined through ratios of correlation functions, can be used to describe the statistical properties of turbulent ensemble, and to close the ensemble averaged NS equations as well.

2 Method

Derived by applying a filter to the NS equation, the momentum equation of LES reads for incompressible flows:

$$\frac{\partial \bar{u}_i}{\partial t} + \bar{u}_j \frac{\partial \bar{u}_i}{\partial x_j} = -\frac{\partial \bar{p}}{\partial x_j} - \frac{\partial \tau_{ij}}{\partial x_j} + \frac{1}{Re} \frac{\partial^2 \bar{u}_i}{\partial x_j \partial x_j}$$

and for compressible flows:

$$\frac{\partial}{\partial t}(\bar{\rho}\tilde{u}_i) + \frac{\partial}{\partial x_j}(\bar{\rho}\tilde{u}_j\tilde{u}_i) = -\frac{\partial P}{\partial x_i} + \frac{\partial P}{\partial x_j}\left[(\mu + \mu_T)\left(\frac{\partial \tilde{u}_j}{\partial x_i} + \frac{\partial \tilde{u}_i}{\partial x_j}\right) - \overline{\rho u_j'' u_j''}\right]$$

We start from the ensemble-averaged LES equations. The ensemble averaged momentum equation can be integrated along the y-direction, yielding:

$$S + T + W = \tau$$

where S is the mean shear rate, W is the resolved part of the Reynolds stress and T is the mean SGS stress. For incompressible channel flow the total stress $\tau = 1 - y$ where y is the wall distance. For compressible boundary layer,

$$\mu \frac{\partial \tilde{u}}{\partial y} + \mu_T \frac{\partial \tilde{u}}{\partial y} - \overline{\rho u_1'' u_2''} = \tau$$

where

$$\tau = \int_0^y \left(\frac{\partial}{\partial x}(\bar{\rho}\tilde{u}_1\tilde{u}_1) + \frac{\partial}{\partial y}(\bar{\rho}\tilde{u}_1\tilde{u}_2) - \frac{\partial \bar{\tau_{11}}}{\partial x} + \frac{\partial \overline{\rho u_1'' u_1''}}{\partial x} \right) dy$$

A length scale, called the mixing length, can be defined from this equation:

$$\ell_M = \frac{\sqrt{W+T}}{S}$$

ℓ_{M} is an order function in the SED theory, which stems from the concept "order parameter" in statistical physics for describing the state transition in the parameter space. Here ℓ_{M} is defined as ratios of terms in mean momentum/energy equations to characterize the flow multi-states and their transitions. The definition of ℓ_{M} is consistent with Prandtl's mixing length definition since:

$$W + T = RS = \nu_{\mathrm{T}} S = (\ell_{\mathrm{M}})^2 |S| S$$

Symmetry analysis [10] reveals a global structure in ℓ_{M}:

$$\ell_M = \ell_c(1 - r^m)$$

where r is the global coordinate centered at the boundary layer edge or channel center and m is determined by geometric constraint: $m = 4$ for channel flow, $m = 5$ for pipe or boundary layer flow. The mixing length removed of the $(1 - r^m)$ structure

$$\ell_{M,ML} = \ell_c = \ell_M / (1 - r^m) \tag{1}$$

is a new order function with multi-layer structure, with different scaling laws between the layers. The layers all have their physical meanings such as viscous sub layer, buffer layer, etc. The buffer layer (approximately $10 < y^+ < 40$) is a region of sharply growing $\ell_{\mathrm{M,ML}}$, which indicates a fast developing of near-wall turbulent fluctuations. A plateau that follows the buffer layer corresponds to a quasi-equilibrium state of the flow and is crucial to the log-law of the mean velocity profile. In fact, the constant $\ell_{\mathrm{M,ML}}$ value in this layer can be directly obtained from the log-law and the Karman constant. Figure 1 shows that $\ell_{\mathrm{M,ML}}$ has much wider constant range than the log-law formulation, covering the whole log-layer region. This indicates that it provides a more precise description of the log layer.

In DES, the fast growing of buffer layer ceases too early, yielding a lower plateau and hence insufficient Reynolds stress. This is due to the excessive damping of small scale resolved vortices. In RANS region, the SA model provides all the Reynolds stress. When the flow has just entered LES region, the eddy viscosity is rapidly reduced, yet still too large to allow the smaller scale motions develop, resulting in insufficient total Reynolds stress which leads to the log layer mismatch.

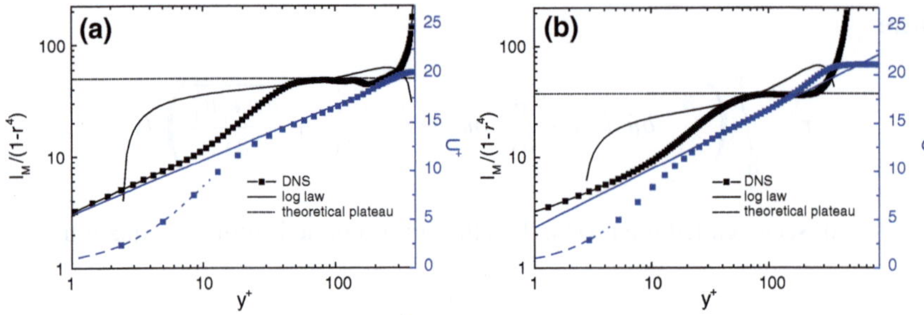

Fig. 1 Comparison of $\ell_{M,ML}$ obtained from log-law with DNS. **a** $Re_\tau = 395$ channel; **b** boundary layer

The rescaling approach to correct the problem follows: at every time step, calculate the spatial averaged Reynolds stress and mean shear rate:

$$U' = \frac{1}{L_x \times L_z} \int_0^{L_z} \int_0^{L_x} u dx dz$$
$$W' = \frac{1}{L_x \times L_z} \int_0^{L_z} \int_0^{L_x} (u - U') v dx dz$$
$$T' = \frac{1}{L_x \times L_z} \int_0^{L_z} \int_0^{L_x} \tau_{12} dx dz$$
$$S' = dU'/dy$$

and the instantaneous mixing length:

$$\ell'_{M,ML} = \frac{\sqrt{W' + T'}}{S'} / (1 - r^m)$$

Assume this $\ell_{\mathrm{M,ML}}$ has correct scaling behavior in the RANS region determined by the SA model:

$$\ell'_{M,ML} = a_1 (y^+)^{n_1}, \text{ in RANS region}$$

Since the correct scaling is violated before reaching the plateau, we have to reconstruct the scaling function by extending the scaling in the RANS region into the LES region. After the function reaches the appointed plateau, it should keep constant. Moreover, the scaling function is decreased quickly in the fully LES region in order to limit the modification to flow field within a narrow range. This can be accomplished by the SED base function [9]:

$$\ell_{M,\text{target}} = a_1 (y^+)^{n_1} \left(1 + \left(\frac{y^+}{y_p}\right)^{-n_1/h}\right)^h \left(1 + \left(\frac{y^+}{y_2}\right)^{-n_1/h}\right)^h$$

which produces a scaling transition at $y^+ = y_p$ into a plateau, where y_p is the solution to:

$$a_1(y_p)^{n_1} = \ell_{M,ML}(y_0^+)$$

and y_2 is the position where the simulation totally enters DES mode. When $y^+ > y_2$ the scaling function $\ell_{\text{M,target}}$ drops quickly and is muted.

The function $\ell_{\text{M,target}}$ is then used to modify the resolved turbulent fluctuations. The required resolved Reynolds stress to get the profile of $\ell_{\text{M,target}}$ is given as:

$$W'_{\text{target}} = (S'\ell_{M,\text{target}}(1 - r^m))^2$$

The ratio between the two Reynolds stress profiles:

$$\beta^2 = \max\left(1, \frac{W'_{\text{target}}}{W'}\right)$$

is used to renormalize the turbulent fluctuation such that the renormalized W' will give the correct $\ell_{\text{M,target}}$. The maximum function is to ensure that the renormalization only inputs energy to the flow field, that is, the fluctuation is only stretched but never compressed by β, mimicking the process of inverse cascade. While W' is made up of two components u' and v', the flow dynamics consideration implies to renormalize v' fluctuations in order to recover the vertical turbulent transportation:

$$v^{(1)} = \beta^2 v$$

The above described procedure is called RNDES (renormalized DES).

3 Results and Discussion

RNDES is tested in both a $Re_\tau = 395$ incompressible channel flow and a compressible boundary layer (CBL) flow with Mach = 2.25. The channel flow is simulated using a domain $2\pi\delta \times 2\delta \times \pi\delta$ in the streamwise, wall-normal and spanwise directions (x, y and z, respectively); δ is the channel half-height. The grid $N_x \times N_y \times N_z$ is $32 \times 129 \times 32$. The CBL flow has the same computational domain as in Gao et al. [11]. and the grids used for the fully developed region ($7 < x < 9$) are $550 \times 85 \times 32$ and $550 \times 85 \times 16$.

Figure 2 compares the velocity profiles. The height of log layer determines the main body of mean velocity because beyond the log layer the balancing terms in NS equations vanish towards the channel center. The log-layer mismatch of DES results in severe overestimate of the mean velocity, in consistent with the results obtained by Nikitin et al. [2]. The IDDES greatly reduces the log-layer mismatch and improves

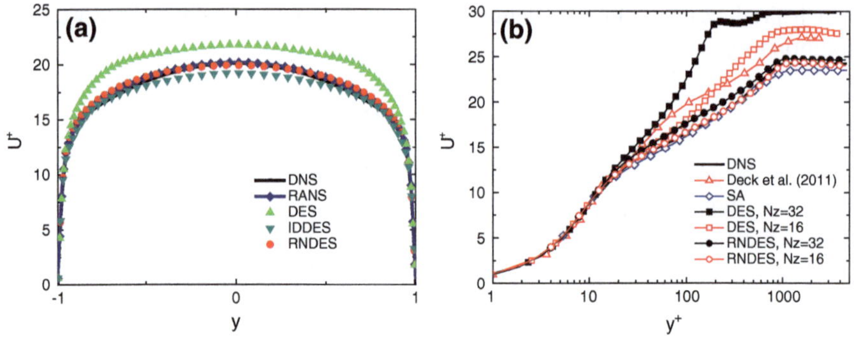

Fig. 2 Comparison of mean velocity profiles. **a** $Re_\tau = 395$ channel; **b** van Driest transformed mean velocity profiles for compressible boundary layer at $x = 11$. The incompressible result of the zonal DES by Deck et al. is also compared

the prediction of DES. On the other hand, RNDES fully eliminates the log-layer mismatch, and results in the best agreement with the DNS result.

For the compressible boundary layer, DES does not yield in two log layers as in channel flow. Instead the log layer is shifted upward as a whole. The zonal DES also has only one log layer. This is also in accordance with the DES result by Caruelle and Ducros [3]. This indicates different behaviors of DES between boundary layer flow and channel flow due to the change of geometrical constraint. However, here we still call the phenomenon a "log layer mismatch" for the mechanism behind the phenomena is the same. RNDES gives a log layer agreeing well with the DNS and also with the empirical log-law. This demonstrates that RNDES solves the LLM problem well, performing superior over DES and IDDES.

Figure 3 compares the r.m.s. profiles of u' component. DES is essentially RANS near the wall and damps too much turbulent energy, as can be seen from the evi-

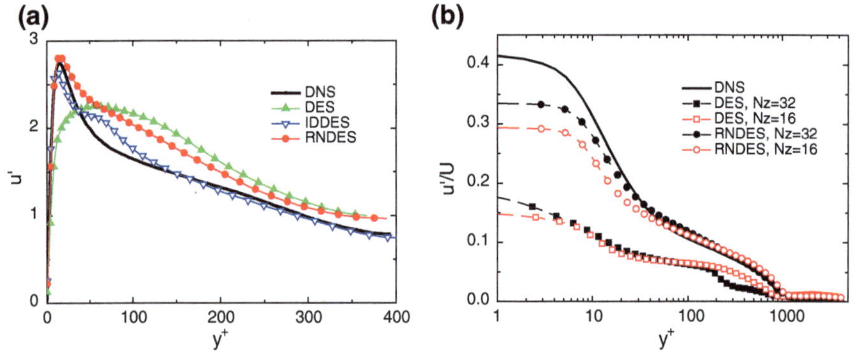

Fig. 3 Comparison for the r.m.s. profiles of u' component. **a** $Re_\tau = 395$ channel; **b** compressible boundary layer at $x = 11$

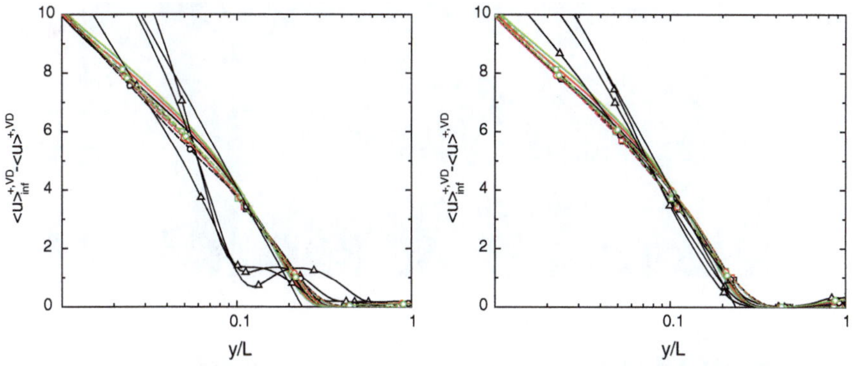

Fig. 4 Velocity defect at several streamwise positions of the compressible boundary layer. *Left* $N_z = 32$; *right* $N_z = 16$

dent lacking of near-wall peak. Both RNDES and IDDES result in good agreement with DNS. This demonstrates that RNDES has correctly recovered the key physical mechanism of vertical turbulent transportation, which is lacking in DES. Similar conclusion can also be drawn for v' and w' r.m.s. profiles (not shown).

The velocity defect at several streamwise positions are compared between DNS, DES, RANS and RNDES in Fig. 4. According to Fernholz [12], the velocity defect only depends on the wall distance scaled by the Rotta integral length

$$L = \frac{1}{u_\tau} \int_0^{\delta_0} (U_\infty^{VD} - U^{VD}) dy$$

where the definition of δ_0 is $U(\delta_0) = 0.995U_\infty$. As shown in Fig. 4, DES performs poorly in the outer flow region and profiles at different positions even do not collapse. On the other hand, both RNDES and RANS yield in good agreement with DNS as well as the empirical formula. It should be noted that RNDES has entered LES mode for long in the outer region, and has also totally left the renormalization range of target function. Therefore, the turbulent flow of RNDES has been fully developed in the outer region, and its defect law is a naturally developed turbulent statistical result.

Figure 5 shows the instantaneous u contour at two different vertical positions: $y^+ = 5$ and 100, at the resolution $N_z = 32$. The field of RNDES is clearly dominated by the alternatively occurring low- and high- speed streaks, which have been proven to be the dominant flow structure by extensive research on wall-bounded turbulences. On the other hand, DES results in unusual near-isotropic or spanwise structures. This demonstrates that RNDES recovers the near-wall streak dynamics violated by DES.

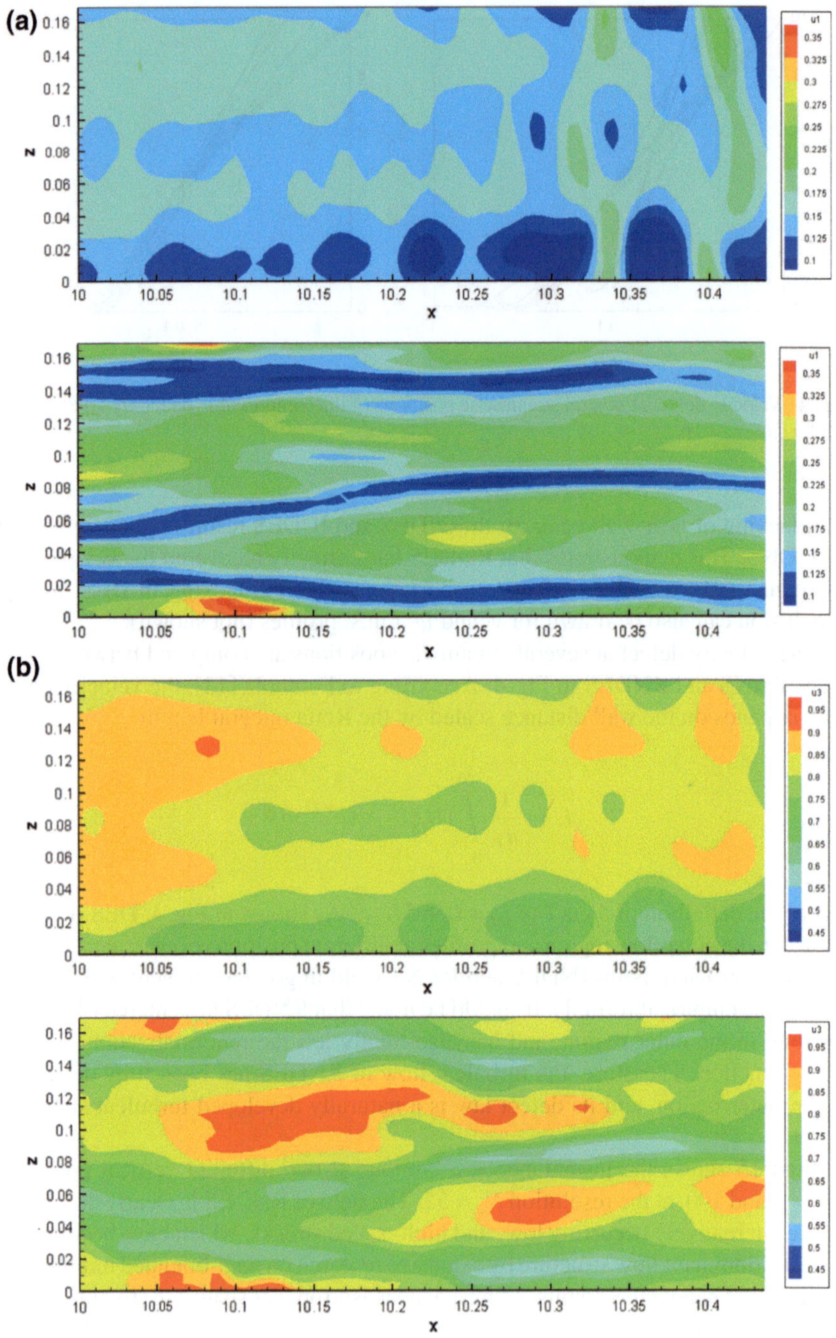

Fig. 5 Instantaneous u contour at two different vertical positions **a** $y^+ = 5$. *Up* DES; *down* RNDES and **b** $y^+ = 100$. *Up* DES; *down* RNDES

4 Concluding Remarks

The reason for log-layer mismatch is analyzed in a statistical way. The scaling of mixing length ℓ_M is violated at RANS/LES interface due to insufficient turbulent fluctuation. $\ell_{M,ML}$ is used as a target function to renormalize the resolved turbulent fluctuations, in order to recover the height of plateau prescribed by the log-law. The renormalization procedure guarantees the height of required by log-law, eliminating log layer mismatch and resulting in good results for both incompressible and compressible flows. RNDES develops correct turbulent structural ensemble under the constraint of statistical quantities. Considering RNDES only needs a rather small averaging sample (only 16 points for the coarsest case), for future simulations of wall-bounded turbulence with no homogeneous direction, statistical order functions can be obtained through local averaging or filtering.

Acknowledgments The Author thanks Prof. Xin-Liang Li for providing the OpenCFD-SC program. This work is supported by the National Natural Science Fund of China (No. 11302213).

References

1. Spalart, P.R., Jou, W.H., Strelets, M., et al.: In: Liu, C., Liu, Z. (eds.) Advances in DNS/LES, pp. 137–147. Greyden Press, Columbus (1997)
2. Nikitin, N.V., Nicoud, F., Wasistho, B., et al.: Phys. Fluids **12**, 1629–1632 (2000)
3. Caruelle, B., Ducros, F.: Int. J. Comput. Fluid Dyn. **17**(6), 433–51 (2003)
4. Piomelli, U., Balaras, E., et al.: Int. J. Heat Fluid Flow **24**, 538–550 (2003)
5. Spalart, P.R., Deck, S., et al.: Theor. Comp. Fluid Dyn. **20**, 181–195 (2006)
6. Travin, A.K., Shur, M.L., Spalart, P.R., et al.: In: Wesseling, P. (ed.) European Conference on Computational Fluid Dynamics, TU Delft, The Netherlands (2006)
7. Deck, S., Weiss P.-E., Pamiès, M., et al.: Comput. Fluids **48**, 1–15 (2011)
8. She, Z.S., Hu, N., Wu, Y.: Acta. Mech. Sin. **25**, 731–736 (2009)
9. She, Z.S., Chen, X., Wu, Y., et al. Acta. Mech. Sin. (2010)
10. She, Z.S., Wu, Y., Chen, X., et al.: New J. Phys. **14**, 093054 (2012)
11. Gao, H., Li, X.L., Fu, D.X., Ma, Y.W.: Chin. Phys. Lett. **22**, 1709 (2005)
12. Fernholz, H.H., Angew, Z.: Math. Mech. **51**, T148 (1971)

Improved Delayed Detached-Eddy Simulation of Transonic and Supersonic Cavity Flows

Kunyu Luo and Zhixiang Xiao

Abstract Transonic and supersonic flows past the M219 cavity with a 5:1:1 aspect ratio are studied using Improved Delayed Detached-Eddy Simulation (IDDES) on a high-quality structured mesh. Comparisons with available experimental and LES data corresponding to the same configuration demonstrate a high level of accuracy at $Ma = 0.85$. The broadband content and the Rossiter modes are captured. Numerical results for a $Ma = 1.5$ inflow are also presented and compared. Typical flow characteristics, such as shear layer instability, vortex impingements and pressure feedback, are well predicted by IDDES.

1 Introduction

A cavity under transonic or supersonic condition often causes highly unsteady and complicated flow phenomena and negative effects on the aircraft and its loads. Thus understanding the characters of transonic or supersonic flow past a cavity is of vital importance for advanced aircraft design. Although a large number of studies have been done about cavity flow under different conditions [4], the accurate prediction of cavity flows still presents a challenge for numerical methods.

The present study investigates the transonic and supersonic flows past the M219 cavity using IDDES method. The M219 cavity wind-tunnel configuration is a three-dimensional cavity mounted into a testing rig. The length of the cavity (L) is 0.508 m (20 in.) and the depth (D) and the width (W) are 0.1016 m (4 in.). The length-to-depth ratio (L/D) is 5, thus presenting an open cavity flow where the shear layer goes straight across the cavity rather than reattaches on the cavity floor. The M219 case has been widely studied since the high Reynolds number subsonic cavity wind-tunnel tests was carried out at QinetiQ [2]. The Rossiter modes [9] have been observed, which are defined as follows:

K. Luo · Z. Xiao(✉)
School of Aerospace Engineering, Tsinghua University, Beijing 100084, China
e-mail: xiaotigerzhx@tsinghua.edu.cn

K. Luo
e-mail: luoky12@mails.tsinghua.edu.cn

S. Girimaji et al. (eds.), *Progress in Hybrid RANS-LES Modelling*,
Notes on Numerical Fluid Mechanics and Multidisciplinary Design 130,
DOI 10.1007/978-3-319-15141-0_13

$$f_n = \frac{U_\infty}{L}\frac{n-\gamma}{Ma_\infty + 1/\kappa} \tag{1}$$

where f_n is the nth mode frequency, U_∞ and Ma_∞ are the velocity and the Mach number of the freestream, κ and γ are empirical parameters, and L is the length of the cavity.

As for the hybrid RANS/LES methods, the authors have chosen the improved delayed detached-eddy-simulation (IDDES) [10]. From our previous studies on the Rudimentary Landing Gear [12], IDDES method has shown its superiority in the prediction of surface flow topology and pressure fluctuation. Though a comprehensive evaluation of performances of different URANS and hybrid RANS/LES methods can be found in EU DESider Project [8], it is still of great interest to see the performance of IDDES in such a case, especially the acoustic related pressure fluctuations, and figure out the advantages and shortcomings in predicting cavity flows.

2 Numerical Approaches

IDDES is an advanced hybrid RANS/LES method. Combining DDES and wall-modelled LES (WMLES), it successfully cures shortcomings like the log-layer mismatch and the separation induced by grid. It has shown considerable improvement in the authors' previous studies. The construction of IDDES method based on the two equation Shear Stress Transport turbulence model [5] is fairly straightforward. In the turbulence kinetic energy (TKE) equation

$$\frac{\partial(\rho k)}{\partial t} + \frac{\partial(\rho U_i k)}{\partial x_i} = \tilde{P}_k - \frac{\rho k^{3/2}}{L_{hybrid}} + \frac{\partial}{\partial x_i}\left[(\mu + \sigma_k \mu_t)\frac{\partial k}{\partial x_i}\right] \tag{2}$$

the length scale, L_{hybrid}, is written as:

$$L_{hybrid} = L_{IDDES} = \tilde{f}_d(1 + f_e) \times L_{RANS} + (1 - \tilde{f}_d) \times L_{LES} \tag{3}$$

Here L_{RANS} and L_{LES} are the turbulence length scales of the RANS part and the filter length scale of the LES part respectively. The grid scale is redefined as

$$\Delta = \min\left[\max\left(C_w \Delta_{max}, C_w d, \Delta_{min}\right), \Delta_{max}\right] \tag{4}$$

where C_w is a constant and d is the distance from the cell center to the nearest wall. The minimum grid scale, Δ_{min} is defined by $\Delta_{min} = \min(\Delta x, \Delta y, \Delta z)$ and similarly the max grid scale, $\Delta_{max} = \max(\Delta x, \Delta y, \Delta z)$. This new definition of grid scale further reduces the sub-grid viscosity in the log layer. Please note that the minimum grid scale Δ_{min} is used here instead of the grid step in the wall-normal direction h_{wn} in the original paper [10] for simplification.

Function $\tilde{f}_d$ is defined as $\max[(1 - f_{dt}), f_B]$, which is determined by both the geometry-related part f_B and the flow-related part $(1 - f_{dt})$.

When $f_e = 0$, L_{IDDES} can be rewritten as

$$L_{hybrid} = L_{IDDES} = \tilde{f}_d \times L_{RANS} + (1 - \tilde{f}_d) \times L_{LES} \quad (5)$$

and IDDES reverts to DDES.

When $f_e > 0$ and $\tilde{f}_d = f_B$, L_{IDDES} can be rewritten as

$$L_{hybrid} = L_{IDDES} = f_B(1 + f_e) \times L_{RANS} + (1 - f_B) \times L_{LES} \quad (6)$$

and IDDES acts in WMLES mode near the wall. The detailed formulations of the functions mentioned above can be found in the original reference [10].

The in-house CFD code UNITs has been invoked in the computations. The UNITs code is a density-based Navier-Stokes solver using the finite volume method on structured grids. For unsteady simulations, a modified fully implicit low-upper symmetric Gauss-Seidel (LU-SGS) method is employed with Newton-like sub-iteration in pseudo time when solving the mean flow and the turbulence model equations.

Furthermore, a spatial scheme with proper dissipation and dispersion levels is important for an unsteady simulation. In this paper, the Roe scheme with 3rd order MUSCL interpolation is applied. The dissipation term is multiplied by an adaptive function ϕ which varies from the minimum cutoff value ϕ_{min} to 1 depending on the flow characteristics [6, 11]. In areas dominated by turbulence, the value of ϕ is small to maintain a low dissipation level in order to resolve the small flow structures. While in non-rotational areas, the adaptive function approaches 1 and provides the scheme with enough dissipation. As the very first step of our research on cavity flows, the dissipation function is set to a constant of 1 in this work.

3 Simulation Configurations

The same multiblock mesh is used for both $Ma = 0.85$ and $Ma = 1.5$ inflows. The geometry configuration includes the cavity, the rig and the wind tunnel but not the support sting. Our structured mesh has about 28.7 million cells in total. We have managed to place 16.6 million (∼58 %) cells in the cavity and its vicinity, by a carefully designed topology. The first grid layers are located 2×10^{-3} mm away from the cavity walls. The grids inside the cavity are kept within 1–1.5 mm in all directions and maintain isotropic and orthogonal as much as possible. Figure 1 shows the surface mesh and its topology. Note that the origin is located right at the middle of the cavity leading edge, with the y axis pointing up.

The inflow/outflow boundaries are set far away enough from the cavity, with a distance more than 5 times the rig length. The rig and the cavity surfaces are set to isothermal no-slip walls and the wind tunnel walls are set to inviscid walls.

Fig. 1 Surface mesh for M219 cavity. Very dense mesh in the cavity and its vicinity

The first computation is performed under freestream conditions of $Ma_\infty = 0.85$, $P_\infty = 6.3 \times 10^4$ Pa, $T_\infty = 266.53$ K and $Re = 1.35 \times 10^7$/m. The non-dimensional time step is 5×10^{-3} corresponding to a physical time step of 1.8×10^{-5} s. The second computation is performed under another wind tunnel conditions of $Ma_\infty = 1.5$, $P_\infty = 4.7 \times 10^4$ Pa, $T_\infty = 202.53$ K and $Re = 2.3 \times 10^7$/m. The non-dimensional time step is still 5×10^{-3}.

4 Results and Discussion

The cavity flow is an ideal case for DES-like modeling because of the fact that the boundary layer detaches from the leading edge of the cavity. The instability of the shear layer over the cavity increases rapidly and breaks down into vortices. Vortex impingements on the cavity aft wall lead to massive unsteady vortex motions and pressure wave propagation. Experimental data by QinetiQ [2] and full LES data [3] are used for validation.

4.1 Mean Flows

Figure 2 compares the streamwise and vertical velocity components with LES data. Very good agreement is achieved. As IDDES actually acts in LES mode in the cavity, this kind of performance is just as predicted. Relatively large differences are limited

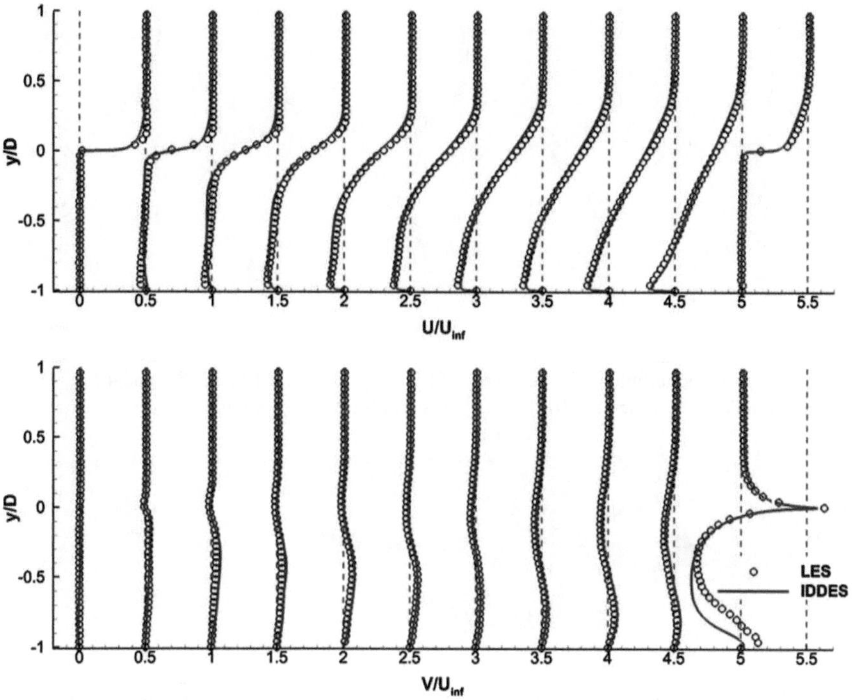

Fig. 2 Mean streamwise (*upper*) and *vertical* (*lower*) velocity profiles at $x/D = 0.003$, 0.5, 1.0, 1.5, 2.0, 2.5, 3.0, 3.5, 4.0, 4.5 and 4.997, $z/D = 0$. The *horizontal* axes are scaled to $(0.5U/U_\infty + x/D)$ and $(V/U_\infty + x/D)$ for better plotting

in the aft wall area for the V component, for RANS simulation is used in the near wall region by IDDES. Other small deviations are due to the different predictions of the shape or size of the recirculation areas.

4.2 Mean Flows

In Fig. 3, the resolved turbulent kinetic energy (TKE) are presented. It can be seen that the predictions of IDDES are very much like the LES results. The modeled part is actually very tiny compared against the resolved part and presents little effect to the TKE profiles. For the shear layer near the leading edge, however, the IDDES method does not fully capture the turbulent fluctuation. This should be blamed on the grey region problem, which is caused by the missing turbulent information when switching from the upstream RANS region to the neighboring LES region. IDDES also under-predictes the TKE at $x/D = 4.5$ compared with LES.

Snapshots of instantaneous spanwise vorticity and iso-surface of Q criterion are displayed in Fig. 4. Small turbulent flow structures are successfully captured by

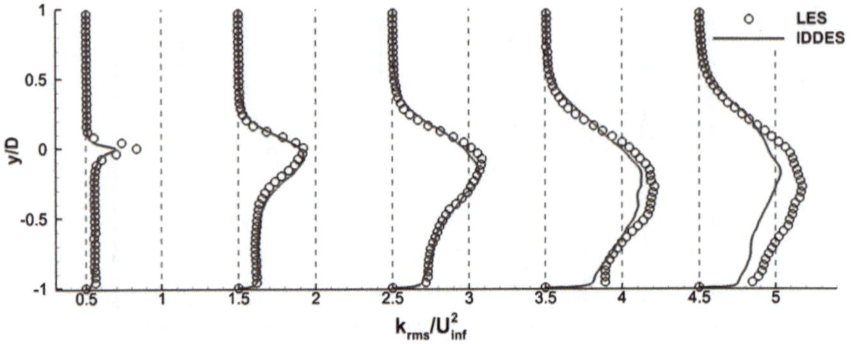

Fig. 3 The resolved turbulent kinetic energy (TKE) at $x/D = 0.5, 1.5, 2.5, 3.5$ and 4.5, $z/D = 0$. The horizontal axis is scaled to $(10\,k_{rms}/U_\infty^2 + x/D)$ for better plotting

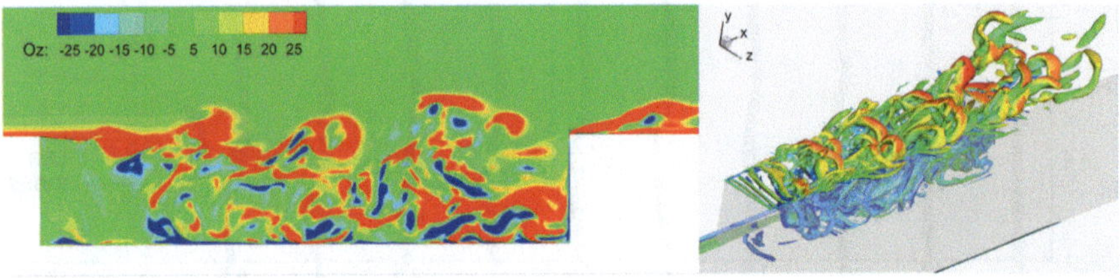

Fig. 4 Snapshots of instantaneous flow structures (*Left* Spanwise vorticity; *Right* Iso-surface of Q criterion, with contours of Mach number)

IDDES. Shear layer instability and breakdown can also be clearly observed. The fixed value of dissipation function ϕ has greatly limited the IDDES capability of resolving even smaller structures. Improvement is expected in our future work.

Pressure fluctuations have been monitored at 10 sampling points along the rig central line on the cavity floor (being denoted K20–K29 respectively), shown in Fig. 5. In the computation, after the flow is fully developed, the pressure at these 10

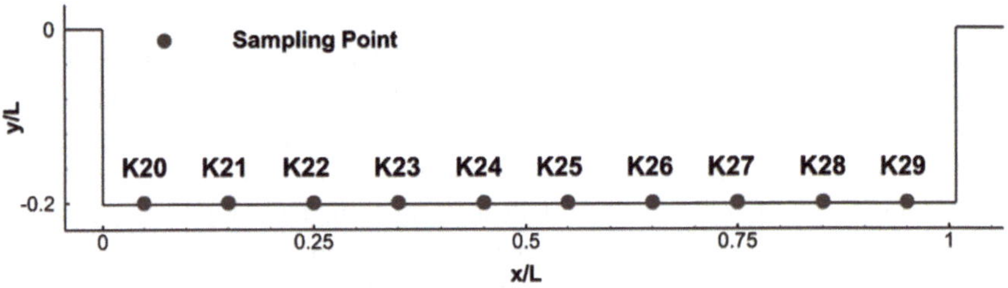

Fig. 5 Locations of 10 sampling points. L is the length of the cavity

Table 1 Frequencies (Hz) for different pressure modes in the cavity

Mode	1st	2nd	3rd	4th
Exp.	135	350	590	820
Rossiter equation	149	360	570	780
IDDES	103	379	601	824

points has been sampled at each time step. The corresponding Nyquist frequency is about 28 kHz. The total sampling time reaches 0.295 s.

The frequencies for different Rossiter modes are listed in Table 1 ($\kappa = 0.57$ and $\gamma = 0.29$ in the Rossiter equation). And the predicted PSD spectra at locations K20, K21, K24, K26, K28 and K29, which are estimated by the Burg algorithm [1], are presented in Fig. 6. The pressure modes are well captured, however with 10–30 Hz errors. Better prediction is achieved for higher frequency. Due to the small time step and fine grid resolution, no high-frequency damping is observed. A longer averaged time might improve the prediction of the 1st mode. The overall sound pressure level (OASPL) distribution along those 10 points is shown in Fig. 7. The predictions of the pattern and the amplitude are satisfactory. The over-prediction reported in previous studies [8] is avoided.

4.3 Comparisons Between $Ma = 0.85$ and $Ma = 1.5$

The flows in the cavity are actually rather complicated as shown in Fig. 8. High pressure region near the aft cavity wall reverses the flow and induced recirculation in the cavity. At $Ma = 0.85$, two sub-recirculation centers are noted, together with a small secondary bubble in the front corner of the cavity and a tiny one in the aft corner. At $Ma = 1.5$, the small sub-recirculation disappears and the bubble in the front corner becomes smaller while the one in the aft corner becomes much bigger. A tiny bubble on the front wall near the leading edge is found only at $Ma = 1.5$, and expansion fans and shockwaves are also observed.

The shear layer is more diffusive at $Ma = 0.85$ than at $Ma = 1.5$, as the shear layer at $Ma = 0.85$ deflects towards the cavity flow earlier than at $Ma = 1.5$. This is further demonstrated in Fig. 9, where the vertical velocity predicted at $Ma = 0.85$ is smaller than at $Ma = 1.5$, at $x/D = 3.5$ and 4.0. The tiny bubble on the front wall also causes discrepancy. However the streamwise velocity shows quite similar profiles.

The high speed flow generates much stronger pressure fluctuations on the cavity floor, which is shown in Fig. 10. Globally, OASPL at $Ma = 1.5$ is 6–7dB higher than at $Ma = 0.85$. In fact, the PSD spectra at some sampling points have indicated

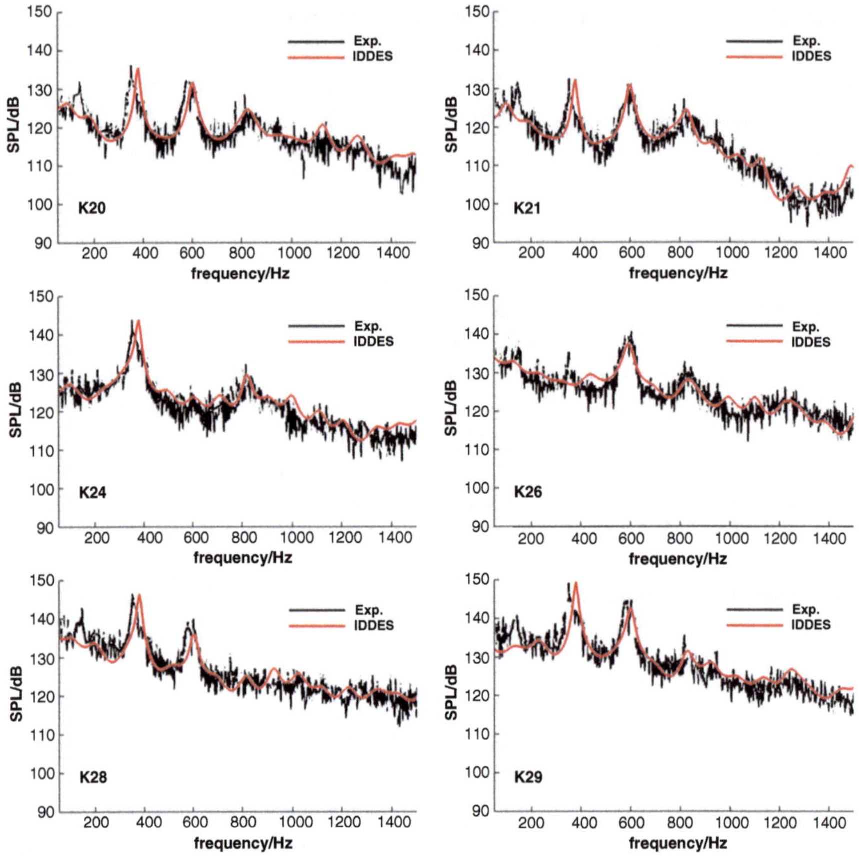

Fig. 6 PSD spectra at K20, K21, K24, K26, K28 and K29, showing the Rossiter modes. Experimental data derived from the paper of Peng [7]

that the fluctuations at $Ma = 1.5$ have more energy in the high frequency range (See Fig. 11, also estimated by the Burg algorithm). It is also noticed that at $Ma = 1.5$, the 2nd and 3rd modes are no longer the dominant modes, as all four modes have close peak amplitudes. The P_{rms}/P_{∞} ratio grows much faster at $Ma = 1.5$ after $x/D = 0.3$. The different tendency near $x/L = 0.9$ is probably caused by the different behaviors of recirculation centers at these two Mach numbers and needs further investigation.

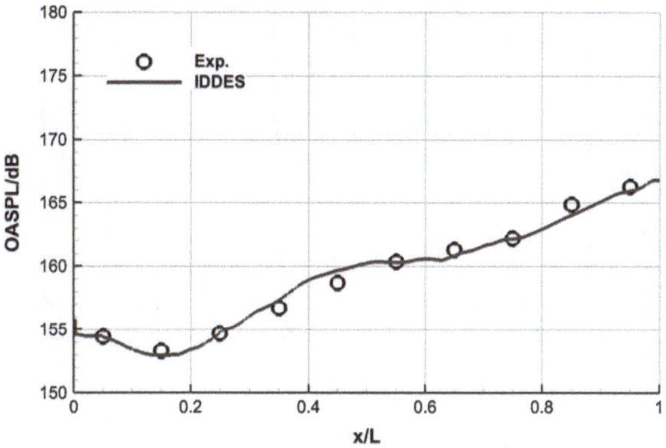

Fig. 7 OASPL distribution along 10 sampling locations. L is the length of the cavity

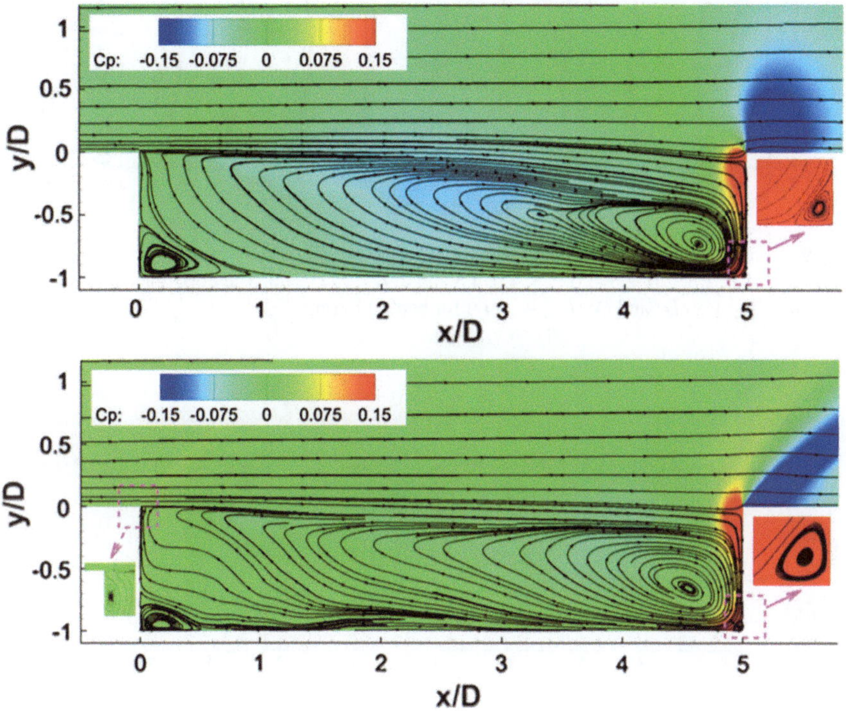

Fig. 8 Pressure contour and streamlines in the central plane of the cavity for the mean flow (*Upper* $Ma = 0.85$; *Lower* $Ma = 1.5$)

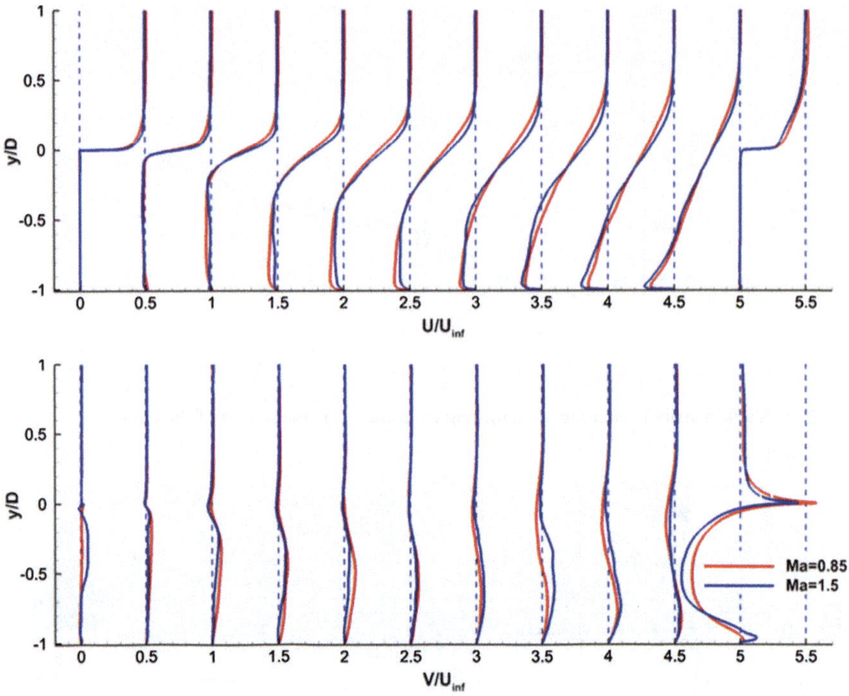

Fig. 9 Comparison of mean streamwise (*upper*) and vertical (*lower*) velocity profiles at $x/D =$ 0.003, 0.5, 1.0, 1.5, 2.0, 2.5, 3.0, 3.5, 4.0, 4.5 and 4.997, $z/D = 0$. The horizontal axes are scaled to $(0.5U/U_\infty + x/D)$ and $(V/U_\infty + x/D)$ for better plotting

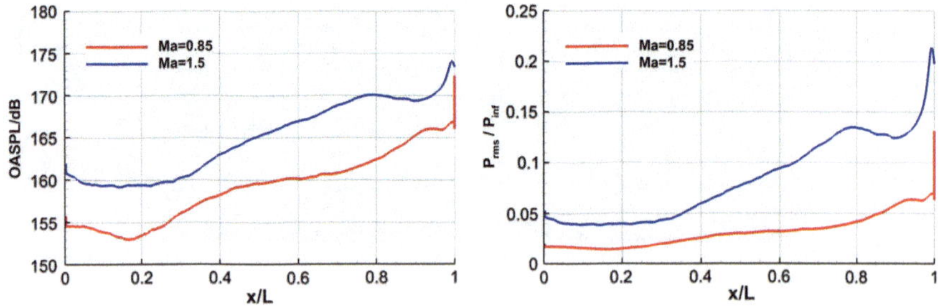

Fig. 10 Comparison of pressure fluctuations on the cavity flow between $Ma = 0.85$ and $Ma = 1.5$

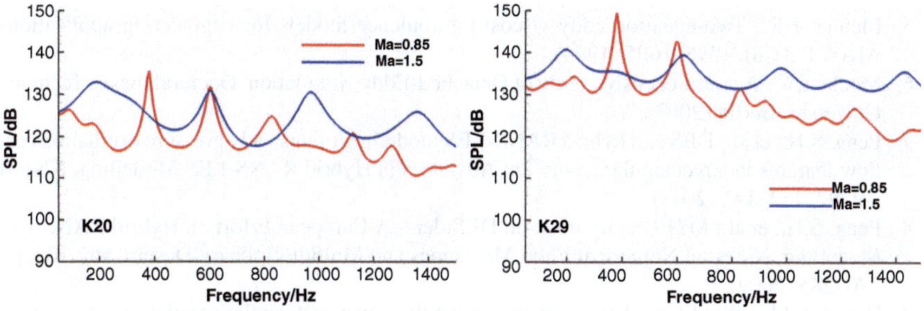

Fig. 11 Comparison of PSD spectra at K20 and K29 between $Ma = 0.85$ and $Ma = 1.5$

5 Conclusions

The present work demonstrates the prediction of the flows past the M219 cavity at $Ma = 0.85$ and $Ma = 1.5$, using IDDES method based on SST turbulence model on a high-quality structured mesh. Numerical results at $Ma = 0.85$ are firstly compared with available experimental and LES data. Time averaged velocity profiles and kinetic turbulent energy show rather good agreement. OASPL distribution on the cavity floor is also in good agreement with the measurements. The predicted PSD spectra at several sampling points are in line with experiment results except for the 1st Rossiter mode.

Flow features at $Ma = 0.85$ and 1.5 are also compared. Time averaged flows in the cavity are quite similar and the differences are caused by the different size and shape of the recirculation and the secondary bubbles. The flow at a higher Mach number presents much stronger pressure fluctuations on the cavity floor, approximately 5–7dB higher at $Ma = 1.5$ than at $Ma = 0.85$. PSD spectra at $Ma = 1.5$ are much higher than those at $Ma = 0.85$ especially in the high frequency range and all four modes have close peak amplitudes and are equally important.

Acknowledgments This work is supported by the National Science Foundation of China (Grant No. 11372159). The authors also thank Tsinghua National Laboratory for Information Science and Technology for the computation resources.

References

1. Burg, J.P.: Maximum entropy spectral analysis. Doctoral thesis, Stanford University (1975)
2. Henshaw, M.J.C.: M219 cavity case. Verification and validation data for computational unsteady aerodynamics. Defense Technical Information Center, ADP010729 (2000)
3. Larcheveque, L., et al.: Large eddy simulation of a compressible flow in a three-dimensional open cavity at high Reynolds number. J. Fluid Mech. **516**, 265–301 (2004)
4. Lawson, S.J., et al.: Review of numerical simulation for high-speed, turbulent cavity flows. Prog. Aerosp. Sci. **47**, 186–216 (2011)

5. Menter, F.R.: Two-equation eddy-viscosity turbulence models for engineering applications. AIAA J. **32**(8), 1598–1605 (1994)
6. Mockett, C.: A comprehensive study of Detached-Eddy-Simulation. Doctoral thesis, Technical University Berlin (2009)
7. Peng, S.H., et al.: DES and hybrid RANS-LES modelling of unsteady pressure oscillations and flow features in a rectangular cavity. In: Advances in Hybrid RANS-LES Modelling, NNFM, vol. 97, 132–141 (2008)
8. Peng, S.H., et al.: M219 cavity flow. In: DESider—A European Effort on Hybrid RANS-LES Modelling, Notes on Numerical Fluid Mechanics and Multidisciplinary Design, vol. 103, pp. 270–285 (2009)
9. Rossiter, J.E.: Wind-tunnel experiments on the flow over rectangular cavities at subsonic and transonic speeds. Rep. Memo. No. 3438, Aeronautical Research Council (1964)
10. Shur, M.L., et al.: A hybrid RANS-LES approach with delayed-DES and wall-modelled LES capabilities. Int. J. Heat Fluid Flow **29**, 1638–1649 (2008)
11. Strelets, M.: Detached eddy simulation of massively separated flows. In: 39th AIAA Aerospace Sciences Meeting and Exhibit, Reno, NV, US. AIAA 2001–0879 (2001)
12. Xiao, Z.X., et al.: Investigation of flows around a rudimentary landing gear with advanced detached-eddy-simulation approaches. AIAA J. **51**(1), 107–125 (2013)

Combining ZDES with Immersed Boundary Conditions Technique for the Treatment of Complex Geometries

L. Mochel, P.-É. Weiss and S. Deck

Abstract The present paper focuses on a numerical strategy called ZIBC consisting of the zonal use of Immersed Boundary Conditions combined with the ability of Zonal Detached Eddy Simulation (ZDES) to simulate high Reynolds number separated flows. The motivation of such a strategy lies in the accurate handling of geometrically complex configurations with validated unsteady tools. A first assessement of the strategy has already been performed by evaluating the introduction of a control device in a form of a short cylindrical serrated skirt into a simplified space launcher afterbody. To go further into validation, this paper focuses on the ability of the ZIBC strategy to reproduce the fluctuating pressure field. The test case corresponds to a simplified space launcher afterbody and consists of a cylinder elongated by another cylinder of smaller diameter (i.e. an extension). Immersed Boundary Conditions are used to handle the introduction of the extension into a structured curvilinear grid fitting the ZDES requirements to treat the blunt body configuration. The governing equations are solved using a standard body-fitted finite volume technique over the whole grid. A direct forcing source term is added when cells are internal to the skirt, i.e. solid, to drive the velocity and the turbulence variables to the chosen values. Numerical simulations are performed at a Reynolds number of 1.2×10^6 and a free stream Mach number of 0.702. The numerical results demonstrate the ability of the "Zonal Immersed Boundary Conditions" to successfully impose the desired values at solid nodes. The first and second order moments illustrate an excellent agreement between the experiment and the numerical simulation. Finally, the "Zonal Immersed Boundary Conditions" appear to successfully reproduce the effect of the extension.

L. Mochel (✉) · P.-É. Weiss · S. Deck
ONERA—The French Aerospace Lab, 92190 Meudon, France
e-mail: loic.mochel@onera.fr

P.-É. Weiss
e-mail: pierre-elie.weiss@onera.fr

S. Deck
e-mail: sebastien.deck@onera.fr

S. Girimaji et al. (eds.), *Progress in Hybrid RANS-LES Modelling*,
Notes on Numerical Fluid Mechanics and Multidisciplinary Design 130,
DOI 10.1007/978-3-319-15141-0_14

1 Introduction

One of the challenges in applied numerical aerodynamics lies in the cababilty to handle accurately geometrically complex configurations with validated unsteady tools. The existence of different flow regions associated to different modelling approaches can induce conflicting demands on the grid. In the framework of scale-resolving unsteady simulations with a classical body-fitted mesh generation approach, strict grid requirements have to be observed not to jeopardize the reliability of the simulation. These requirements can become prohibitive making the grid generation dramatically time-consuming or even unfeasible.

In such a framework, a strategy based on the Immersed Boundary Conditions (IBC) appears well-suited to avoid these grid limitations for complex geometries. Many applications of this approach can be found in the literature [9, 12]. Nevertheless, the IBC are generally applied on the whole geometry. In this paper, we propose another strategy to broaden the application range of structured solvers in term of complexity of the geometry. This approach is called Zonal Immersed Boundary Condition (ZIBC) (see Fig. 1) and is combined with the Zonal Detached Eddy Simulation (ZDES) [5, 6] which is based on a large validation base and is implemented in several industrial solvers commonly used by the aeronautical industry [1, 2, 18, 19, 24].

The ZIBC strategy has already been evaluated on a first validation test case [13]. Nevertheless, its ability to properly reproduce the fluctuating pressure field at the wall of the immersed object has not been assessed yet.

Thus, the present paper aims at investigating the ability of ZIBC combined with ZDES to properly reproduce the fluctuating pressure field at the wall of the

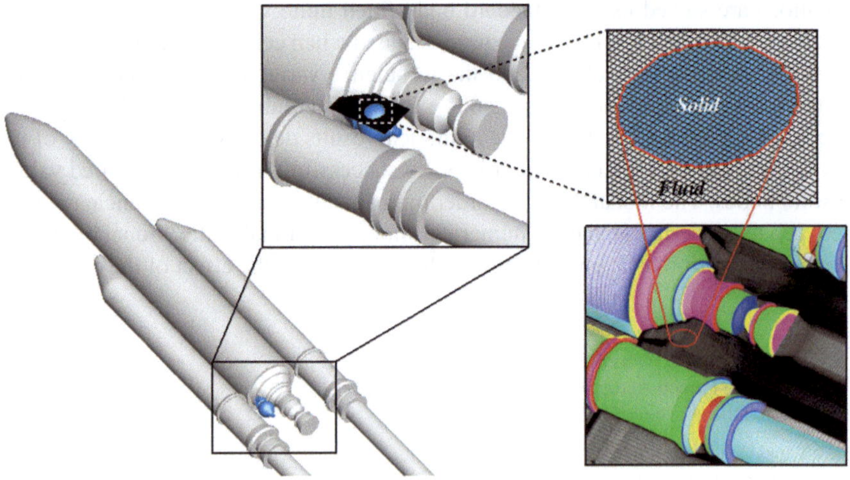

Fig. 1 Principle of the zonal immersed boundary conditions (ZIBC) methodology [13]

immersed object. To achieve this purpose, the paper is organized as follows. Section 2 describes the test case. Then, Sect. 3 is dedicated to the description of the ZIBC with the governing equations and the associated numerical method. Finally, in Sect. 4, the immersion procedure is described and the instantaneous, statistical and fluctuating properties of the flow are presented and assessed in comparison with the available experimental and numerical data.

2 Test Case

To assess the ability of the strategy combining the ZDES and the zonal use of the IBC to reproduce the fluctuating pressure field, a numerical configuration is considered and designed to fit with the experiments of the ONERA's S3Ch continuous research wind tunnel described by Deprés et al. [7] and Meliga and Reijasse [11] and the numerical simulation of Weiss et al. [25]. The configuration, shown in Fig. 2, consists of a cylinder with a diameter D equal to 100 mm elongated by another cylinder of smaller diameter d equal to 40 mm. This extension is 120 mm long which brings a L/D ratio of 1.2, where L is the emergence length. The height h for the axisymmetric step is equal to 30 mm. The numerical simulations are performed with a free stream Mach number of 0.702 leading to a Reynolds number based on a free stream velocity U_∞ and on the diameter D of 1.2×10^6. The initial external boundary layer thickness δ was obtained by modeling the necessary length for the forebody to have a δ/D ratio equal to 0.2.

The strategy is employed to immerse the rear body into a curvilinear grid fitting the ZDES grid requirements to treat the blunt body configuration. That underlying grid was obtained by adapting the pre-existing mesh from the numerical study of Weiss et al. [25]. From a numerical point of view, the case is computed using 14 million hexaedric cells (Fig. 2, middle). A particular care is taken to discretize the azimuthal direction with 240 planes (i.e. 1.5? per plan) considering the significance of the azimuthal wavelength evidenced by Deck and Thorigny [3], Weiss et al. [25], Weiss and Deck [26]. Downstream the blunt body, an O–H topology has been set to avoid singularity problem near the axis. It should be noticed that the external grid

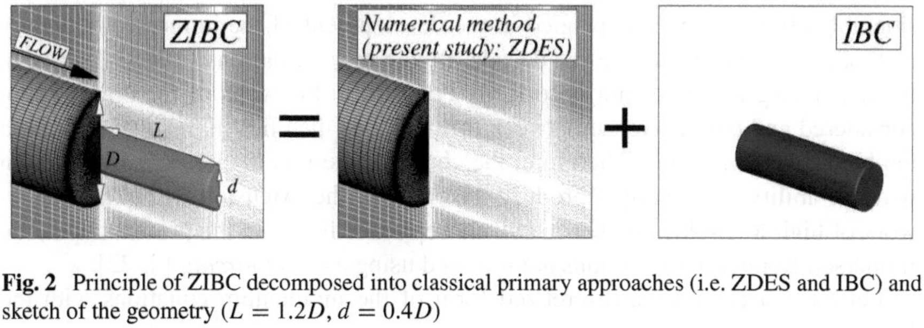

Fig. 2 Principle of ZIBC decomposed into classical primary approaches (i.e. ZDES and IBC) and sketch of the geometry ($L = 1.2D$, $d = 0.4D$)

(every cells outside the rear body) is exactly the same as the classical body-fitted mesh employed by Weiss et al. [25]. This was made on purpose in order to evidence the influence of the forcing source term on the fluctuating pressure field prediction. In particular, the effect of a non coincidence between the wall of the rear body and the underlying grid is not assessed in the present paper. It should be mentioned that the flow dynamics in the neighborhood of the extension is mainly ruled by the recirculation zone. This implies that the cells have been designed to fit with the LES requirements in the detached area in terms of number of points and cell isotropy.

3 Numerical Strategy: Zonal IBC Combined with ZDES

Numerical simulation is performed using the finite-volume solver FLU3M which has been developed by Onera to perform simulations with the compressible Navier-Stokes equations on multiblock structured grids. The time integration is carried out using the second-order-accurate backward scheme of Gear. A modified AUSM+ scheme proposed by Liou [10] is employed for the advective fluxes. The accuracy of the solver has been assessed in various applications including base afterbody flows for several regimes and in particular on launcher geometries [3, 25, 26]. In these last references, the numerical results are thoroughly compared to the available experimental data including single and two-point spectral analysis. Further details concerning the numerical method and implementation of turbulence models can be found in references [4, 16].

The approach selected for the present study is the Zonal Detached Eddy Simulation (ZDES) proposed by Deck [5, 6] which belongs to the RANS/LES approaches. ZDES has proven to be suitable for the simulation of complex turbulent phenomena at high Reynolds number configurations [3, 20, 25, 26]. Contrary to the classical Detached Eddy Simulation (DES97) introduced by Spalart et al. [21] which is a non-zonal approach, the user has to select RANS and DES areas according to a classification of typical flow problems indicated in Fig. 3. In the present case, the area upstream from the separated point is computed using a URANS (Unsteady Reynolds Averaged Navier-Stokes) model while the downstream part is modelled with ZDES mode I, since the separation is clearly fixed by the geometry as illustrated by Fig. 3.

Since the introduction of the first IB formulation by Peskin in 1972 [17], a considerable effort has been expended in the development of the immersed boundary method, particularly over the last decade. By the very definition of the IB approach, the underlying grid does not necessarily conform to the walls of the object being considered and the difference between formulations then lies in the way to impose the boundary conditions. Therefore, the challenge here is to choose a formulation with the ability to properly reproduce the effect of the extension within the framework of high Reynolds number massively separated flows. In the present paper, the Immersed Boundary Conditions are imposed using a *direct forcing* [8, 22].

Let us first consider a discretized form of the momentum equations with the forcing function:

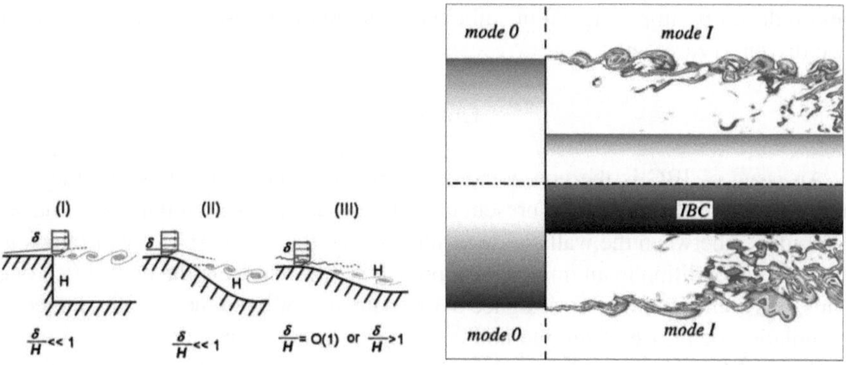

Fig. 3 *Left* Classification of typical flow problems [6]. I: separation fixed by the geometry, II: separation induced by a pressure gradient on a curved surface, III: separation strongly influenced by the dynamics of the incoming boundary layer. *Right* ZDES zones for both the body-fitted (*top*) and the IB (*bottom*) approaches. Mode 0 (i.e. RANS) and mode I correspond to the flow classification

$$\frac{\rho u_i^{n+1} - \rho u_i^n}{\Delta t}\Omega = RHS_{\rho u_i} + f_{\rho u_i},\ i = 1, 3 \tag{1}$$

where Δt denotes the time step, Ω the cell volume, u_i the velocity component in the i direction, ρ the density and $RHS_{\rho u_i}$ the discretized form of the convective and diffusive fluxes and eventually source terms in the LES and RANS equations. Assuming that the face of the cell being considered coincides with the immersed boundary and that the Dirichlet boundary condition $(\rho u_i) = (\rho u_i)_S$ is imposed $(\rho u_i)_S$ stands for the desired value of the conservative velocity component at solid walls), the forcing function in Eq. (1) to ensure direct enforcement of the boundary conditions is given by:

$$f_{\rho u_i} = \frac{(\rho u_i)_S - \rho u_i^n}{\Delta t}\Omega - RHS_{\rho u_i},\ i = 1, 3 \tag{2}$$

The combination of Eqs. (1) and (2) then yields:

$$\rho u_i^{n+1} = (\rho u_i)_S \tag{3}$$

Concerning the internal treatment of the body, the forcing is applied on each solid cell. As the no-slip condition is desired, velocity variables are enforced to be zero when the forcing is applied, i.e.:

$$(\rho u_i)_S = 0 \tag{4}$$

What is more, in the framework of the use of RANS with IBC, Iaccarino and Verzicco [9] indicate that the same treatment can be applied for the velocity variables and the conservative modified pseudo eddy-viscosity $\rho\tilde{\nu}$ (from the Spalart-Allmaras model). In order to prevent the generation of numerical oscillations into the flow field induced by imposing $\rho\tilde{\nu}$ in solid cells, the modified pseudo eddy-viscosity is also driven to zero, i.e.:

$$(\rho\tilde{\nu})_S = 0 \tag{5}$$

An issue of IBC is the non coincidence between the grid points and the solid interface. Nevertheless, in the present case it has been chosen on purpose to have a coincidence between the walls of the extension and the mesh. The way of imposing the no-slip condition in an immersed approach is different from the body-fitted one. Thus, this configuration was designed to evaluate the ability of the specific immersed formulation being free from the non-coincidence issue. In practical term, contrary to classical boundary condition, the source term is designed to drive the velocity variables to the desired value at the center of a cell. This means that even if the grid is perfectly coincident with the wall of the extension, the source term is not able to impose the no-slip condition exactly at the wall. This leads to a relocation of the wall from the vertexes to the centers of the cells. The idea was to assess the results despite this approximation. A non-coincident grid in azimuth has been tackled in a previous study [13]. In this study, the coupled approach (ZDES + IBC) has been applied to introduce a short cylindrical serrated skirt on the same simplified space launcher afterbody. The results illustrate a fairly good agreement with the experiment and despite the fact that the skirt is described in a stepwise manner due to the absence of velocity reconstruction, the method still allows the development of a wide variety of turbulent scales. In such conditions, the major difficulty appears to find an appropriate way to reconstruct the velocity [8, 9, 12]. However the use of fully non-coincident grids in the framework of ZIBC remains an open issue.

For more detailed discussions of the various elements of the IB method, the reader is invited to see reviews from Iaccarino and Verzicco [9] and Mittal and Iaccarino [12].

4 Application and Validation

The application of the ZIBC is composed with two steps summarized in Fig. 4, i.e. the ray tracing method [15] and the wall distance assessment:

1. The aim of the ray tracing technique is to distinguish the cells of the underlying computational grid as *solid* and *fluid*, whether they are inside or outside the object of interest. This operation needs the description of such an object in STereo-Lithography (STL) format (for more information see O'Rourke [15]).
2. The wall distances are mandatory to compute for the use of both the Spalart-Allmaras model of turbulence and the ZDES. First, the wall distances d_w^B due to the object are computed. During this procedure, the internal cells are treated as if they were infinitely far from the object. This is achieved in order to prevent the destruction term of the transport equation from driving the modified pseudo

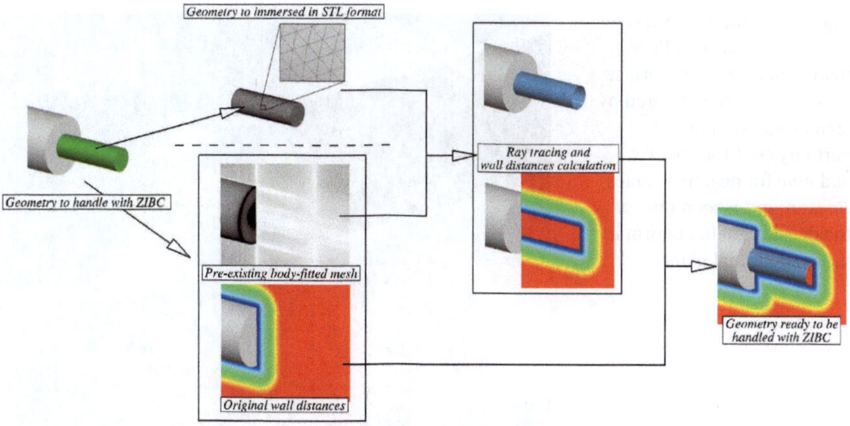

Fig. 4 Decomposition of the procedure used to immerse the rear-body

eddy-viscosity to negative values. In practical terms, the infinite distance expressed in meters is set to 10^9. Then, the algorithm reads the wall distances d_w^A corresponding to the geometry without the immersed object. Finally, the minimum between both distances becomes the new wall distances of the whole configuration $\left(d_w = min\left(d_w^A, d_w^B\right)\right)$ except for the internal cells which keep their value d_w^B.

Once this step has been performed, the solver is able to apply the specific source term (Eq. 2) on each solid cell. A flag variable computed by the ray tracing algorithm and denoted as $tagibc$ is equal to 1 if the cell is *solid* and 0 otherwise which allows to apply the forcing function or not. First, the simulation runs over 200 ms to ensure any spurious numerical oscillation has been evacuated. After this transient step, the simulation is running during another 200 ms and the unsteady aerodynamic field is time-averaged during the calculation. The time step is equal to 2×10^{-6} s leading to 50,000 time steps.

A first visual impression of the turbulent content is provided by Fig. 5. In the foreground, an isosurface of the $\mathcal{Q}$ criterion is plotted colored by the sign of the streamwise vorticity (red for positive and blue for negative values). One can observe that the forcing source term does not seem to corrupt the flow since it appears to be representative of this particular type of configuration. Indeed, close to the abrupt change of geometry, one can clearly identify a sequence of toroïdal structures denoting the roll-up process of the shear-layer. As analysed by Deck and Thorigny [3] and Weiss et al. [25], these coherent structures merge together as they convect downstream (pairing phenomenon), grow in size, then distort into 3D structures with a hairpin shape near the solid reattachment and finally adopt a fully three dimensional organization in the wake region.

The background of Fig. 5 also proposes a qualitative overview of the instantaneous flow fields with the plot of iso-contours of the density gradient norm ($||\overrightarrow{grad}\rho||$) in

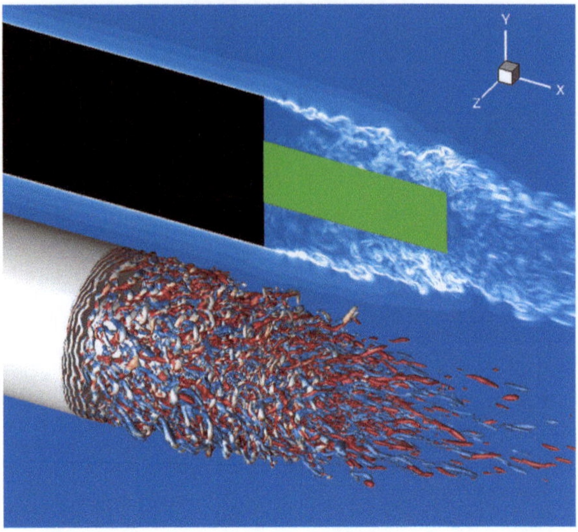

Fig. 5 Qualitative overview of the instantaneous flow field. *Foreground* isosurface of the $\mathcal{Q}$ criterion colored by sign of the streamwise vorticity (*red* for positive and *blue* for negative values). *Background* isocontours of the density gradient norm in the longitudinal plane

the longitudinal plane. The Kelvin-Helmholtz instabilities of the mixing layer appear right after the separation point and the growth and pairing process of the shear layer instabilities can also be observed. Moreover the complex three-dimensional turbulent dynamics is apparently not affected by the application of the forcing source term. What is more, it is worth mentioning that no delay occurs in the formation of instabilities in the mixing layer as expected by the use of the adequate ZDES mode (see discussion in [6]).

To go deeper into the validation of the rear body immersion, the statistical data field associated with both approaches (body fitted and immersed boundary) is investigated. According to the streamlines plotted in the mean field for both approaches (see Fig. 6a), the topology of the flow appears to be well reproduced by the application of the forcing source term. In particular, the recirculation bubble sizes and the location of the reattachment point are the same. This is confirmed by the isoline of velocity $U/U_\infty = 0.01$ in Fig. 6b where one can also observed that the velocity profiles give a streamwise velocity equal to zero for the cells inside the rear-body. As a consequence, the forcing function properly drives the velocity variable to the desired value. Concerning the iso-Cp, the fields appear to be very similar which is confirmed by Fig. 7 which shows the evolution of the mean and Root Mean Square pressure coefficients ($Cp = \left(\overline{P} - P_\infty\right)/0.5\rho U_\infty^2$; $Cp_{rms} = P_{rms}/0.5\rho U_\infty^2$ with $P_{rms} = \overline{p'^2}$) at the wall along the streamwise direction. The streamwise distributions are averaged in the azimuthal direction. Concerning the evolution of the Cp, it is worth mentioning that the values obtained with the IB approach are in excellent agreement with both the numerical data of Weiss et al. [25] and the experimental data of Deprés et al. [7] and Meliga and Reijasse [11] on analogous configurations. The evolution of Cp_{rms} illustrates also a good agreement between both the available

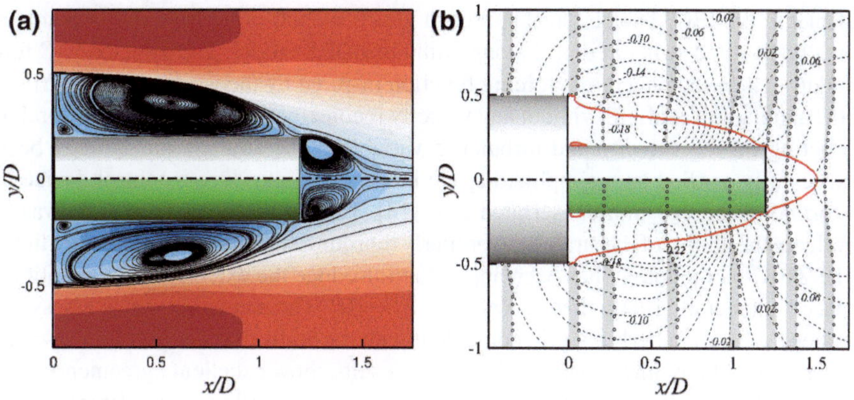

Fig. 6 Time-averaged flow fields. *Top* classical body-fitted approach. *Bottom* IB approach. **a** Streamlines in the mean field. **b** (- -) Iso-Cp with $\Delta C_p = 0.02$ between two lines (–) isoline of velocity $U/U_\infty = 0.01$, and (◯) velocity profiles at several streamwise locations

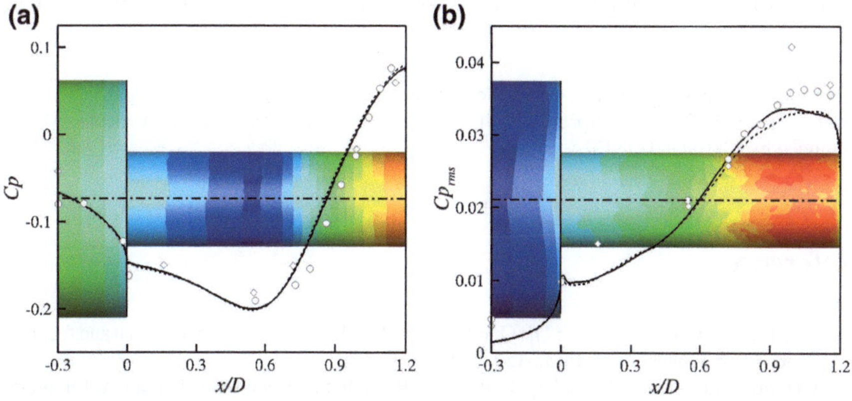

Fig. 7 Streamwise distribution of the mean pressure coefficient and the rms pressure coefficient with the corresponding isocontours at the afterbody wall plotted in the background (*top* body-fitted approach, *bottom* IB approach). Numerical simulations (ZDES): (—) body-fitted approach (- -) IB approach. Experiments: (◇) [7] (○) [11]

experimental and ZDES numerical data (classical approach without IBC) and the current ZIBC simulation.

5 Conclusion

A new validation of the ZIBC/ZDES strategy has been presented in this paper. The main idea relies on the treatment of a simplified geometry with a classical approach thoroughly validated and on the local addition of geometrically complex elements thanks to the IBC. The classical approach uses a second-order finite-volume solver on a body-fitted structured grid along with the ZDES model. The selected IBC formulation is based on a direct forcing function inspired by the works of Mohd-Yusof [14] and Fadlun et al. [8] without any velocity reconstruction procedure. The forcing is acting on the velocity and turbulence variables. The strategy has already been succesfully applied on a simplified space launcher to introduce a control device in a form of a short cylindrical serrated skirt [13]. The aim of the present study was to evaluate the ability of the strategy to properly reproduce the fluctuating pressure field. The ZIBC methodology has been applied to introduce the extension of a simplified space launcher afterbody flow at $Re_D = 1.2 \times 10^6$. The main idea was to assess the strategy with an underlying grid coincident with the walls of the object immersed in order to avoid any interpolation effect. The results show excellent agreements with the available experimental and numerical data for the first and second order moments. In particular, the salient features of the flow topology are properly reproduced with the zonal IB approach compared to the validated body fitted approach.

Future work will be dedicated to the spectral analysis of the fluctuating pressure field. Particular attention will be paid on the influence of the strategy on the sideloads prediction.

Acknowledgments The authors would like to thank the Centre National d'Études Spatiales (CNES) and the ALLIGATOR research project (ONERA) for financial support. The Ph.D. work of L. Mochel is funded by CNES and ONERA.

References

1. Cambier, L., Heib, S., Plot, S.: The Onera elsA CFD software: input from research and feedback from industry. Mech. Ind. **14**(3), 159–174 (2013)
2. Chalot, F., Levasseur, V., Mallet, M., Petit, G., Reau, N.: LES and DES simulations for aircraft design. AIAA Paper 2007–0723, 45th AIAA Aerospace Sciences Meeting and Exhibit, Reno, Nevada (2007)
3. Deck, S., Thorigny, P.: Unsteadiness of an axisymmetric separating-reattaching flow: numerical investigation. Phys. Fluids. **19**(065103) (2007)
4. Deck, S., Duveau, P., d'Espiney, P., Guillen, P.: Development and application of spalart allmaras one equation turbulence model to three-dimensional supersonic complex configurations. Aerosp. Sci. Technol. **6**(3), 171–183 (2002)
5. Deck, S.: Zonal-Detached-Eddy simulation of the flow around a high-lift configuration. AIAA J. **43**, 2372–2384 (2005)
6. Deck, S.: Recent improvements in the zonal detached eddy simulation (ZDES) formulation. Theor. Comput. Fluid Dyn. **26**(6), 523–550 (2012)
7. Deprés, D., Reijasse, P., Dussauge, J.-P.: Analysis of unsteadiness in afterbody transonic flows. AIAA J. **42**(12), 2541–2550 (2004)
8. Fadlun, E.A., Verzicco, R., Orlandi, P., Mohd-Yusof, J.: Combined immersed-boundary/finite-difference methods for three-dimensional complex flow simulations. J. Comp. Phys. **161**(1), 35–60 (2000)
9. Iaccarino, G., Verzicco, R.: Immersed boundary technique for turbulent flow simulations. Appl. Mech. Rev. **56**(3), 331–347 (2003)
10. Liou, M.S.: A sequel to AUSM, AUSM+. J. Comp. Phys. **129**(0256), 364–382 (1996)

11. Meliga, P., Reijasse, P.: Unsteady transonic flow behind an axisymmetric afterbody equipped with two boosters. In: 25th AIAA applied aerodynamics conference, Miami, FL, AIAA Paper 2007–4564, 103–115 (2007)
12. Mittal, R., Iaccarino, G.: Immersed boundary methods. Annu. Rev. Fluid Mech. **37**, 229–261 (2005)
13. Mochel, L., Weiss, P.-É., Deck, S.: Zonal immersed boundary conditions: application to a high reynolds number afterbody flow. AIAA J. (2014). doi:10.2514/1.J052970
14. Mohd-Yusof, J.: Combined immersed-boundary/B-spline methods for simulations of flows in complex geometries, pp. 317–328. Annual Research Briefs, Center for Turbulence Research (1997)
15. O'Rourke, J.: Computational Geometry in C. Cambridge University Press, Cambridge (1998)
16. Péchier, M., Guillen, P., Caysac, R.: Magnus effect over finned projectiles. J. Spacecr. Rockets **38**(4), 542–549 (2001)
17. Peskin, C.S.: Flow patterns around heart valves: a numerical method. J. Comp. Phys. **10**, 252–271 (1972)
18. Roux, A., Reichstadt, S., Bertier, N., Gicquel, L., Vuillot, F., Poinsot, T.: Comparison of numerical methods and combustion models for LES of a ramjet. Combust. Aerosp. Propul. **337**(6–7), 313–572 (2009)
19. Sainte-Rose, B., Bertier, N., Deck, S., Dupoirieux, F.: Numerical simulations and physical analysis of an overexpanded reactive gas flow in a planar nozzle. Combust. Flame **159**, 2856–2871 (2012)
20. Simon, F., Deck, S., Guillen, P., Sagaut, P., Merlen, A.: Numerical simulation of the compressible mixing layer past an axisymmetric trailing edge. J. Fluid Mech. **591**, 215–253 (2007)
21. Spalart, P., Jou, W.H., Strelets, M., Allmaras, S.R.: Comments on the feasibility of LES for wings and on a hybrid RANS/LES approach. In: Proceedings of the 1st AFSOR International Conference on DNS/LES, Ruston, 137–147 (1998)
22. Verzicco, R., Mohd-Yusof, J., Orlandi, P., Haworth, D.C.: Large Eddy Simulation in complex geometries using boundary body forces. AIAA J. **38**, 427–433 (2000)
23. Verzicco, R., Fatica, M., Iaccarino, G., Moin, P., Khalighi, B.: Large Eddy Simulation of a road-vehicle with drag reduction devices. AIAA J. **40**, 2447–2455 (2002)
24. Vuillot, F., Houssen, F., Manoha, E., Redonnet, S., Jacob, J.: Applications of the CEDRE unstructured flow solver to landing gear unsteady flow and noise predictions. In: AIAA Paper 2011–2944, 17th AIAA/CEAS Aeroacoustics Conference, Portland, Oregon (2011)
25. Weiss, P.-É., Deck, S., Robinet, J.-C., Sagaut, P.: On the dynamics of axisymmetric turbulent separating/reattaching flows. Phys. Fluids. **21**(075103) (2009)
26. Weiss, P.-É., Deck, S.: Control of the antisymmetric mode (m = 1) for high Reynolds axisymmetric turbulent separating/reattaching flows. Phys. Fluids. **23**(095102) (2011)
27. Yang, J., Balaras, E.: An embedded-boundary formulation for large-eddy simulation of turbulent flows interacting with moving boundaries. J. Comp. Phys. **215**, 12–40 (2006)
28. Zhu, W.J., Behrens, T., Shen, W.Z., Sørensen, J.N.: Hybrid immersed boundary method for airfoils with a trailing-edge flap. AIAA J. **51**(1), 30–41 (2013)

Two Non-zonal Approaches to Accelerate RANS to LES Transition of Free Shear Layers in DES

Charles Mockett, Marian Fuchs, Andrey Garbaruk, Michael Shur, Philippe Spalart, Michael Strelets, Frank Thiele and Andrey Travin

Abstract We present two novel approaches to improve the behaviour of DES in the region where an attached boundary layer (handled with RANS) flows into a separated shear layer (to be resolved using LES). The approaches aim to be generally-applicable and retain the non-zonal nature of DES. Furthermore, the formulations are local and can be readily implemented in general-purpose solvers. One approach introduces an adaptive grid scale definition, sensitised to the local vorticity orientation. The second approach, which can be combined with the first, involves the incorporation of alternative SGS model formulations that discern between quasi 2D and developed 3D flow states. Both modifications lead to a strong reduction of eddy viscosity in the early shear layer. Consequently, a significant acceleration of RANS to LES transition is demonstrated for a plane shear layer, a backwards-facing step and a round jet, with results from two different flow solvers shown. The greatest improvement is seen when the approaches are applied in combination.

1 Introduction

Since its conception in 1997 [15], the Detached-Eddy Simulation (DES) method, with later extensions [10, 14] increasing its robustness and widening its applicability, has become a powerful computational tool. Particularly for massively-separated flows at

C. Mockett (✉) · M. Fuchs · F. Thiele
CFD Software Entwicklungs- und Forschungsgesellschaft mbH, Berlin, Germany
e-mail: charles.mockett@cfd-berlin.com

A. Garbaruk · M. Shur · M. Strelets · A. Travin
New Technologies and Services, St. Petersburg State Polytechnic University,
St. Petersburg, Russian Federation

M. Strelets
e-mail: strelets@mail.rcom.ru

P. Spalart
Boeing Commercial Airplanes, Seattle, USA
e-mail: philippe.r.spalart@boeing.com

S. Girimaji et al. (eds.), *Progress in Hybrid RANS-LES Modelling*,
Notes on Numerical Fluid Mechanics and Multidisciplinary Design 130,
DOI 10.1007/978-3-319-15141-0_15

high Reynolds number, DES has delivered considerably improved accuracy relative to RANS at a fraction of the computational expense of pure LES (see e.g. [12] for a review of DES progress). However, a particular manifestation of the "grey area" problem anticipated from the outset has been shown to cause significant problems for flows such as shallow recirculation regions or jets (unless the grid is fine in all directions). Improvement on this front is the target of the current contribution.

We focus more precisely on problems in the "RANS to LES transition" (RLT) region of the early shear layer following separation of the attached boundary layer treated by RANS. Since, except in a few limited research studies, no fluctuations are explicitly introduced in the RLT zone, DES relies on the natural instability of the shear layer to generate resolved turbulence. This process is hampered by the production of high levels of eddy viscosity in standard DES formulations, particularly with typical grids that have loose spacing in the lateral direction compared with the shear layer thickness. This issue has led to the use of Implicit LES (ILES) for much jet-noise work (see e.g. [9]), but ILES is not satisfactory for flows in which the turbulence again interacts with bodies. Examples are jet-flap interaction on an airplane, or rotor-wake/body interaction on a helicopter. Pioneering work on the RLT problem has been carried out by Kok et al. [5, 6], with a temporal high-pass filter effectively reducing eddy viscosity in the early shear layer. This remedy however has its drawbacks, e.g. being incompatible with unsteady geometries. The strategies we pursue likewise seek to counteract excess eddy viscosity without compromising the model behaviour in other regards. In other words, we aim to retain generality of the method, not to derive case-specific fixes.

It is of course clear that despite these efforts, the turbulence in the RLT zone will remain somewhat pathological, being overwhelmingly two-dimensional and dominated by the Kelvin–Helmholtz instability as a necessary precursor to full, three-dimensional turbulence further downstream. Therefore, our goal is not a complete elimination of this "pseudo transition" region and corresponding flow artefacts but reduction of its length as much as allowed by the grid.

In the following, proposed approaches to mitigate the RLT problem are outlined. Due to space constraints, reference will be made to other publications for full formulations of some model expressions.

2 Proposed Approaches

For convenience we adopt a generalised notation for the LES models considered. For DES models in LES mode, the following form can be derived under the assumption of local equilibrium (i.e. equality of the dissipation and generation terms of the underlying transport equations):

$$\nu_{sgs} = \left(C_{sgs}\Delta\right)^2 \mathscr{D}_{sgs}(u), \tag{1}$$

Table 1 Summary of model formulations in LES mode

Model	C_{sgs}	$\mathscr{D}_{sgs}(u)$
Smagorinsky [11]	C_S	$\sqrt{2S_{ij}S_{ij}}$
WALE [7]	C_W	S^*_W
σ [8]	C_σ	S^*_σ
DES [14, 15]	$\sqrt{A}\ C_{DES}\ \Psi$	S^*_{RANS}
WALE-DES	$\sqrt{A}\ C_{DES}\ \Psi$	$B_W\ S^*_W$
σ-DES	$\sqrt{A}\ C_{DES}\ \Psi$	$B_\sigma\ S^*_\sigma$

For definitions of S^*_W and S^*_σ see Refs. [7, 8], respectively. Coefficient A depends on the underlying RANS model and may or may not be constant

where different choices of C_{sgs} (the calibrated model parameter), Δ (the grid scale measure) and $\mathscr{D}_{sgs}(u)$ (a differential operator acting on the resolved velocity field) give different SGS model formulations (see e.g. Table 1).

Two main strategies are pursued, which can be applied separately or in combination. The first proposes a modification of Δ and is described in Sect. 2.1. The second involves adopting different SGS model formulations (i.e. modification of $\mathscr{D}_{sgs}(u)$ and C_{sgs}) and is covered in Sect. 2.2.

2.1 Modification of the Grid Scale, Δ

In the following, we consider e.g. jet shear layers or trailing edges and define x, y and z to be aligned with the streamwise direction, across the shear layer and in the spanwise or azimuthal directions, respectively. In such situations, efficient grids are fine in the y direction and perhaps also in x, but coarse in z. This creates "pencil" or "ribbon" grid cells. The maximum cell diameter definition of Δ normally used with DES is

$$\Delta_{max} = \max(\Delta_x, \Delta_y, \Delta_z). \tag{2}$$

Although this is a rational and robust choice for archetypal LES in the inertial range with near-cubic cells, it turns out to be too "conservative" in the initial region of shear layers resolved by such anisotropic grids.

Chauvet et al. [1] introduced the promising concept of sensitising Δ to the orientation of the vorticity vector with the grid. The formulation was subsequently generalised for unstructured meshes by Deck [3]. In regions where the flow is essentially 2D with the vorticity axis aligned with the coarse z direction, their Δ_ω quantity reduces to $\sqrt{\Delta_x\Delta_y}$. Although removing the dominance of Δ_z here is helpful, we consider the strong influence of the smallest grid direction in this formulation troublesome. This is the same as our objection to the commonplace use of the cubic root of the cell volume, which was introduced by Deardorff [2] without logical justification. We therefore propose an alternative concept that reduces to $\max(\Delta_x, \Delta_y)$.

Considering a cell with its centre at $\mathbf{r}$ and vertices at $\mathbf{r}_n$ ($n = 1 \ldots 8$ for hexahedra), the proposed definition reads as:

$$\tilde{\Delta}_\omega = \frac{1}{\sqrt{3}} \max_{n,m=1,8} |(\mathbf{l}_n - \mathbf{l}_m)|, \tag{3}$$

where $\mathbf{l}_n = \mathbf{n}_\omega \times (\mathbf{r}_n - \mathbf{r})$ and $\mathbf{n}_\omega$ is the unit vector aligned with the vorticity vector. Thus, the quantity $\tilde{\Delta}_\omega$ is the diameter of the set of cross-product points $\mathbf{l}_n$ divided by $\sqrt{3}$.

As intended, in the shear layer situations outlined above it reduces to $\frac{1}{\sqrt{3}}(\Delta_x^2 + \Delta_y^2)^{1/2}$, i.e. is $O(\max\{\Delta_x, \Delta_y\})$. In 3D cases, $\tilde{\Delta}_\omega$ is of the order of Δ_{max} except for the situation when the vorticity vector is aligned with one of the grid coordinate directions (say, k), when it reduces to $O(\max\{\Delta_i, \Delta_j\})$. Therefore, the smallest grid-spacing never rules.

Another improvement over the original proposal [1, 3] occurs when the shear layer is skewed so that the vortex cores are not aligned with the z direction. In such a case, the Kelvin–Helmholtz instability cannot be resolved well, yet the original definition does not recognise this situation rapidly enough and keeps Δ excessively small.

Testing for isotropic turbulence on an isotropic grid showed that $\tilde{\Delta}_\omega$ indeed adopts on average 97.5 % of the value of Δ_{max}. To balance this, $\tilde{\Delta}_\omega$ can be multiplied by a factor 1.025 as shown in Sect. 3.2. The influence of this is however very minor.

2.2 Alternative Model Formulations in LES Mode

DES is formulated as a modification to an existing RANS model whereby the model's length scale, L_{RANS} is substituted by a DES length scale. In its original formulation [15], referred to as DES97, the DES length scale reads

$$L_{DES97} = \min(L_{RANS}, L_{LES}), \text{ with } L_{LES} = \Psi C_{DES} \Delta. \tag{4}$$

For the Spalart–Allmaras (SA) RANS model [13] employed here, the RANS length scale is defined by the wall-normal distance. C_{DES} is a model parameter analogous to the Smagorinsky constant and Ψ is a term that was added later [14] to compensate for unwanted activity of low-Re terms in LES mode. DES97 hence blends between RANS-mode near the wall and LES-mode far away from the wall, with the interface location determined by the grid resolution.

The subsequent "delayed DES" (DDES) formulation [14] extends this to include a shield function that detects attached turbulent boundary layers and aims to ensure RANS mode there irrespective of the grid resolution. The DDES length scale is formulated as

$$L_{DDES} = L_{RANS} - f_d \max(0, \; L_{RANS} - L_{LES}), \tag{5}$$

where the function f_d is close to 0 inside the boundary layer and blends rapidly to 1 near the boundary layer edge. Furthermore, in free shear flows $f_d = 1$ leading to $L_{DDES} = L_{LES}$.

As mentioned, DES can be mapped to a Smagorinsky model form in LES mode, for which several well-established shortcomings are known. Of greatest relevance here is the Smagorinsky model's inability to correctly handle laminar-to-turbulent transition, where its sensitivity to mean flow shear gives rise to high levels of eddy viscosity that attenuate the (resolved) transition process. Turning to DES, the same mechanism contributes to the RLT problem, hampering the development of resolved turbulence arising from the natural shear layer instability.

Several models without this shortcoming have been formulated by the LES research community, however many of these (e.g. dynamic and high-pass filtered models) involve non-local terms that are impractical for industrial CFD solvers. The WALE [7] and σ [8] models proposed by Nicoud and co-workers however seem particularly promising for further consideration in the context of this study. In these approaches local velocity gradient information is processed to distinguish between essentially two-dimensional situations such as plane shear, for which very low eddy viscosity is generated, and three-dimensional turbulence, where regular SGS model activity is recovered. This should offer a highly effective measure for accelerating RANS to LES transition whilst maintaining a practical and robust local formulation. Note that our goal in adopting the WALE and σ approaches in the LES mode of DES is exclusively targeted at such RLT improvement—the use of these models in a DES framework renders issues regarding their near-wall behaviour irrelevant, since the RANS branch of DES is active there. Finally, a further potential feature of this approach is that its activity is independent of the grid cell shape, whereas the $\tilde{\Delta}_\omega$ countermeasure is only effective for anisotropic meshes.

The key changes relative to the Smagorinsky model involve the differential operator acting on the velocity field ($\mathscr{D}_{sgs}(u)$ in Eq. 1). For brevity, these terms will be denoted simply as S_W^* and S_σ^* for the WALE and σ models respectively with reference made to the original publications for full details. Although derived with very different considerations, both return very low values of SGS viscosity in plane shear flows and involve negligible computational overhead. The analysis in Ref. [8] however indicates that the σ model possesses greater generality. We have decided to test both approaches for a range of complex flows before deciding for one formulation over the other.

To modify DES to behave like the WALE and σ models in LES mode, we leave the length scale substitution unmodified (Eq. 5 for DDES) and introduce an additional function to substitute the corresponding term for $\mathscr{D}_{sgs}(u)$ in the LES mode region only. The velocity gradient invariant in the underlying RANS model, S_{RANS}^* [1] is substituted by

$$S^*_{(W,\sigma)-DDES} = S^*_{RANS} - f_d \ \mathrm{pos}(L_{RANS} - L_{LES}) \ (S^*_{RANS} - B_{W,\sigma} \ S^*_{W,\sigma}), \qquad (6)$$

[1] For the Spalart–Allmaras model, $S_{RANS}^* = \sqrt{2\Omega_{ij}\Omega_{ij}}$ is substituted (not $\tilde{S}$).

where the operator used to detect DES97 RANS and LES mode acts as

$$\text{pos}(a) = \begin{cases} 0, & \text{if } a \leq 0 \\ 1, & \text{if } a > 0 \end{cases} . \tag{7}$$

Where the DDES shield function f_d is active, the values are blended smoothly according to the value of f_d. As such, the blending occurs at the same location as the length scale blending of DDES. Note that if the grid is sufficiently coarse that the interface between RANS and LES would occur outside the boundary layer (according to DES97), Eq. 6 gives a discontinuous switch between S^*_{RANS} on the RANS side and $B_{W,\sigma}\ S^*_{W,\sigma}$ on the LES side of the interface. On the other hand, if we would not apply the discontinuous pos-function in Eq. 6, the LES mode of the $S^*_{(W,\sigma)-DDES}$ term would be activated as soon as f_d deviates from zero, which would not be desirable considering the coarseness of the grid.

The resulting form of the model in LES mode is given in Table 1 with reference to Eq. 1. The factor $B_{W,\sigma}$ is included in $\mathcal{D}_{sgs}(u)$ such that C_{sgs} is equivalent between DES variants, since C_{DES} contributes to the RANS-LES interface determination and the pure LES models have widely different values of C_{sgs}. In principle the value of this parameter can be derived as $B_{W,\sigma} = C^2_{W,\sigma}/C^2_S$, which is checked for isotropic turbulence as shown in Sect. 3.2. For a given underlying RANS model and numerical method, a single value of C_{DES} is hence maintained for all variants. The expression for Ψ is unaltered by the WALE/σ modification.

3 Results

Both developed approaches are compared in isolation and in combination in two different flow solvers, and for a range of test cases.

3.1 Numerical Methods Applied

3.1.1 OpenFOAM-Based Approach

A customised version of the open source CFD environment OpenFOAM[2] was used, which is a cell-centred, unstructured, finite-volume based code of second order accuracy in space and time. An incompressible solver has been applied for all test cases presented here. Relevant customised features include standardised implementations of DDES and the underlying SA model as well as the hybrid convection scheme

[2] www.openfoam.org.

of Travin et al. [16]. The latter is used to blend between low-dissipation second order central differencing in well-resolved turbulent regions and robust second order upwind differencing in coarse or irrotational regions.

3.1.2 NTS Code

The NTS in-house code [16] is a cell-vertex, finite-volume based code for structured multiblock and overset meshes. It is of 2nd order accuracy in time and of higher (4th/5th) order accuracy in space and employs dual time-stepping with implicit Gauss-Seidel relaxation by planes. The Rogers and Kwak or Roe flux difference splitting schemes are used. Except for the jet case, the incompressible branch of the code has been employed here. As for the OpenFOAM approach, the hybrid convection scheme of Travin et al. is used, however in the NTS code the blending is between 4th order central differencing and 3rd or 5th order upwind schemes.

3.2 Calibration and Verification of Basic Functionality

Whilst improving RLT behaviour, the proposed modifications give unchanged functionality in "fully-developed" LES turbulence. This is demonstrated using decaying isotropic turbulence, for which model and code-specific values of C_{sgs} have been calibrated. Following calibration, the strong similarity of turbulent spectra from different models and Δ formulations is evident in Fig. 1.

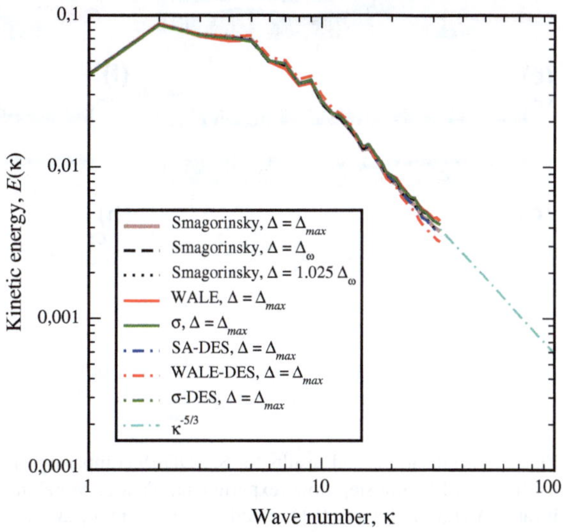

Fig. 1 Comparison of spectra for decaying isotropic turbulence obtained on a 64^3 grid from all model variants using calibrated constant values. Also shown are spectra obtained using $\tilde{\Delta}_\omega$, with and without a coefficient of 1.025 (OpenFOAM results)

Using a suitable flat plate grid, it was also found that the DDES shield function f_d needs adjustment for the SA-σ-DDES and SA-WALE-DDES variants. Recalibrating the value of C_{d1} to 10 gave equivalent shielding behaviour to standard SA-DDES in both cases.

3.3 Backward-Facing Step

DDES of the flow over a backward-facing step has been conducted to compare the Δ_{max} and $\tilde{\Delta}_\omega$ grid scale definitions using the NTS code. The grid resolution in the homogeneous spanwise direction has furthermore been varied between $\Delta_z/H = 0.1$ and $\Delta_z/H = 0.05$, where H is the step height.

Visualisations of the grid and the activity of the $\tilde{\Delta}_\omega$ quantity are shown in Fig. 2a, b, respectively. In the very early shear layer, where the grid cells are highly anisotropic and the vorticity vector is aligned with the z direction, $\tilde{\Delta}_\omega$ gives significantly reduced values relative to Δ_{max}. Further downstream, where three-dimensional resolved turbulence has developed, values of between around 70 and 85 % of Δ_{max} are seen. Correspondingly, eddy viscosity levels (Fig. 2c, d) are strongly reduced in the early

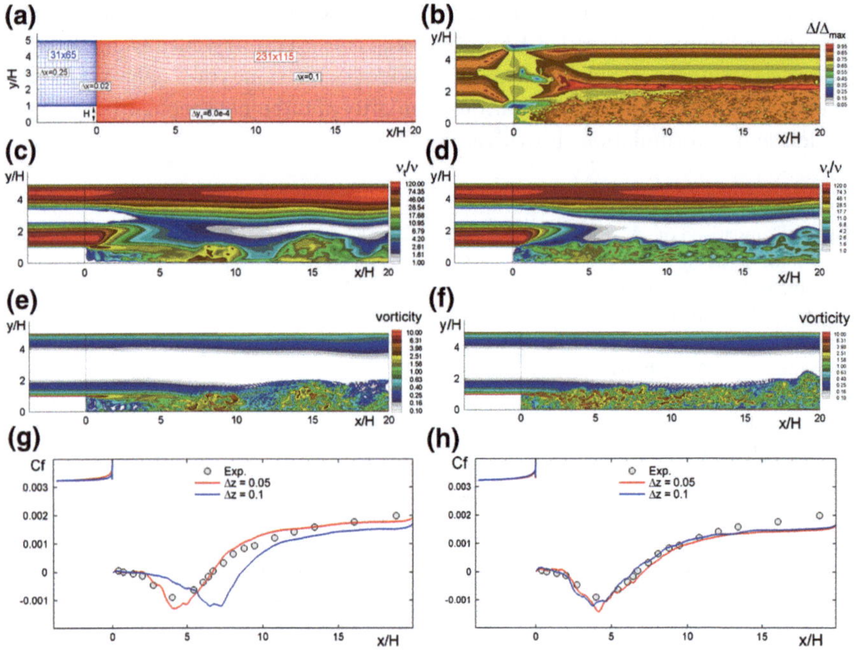

Fig. 2 Visualisations and results for SA-DDES comparing the Δ_{max} and $\tilde{\Delta}_\omega$ grid scale measures for backward-facing step flow (experimental data of Vogel and Eaton [17]). **a** x/y plane of grid, **b** ratio $\tilde{\Delta}_\omega/\Delta_{max}$ ($\Delta_z = 0.1$), **c** eddy viscosity ratio, Δ_{max} ($\Delta_z = 0.1$), **d** eddy viscosity ratio, $\tilde{\Delta}_\omega$ ($\Delta_z = 0.1$), **e** vorticity magnitude, Δ_{max} ($\Delta_z = 0.1$), **f** vorticity magnitude, $\tilde{\Delta}_\omega$ ($\Delta_z = 0.1$), **g** mean skin friction, Δ_{max}, **h** mean skin friction, $\tilde{\Delta}_\omega$

shear layer, enabling a significantly accelerated development of resolved structures (Fig. 2e, f). Owing to the DDES shield function, the eddy viscosity levels in the attached boundary layers (treated with RANS) are unaffected, as intended.

As seen from the profiles of mean skin friction coefficient on the lower wall (Fig. 2g, h), the agreement with experiment is improved significantly by the $\tilde{\Delta}_\omega$ formulation for the coarser $\Delta_z/H = 0.1$ mesh. Furthermore, the strong sensitivity of the Δ_{max} results to the spanwise mesh resolution is dramatically reduced using the $\tilde{\Delta}_\omega$ expression, which is a highly desirable result.

3.4 Plane Shear Layer

A plane shear layer issuing from the training edge of a flat splitter plate is used as a fundamental test case for examining the RLT problem. Indeed, this flow has been employed by Deck [3] to demonstrate the effectiveness of the original Δ_ω formulation and by Kok and van der Ven [5, 6] to investigate their high-pass filtered and stochastic SGS model approaches for RLT acceleration.

OpenFOAM simulations comparing all combinations of the proposed methods have been carried out on the same coarse structured mesh used by Kok and van der Ven, consisting of some 1.3 m cells. The region of interest extends 1 m downstream of the trailing edge (located at $x = 0$ m) and has a constant streamwise and spanwise resolution of 3.125 mm (gradual streamwise coarsening is applied for $x > 1$ m). The initial shear layer vorticity thickness of $\delta_\omega \approx 7.5$ mm is therefore highly under-resolved in these directions. It is however well-resolved by around 40 cells in the direction traversing the shear layer. The point distribution across the shear layer is gradually fanned out downstream to track the shear layer growth, measurements of which are plotted in Fig. 6. The spanwise domain size is 0.15 m. The running lengths of the (shielded RANS) boundary layers have been adjusted to give velocity profiles at the plate trailing edge in agreement with those measured in the experiment of Delville [4].

A good impression of the effectiveness of the proposed measures can be obtained from Fig. 3, which shows snapshots of resolved vortical structures identified using the Q criterion (which was expressly conceived to distinguish vortices from a vorticity layer). For standard DDES with Δ_{max}, no resolved vortices can be observed within the focus region downstream of the trailing edge. The WALE-DDES and σ-DDES approaches bring a strong improvement (with the σ formulation proving most effective), and the $\tilde{\Delta}_\omega$ definition delivers further RLT acceleration to all three models.

Snapshots of eddy viscosity ratio showing the entire shear layer focus region (from $0 < x < 1$) are contrasted between the standard formulation and the most effective combined approach in Fig. 4. The strong initial reduction of eddy viscosity as well as the activation of regular SGS levels following the emergence of 3D resolved structures can be seen in the latter case.

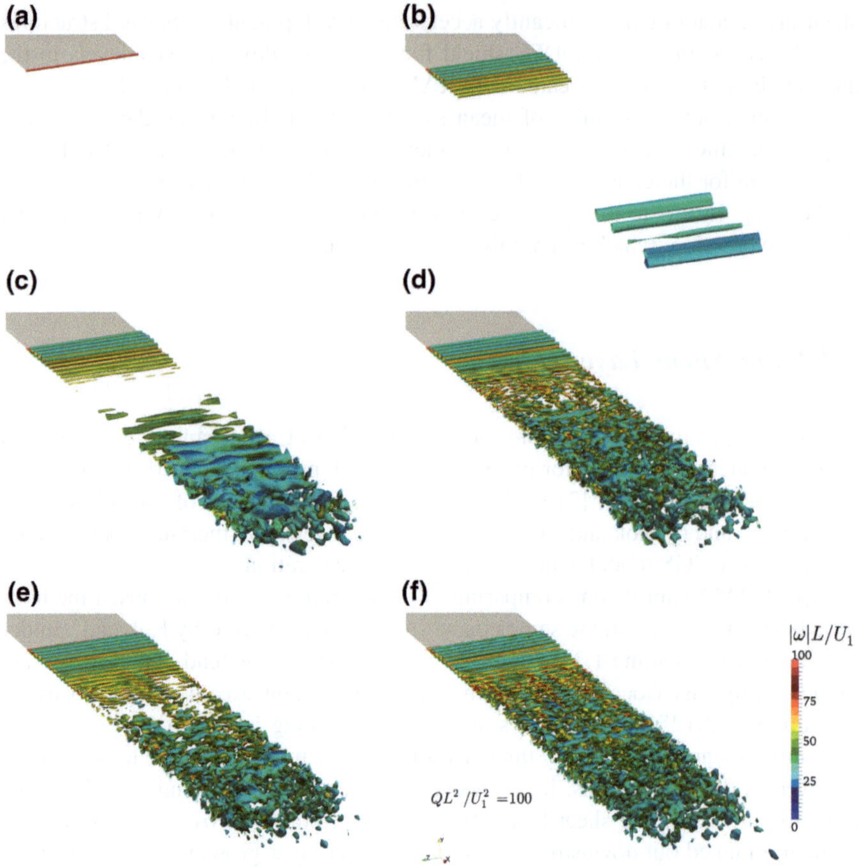

Fig. 3 Resolved vortical structures from different model variants for the plane shear layer test case. **a** DDES, Δ_{max}, **b** DDES, $\tilde{\Delta}_\omega$, **c** WALE-DDES, Δ_{max}, **d** WALE-DDES, $\tilde{\Delta}_\omega$, **e** σ-DDES, Δ_{max}, **f** σ-DDES, $\tilde{\Delta}_\omega$

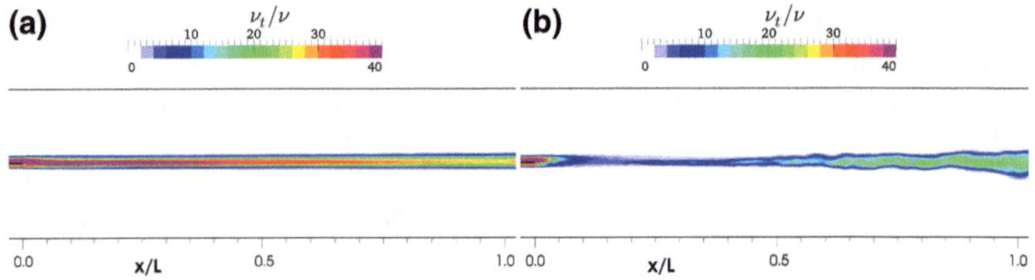

Fig. 4 Instantaneous eddy viscosity ratio for selected simulations of the plane shear layer. **a** DDES, Δ_{max}, **b** σ-DDES, $\tilde{\Delta}_\omega$

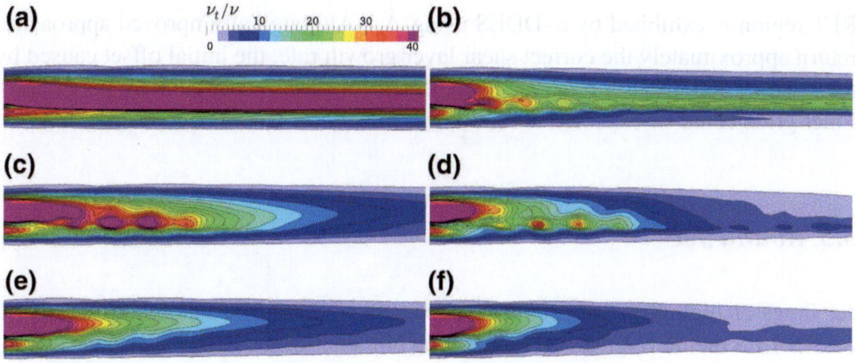

Fig. 5 Instantaneous eddy viscosity ratio for the plane shear layer (zoom in region $0 < x < 0.2$). **a** DDES, Δ_{max}, **b** DDES, $\tilde{\Delta}_\omega$ **c** WALE-DDES, Δ_{max} **d** WALE-DDES, $\tilde{\Delta}_\omega$ **e** σ-DDES, Δ_{max} **f** σ-DDES, $\tilde{\Delta}_\omega$

To shed more light on the differences between the approaches, the instantaneous eddy viscosity is plotted in Fig. 5 with magnification of the initial shear layer region. The reduction of the length scale achieved by the $\tilde{\Delta}_\omega$ definition on this anisotropic grid gives rise to overall lower eddy viscosity levels and a faster attenuation of the eddy viscosity convected from the RANS boundary layers. The effective elimination of eddy viscosity production caused by the alternative formulations of $\mathscr{D}_{sgs}(u)$ is also clearly visible. A differentiation between WALE-DDES and σ-DDES is seen for the initial Kelvin–Helmholtz vortices; WALE-DDES applies high levels of eddy viscosity to these whereas σ-DDES does not, owing to their two-dimensionality. The effect of this can be clearly seen by comparison e.g. of Fig. 3c, e.

The overall picture is confirmed by Fig. 6, in which the time and spanwise-averaged vorticity thickness of the shear layer is compared between approaches and with experimental measurements and RANS. The most rapid recovery from the

Fig. 6 Comparison of shear layer vorticity thickness δ_ω between DDES variants, with RANS and with measurements [4]

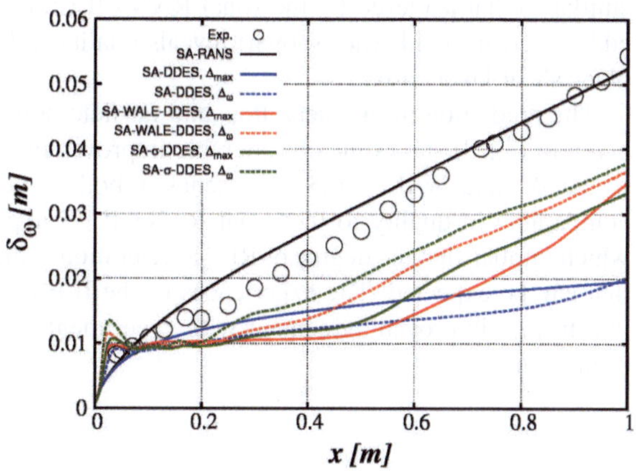

RLT region is exhibited by σ-DDES using $\tilde{\Delta}_\omega$. Although all improved approaches return approximately the correct shear layer growth rate, the initial offset caused by the residual RLT region remains, as expected (cf. Sect. 1, penultimate paragraph). Finer grids could be expected to reduce this offset however.

3.5 Round Jet

Jet flows at high Reynolds numbers represent a significant challenge for hybrid RANS-LES methods due to the dominating influence of the RLT issue on the overall prediction. Furthermore, turbulence-resolving simulation of such flows is key for complex jet noise applications. A zonal coupling of RANS (for the nozzle boundary layers) with ILES (in the jet plume) has previously been employed [9] as a means to handle the RLT area. New simulations are presented in which the effectiveness of the $\tilde{\Delta}_\omega$ and WALE-DES measures are assessed individually and in combination. The compressible simulations of an unheated, single-stream, static round jet at $M = 0.9$, $Re_D = 1.1 \times 10^6$ were conducted on a fairly coarse mesh of 4.2 m cells (mesh "G2" from Ref. [9]) using the NTS code. As for the plane shear layer case, the initial shear layer of the jet is strongly under-resolved and the grid cells are highly anisotropic. Unlike the shear layer however, the initial grid cells are "ribbon" rather than "book" shape, with streamwise, minimum-radial and azimuthal resolutions of respectively 0.009, 0.003 and 0.04 (normalised by the jet diameter, D). The mesh has 80 cells in the azimuthal direction. Steady RANS profiles from a precursor simulation are applied at the nozzle exit plane. These include the boundary layer eddy viscosity and the approach therefore effectively mimics a non-zonal coupled DDES of the nozzle and jet plume flows.

As seen in snapshots (Fig. 7), the development of turbulent structures in the jet shear layer is significantly promoted by both the $\tilde{\Delta}_\omega$ and WALE-DES measures. Furthermore, as seen for the plane shear layer in Sect. 3.4, the combination of both approaches yields further improvement and the speed of shear layer development is similar to that achieved by the zonal RANS-ILES of Ref. [9]. Examination of the eddy viscosity fields (not shown) reveals qualitatively similar behaviour as in the plane shear layer case.

The prediction of the mean flow and far-field noise (selected plots of which are given in Fig. 8) shows the considerable improvement achieved by the combination of the $\tilde{\Delta}_\omega$ and WALE-DES approaches in both respects. These results are highly comparable in quality to the zonal RANS-ILES approach for this sensitive flow, which implies that the degree of RLT-acceleration achieved is close to the maximum that can be expected. The experiences of the ILES study [9] furthermore indicate that further improved agreement with experiment could be expected through grid refinement.

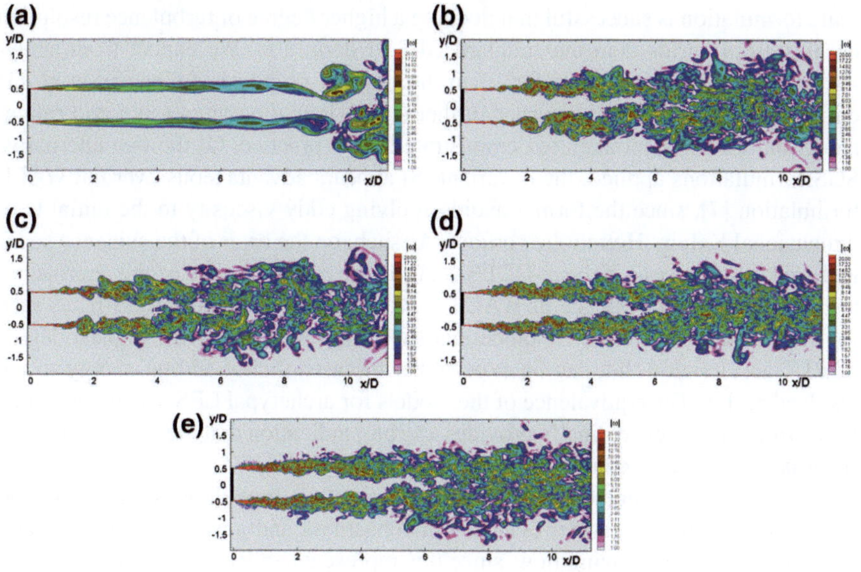

Fig. 7 Instantaneous contours of vorticity magnitude for the round jet case. **a** DES, Δ_{max}, **b** DES, $\tilde{\Delta}_{\omega}$ **c** WALE-DES, Δ_{max}, **d** WALE-DES, $\tilde{\Delta}_{\omega}$, **e** zonal RANS-ILES, from [9]

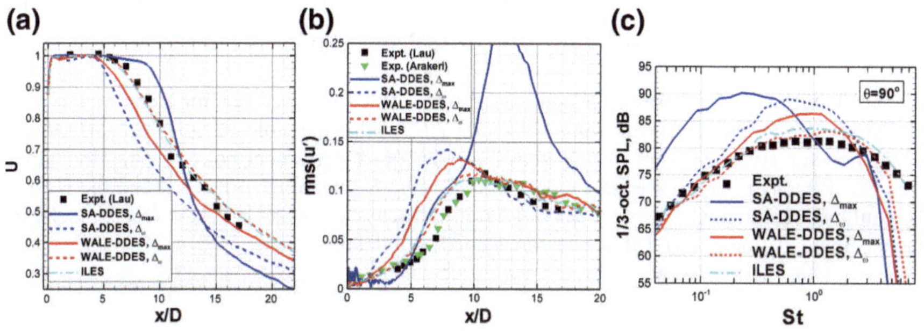

Fig. 8 Mean flow, turbulence and far-field noise for round jet case, comparing current results with the zonal RANS-ILES of Shur et al. [9]. **a** Mean centreline velocity, **b** centreline $\overline{u'u'}$, **c** 1/3-octave far-field spectra

4 Conclusion and Outlook

Practical remedies to the RANS to LES transition (RLT) problem have been derived, which maintain both the non-zonal nature as well as the local formulation of DDES. The effectiveness of these has been demonstrated for a range of flows exhibiting an influential RLT zone with significant improvements over standard DDES seen. The greatest RLT acceleration is furthermore achieved when the approaches are applied in combination. Results indicate that the proposed vorticity-adaptive grid scale formulation is successful in unlocking a higher degree of turbulence resolution on anisotropic grids than the standard DES Δ definition. We expect from analysis that our formulation is "safer" than the original proposal of Chauvet et al. [1] on very ribbon-like grids, however further study (including aggressive grid refinement) should be undertaken to demonstrate this in practice. Of the two alternative SGS formulations applied, the σ variant [8] appears advantageous over the WALE formulation [7], since the former avoids applying eddy viscosity to the initial two-dimensional Kelvin–Helmholtz vortices. As such, on the basis of the evidence so far gathered, it appears as if the σ-DDES + $\tilde{\Delta}_\omega$ is the most effective of the approaches tested. The near equivalence of WALE + $\tilde{\Delta}_\omega$ results with zonal RANS-ILES seen for the jet flow leads us to conclude that such approaches offer the highest degree of RLT acceleration that can be expected without explicitly adding energy to the resolved scales. The equivalence of the models for archetypal LES, as demonstrated for isotropic turbulence, finally provides a strong indication that the improvement of shear layer behaviour comes at no cost to generality.

Future work will include testing for more complex, three-dimensional flows and geometries to provide further evidence of robustness and generality. Of particular interest will be a delta wing flow, since this represents an RLT-dominated situation (as shown e.g. in [6]) with a significant topological departure from planar shear layers. Further comparison of the WALE and σ variants will furthermore serve to confirm the initial conclusion arising from this work of a preference for the latter. Finally, testing of these new approaches within the framework of IDDES [10] for wall-modelled LES (e.g. channel flow) will be carried out.

Acknowledgments The research leading to these results has received funding from the European Union Seventh Framework Programme *FP7/2007-2013* within the project Go4Hybrid ("Grey Area Mitigation for Hybrid RANS-LES Methods") under grant agreement no. 605361. The authors from St. Petersburg acknowledge support from Boeing Commercial Airplanes and the Russian Scientific Foundation (Grant 14-11-00060). Fruitful discussions with Dr. J. Kok (NLR), including the recommendation of Ref. [8], are acknowledged with thanks.

References

1. Chauvet, N., Deck, S., Jacquin, L.: Zonal detached eddy simulation of a controlled propulsive jet. AIAA J. **45**(10), 2458–2473 (2007)
2. Deardorff, J.: A numerical study of three-dimensional turbulent channel flow at large Reynolds numbers. J. Fluid Mech. **41**(2), 453–480 (1970)
3. Deck, S.: Recent improvements in the zonal detached-eddy simulation (ZDES) formulation. Theoret. Comput. Fluid Dyn. **26**, 523–550 (2012)
4. Delville, J.: La décomposition orthogonal aux valeurs propres et l'analyse de l'organisation tridimensionnelle de écoulements turbulents cisaillés libres. PhD thesis, Université de Poitiers (1995)
5. Kok, J., van der Ven, H.: Destabilizing free shear layers in X-LES using a stochastic subgrid-scale model. Technical Report NLR-TP-2009-327, National Aerospace Laboratory NLR (2009)

6. Kok, J., van der Ven H.: Capturing free shear layers in hybrid RANS-LES simulations of separated flow. Technical Report NLR-TP-2012-333, National Aerospace Laboratory NLR, (2012)
7. Nicoud, F., Ducros, F.: Subgrid-scale stress modelling based on the square of the velocity gradient tensor. Flow Turbul. Combust. **62**(3), 183–200 (1999)
8. Nicoud, F., Toda, H., Cabrit, O., Bose, S., Lee, J.: Using singular values to build a subgrid-scale model for large eddy simulations. Phys. Fluids **23**(8), 5106 (2011)
9. Shur, M., Spalart, P., Strelets, M.: LES-based evaluation of a microjet noise reduction concept in static and flight conditions. J. Sound Vib. **330**(17), 4083–4097 (2011)
10. Shur, M., Spalart, P., Strelets, M., Travin, A.: A hybrid RANS-LES approach with delayed-DES and wall-modelled LES capabilities. Int. J. Heat Fluid Flow **29**(6), 1638–1649 (2008)
11. Smagorinsky, J.: General circulation experiments with the primitive equations. Mon. Weather Rev. **91**(3), 99–164 (1963)
12. Spalart, P.: Detached-eddy simulation. Annu. Rev. Fluid Mech. **41**, 181–202 (2009)
13. Spalart, P., Allmaras, S.: A one-equation turbulence model for aerodynamic flows. La recherche aérospatiale **1**(1), 5–21 (1994)
14. Spalart, P., Deck, S., Shur, M., Squires, K., Strelets, M., Travin, A.: A new version of detached-eddy simulation, resistant to ambiguous grid densities. Theoret. Comput. Fluid Dyn. **20**(3), 181–195 (2006)
15. Spalart, P., Jou, W., Strelets, M., Allmaras, S.: Comments of feasibility of LES for wings, and on a hybrid RANS/LES approach. In: Advances in DNS/LES, vol. 1 (1997)
16. Travin, A., Shur, M., Strelets, M., Spalart, P.: Detached-eddy simulations past a circular cylinder. Flow Turbul. Combust. **63**(1), 293–313 (2000)
17. Vogel, J., Eaton, J.: Combined heat transfer and fluid dynamic measurements downstream of a backward-facing step. J. Heat Transfer **107**(4), 922–929 (1985)

On the Interface Positioning in a Zonal Detached Eddy Simulation (ZDES) of a Spatially Developing Flat Plate Turbulent Boundary Layer

Nicolas Renard and Sébastien Deck

Abstract Hybrid RANS/LES methods may be used as Wall-Modeled Large Eddy Simulations to predict wall-bounded turbulent dynamics at a computational cost compatible with complex applications. The present study investigates the sensitivity to the RANS/LES interface position in one such method, the mode III of the ZDES technique (see Deck, Theor Comput Fluid Dyn 26(6):523–550, 2012 [2]). A canonical spatially developing flat plate boundary layer is simulated over a wide range of Reynolds numbers $3{,}150 \leq Re_\theta \leq 14{,}000$ at a low Mach number enabling comparisons with the incompressible case. The method is assessed on relatively coarse meshes and is compared to a Wall-Resolved Large Eddy Simulation together with experimental data. The influence of the interface position on skin friction prediction and the ratio of the resolved to modeled turbulent friction is discussed in the framework of the FIK identity (Fukagata, K., Iwamoto, K., Kasagi, N.: Contribution of Reynolds stress distribution to the skin friction in wall-bounded flows. Phys. Fluids **14**, L73 (2002) [7]). The mean velocity and Reynolds stresses profiles are analyzed, including spectral analysis of the streamwise velocity and Reynolds shear stress.

1 Hybrid RANS/LES Methods as Wall-Modeled LES

Mean flow quantities in attached turbulent boundary layers may be predicted by RANS simulations, but resolving the most energetic turbulent fluctuations is useful in cases such as mild flow separation, dynamic loading or noise prediction. A DNS and even a Wall-Resolved LES (WRLES) at high Reynolds numbers is extremely expensive, mainly because of the very small structures to be resolved in the inner layer [3, 19]. This motivates Wall-Modeled LES (WMLES), where the inner layer

N. Renard (✉) · S. Deck
ONERA, The French Aerospace Lab, 92190, Meudon, France
e-mail: nicolas.renard@onera.fr

S. Deck
e-mail: sebastien.deck@onera.fr

S. Girimaji et al. (eds.), *Progress in Hybrid RANS-LES Modelling*,
Notes on Numerical Fluid Mechanics and Multidisciplinary Design 130,
DOI 10.1007/978-3-319-15141-0_16

is modeled and most of the turbulent kinetic energy in the outer layer is resolved. Reviews of wall models may be found in [20] and [19].

Some WMLES techniques may resort to dynamical wall models or to a set of simplified equations solved in the vicinity of the wall to provide the proper wall fluxes (see e.g. [18]). Such methods may lack universality, and their use in complex geometries is not straightforward, contrary to the promising unified way of dealing with complex flows that is provided by hybrid RANS/LES methods. These methods, very efficient in massively separated flows, may be used as WMLES, where the inner layer is represented by a RANS zone, while the outer layer is resolved in LES.

Only a few examples from the vast literature devoted to this topic are mentioned here. One major issue is the role played by the RANS/LES interface lying within the boundary layer, where the solution transitions from LES-filtered values (outer layer) to the Reynolds average (inner layer). Detached Eddy Simulations with a passive interface by [16] led to an unphysical phenomenon called *log-layer mismatch*. Many corrections have been studied, whereby the interface is made active: stochastic forcing by [10], additive filtering by [22] and [21], but also additional filtering from a heterogeneous LES filter which reduces to the Reynolds average near the wall by [8].

The RANS/LES interface is sometimes treated in a passive way instead. RANS and LES models have been blended by [1] in the case of a supersonic boundary layer. Another example is the Improved Delayed Detached Eddy Simulation of [23], which switches automatically to a WMLES behavior if turbulent inflow is provided and if the grid resolution can support resolved fluctuations. Tests in both channel flows [23] and spatially developing boundary layers [24] suggest that the *log-layer mismatch* issue is mitigated, but deviations from the logarithmic law are still present. Moreover, the method automatically switches between the WMLES and DDES branches, but in order to perform a WMLES simulation using IDDES, imposing turbulent fluctuations at the inflow may be needed in attached boundary layers, leading to a zonal approach to some of the applications (embedded WMLES).

2 Mode III of Zonal Detached Eddy Simulation (ZDES)

The present study focuses on Zonal Detached Eddy Simulation, which provides a broad framework to treat complex flows by defining zones of interest as detailed in [2]. The ZDES mode III used here is a WMLES technique able to perform not only WMLES [3] but also WRLES [4] simulations of spatially developing boundary layers.

The turbulence model is the Spalart-Allmaras model [25], used as a RANS model in the inner layer where the characteristic length $\tilde{d}^{III}_{ZDES}$ is the wall distance d_w, and as a subgrid scale model in the outer layer, where the characteristic length depends on the mesh size $\Delta = (\Delta x \Delta y \Delta z)^{1/3}$. The user defines the height of the RANS/LES interface $d_w^{\text{interface}}$:

$$\tilde{d}^{III}_{ZDES} = \begin{cases} d_w \; if \, d_w < d_w^{interface} \\ \min(d_w, C_{DES}\Delta) \text{ otherwise} \end{cases} \quad (1)$$

with $C_{DES} = 0.65$. The RANS/LES interface is treated in a passive way (no extra filtering or forcing), which is simple, robust, numerically stable, does not require calibration of a forcing or filtering, and generates no spurious acoustics other than the turbulent inflow generation. However, since the difference between RANS (inner) and LES (outer) variables is neglected, some of the physics may be missing. The height of the interface has to be carefully optimized to minimize the error associated to the neglected terms.

It has been shown in [5] that the interface height should be imposed independently from the grid resolution, so that mesh refinement procures a monotonous decay of the error. This is facilitated by the zonal framework of ZDES. Moreover, Deck et al. [3] concluded that the interface seems to be better positioned at a constant outer-scaled rather than inner-scaled wall distance. The interface positioning is further investigated in the present study, especially regarding the resolved physics. ZDES calculations on relatively coarse meshes are compared with a WRLES made on a much finer mesh (from [4]).

3 Case Definition and Flow Visualization

A ZDES mode III simulation of a canonical spatially developing zero pressure gradient flat plate turbulent boundary layer over a wide range of Reynolds numbers $3{,}150 \leq Re_\theta \leq 14{,}000$ is made. The computational domain is $350\delta_0$ long, with δ_0 the inflow boundary layer thickness. Spatial development increases the demand on the method compared with channel flow tests, and is mandatory for validation of the technique before use in applied external aerodynamics. The finite-volume compressible flow solver FLU3M is used with a low-dissipation version of the AUSM+P numerical scheme by [14], and second-order time integration. The compressibility effects are negligible ($M_\infty = 0.21$). The inflow is provided by the Synthetic Eddy Method of [9] adapted by [17] and [11].

Three interface positions are considered: $y_{\text{interface}} = 0.1\delta$, $y^+_{\text{interface}} = 200$ and $y^+_{\text{interface}} = 3.9\sqrt{Re_\tau}$. The mesh has a $\Delta x^+ = 200$ ($\Delta x/\delta_0 = 0.212$) streamwise- and $\Delta z^+ = 100$ ($\Delta z/\delta_0 = 0.107$) spanwise spacing. The first mesh cell is located within the first wall unit from the wall. On the coarse mesh ($N_{xyz} = 30 \times 10^6$), the ZDES mode III technique performs a WMLES. The interface positions are compared in Fig. 1. Comparisons are made with a ZDES simulation on a much finer mesh (50^+ streamwise- and 12^+ spanwise spacing, $N_{xyz} = 800 \times 10^6$), which provides WRLES results with most of the near-wall dynamics resolved.

The flow fields from both meshes may be compared in Fig. 2, showing an isosurface of the instantaneous streamwise velocity $u/U_\infty = 0.8$ near the $Re_\theta = 13{,}000$ station, colored by the distance from the wall y/δ. Two numerical Schlieren (density gradient magnitude) are also provided in the (x, y) and (y, z) planes, where

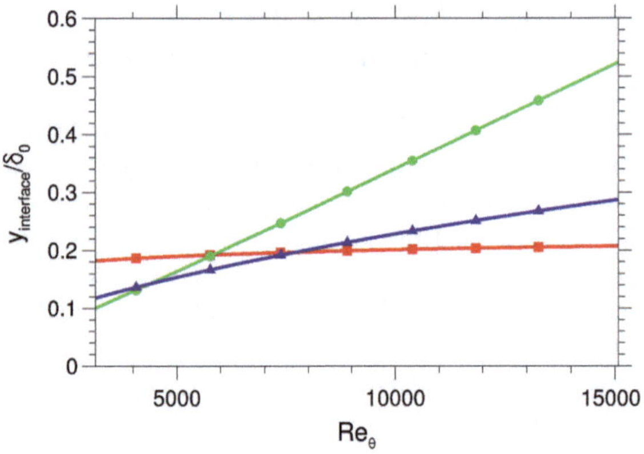

Fig. 1 RANS/LES interface positioning as a function of the Reynolds number Re_θ, in terms of the incoming boundary layer thickness δ_0. —■— $y^+_{\text{interface}} = 200$; —●— $y_{\text{interface}} = 0.1\delta$; —▲— $y^+_{\text{interface}} = 3.9\sqrt{Re_\tau}$.

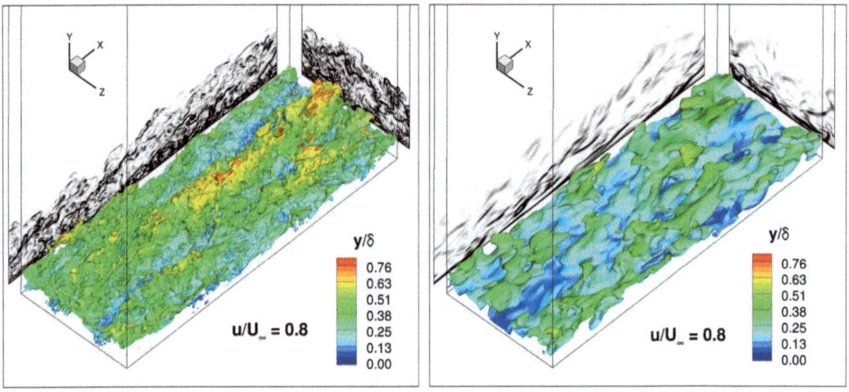

Fig. 2 Isosurface of the streamwise velocity $u/U_\infty = 0.8$ colored by the wall distance y/δ, near $Re_\theta = 13{,}000$, and numerical Schlieren (density gradient magnitude). *Left* fine mesh $(50^+/12^+)$ ZDES. *Right* coarse mesh $(200^+/100^+)$ ZDES, $y_{\text{interface}} = 0.1\delta$

x, y, z refer to the streamwise, wall-normal and spanwise directions respectively. Even though the simulation on the coarse mesh does not provide the same degree of resolution of small-scale structures, it should be noted that the large-scale structures that are resolved have common features with those resolved on the finer mesh. Especially the inclination angle in the (x, y) plane of the hairpin packets, but also the tendency of hairpin packets to adopt preferential spatial organizations in the form of very long and meandering streets may be recognized to some extent in the simulation on the coarser mesh Fig. 2.

4 Skin Friction Prediction

The skin friction coefficient defined by $C_f = \frac{\tau_w}{\frac{1}{2}\rho U_\infty^2}$ with $\tau_w = \mu \frac{\partial \langle u \rangle}{\partial y}(y = 0)$, is compared for all ZDES simulations with a RANS simulation using the Spalart-Allmaras model in Fig. 3. The Coles-Fernholz empirical correlation fitted to experimental data by [15] is also shown. This reveals that the $y^+_{\text{interface}} = 200$ interface tends to lead to the underestimation of C_f at high Reynolds numbers, compared with the $y_{\text{interface}} = 0.1\delta$ interface. The $y^+_{\text{interface}} = 3.9\sqrt{Re_\tau}$ interface features an intermediate behavior, which is not surprising if Fig. 1 is considered. All simulations are within a 5 % tolerance margin about the Coles-Fernholz correlation for Reynolds numbers higher than $Re_\theta = 5{,}200$. In terms of C_f versus Reynolds number trend, the $y_{\text{interface}} = 0.1\delta$ coarse mesh ZDES and the fine mesh ZDES are the closest to the trend of the Coles-Fernholz correlation.

The skin friction may be analyzed in the framework of the FIK identity [7], as exercised in [4] and [3]. In the case of an incompressible zero pressure gradient flat plate boundary layer (with the boundary layer hypotheses), it reads:

$$C_f = C_{f,1} + C_{f,2} + C_{f,3} \tag{2}$$

$$\begin{aligned} C_{f,1} &= \frac{4(1-\delta_1/\delta)}{Re_\delta} \\ C_{f,2} &= -4\int_0^1 \frac{\langle u'v' \rangle}{U_\infty^2}\left(1 - \frac{y}{\delta}\right) d\left(\frac{y}{\delta}\right) \\ C_{f,3} &= -2\int_0^1 \left(1 - \frac{y}{\delta}\right)^2 \left(\overline{I_x} + \frac{\partial \langle u \rangle}{\partial t}\right) \frac{\delta}{U_\infty^2} d\left(\frac{y}{\delta}\right) \end{aligned} \tag{3}$$

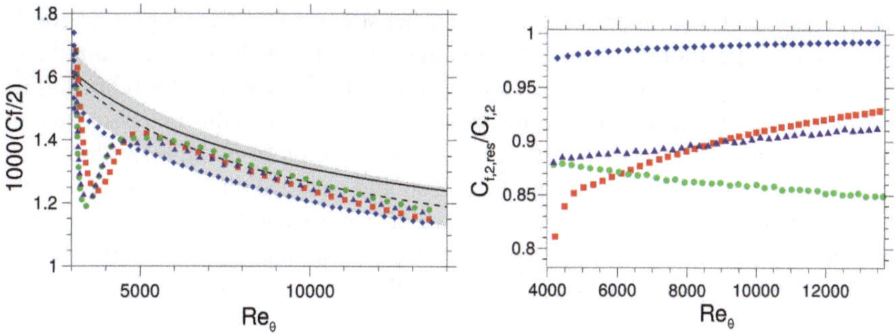

Fig. 3 Friction coefficient $1000\frac{C_f}{2}$ (*left*) and resolved fraction of the turbulent friction $\frac{C_{f,2,res}}{C_{f,2}}$ (*right*) according to the FIK identity [7] as a function of the Reynolds number Re_θ. – – – Coles-Fernholz correlation fitted to experimental data in [15]; —— S-A RANS; ♦ ZDES, fine mesh ($50^+/12^+$); ■ ZDES, coarse mesh ($200^+/100^+$), interface $y^+_{\text{interface}} = 200$; • ZDES, coarse mesh ($200^+/100^+$), interface $y_{\text{interface}} = 0.1\delta$; ▲ ZDES, coarse mesh ($200^+/100^+$), interface $y^+_{\text{interface}} = 3.9\sqrt{Re_\tau}$. The *shaded area* depicts a 5 % tolerance margin with respect to the Coles-Fernholz correlation

where $\overline{I_x} = \frac{\partial}{\partial x}\left(\langle u\rangle^2\right) + \frac{\partial}{\partial y}\left(\langle u\rangle\langle v\rangle\right)$, $Re_\delta = \frac{\delta U_\infty}{\nu}$, $\langle u\rangle(y=\delta) = 0.99\,U_\infty$, and $\delta_1 = \int_0^\delta\left(1-\frac{\langle u\rangle}{U_\infty}\right)dy$. Most importantly, $C_{f,2}$ represents the weighted contribution of the Reynolds shear stress to mean friction.

It is possible to decompose $C_{f,2}$ into its resolved and modeled parts, because $\langle u'v'\rangle$ is partially resolved and partially reconstructed by the turbulence model. Figure 3 presents the ratio of the resolved part of $C_{f,2}$, noted $C_{f,2,res}$ to the total $C_{f,2}$. This emphasizes the negligible role played by the model on the fine mesh. On the coarse mesh, the resolved fraction decreases with the Reynolds number in the $y_{\text{interface}} = 0.1\delta$ case, while it increases in the $y^+_{\text{interface}} = 200$ case. However, as shown in Fig. 3, the overall C_f prediction in the $y^+_{\text{interface}} = 200$ case is not fully satisfying, because C_f seems to decrease too quickly with the Reynolds number. On the other hand, the $y_{\text{interface}} = 0.1\delta$ case provides a decreasing ratio $\frac{C_{f,2,res}}{C_{f,2}}$ with Re_θ, which is not fully satisfying either, because the higher this ratio, the smaller the role played by the model and the more important the physics actually resolved by the simulation. The third interface is positioned at $y^+_{\text{interface}} = 3.9\sqrt{Re_\tau}$. This provides a slightly increasing ratio $\frac{C_{f,2,res}}{C_{f,2}}$, which is satisfying because a proper interface positioning would be expected to lead to a behavior that does not depend on the Reynolds number, so that it can be used over different Reynolds number ranges. Moreover, it has already been seen in Fig. 3 that the prediction of the total C_f is reasonably good with this interface position.

5 Velocity Field Study

The velocity field is investigated using first and second order statistical moments, and spectral analysis. Figure 4a shows the mean streamwise velocity profiles for the three interface positions considered in this study, together with the RANS simulation profile and experimental data. The discrepancy in the outer velocity is related to C_f prediction, which directly impacts the value of U_∞/u_τ seen in this inner-scaled plot. Besides, the most remarkable feature is a slight deviation from the logarithmic law, especially near the RANS/LES interface. It should be noted that the deviation is the largest with the $y^+_{\text{interface}} = 200$ interface, whereas it is small for the $y_{\text{interface}} = 0.1\delta$ interface. The results from the $y^+_{\text{interface}} = 3.9\sqrt{Re_\tau}$ simulation are in between.

The normal Reynolds stresses from the ZDES simulations on the fine mesh and on the coarse mesh with the $y_{\text{interface}} = 0.1\delta$ and $y^+_{\text{interface}} = 3.9\sqrt{Re_\tau}$ interfaces are compared in Fig. 4b with experimental data. The ZDES simulation on the fine mesh resolves very well the outer layer, whereas the coarse mesh ZDES underestimates the stresses in the outer layer, especially the streamwise turbulent intensity u^+_{rms} at an intermediate height. It should also be noted that there are significant turbulent fluctuations in the RANS zone, so that this zone is only formally RANS.

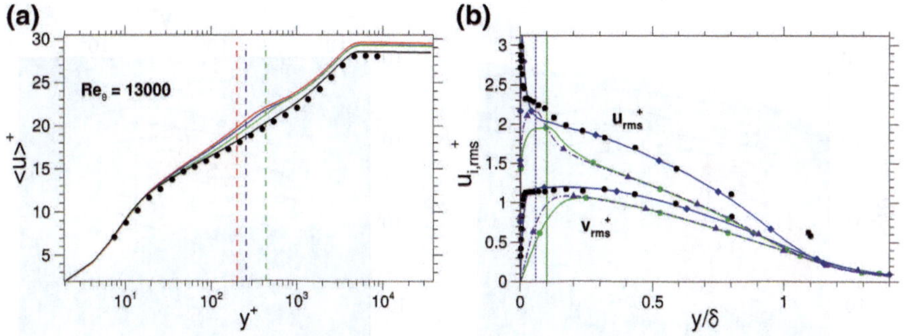

Fig. 4 Velocity and normal Reynolds stresses profiles at $Re_\theta = 13{,}000$. **a** Mean velocity profile. • Experimental data from [6]. — S-A RANS. ZDES, coarse mesh ($200^+/100^+$): — $y^+_{\text{interface}} = 200$; — $y_{\text{interface}} = 0.1\delta$; — $y^+_{\text{interface}} = 3.9\sqrt{Re_\tau}$. *Dashed lines* RANS/LES interfaces. **b** Normal Reynolds stresses. • Experimental data from [6]. —♦— ZDES, fine mesh ($50^+/12^+$). ZDES, coarse mesh ($200^+/100^+$): —•— $y_{\text{interface}} = 0.1\delta$, —▲— $y^+_{\text{interface}} = 3.9\sqrt{Re_\tau}$. *Dashed lines* RANS/LES interfaces

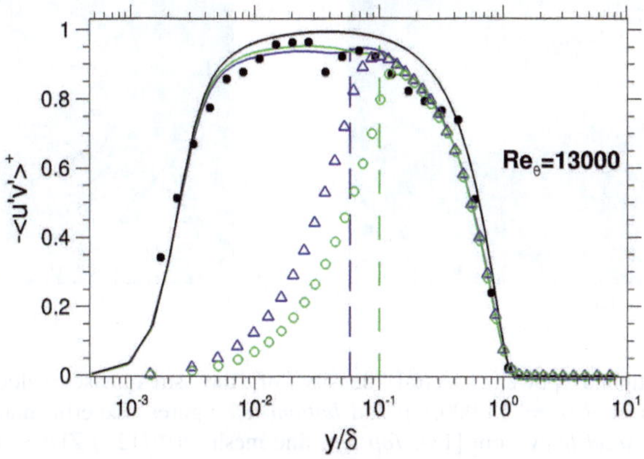

Fig. 5 Reynolds shear stress at $Re_\theta = 13{,}000$. • Experimental data from [6]. *Solid lines* total stress. — S-A RANS; — ZDES, coarse mesh ($200^+/100^+$), $y_{\text{interface}} = 0.1\delta$; — ZDES, coarse mesh ($200^+/100^+$), $y^+_{\text{interface}} = 3.9\sqrt{Re_\tau}$. *Symbols* resolved stress. ○ ZDES, coarse mesh ($200^+/100^+$), interface $y_{\text{interface}} = 0.1\delta$; △ ZDES, coarse mesh ($200^+/100^+$), interface $y^+_{\text{interface}} = 3.9\sqrt{Re_\tau}$. *Dashed lines* RANS/LES interfaces

Figure 5 further illustrates the impact of the interface positioning on the amount of resolved stresses in the inner layer, emphasizing the benefit of the $y^+_{\text{interface}} = 3.9\sqrt{Re_\tau}$ interface over the $y_{\text{interface}} = 0.1\delta$ interface. Furthermore, it confirms that the total Reynolds shear stress magnitude is underestimated in the outer layer. Since the modeled stress is negligible in this zone, this lack of Reynolds shear stress may be attributed to the resolved turbulent fluctuations.

The lack of resolved Reynolds stress in the outer layer is investigated by spectral analysis. Time signals are used to estimate the streamwise (co-)spectra of the

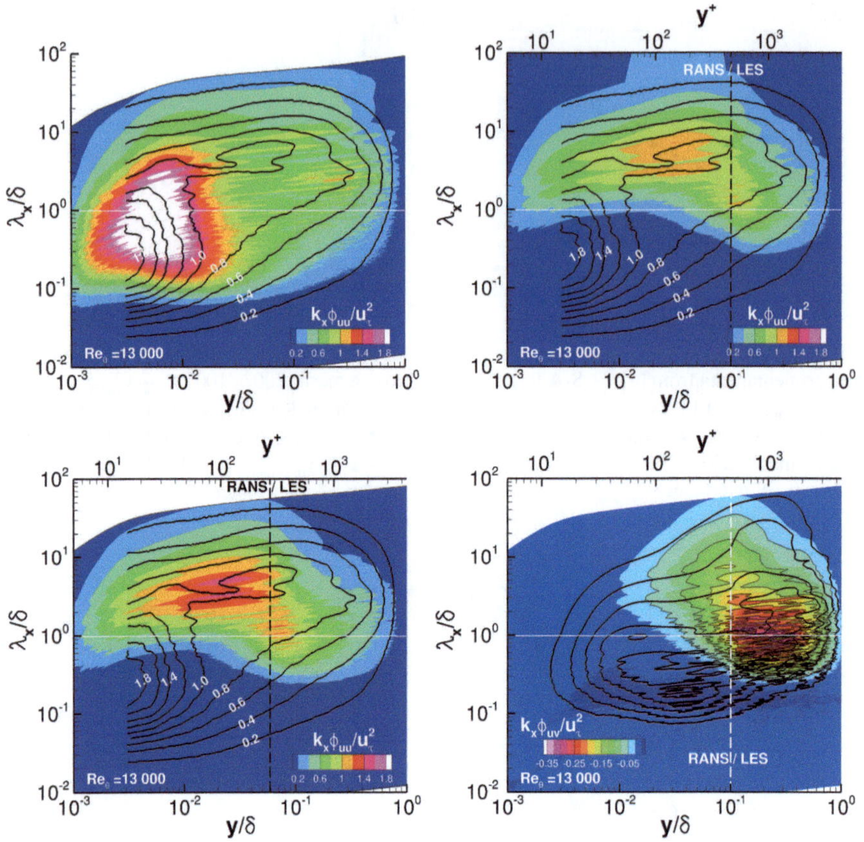

Fig. 6 Premultiplied power spectral density of the streamwise velocity fluctuations $k_x \Phi_{uu}/u_\tau^2(\lambda_x/\delta)$ at $Re_\theta = 13{,}000$, *top* and *bottom left* figures. Experimental data at $Re_\tau = 3{,}900$ shown by *solid lines* from [13]. *Top left*, fine mesh $(50^+/12^+)$ ZDES; *Top right*, coarse mesh $(200^+/100^+)$ ZDES, $y_{\text{interface}} = 0.1\delta$; *Bottom left*, coarse mesh $(200^+/100^+)$ ZDES, $y^+_{\text{interface}} = 3.9\sqrt{Re_\tau}$. *Bottom right* premultiplied co-spectrum of the resolved Reynolds shear stress $k_x \Phi_{uv}/u_\tau^2(\lambda_x/\delta)$ at $Re_\theta = 13{,}000$, coarse mesh $(200^+/100^+)$ ZDES, $y_{\text{interface}} = 0.1\delta$. Fine mesh $(50^+/12^+)$ ZDES shown by thick *solid lines* (matching the thin *solid lines* corresponding to the colored contour levels)

streamwise velocity and of the Reynolds shear stress, using Taylor's hypothesis with the Reynolds-averaged streamwise velocity as the convection velocity. The co-spectrum of the resolved Reynolds shear stress shown in Fig. 6 reveals that the deficit of resolved shear stress in the outer layer is mainly located at the largest wavelengths. The streamwise velocity spectra, compared with experimental data in Fig. 6, confirm that the ZDES simulations on the coarse mesh are 'under-resolving' in the outer layer both the smallest scales (which is the very principle of Large Eddy Simulation) and the largest scales (which is surprising, and might be explained by the impact of the RANS/LES interface on these very large scale motions, or maybe by a role played by either the smaller scales or the fluctuations close to the wall in the generation of the largest scales in the outer layer).

Another interesting feature is the presence in the $y_{\text{interface}} = 0.1\delta$ case of coherent structures seen as very large scale energy in the streamwise velocity spectrum. Both their location, size and level are in fair agreement with the experimental data, where this site is known to correspond to the superstructures that become stronger as the Reynolds number increases. This means that the coarse mesh ZDES simulation might be resolving the so-called very large scale motions (VLSMs), as also suggested by the flow visualization of Fig. 2. However, the comparison with the $y^+_{\text{interface}} = 3.9\sqrt{Re_\tau}$ interface simulation reveals that the strength of these coherent motions is not properly predicted if the interface position is different from $y_{\text{interface}} = 0.1\delta$. Even though these structures are stronger with $y^+_{\text{interface}} = 3.9\sqrt{Re_\tau}$, the lack of resolved Reynolds shear stress in the outer layer is still present (the Reynolds shear stress co-spectrum is similar to the one shown here for the $y_{\text{interface}} = 0.1\delta$ interface).

6 Outlook

The present study reveals the overall good prediction capability of the 3rd mode of the ZDES technique applied to zero pressure gradient boundary layers over a wide range of Reynolds numbers $3{,}150 \le Re_\theta \le 14{,}000$. The precision on C_f is better than 5 % after the initial flow adaptation. It should be noted that the present study features a rather long flow relaxation length downstream the turbulent inflow condition, which could be significantly reduced by dynamic forcing (see [12]). This was not done here in order to isolate the impact of the interface positioning on the high Reynolds number asymptotic trend of skin friction prediction from any possible effect of a dynamic forcing. Around 90 % of the turbulent skin friction may be resolved rather than modeled if the interface is properly positioned by the user, as analyzed in the framework of the FIK identity [7]. The study gives hope for an invariant behavior of the method towards higher Reynolds numbers.

The spectral analysis of the resolved velocity field reveals that some of the missing Reynolds stresses in the outer layer are surprisingly associated to the largest length scales, as compared with the simulation performed on a much finer mesh and with experiment. Several ways of improving these results will be considered in a later study. Besides, the spectral analysis suggests that a proper interface positioning enables the resolution of coherent motions with energy level and location that are in fair agreement with the so-called Very Large Scale Motions observed experimentally. Unfortunately, the various criteria addressed in this study do not lead yet to a straightforward optimal interface positioning rule.

A study of the sensitivity to non-optimal interface positioning might be of high practical interest, as the optimum location should not be expected to be exactly reached in the case of a complex curvilinear geometry with pressure gradient and cross flow. Treating the interface in an active way, i.e. resorting to additional filtering or forcing in contrast to the present study, would probably improve this aspect, as well as reduce the mitigated but still present deviation from the log law, but at the cost of lack of universality and spurious acoustic signature.

Acknowledgments The authors wish to thank all the people involved in the past and present evolution of the FLU3M code. Pierre Sagaut, Romain Laraufie and Pierre-Élie Weiss are warmly acknowledged for very stimulating discussions. The fine mesh WRLES computation was made thanks to the HPC resources from GENCI-CINES (Project ZDESWALLTURB, Grant 2012-[c2012026817]). The thesis of Nicolas Renard is partly funded by the French defense procurement agency DGA.

References

1. Choi, J.I., Edwards, J., Baurle, R.: Compressible boundary-layer predictions at high Reynolds number using hybrid LES/RANS methods. AIAA J. **47**(9), 2179–2193 (2009)
2. Deck, S.: Recent improvements of the zonal detached eddy simulation (ZDES) formulation. Theoret. Comput. Fluid Dyn. **26**(6), 523–550 (2012). doi:10.1007/s00162-011-0240-z
3. Deck, S., Renard, N., Laraufie, R., Sagaut, P.: Zonal detached eddy simulation (ZDES) of a spatially developing flat plate turbulent boundary layer over the Reynolds number range $3{,}150 \leq Re_\theta \leq 14{,}000$. Phys. Fluids **26**, 025–116 (2014)
4. Deck, S., Renard, N., Laraufie, R., Weiss, P.E.: Large scale contribution to mean wall shear stress in high Reynolds number flat plate boundary layers up to Re_θ =13,650. J. Fluid Mech. **743**, 202–248 (2014). doi:10.1017/jfm.2013.629
5. Deck, S., Weiss, P., Pamiès, M., Garnier, E.: Zonal detached eddy simulation of a spatially developing flat plate turbulent boundary layer. Comput. Fluids **48**, 1–15 (2011). doi:10.1016/j.compfluid.2011.03.09
6. DeGraaff, D., Eaton, J.: Reynolds number scaling of the flat-plate turbulent boundary layer. J. Fluid Mech. **422**, 319–346 (2000)
7. Fukagata, K., Iwamoto, K., Kasagi, N.: Contribution of Reynolds stress distribution to the skin friction in wall-bounded flows. Phys. Fluids **14**, L73 (2002)
8. Hamba, F.: Log-layer mismatch and commutation error in hybrid RANS/LES simulation of channel flow. Int. J. Heat Fluid Flows **30**, 20–31 (2009)
9. Jarrin, N., Benhamadouche, S., Laurence, D., Prosser, R.: A synthetic-eddy-method for generating inflow conditions for large eddy simulation. Int. J. Heat Fluid Flows **27**, 585–593 (2006)
10. Keating, A., Piomelli, U.: A dynamic stochastic forcing method as a wall-layer model for large-eddy simulation. J. Turbul. **7**(12), 1–24 (2006)
11. Laraufie, R., Deck, S.: Assessment of Reynolds stresses tensor reconstruction methods for synthetic inflow conditions. Application to hybrid RANS/LES methods. Int. J. Heat Fluid Flow **42**, 68–78 (2013)
12. Laraufie, R., Deck, S., Sagaut, P.: A dynamic forcing method for unsteady turbulent inflow conditions. J. Comput. Phys. **230**(23): 8647–8663 (2011). doi: 10.1016/j.jcp.2011.08.012. http://www.sciencedirect.com/science/article/pii/S0021999111100484
13. Marusic, I., Mathis, R., Hutchins, N.: High Reynolds number effects in wall turbulence. Int. J. Heat Fluid Flow **31**, 418–428 (2010)
14. Mary, I., Sagaut, P.: Large eddy simulation of flow around an airfoil near stall. AIAA J. **40**(6), 1139–1145 (2002)
15. Nagib, H., Chauhan, K., Monkewitz, P.: Approach to an asymptotic state for zero pressure gradient turbulent boundary layers. Philos. Trans. R. Soc. A **365**, 755–770 (2007)
16. Nikitin, N.V., Nicoud, F., Wasistho, B., Squires, K.D., Spalart, P.R.: An approach to wall modeling in large-eddy simulations. Phys. Fluids **12**(7), 1629–1632 (2000)
17. Pamiès, M., Weiss, P., Garnier, E., Deck, S., Sagaut, P.: Generation of synthetic turbulent inflow data for large eddy simulation of spatially evolving wall-bounded flows. Phys. Fluids **21**, 045103 (2009)

18. Park, G.I., Moin, P.: An improved dynamic non-equilibrium wall-model for large eddy simulation. Phys. Fluids **26**, 015–108 (2014)
19. Piomelli, U.: Wall-layer models for large-eddy simulations. Prog. Aerosp. Sci. **44**, 437–446 (2008)
20. Piomelli, U., Balaras, E.: Wall-layer models for large-eddy simulations. Annu. Rev. Fluid Mech. **34**, 348–374 (2002)
21. Rajamani, B., Kim, J.: A hybrid-filter approach to turbulence simulation. Flow Turbul. Combust. **85**, 421–441 (2010)
22. Sánchez-Rocha, M., Menon, S.: An order-of-magnitude approximation for the hybrid terms in the compressible hybrid RANS/LES governing equations. J. Turbul. **12**(16), 1–22 (2011)
23. Shur, M., Spalart, P., Strelets, M., Travin, A.: A hybrid RANS-LES approach with delayed-DES and wall-modelled LES capabilities. Int. J. Heat Fluid Flows **29**(6), 1638–1649 (2008)
24. Shur, M., Spalart, P., Strelets, M., Travin, A.: A rapid and accurate switch from RANS to LES in boundary layers using an overlap region. Flow Turbul. Combust. **86**, 179–206 (2011)
25. Spalart, P., Allmaras, S.: A one equation turbulence model for aerodynamic flows. La Recherche Aérospatiale **1**, 5–21 (1994)

18. [illegible] An improved [illegible]
19. [illegible]
20. [illegible]
21. [illegible]
[illegible]

Simple Improvements in the SST-DES Formulation for Mild Aerofoil Trailing-Edge Separation

Xiangyu Wang and Dong Li

Abstract We present an improved Detached Eddy Simulation (DES) based on the SST turbulence model, WAD SST-DES, for accelerating the Large Eddy Simulation (LES) development in mixing layers. Similar to WAD-DES based on the SA turbulence model proposed by Wang and Qin, the formulation of Yoshizawa's k-equation-like subgrid model in LES zones in the original SST-DES is rewritten as a standard one to end the previous turbulence transfer in the interface between the RANS and LES regions, and the numerical results for flow around A-aerofoil show that these improvements efficiently weaken the influence of the grey area while delaying the RANS at the boundary, in a manner similar to DDES. In addition, we discuss the influence of several coefficients in the improved SST-DES; these coefficients may play a very important role in the final simulation results.

1 Introduction

Delayed Detached Eddy Simulation (DDES) [11] has successfully overcome Grid Induced Separation (GIS), which may occur under some inaccurate grid distributions in the original DES97 by delaying RANS to cover the entire boundary layer. However, this simple shielding may increase the problems associated with the grey area and further decelerate the development of downstream resolved turbulence. Previous studies addressing this issue usually focused on the grid spacing and rewriting the formulation of the original length scale (the maximum edge of a cell volume) [12] to reduce the turbulent viscosity in specific areas, which actually results in an opposite DDES modification; there also have been studies involving some advanced but complicated zonal/embedded hybrid methods, which usually need extra resolved turbulence near the interface. It has been shown that with this improvements, irrespective

X. Wang (✉) · D. Li
School of Aeronautics, Northwestern Polytechnical University, Xi'an, China
e-mail: xiangyu.wang@mail.nwpu.edu.cn

D. Li
e-mail: ldgh@nwpu.edu.cn

S. Girimaji et al. (eds.), *Progress in Hybrid RANS-LES Modelling*,
Notes on Numerical Fluid Mechanics and Multidisciplinary Design 130,
DOI 10.1007/978-3-319-15141-0_17

of whether they involve ZDES [2] embedded LES based on synthetic turbulence [1], synthetic turbulence generator [5], or recycling methods [8], they all lead to a better simulation.

Recently, Wang and Qin [14] proposed a new improved DES based on the SA turbulence model called WAD-DES for mild A-aerofoil trailing-edge separation, which focuses on the equation formulation and the integral DES method itself, and does not entirely follow previous results to restructure a new complex length scale or adopt any zonal/embedded thoughts. This improvement is interesting, and rewriting a mature CFD code at the same time is very simple. Therefore, in the present paper, an improvement similar to the WAD-DES but based on the SST turbulence model, called WAD SST-DES is presented and mild A-aerofoil trailing-edge separation is also used as a numerical test case. Because Qin and his co-workers have performed much research [4] with respect to the simulation of A-aerofoil by SA-DES and obtained a variety of meaningful results, the authors aimed to not simply repeat their work, showing the comprehensive behaviours of the WAD SST-DES, but to provide some discussion on the influence of several coefficients in this approach.

2 Review of WAD-DES

The DES97 and DDES based on the SA turbulence model [10] are formulated as

$$\mu_T = f_{v1}\rho \upsilon \tag{1}$$

$$\frac{\partial \upsilon}{\partial t} + \frac{\partial (\upsilon v_j)}{\partial x_j} = C_{b1} S \upsilon + \frac{1}{\sigma}\left\{\frac{\partial}{\partial x_j}\left[(\upsilon_L + \upsilon)\frac{\partial \upsilon}{\partial x_j}\right] + C_{b2}\left(\frac{\partial \upsilon}{\partial x_j}\right)^2\right\} - C_{\omega 2} f_\omega \left(\frac{\upsilon}{d}\right)^2 \tag{2}$$

$$d_{DES} = \min\left(d_{wall}, C_{DES}\Delta\right) \tag{3}$$

$$d_{DDES} = \min\left(d_{wall}, (1 - f_d)\, d_{wall} + f_d C_{DES}\Delta\right) \tag{4}$$

It is known that when DES operates far from the wall, and if the production and destruction terms are balanced,

$$C_{b1} S \upsilon = C_{\omega 2} f_\omega \left(\frac{\upsilon}{C_{DES}\Delta}\right)^2 \tag{5}$$

Then, the final turbulence viscosity can be deduced as

$$\mu_T^{SA-DES} = \frac{c_{b1} f_{v1}}{c_{w1} f_w} (C_{DES}\Delta)^2 S \tag{6}$$

On one hand, Eq. (6) is similar to the turbulence viscosity formulation in the Smagorinsky subgrid model [9]

$$\mu_T^{SGS} = (C_S\Delta)^2 S \tag{7}$$

It may also function as a LES mode in the area far from the wall, which is the core concept of the DES. On the other hand, Eq. (6) is simple a Smagorinsky-like model, which is not entirely the same as the formulation in the standard Smagorinsky subgrid model. In Wang and Qin's opinion, Eq. (6) implies that when the S-A model works as a SGS model, the original small value of the wall damping function f_w will lead to a large turbulent viscosity in the LES mode, suppressing the development of resolved flow fluctuations in regions further away from the wall. By comparison with DES, DDES triggers the LES mode and depends on the wall distance, the grid spacing and the modeled turbulent viscosity, rather than only the grid spacing and the wall distance as DES does, which may "delay" the RANS mode in thick boundary layers, generating higher turbulent viscosities than DES.

Therefore, a new destruction term can be derived using Eqs. (5) and (7) to balance the production term in the SA turbulence model in Wang and Qin's work

$$D^{Smag} = \frac{c_{b1} f_{v1}}{C_s^2} \left(\frac{\upsilon}{\Delta}\right)^2 \tag{8}$$

To maintain the advantage of DDES for thick boundary layers and simultaneously increase the resolved turbulence in the wake, a shielding function is used and the final formulation can be rewritten as

$$\begin{aligned} D_{WAD-DES} &= (1 - f_d)\, D^{SA-DES} + f_d D^{Smag} \\ &= (1 - f_d)\, c_{w1} f_{w1} \left(\frac{\upsilon}{d}\right)^2 + f_d \frac{c_{b1} f_{v1}}{C_s^2} \left(\frac{\upsilon}{\Delta}\right)^2 \end{aligned} \tag{9}$$

We skip the derivation process and summarizing Wang and Qin's conclusion that "a subgrid-like model in DES is different from a standard one" -because the extra variables involved in the subgrid-like model may confuse the simulation results.

3 Similar Improvements in SST-DES

DES based on Menter's SST turbulence model [13] is formulated as

$$\mu_T = \min\left[\frac{\rho k}{\omega}, \frac{a_1 \rho k}{S F_2}\right] \tag{10}$$

$$P_k = \mu_T S^2, \quad P_\omega = \gamma \rho S^2 \tag{11}$$

$$\frac{\partial k}{\partial t} + u_j \frac{\partial k}{\partial x_j} = \frac{1}{\rho} P_k - \beta' k \omega F_{DES} + \frac{1}{\rho} \frac{\partial}{\partial x_j} \left[\left(\mu_L + \frac{\mu_T}{\sigma_k} \right) \frac{\partial k}{\partial x_j} \right] \tag{12}$$

$$\begin{aligned} \frac{\partial \omega}{\partial t} + u_j \frac{\partial \omega}{\partial x_j} &= \frac{1}{\rho} P_\omega - \beta \omega^2 + \frac{1}{\rho} \frac{\partial}{\partial x_j} \left[\left(\mu_L + \frac{\mu_T}{\sigma_\omega} \right) \frac{\partial \omega}{\partial x_j} \right] \\ &+ 2(1 - F_1) \frac{1}{\sigma_\omega} \frac{1}{\omega} \frac{\partial k}{\partial x_i} \frac{\partial \omega}{\partial x_j} \end{aligned} \tag{13}$$

$$\begin{aligned} F_{DES} &= \max\left(\frac{l_{RANS}}{C_{DES}\Delta}, 1 \right), \quad F_{DDES} = \max\left(\frac{l_{RANS}}{C_{DES}\Delta} (1 - f_d), 1 \right), \\ l_{RANS} &= \frac{\sqrt{k}}{\beta_k \omega} \end{aligned} \tag{14}$$

$$f_d = \tanh\left([8 r_d]^3 \right), r_d = (\mu_L + \mu_T) / \left(\sqrt{S_{ij} S_{ij}} \kappa^2 d^2 \right) \tag{15}$$

The k equation SGS model of Yoshizawa [15] reads

$$\begin{aligned} \mu_T &= C_k \Delta \sqrt{k}, \frac{\partial k}{\partial t} + u_j \frac{\partial k}{\partial x_j} \\ &= 2 P_k - C_d \frac{k^{3/2}}{\Delta} + \frac{\partial}{\partial x_j} \left[(\mu_L + \mu_T) \frac{\partial k}{\partial x_j} \right] \end{aligned} \tag{16}$$

It is obvious that when the SST-DES functions as an LES mode out of the boundary layer, destruction term in the k equation changes and the original SST becomes a Yoshizawa-like SGS model. However, Eq. (10) remains the same, and in fact, the final turbulence viscosity is still obtained in the manner of a k-ε turbulence model:

$$\mu_T = \rho k / \omega = \rho \left(\sqrt{k} / \omega \right) \sqrt{k} = \rho l_{RANS} \sqrt{k} / \beta_k \tag{17}$$

Comparing Eq. (15) with Eq. (16), it is the RANS length scale l_{RANS}, rather than the LES length scale $\boldsymbol{\Delta}$, that controls the turbulence viscosity far from the wall; there is something similar to what happens in the original SA-DES above, and the extra l_{RANS}, which is usually much larger than the LES length scale, may confuse the LES mode and the resolved turbulence. Furthermore, because of an unchanged ω equation such as Eq. (12) even if the turbulence energy k could decrease rapidly at the interface between RANS and LES, the low turbulence dissipation ω determined from the upstream RANS region may be insufficient to significantly resolve the turbulence downstream. It seems necessary to rewrite Eq. (10) in a standard Yoshizawa SGS model, as similar to what has been done in WAD-DES. To avoid a potential GIS such as DES97, the same weighting function is also adopted as

$$\mu_T = \min\left(f_d\frac{\rho k}{\omega} + (1 - f_d)\, C_k\sqrt{k}\Delta,\ \frac{a_1\rho k}{SF_2}\right) \tag{18}$$

At the same time Eq. (12) has a new formulation for consistency with Eq. (17) in the same manner:

$$\begin{aligned}\frac{\partial k}{\partial t} + u_j\frac{\partial k}{\partial x_j} = \frac{1}{\rho}P_k - \left(f_d\beta' k\omega + (1 - f_d)\, C_d\frac{k^{3/2}}{\Delta}\right)\\ + \frac{1}{\rho}\frac{\partial}{\partial x_j}\left[\left(\mu_L + \frac{\mu_T}{\sigma_k}\right)\frac{\partial k}{\partial x_j}\right]\end{aligned} \tag{19}$$

That is, a Yoshizawa-like SGS model has been translated to a standard one by rewriting its dissipation term, although a slight difference still exists between those two production terms. As a result, the previous turbulence transfer near the interface with the variable ω in the k equation is completely eliminated. Also, it would be simple to rewrite a mature CFD code from the original SST-DES to the WAD SST-DES similar to Eqs. (17) and (18).

From another perspective, when the new method works as an LES mode, a relationship $\mu_T = C_k\rho\sqrt{k}\Delta$ can be obtained as $f_d \approx 0$ and inserted into Eq. (19); then, the final turbulence viscosity equation can be deduced as

$$\frac{\partial \mu_T}{\partial t} + u_j\frac{\partial \mu_T}{\partial x_j} = \frac{C_k^2}{2}\Delta^2 S^2 - \frac{C_k C_d k}{2} + \mathit{diffusion} \tag{20}$$

In contrast to the SA-DES, the length scale is involved in the production term in Eq. (20). In Wang and Qin's work, the discussion of the choice of length scale is skipped and a traditional length scale $\Delta = \max(\Delta x, \Delta y, \Delta z)$ is also adopted; however, it is known that in the diffusion-convection equations, production terms usually play the decisive role in comparison with the other source terms. Therefore, the original formulation of the length scale in Yoshizawa's model is used in this improved SST-DES method:

$$\Delta = \sqrt[3]{\Delta x\Delta y\Delta z} \tag{21}$$

It may be confusing that Eq. (21) itself also can contribute to the decrease of high turbulence viscosity at the interface. To clarify this matter in all of the following test cases, the formulation in Eq. (21) is adopted, irrespective of whether the new improved method or the traditional SST-DES and SST-DDES are used for comparison.

Moreover, although constant values for both C_k and C_d are recommended in the standard Yoshizawa's k equation SGS model, it is necessary but complicated to recalibrate both of them in the improved method. When the production term and dissipation term are balanced in Eq. (19), as is the practice in WAD-DES,

$$\frac{C_k^2}{2}\Delta^2 S^2 = \frac{C_s C_d k}{2} \tag{22}$$

Through the derivation, an ideal turbulence viscosity can be written as

$$\mu_T = \sqrt{\frac{C_k^3}{C_d}}\Delta^2\Omega \tag{23}$$

Equation (22) seems quite similar to the form of the Smagorinksy SGS model $\mu_T = C_s\Delta^2\Omega$. Because $C_s = 018^2$ has been commonly used by many researchers, an extra and new relationship between C_k and C_d can be obtained from the following:

$$\sqrt{\frac{C_k^3}{C_d}} = C_s^2 = 0.18^2 \tag{24}$$

This means if the value of either C_k or C_d is chosen, the other one is fixed. Using the decaying homogeneous isotropic turbulence simulation, suitable values of $C_k = 0.1$ and $C_d = 1.0$ are adopted in this paper.

4 Simulation Approach

In this paper, flow around an A-aerofoil at the maximum lift condition, identical to that in Wang and Qin's work, is simulated to test the improvements in the WAD SST-DES method as well as the behaviours of the original SST-DES and SST-DDES. The C-type grid in the stream-wise slice is consists 1,960,000 points and the final mesh is extruded from the XY plane in the Z direction with 50 elements distributed in a span-wise length of 0.5, which is longer than the span-wise space in the mesh Wang and Qin used.

The computation presented has been performed using a Chinese open-source CFD code, OpenEC-3d, first written by Xinliang Li, a researcher at the Institute of Mechanics (IM) of the Chinese Academy of Sciences (CAS). He hopes his code could provide a free, basic, and initiative framework like LINUX, so that other CFD researchers could be able to rewrite it for different needs or just take it as a platform to add some other useful modules. Without commercial purpose anyone focusing on CFD can contact him at lixl@imech.ac.cn for an original code.

OpenEC-3d is a compressible, finite volume code based on structured mesh and Fortran 90. The authors themselves have rewritten this code in many places; and in this paper, a dual time stepping method combining implicit physical time marching, and LU-SGS pseudo time marching is employed and the spatial discretization scheme is as second order accurate as most other finite volume codes.

5 Results and Discussions

The WAD SST-DES test results are compared with the original SST-DES and SST-DDES results. Figure 1 shows the modeled turbulent viscosity and vorticity magnitude of the three cases; unless stated otherwise all the simulation tests are based on the length scale from Eq. (20). Obviously, SST-DDES models higher turbulent viscosity than the other two, especially in the wake region, because of the delayed RANS mode in Fig. 1b1. Although this shielding may complement the insufficiently modeled and resolved Reynolds stress, it also limits the development of turbulence downstream, and a "cloudy" vortex shedding is obtained as a result. The length scale in Yoshizawa's SGS model has such a significant effect on SST-DES that unexpectedly low turbulent viscosity appears in both the area near the wall and the wake region; finally, severe GIS occurs; the vorticity contour in Fig. 1a2 is not a mild separation but a massive one. The weighting function f_d can mask the influence of Yoshizawa's length scale, as in the DDES results. WAD SST-DES predicts higher turbulent viscosity than SST-DES, which is suddenly "cut off" as in ZDES [3]

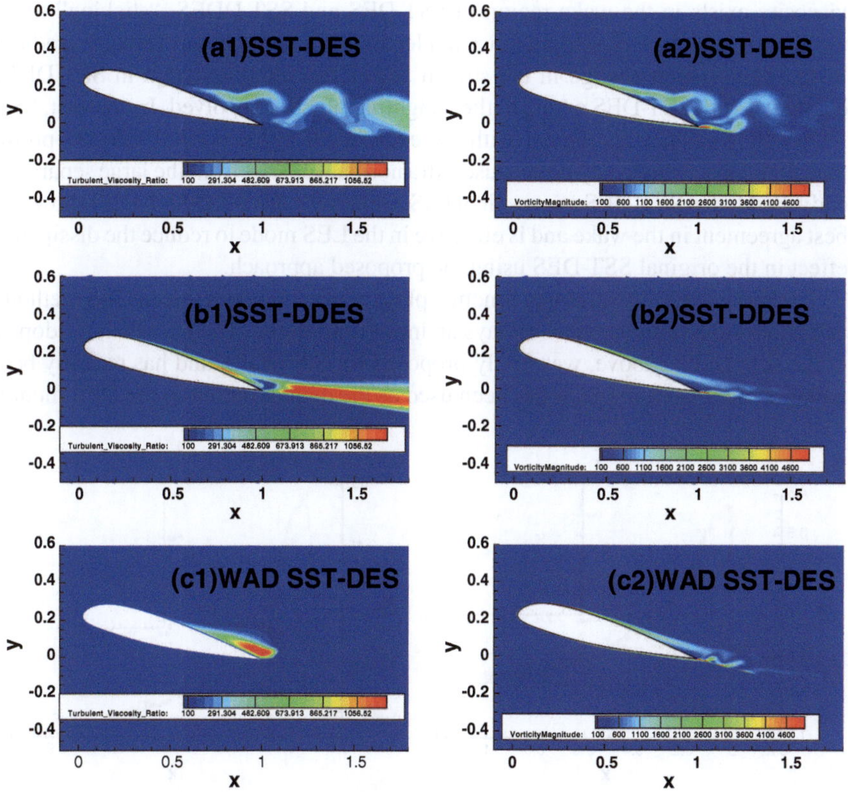

Fig. 1 Turbulent viscosity ratio (*left column*) and vorticity magnitude (*right column*)

functions, and is contradictory to SST-DDES in the near-wall separation region in Fig. 1c1. This implies that WAD SST-DES inherits the advantages of DDES to some extent in delaying RANS in regions with thick boundary layers, and in the wake, it obtains much smaller modeled turbulent viscosity, encouraging the development of turbulence fluctuations. This viewpoint can be proved by the distinct and mild vortex shedding in the wake, as shown in Fig. 1c2.

Figure 2 shows the pressure coefficient along the aerofoil and skin friction coefficient in the suction side. Compared with the trailing edge separation point, the 0.82 obtained from experiments, is under-predicted in SST-DES at no more than 0.7 owing to the insufficiently modeled stress in the near-wall boundary layer. Both SST-DDES and WAD SST-DES predict the separation point at about 0.8, which means the weighting function in WAD SST-DES plays a similar role as in SST-DDES to delay the RANS in the boundary layer. As apparent from the pressure coefficient, WAD SST-DES obtains results similar to those with DDES before the separation point, but after the separation point, it gives a better agreement with the experimental data.

Figures 3 and 4 plot the parallel and normal velocity profile in the direction of the free stream velocity. Intuitively, because of the upstream RANS region high turbulent viscosity exists in the wake region in SST-DES and SST-DDES, "sticking" to the turbulence, which be developing fast and leading to a large stream velocity gradient. In fact this sudden change in the stream velocity is more apparent in SST-DDES compared with SST-DES owing to the weighting function involved. In contrast, SST-DDES shows better agreement with the experiment in the normal velocity profile than SST-DES, which may be because extra normal velocity from the large separation eddy is added in SST-DES as a result of GIS. Generally, the WAD SST-DES shows the best agreement in the wake and is effective in the LES mode to reduce the dissipation effect in the original SST-DES using the proposed approach.

It is known that the weighting function plays a very important role in DES methods, and choosing a suitable one is always an interesting tropic. The f_d, which is adopted in the simulations above, was firstly proposed for SA-DDES and has recently been widely used. For example, it has been used as the default option in the DES module

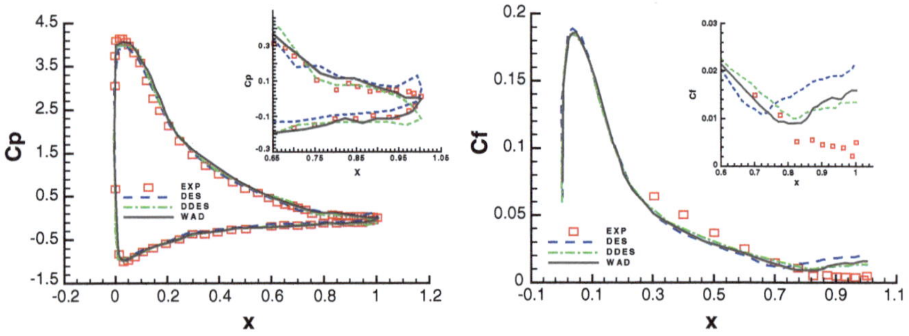

Fig. 2 Pressure coefficient (*left*) and skin friction coefficient (*right*)

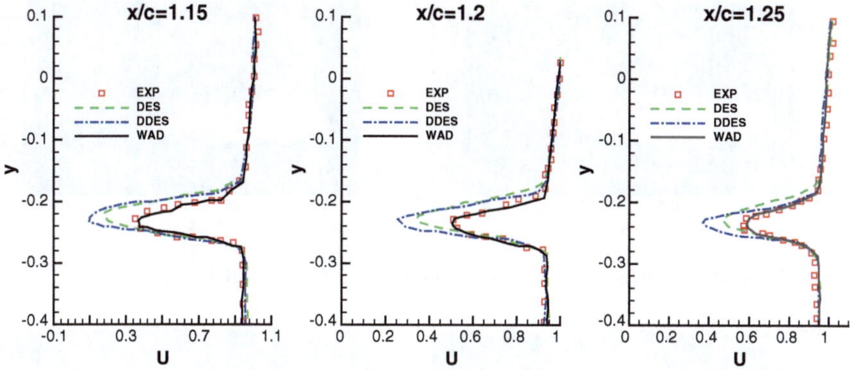

Fig. 3 Velocity parallel to the direction of the free stream velocity at x = 1.15, 1.2 and 1.25

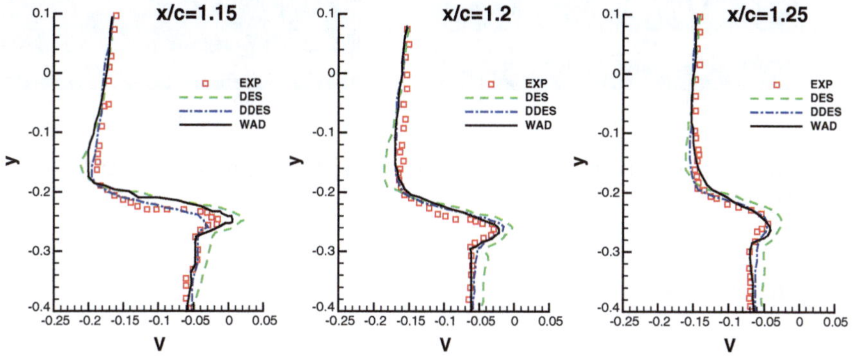

Fig. 4 Velocity normal to the direction of the free stream velocity at x = 1.15, 1.2 and 1.25

of Ansys Fluent. In addition, f_2, the function involved in the SST turbulence model itself, is recommended in some early works [6, 7]. Figures 5 and 6 show a comparison between f1 and f2 in WAD SST-DES; obviously, f_d covers a larger flow area and yields higher values in the near-wall separation, which may delay the RANS mode strongly and further confirm the previous points about compensating for the new small length scale. Conversely, f_2 covers only a thin layer near the wall, and from Fig. 6, it cannot remedy the low turbulent viscosity near the trail region resulting from both the new length scale and the formulation itself, although the result may be better than that with the original SST-DES seen in Fig. 1a.

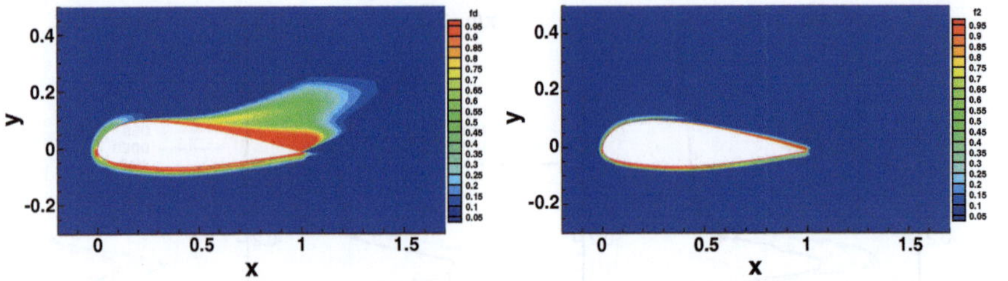

Fig. 5 Comparison between the two weighting functions, f_d (*left*) and f_2 (*right*)

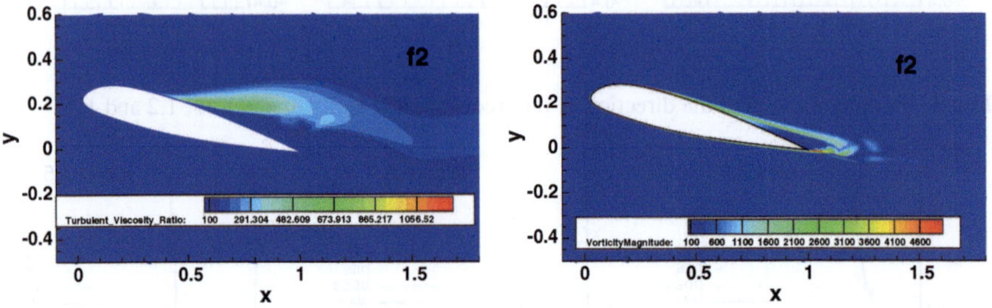

Fig. 6 Turbulent viscosity ratio (*left*) and vorticity magnitude (*right*) in WAD SST-DES under the weighting function f_2

6 Conclusions

We presented simple improvements in the SST-DES, which we call WAD SST-DES, for accelerating the development of turbulence flow in shear layers. As a generalization of Wang and Qin's work, the original SST-DES has been rewritten in a new formulation in which a standard SGS model, rather than a SGS-like one, functions in the LES mode. Numerical investigation showed that the new formulation retains the advantages of DDES in the boundary layer while increasing the prediction accuracy in the grey area. It is also provide a more distinct trailing-edge vortex shedding. Moreover, some coefficients involved in the WAD SST-DES have been discussed, and a brief test of different weighting functions has also been performed.

References

1. Davidson, L., Peng, S.H.: Embedded large-eddy simulation using the partially averaged Navier-Stokes model. AIAA J. **51**(5), 1066–1079 (2013)
2. Deck, S.: Zonal-detached eddy simulation of the flow around a high-lift configuration. AIAA J. **43**(11), 2372–2384 (2005)
3. Deck, S.: Recent improvements of the zonal detached eddy simulation (ZDES) formulation. Theoret. Comput. Fluid Dyn. **26**, 523–550 (2012)

4. Durrani, N., Qin, N.: Behaviour of detached-eddy simulations for mild trailing-edge separation. J. Aircr. **48**(1), 193 (2011)
5. Jarrin, N., Cokljat, D., Bertoglio, J.P., Sergent, E.: Reconstruction of turbulent fluctuations for hybrid RANS/LES simulations using a synthetic-eddy method. Int. J. Heat Fluid Flow **30**, 435–442 (2009)
6. Lynch, C.E., Smithy, M.J.: Hybrid RANS-LES turbulence models on unstructured grids. In: AIAA Paper, 2008–3854 (2008)
7. Sanchez-Rocha, M., Kirtas, M., Menon, S.: Zonal hybrid RANS-LES method for static and oscillating airfoils and wings. In: AIAA Paper, 2006-1256 (2006)
8. Shur, M.L., Spalart, P.R., Strelets, A., Travin, A.: A rapid and accurate switch from RANS to LES in boundary layers using an overlap region. Turbul. combust. **86**, 179–206 (2011)
9. Smagorinsky, J.: General circulation experiments with the primitive equations. Mon. Weather Rev. **91**(3), 99–164 (1963)
10. Spalart, P.R.: Strategies for turbulence modelling and simulations. Int. J. Heat Fluid Flow **21**, 252–263 (2000)
11. Spalart, P.R., Deck, S., Shur, M., Squares, K.D., Strelets, M., Travin, A.: A new version of detached-eddy simulation, resistant to ambiguous grid densities. Theor. Comput. Fluid Dyn. **20**, 181–195 (2006)
12. Spalart, P.R.: Detached-eddy simulation. Annu. Rev. Fluid Mech. **41**, 181–202 (2009)
13. Strelets, M.: Detached eddy simulation of massive separated flow. In: AIAA Paper, 2001-0879 (2001)
14. Wang, W., Qin, N.: Balancing destruction and production in S-A model-based hybrid RANS-LES for flow around an aerofoil with mild separation. Progress in Hybrid RANS-LES Modelling, NNFM **117**, 379–388 (2012)
15. Yoshizawa, A., Horiuti, K.: A statistically-derived subgrid-scale kinetic energy model for the large-eddy simulation of turbulent flows. J. Phys. Soc. Jpn. **54**(8), 2834–2839 (1985)

4. [illegible] detached eddy simulation [illegible]
5. [illegible] synthetic-eddy method [illegible]
6. [illegible] RANS [illegible] AIAA Paper [illegible]
7. [illegible]
8. [illegible]
9. [illegible]
10. [illegible]
11. [illegible]
12. [illegible]
13. [illegible]
14. [illegible]
15. [illegible]

Part IV
Hybrid and Zonal Methods

Prediction of Transonic Duct Flow Using a Zonal Hybrid RANS-LES Modeling Approach

Sebastian Arvidson, Shia-Hui Peng and Lars Davidson

Abstract Transonic duct flow with shock/boundary-layer interaction (SBLI) was analyzed using a zonal hybrid RANS-LES approach. The proposed zonal approach simulates the attached boundary layer flow, prior to the SBLI region in RANS mode. At a prescribed streamwise location, upstream of the SBLI region, the model switches to its hybrid RANS-LES mode over a buffer zone in order to avoid possible discontinuities. The corner separation bubbles, induced by the shock at M = 1.4, were mostly simulated in LES mode in order to improve the SBLI flow prediction. In addition to comparisons with experimental data, the zonal approach is compared to simulations using the SA-DDES and SA-IDDES model. The zonal approach predicted a corner separation bubble and a λ-shape shock which is in good agreement with experimental data. Furthermore, the predicted pressure rise across the shock agrees reasonably well with the experiment.

1 Introduction

Hybrid RANS-LES modeling approaches have been increasingly used in aeronautic applications for better predictions of unsteady and complex aerodynamic flows. Facilitated by this, hybrid RANS-LES methods have been extensively studied and verified in simulations of a large variety of flows. At an early stage, the development

S. Arvidson (✉) · S.-H. Peng · L. Davidson
Division of Fluid Dynamics, Chalmers University of Technology,
SE-412 96 Gothenburg, Sweden
e-mail: sebastian.arvidson@chalmers.se; sebastian.arvidson@saabgroup.com

S. Arvidson
Saab Aeronautics, SE-581 88 Linköping, Sweden

S.-H. Peng
Swedish Defence Research Agency (FOI), SE-164 90 Stockholm, Sweden
e-mail: peng@foi.se; peng@chalmers.se

L. Davidson
e-mail: lada@chalmers.se

S. Girimaji et al. (eds.), *Progress in Hybrid RANS-LES Modelling*,
Notes on Numerical Fluid Mechanics and Multidisciplinary Design 130,
DOI 10.1007/978-3-319-15141-0_18

of hybrid RANS-LES methods focused most on high-Re number turbulent flows with massive separation, such as the flow over an aircraft wing at high incidence or over other bluff bodies. With increasing computer capacities and refined modeling approaches, hybrid RANS-LES methods have shown promising advantages in the computational analysis of unsteady boundary layer separation. This has motivated the present study, verifying hybrid RANS-LES methods for a SBLI flow.

SBLI flows have been commonly encountered in aeronautical applications and appears in such phenomena as buffeting on aircraft wings and in high-speed inlet configurations. For high-speed inlets, SBLI can typically take place inside the inlet duct, which can have a severe impact on the engine/inlet stability if it is not controlled. SBLI flow in a duct often causes flow separation in the corners of the duct. However, predicting SBLI in duct flows using CFD methods has been found to be challenging, especially using hybrid RANS-LES and DES methods [1, 2].

This paper analyzes a transonic duct flow involving SBLI [4]. A rectangular duct with a convergent-divergent nozzle was used to accelerate the flow from subsonic speed to supersonic speed, where a shock takes place at Mach 1.4. The λ-shape shock wave interacts with the boundary layer and consequently leading to local recirculation bubbles in the duct corners as shown in Fig. 1. The walls are parallel in the shock region and the cross-sectional area is constant with $H = 178$ mm and $W = 117$ mm in height and width, respectively. The center line length of the duct is $L = 1{,}030$ mm. The computational domain can be seen in Fig. 2.

Bottom wall pressures and streamwise velocities in the duct center plane are available from the experiment [4]. A comparison is made of profiles from the simulations and the experiment at a distance Δx relative to the shock location, as illustrated in Fig. 1. Schlieren photographs of the shock wave and oil-flow visualization of the bottom wall flow are also available and used for comparison. These are shown in Fig. 1. Measured at the center line of the duct, the shock wave in the experiment occurs at $x = 659$ mm ($\Delta x = 0$), i.e. 205 mm downstream of the nozzle exit. The Reynolds number, based on the displacement thickness and the local freestream velocity, U_0, at $\Delta x = -30$ mm is $Re_{\delta*} = 13{,}600$.

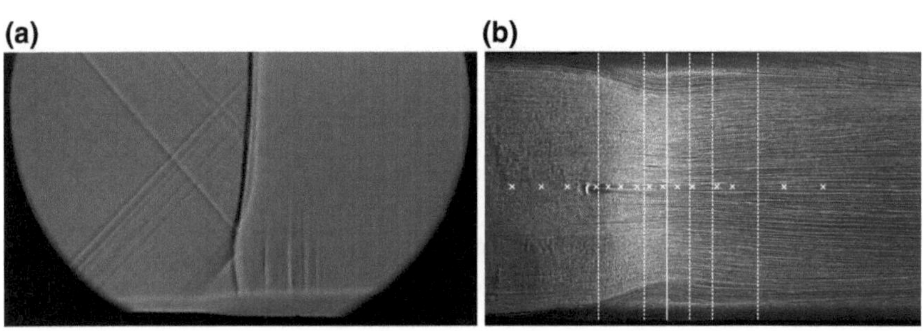

Fig. 1 Experimental visualization of **a** shock wave with λ-foot and **b** oil-flow visualization of boundary layer separation. In (**b**), *lines* indicate the locations of the velocity profiles from the shock at $\Delta x = -30, -10, 0, 10, 20$ and 40 mm. *Crosses* indicate locations where wall pressures were measured

For the transonic duct flow involving SBLI analyzed in this paper, conventional hybrid RANS-LES, DES and RANS approaches fail to predict the SBLI flow field [2, 5]. It is shown that the response of modeling to the onset of the corner separation bubble is one of the key issues for an accurate prediction of the shock wave and the subsequent SBLI flow. It is further revealed that an accurate prediction of the shock intensity is closely associated with the modeling capability and gives significant improvements on the prediction of the overall SBLI flow properties.

This work is a continuation of a previous work presented in [2] to improve the modeling and further to explore the underlying physics of SBLI with an improved prediction. The paper presents a zonal hybrid RANS-LES modeling approach based on a low-Re number $k - \omega$ model, which was applied to the transonic duct flow. Moreover, an extensive analysis is given of the simulations using the zonal approach. In addition to the zonal approach, SA-model [10] based DDES [12] and IDDES [11] were also used to simulate the flow. RANS results with the low-Re number $k - \omega$ model are also included for reference.

2 Zonal Hybrid RANS-LES Modeling

The PDH-LRN $k - \omega$ model by Peng et al. [9] (hereafter PDH-LRN) was shown in [2] to give reasonably accurate RANS results for the flow analyzed. Moreover, it was shown in [2] that RANS simulations using the Spalart-Allmaras model [10] and the Menter SST $k - \omega$ model [7] fail to give symmetric corner separation bubbles leading to a collapsed shock. The PDH-LRN is therefore used as the base model in the zonal hybrid RANS-LES modeling approach (hereafter ZHYBRID) used in this paper. The k and ω transport equations for the PDH-LRN model read:

$$\frac{D\rho k}{Dt} = \tau_{ij}\frac{\partial u_i}{\partial x_j} - D^k + \frac{\partial}{\partial x_j}\left[\left(\mu + \frac{\mu_t}{\sigma_k}\right)\frac{\partial k}{\partial x_j}\right] \tag{1}$$

$$\begin{aligned}\frac{D\rho\omega}{Dt} &= C_{\omega 1} f_\omega \frac{\omega}{k}\tau_{ij}\frac{\partial u_i}{\partial x_j} - C_{\omega 2}\rho\omega^2 \\ &+ \frac{\partial}{\partial x_j}\left[\left(\mu + \frac{\mu_t}{\sigma_\omega}\right)\frac{\partial \omega}{\partial x_j}\right] + C_\omega \frac{\mu_t}{k}\frac{\partial k}{\partial x_j}\frac{\partial \omega}{\partial x_j}\end{aligned} \tag{2}$$

$$\mu_t = C_\mu f_\mu \frac{\rho k}{\omega} \tag{3}$$

The turbulence-resolving LES mode, based on the PDH-LRN model, is incorporated through the dissipation term in the k-equation, see further [1, 3].

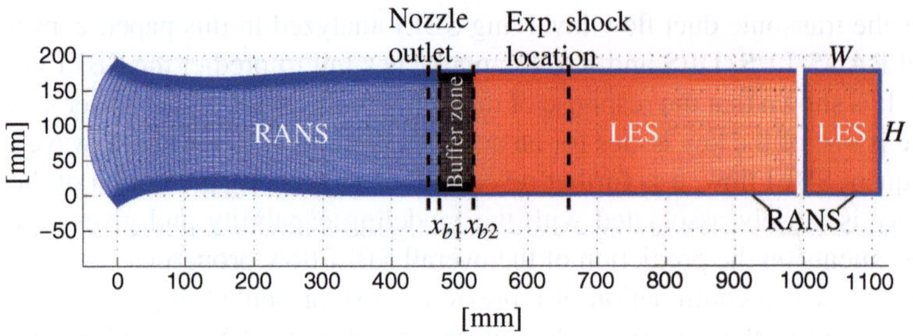

Fig. 2 Computational grid with RANS and LES zones used in the ZHYBRID simulations. *Blue*: RANS; *black*: buffer zone; *red*: LES. *Left grid*: xy-plane, flow from *left* to *right*. *Right grid* yz-plane, grid cross-section

$$D^k = C_k f_k \rho k \omega = \rho f_k \frac{k^{3/2}}{l_{turb}} \tag{4}$$

The RANS and LES zones are prescribed for this transonic duct flow as shown in Fig. 2. Three zones have been accordingly defined. The upstream attached boundary layer flow is treated in RANS mode, via a buffer zone, hybrid RANS-LES modeling is then used to accommodate the flow in the SBLI region and further downstream to the outflow section. In previous works [1, 3] on channel flow, the switch from RANS to LES was made at a prescribed interface plane orthogonal to the streamwise direction. Moreover, the RANS turbulent kinetic energy was manipulated across the interface to match the LES subgrid scale (SGS) turbulent kinetic energy level. Thus, the turbulent viscosity at the interface was reduced from a RANS level to a SGS level. However, in this work we instead use a buffer zone for the turbulent length scale according to Eqs. (5)–(7), which has additionally been defined in between the RANS and the hybrid RANS-LES zones. The buffer zone helps to eliminate possible discontinuities in the transition from RANS to LES mode and is patched between x_{b1} and x_{b2} in the streamwise direction upstream of the shock. In this paper $x_{b1} = 470$ mm and $x_{b2} = 520$ mm, see Figs. 2 and 4.

$$l_b = f_b l_{LES} + (1 - f_b)\, l_{RANS} \tag{5}$$

$$f_b = \tanh\left([C_{b1} r_b]^{C_{b2}}\right), \; C_{b1} = 2, \; C_{b2} = 3 \tag{6}$$

$$r_b = \frac{x - x_{b1}}{x_{b1} - x_{b2}}, \; x_{b1} \le x \le x_{b2} \tag{7}$$

The RANS-LES interface in the near-wall layer for hybrid RANS-LES modeling is predefined at a distance of $d_w = d_{ws}$ from the wall, where d_w is the wall distance and d_{ws} is the distance from the wall at which the RANS-LES switch is located.

In this paper $d_{ws} = 5\,\text{mm}$ ($d_{ws}/H = 0.028$). The turbulent length scale used in the zonal approach is defined as:

$$l_{turb} = \begin{cases} l_{RANS} & \text{if } x < x_{b1}, \\ l_{RANS} & \text{if } d_w \le d_{ws}, \\ l_b & \text{if } x_{b1} \le x \le x_{b2} \text{ and } d_w > d_{ws}, \\ l_{LES} & \text{if } x > x_{b2} \text{ and } d_w > d_{ws}. \end{cases} \quad (8)$$

$$l_{RANS} = \frac{k^{1/2}}{c_k \omega}, \; l_{LES} = \Psi_{PDH} C_{LES} \Delta_{dw} \quad (9)$$

$$\Delta_{dw} = \min\left(\max\left[c_w d_w, c_w \Delta_{max}\right], \Delta_{max}\right), \quad (10)$$
$$\Delta_{max} = \max\left(\Delta_x, \Delta_y, \Delta_z\right), \; c_w = 0.15$$

The LES length scale, Δ_{dw}, was originally used in the IDDES formulation [11]. However, compared to the original formulation, the grid step in the wall normal direction is omitted in the ZHYBRID model in the max-argument in Eq. (10). This modification was also used in [8]. The details in the hybrid RANS-LES model can be found in [3].

3 Computational Setup

The simulations presented were performed with the unstructured compressible Navier-Stokes solver Edge [6] on a hexahedral grid using $(n_{xmax}, n_{ymax}, n_{zmax}) = (306, 180, 136)$ grid points. The time-marching was done using an implicit second-order backward Euler scheme with a time step of $\Delta t = 4 \times 10^{-6}$ s. This gave a maximum convective CFL number of approximately 1. At each time step, 100–200 sub-iterations have been conducted to converge the solution using a three-stage Runge-Kutta scheme and an algebraic multigrid method. The momentum and turbulent transport equations were discretized using a second-order central differencing scheme.

On the inlet boundary, total pressure ($p_{0,in} = 147.5$ kPa), total temperature ($T_{0,in} = 293$ K) and the flow direction were specified according to the experiment. A turbulence intensity of one percent and $\mu_t/\mu = 1$ were specified to set the inlet boundary conditions for the turbulent quantities. On the outlet boundary a static pressure, p_{out}, was specified. It is noted here that the predicted location of the shock wave is very sensitive to the specified outlet pressure as seen in Table 1. As expected, a relatively large outlet pressure often leads to an early formation of the shock wave. The outlet pressure is not available from the experiment.

Table 1 Summary of simulations. Δx_{shock} and Δx_{nozzle} indicate time-averaged values

Case	P_{out} (Pa)	Δx^{a}_{shock} (mm)	Δx^{b}_{nozzle} (mm)
ZHYBRID-a	84,000	−84	121
ZHYBRID-b	78,500	0	205
SA-DDES	83,000	+61	266
SA-IDDES	83,250	+26	231
PDH-LRN (RANS)	84,000	+2	207

[a] Simulated shock location in relation to experimental location $\Delta x_{shock} = x_{shock,sim} - x_{shock,exp.}$
[b] Simulated shock location in relation to nozzle outlet location $\Delta x_{nozzle} = x_{shock,sim} - x_{nozzle}$.

In the simulations 8,000–16,000 time steps were needed to establish a fully-developed turbulence-resolving SBLI flow field from the initial RANS flow field. Another 8,000 time steps, which represents a fluid particle passing through the duct approximately ten times, were used for the statistical analysis.

In relation to the modeling approach and the outlet pressure, p_{out}, specified, the offset, Δx_{shock}, between the predicted and the measured shock location is presented in Table 1.

Two simulations using the ZHYBRID model are presented, one with the same outlet pressure used as in the PDH-LRN RANS simulation and another with the outlet pressure being adjusted to agree with the experimental shock location. The PDH-LRN RANS simulation is also included in Table 1 for reference. For the SA-DDES and SA-IDDES computations $p_{out} = 83{,}000$ and 83,250 Pa were used, respectively. Applying the SA-RANS model to this transonic duct flow gives an asymmetric flow field and a weak shock [2]. However, the shock location at the duct center line agrees reasonably well with experimental data for $p_{out} = 83{,}000$ Pa, which has motivated the choice of outlet pressure in the SA-DDES and SA-IDDES computations.

4 Results and Discussion

As seen in Table 1 and in Fig. 3a, the PDH-LRN RANS simulation using $p_{out} = 84{,}000$ Pa reproduces well the shock location and the shock intensity, i.e. the pressure rise across the shock. Applying the outlet pressure $p_{out} = 84{,}000$ Pa to the ZHYBRID model (denoted ZHYBRID-a), the simulated shock location is shifted upstream as compared to the PDH-LRN RANS simulation and to the experiments. Moreover, a weak secondary shock is observed. Nevertheless, the shock intensity in terms of the pressure distribution over the shock is reasonably comparable to the experimental data as shown in Fig. 3.

As observed in Fig. 4, which shows surface streamlines on the bottom wall in the SBLI region, the predicted corner separation bubbles using the ZHYBRID model are larger than in the PDH-LRN RANS simulation. Moreover, in both ZHYBRID simulations, cross flow is observed downstream of the shock, moving from the side

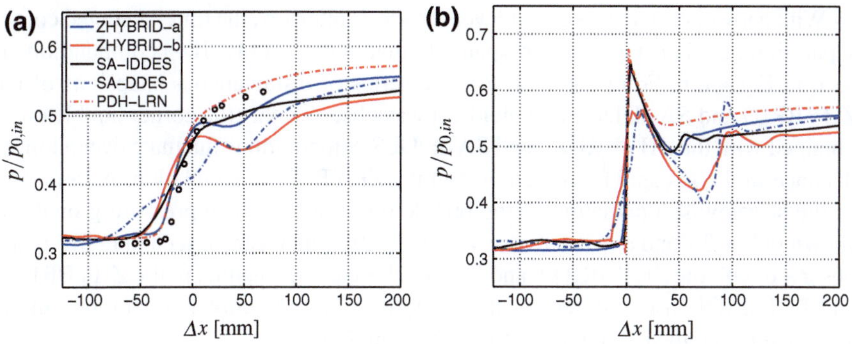

Fig. 3 Time-averaged static pressure across the shock. **a** Surface pressure on *bottom* wall. **b** Pressure along the duct *center line*. Markers are experimental data

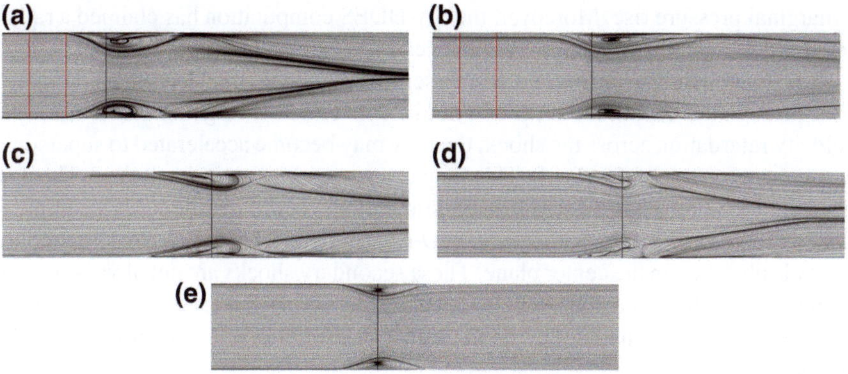

Fig. 4 Time-averaged surface streamlines on the *bottom* wall around the shock. The *black line* indicates shock location, and the *red lines* indicate the buffer zone used in ZHYBRID. **a** ZHYBRID-a. **b** ZHYBRID-b. **c** SA-DDES. **d** SA-IDDES. **e** PDH-LRN

walls into the duct center part, which is not present in the PDH-LRN RANS simulation. This is particularly the case for the ZHYBRID-a simulation. The larger corner separation bubble and the subsequent cross flow observed in the ZHYBRID simulations have reduced the skin friction as compared to the PDH-LRN simulation. Since the same outlet pressure was used in the ZHYBRID-a and PDH-LRN RANS simulations, the shock moved upstream in the ZHYBRID-a case.

The second ZHYBRID simulation presented in Table 1 (denoted ZHYBRID-b) was given a lower outlet pressure in order to match the experimental shock location. As shown in Fig. 3a, a weaker shock intensity is obtained as expected. Figure 3b, which shows the time-averaged static pressure across the shock along the duct center line, indicates further that the main pressure peak is much reduced and much less sharp as compared to the ZHYBRID-a result. Moreover, the secondary shock, downstream of the main shock, is more intensive in the ZHYBRID-b simulation.

With a lower outlet pressure and weaker shock intensity, the time-averaged corner separation bubble is longer as compared to that in the ZHYBRID-a computation, as seen in Fig. 4a, b. From the analyses of a series of instantaneous snapshots of the ZHYBRID and SA-IDDES flow fields, it was observed that the corner bubble onset obtained with the ZHYBRID-b and SA-IDDES computations fluctuate over a longer distance in the streamwise direction than the ZHYBRID-a simulation. As an effect of the streamwise fluctuation of the bubble onset, a less steep pressure growth, as shown in Fig. 3a, and a reduced near wall velocity, which can be seen in in Fig. 6, are observed with the ZHYBRID-b and SA-IDDES cases compared to the ZHYBRID-a and PDH-LRN cases. Moreover, it was noticed that the corner separation bubble shifts from side to side in the ZHYBRID-b simulation.

A strong fluctuation in the onset of the corner separation bubbles was also observed in the SA-DDES case. The time-averaged pressure distribution in Fig. 3 for the SA-DDES computation indicates an obvious shock collapse near the wall with only a marginal pressure rise. Moreover, the SA-DDES computation has claimed a rather intensive secondary shock over the duct center, as shown in Fig. 3b.

It is noted that an exaggerated corner separation bubble may lead to a contracted flow through the duct in the form of a fluidic convergent-divergent nozzle. After the velocity retardation across the shock, the flow may become accelerated to supersonic speed downstream of the shock. This may cause secondary shock patterns. This has been the case in the hybrid RANS-LES simulations, but not in the RANS computation, as shown in Fig. 3b. With ZHYBRID-a and SA-IDDES only a weak secondary shock is observed in the center plane. These secondary shocks are not always present in the resolved instantaneous flow field, but appears when the cross flow towards the duct center is strong and contracts the effective duct flow area downstream of the shock.

The difference seen in the pressures at $\Delta x = 200$ in Fig. 3 is partly due to the different back pressures applied at the outlet boundary. Since the cross flow from the side walls towards the duct center is over-predicted, the contraction of the flow contributes to a higher velocity in the duct center region and a lower static pressure than in the experiments. Further downstream, however, it was found that the pressure recovers to the level specified at the boundary.

The switch from RANS to LES can be estimated from the turbulent viscosities in the center plane of the duct shown in Fig. 5. The ZHYBRID RANS-LES switch was prescribed and occurred at the same distance from the wall in both simulations. For the SA-DDES and the SA-IDDES model, the RANS-LES switch is regulated by the local flow properties in the near-wall region. As shown in the profiles of the SA-DDES turbulent viscosity and the streamwise velocity in the center plane, most of the boundary layer has been treated in RANS mode as expected. In the center plane, the RANS-LES interface with the SA-IDDES model occurs closer to the wall as compared to the SA-DDES and the ZHYBRID model. This may suggest that SA-IDDES is in wall-modeled LES mode and has resolved a part of the boundary layer.

A more rapid reduction of the turbulent viscosity is observed in the SA-IDDES model than in the ZHYBRID model. The RANS-LES switch occurs somewhat farther

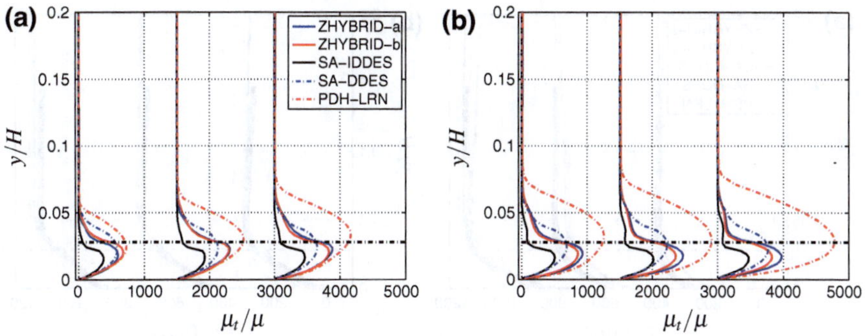

Fig. 5 Time-averaged turbulent viscosity in the SBLI region. *Dash-dotted horizontal black line* indicates ZHYBRID switch from RANS to LES. **a** $\Delta x = -30, -10, 0$ mm. **b** $\Delta x = 10, 20, 40$ mm

out from the wall with the ZHYBRID compared to the SA-IDDES and the turbulent viscosity produced by the ZHYBRID model in the near-wall RANS layer is higher, which contributes to the higher level of turbulent viscosity observed in the LES region. Furthermore, the near-wall switch from RANS to LES in the ZHYBRID model does not involve any empirical functions to manipulate the turbulent viscosity across the RANS-LES interface in order to mitigate the log-layer mismatch as in the SA-IDDES model. On the other hand, in [1, 3] it was shown in channel flow that the zonal approach based on PDH-LRN has only a weak dependency of the RANS-LES switch location with respect to the log-layer mismatch.

Far upstream at $x = 400$ mm (not shown here though), where the effect of the shock is negligible, the velocity profiles of the PDH-LRN and the two ZHYBRID simulations coincide as expected, since the ZHYBRID model is in RANS mode in this flow region. Comparing the SA based models with the PDH-LRN based models at $x = 400$ mm, the SA based models predicted a slightly fuller profile than the PDH-LRN based models.

The shape and width of the corner separation bubbles, as well as the shock foot shape affect significantly the velocity profiles in the center plane of the duct as shown in Fig. 6. The ZHYBRID-a simulation predicts a velocity profile at $\Delta x = -30$ mm which is in reasonably good agreement with experimental data, indicating a well predicted shock foot and a bubble onset relative to the shock. Moving downstream to $\Delta x = -10$ mm, all simulations predict too low a velocity in the near wall region, indicating too early a boundary-layer growth caused by the λ-shape shock. At the shock, the profiles are close to each other in the near-wall region, except for the SA-DDES simulation, and compare fairly well with the experimental velocity profile.

The near-wall flow after the shock is best predicted in the SA-IDDES and the PDH-LRN simulations with a slight over-prediction of the velocity in the off-wall region for the SA-IDDES model. At $\Delta x = 20$ and 40 mm, both ZHYBRID simulations over-predicts the flow velocity. Moreover, the different shape of the SA-DDES velocity profile at $\Delta x = 10$ mm, compared to the experiment and to the other simulations,

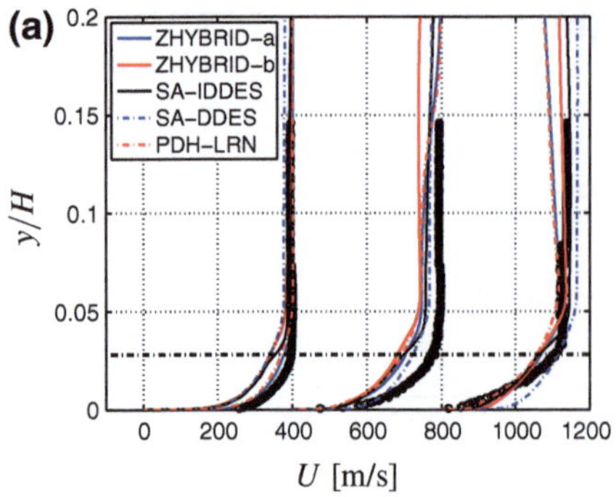

Fig. 6 Time-averaged streamwise velocity, U, in the SBLI region. *Dash-dotted horizontal black line* indicates ZHYBRID switch from RANS to LES. Markers are experimental data. **a** $\Delta x = -30$, -10, 0 mm. **b** $\Delta x = 10$, 20, 40 mm

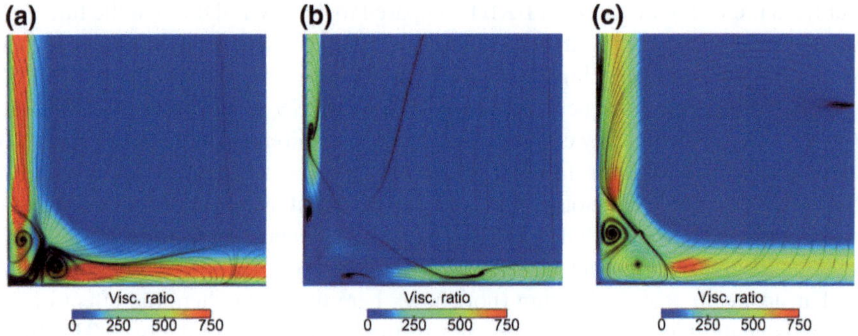

Fig. 7 Time-averaged streamlines and turbulent viscosity field normalized by molecular viscosity in a cross-sectional cut at $\Delta x = 20$ mm in the *lower left* (aft looking forward) corner separation bubble. **a** ZHYBRID-a. **b** SA-IDDES. **c** SA-DDES

is an effect of time-averaging the shock motion, which is more pronounced for this simulation.

Large differences in modeling/resolving the flow have been observed when the corner separation flow is analyzed more in details. In Fig. 7, the time-averaged field of turbulent viscosity is overlayed with streamlines of the recirculating corner flow for the models used. In the corner, the SA-DDES model has produced a high level of turbulent viscosity and only very large structures have been resolved as shown in Fig. 8.

The SA-IDDES model, on the other hand, adapts nicely to the recirculating unsteady flow in the corner separation bubble, switching to LES and resolving reasonably well the core of the recirculations. In the ZHYBRID-a case, the "eye" of the recirculation is located in the prescribed RANS region. However, the outer part of the recirculation region predicted in the ZHYBRID-a simulation is resolved with LES, which gives a higher level of resolved turbulent structures compared

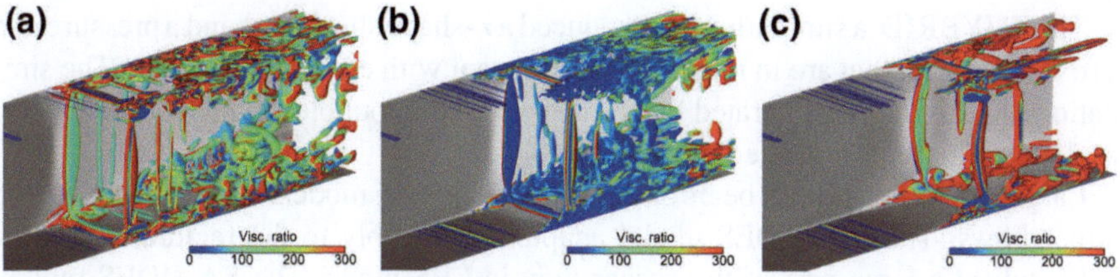

Fig. 8 Resolved turbulent structures displayed using iso-surfaces of Q-criterion, $QH^2/U_0^2 = 1$. Skin friction displayed on the duct walls **a** ZHYBRID-a. **b** SA-IDDES. **c** SA-DDES

to the SA-DDES model as shown in Fig. 8. In an additional simulation using the ZHYBRID model with $d_{ws} = 2.5$ mm and $p_{out} = 78{,}500$ Pa (not included in this paper), more resolved turbulence was observed in the corner separation bubble compared to $d_{ws} = 5$ mm. It is also observed in all simulations that the turbulent structures are concentrated in the corner separation bubbles and the cross flow moving from the side walls towards the duct center. Moreover, no resolved turbulence is present in the upstream attached boundary layer in any of the simulations since the attached boundary layer is simulated in RANS mode and the strong fluctuations imposed by the shock have a very weak effect on the flow field upstream of the SBLI region. The concentration of resolved turbulent structures in the SBLI region is thus expected and the aim of the modeling techniques chosen.

Analyzing the ZHYBRID simulations, the blending function used in the buffer zone rapidly forces the ZHYBRID model to switch from RANS to LES without any discontinuities in the turbulent quantities. However, in the ZHYBRID-a simulation, the buffer zone, which is located close to the onset of the corner bubble, delays the formation of resolved turbulence compared to the SA-IDDES computation.

5 Summary and Conclusions

A zonal hybrid RANS-LES approach (ZHYBRID), based on the PDH-LRN $k - \omega$ model (PDH-LRN), has been applied to a transonic duct flow with shock/boundary-layer interaction. The ZHYBRID simulations have been compared to simulations performed with the Spalart-Allmaras (SA) based DDES and IDDES computations. Moreover, a RANS simulation using the PDH-LRN model has been included as a reference. ZHYBRID simulations with two different outlet pressures have been presented in order to investigate the effect of shock intensity on the corner separation bubbles and the flow downstream of the shock in the SBLI region.

The simulations clearly indicate that the onset of the corner separation bubble relative to the shock location is one of the key issues for accurately predicting the SBLI flow. Moreover, the incoming boundary layer, the shock intensity and the shock foot shape are closely interconnected and an accurate prediction of these SBLI flow properties is challenging using hybrid RANS-LES simulations.

The ZHYBRID-a simulation has produced a λ-shape shock foot and a pressure rise across the shock, that are in reasonable agreement with experimental data. The simulation has slightly exaggerated the corner separation bubbles, which subsequently has led to a more pronounced cross flow.

Large differences have been observed on how the models have simulated the corner flows. The SA-IDDES model adapted reasonably to the recirculating flow and resolved a large part of the corner flow in LES mode. The SA-DDES model, on the other hand, produced high levels of turbulent viscosity in the corner separation bubbles and has much less turbulent structures resolved. As compared to the ZHYBRID-a and the experimental data, the SA-IDDES predicted a weaker λ-foot, which caused a slightly less distinct pressure rise in the wall pressure distribution across the shock. In the ZHYBRID simulations, the RANS and LES zones were prescribed and the model has resolved a part of the corner separation bubbles in LES mode, leading to an improved SBLI flow prediction.

As a continuation of this paper, a grid resolution study will be carried out in order to analyze the effect on the SBLI region. Moreover, an extended duct will be used in order to locate the outlet boundary farther downstream from the SBLI region.

Acknowledgments This work was funded by the Swedish Governmental Agency for Innovation Systems (VINNOVA) in the Swedish National Flight Research Program (NFFP, Contract No. 2009-01346 and 2013-01209) and Saab Aeronautics with support from the EU project ATAAC, Contract No. 233710. Computational resources were supported by the Swedish National Infrastructure for Computing (SNIC).

References

1. Arvidson, S.: Assessment and some improvements of hybrid RANS-LES methods. Licenciate thesis, Applied Mechanics, Chalmers University of Technology (2013). ISSN:1652–8565, 2013:01
2. Arvidson, S., Peng, S.H., Davidson, L.: Feasibility of hybrid RANS-LES of shock/boundary-layer interaction in a duct. In: Fu, S., et al. (eds.) Progress in Hybrid RANS-LES Modelling, NNFM, vol. 117, pp. 245–256. Springer, Heidelberg (2012)
3. Arvidson, S., Davidson, L., Peng, S.H.: Hybrid RANS-LES modeling using a low-Reynolds-number $k - \omega$ based model. In: AIAA paper 2014–0225, National Harbour, Maryland (2014)
4. Bruce, P.J.K., Babinsky, H.: Unsteady shock wave dynamics. J. Fluid Mech. **603**, 463–473 (2008)
5. Bruce, P.J.K., Babinsky, H., Tartinville, B., Hirsch, C.: Corner effect and asymmetri in transonic channel flows. AIAA J. **49**, 2382–2392 (2011)
6. Eliasson, P.: EDGE, a Navier-Stokes solver for unstructured grids. Scientific report FOI-R-0298-SE, Computational Aerodynamics Department, Aeronautics Division, FOI (2001)
7. Menter, F.R.: Two-equation eddy-viscosity turbulence models for engineering applications. AIAA J. **32**, 1598–1605 (1994)
8. Gritskevich, M.S., Garbaruk, A.V., Schütze, J., Menter, F.: Development of DDES and IDDES formulations for the $k - \omega$ shear stress transport model. Flow Turbul. Combust. **88**, 431–449 (2012)
9. Peng, S.H., Davidson, L., Holmberg, S.: A modified low-Reynolds-number $k - \omega$ model for recirculating flows. J. Fluids Eng. **119**, 867–875 (1997)

10. Spalart, P.R., Allmaras, S.R.: A one-equation turbulence model for aerodynamic flows. La Recherche Aérospatiale **1**, 5–21 (1994)
11. Shur, K.L., Spalart, P.R., Strelets, M.K., Travin, A.K.: A hybrid RANS-LES approach with delayed-DES and wall-modelled LES capabilities. Int. J. Heat Fluid Flow **29**, 1638–1649 (2008)
12. Spalart, P.R., Deck, S., Shur, M.L., Squires, K.D., Strelets, M.K., Travin, A.: A new version of detached-eddy simulation, resistant to ambiguous grid densities. Theory Comput. Fluid Dyn. **20**, 181–195 (2006)

Hybrid RANS-LES Methods Applied to Acoustic Problems

Abdelkader Frendi

Abstract In this paper, two acoustic problems of importance to the engineering community are solved using hybrid RANS-LES methods. More specifically, results from the problems of supersonic flow over a surface mounted protuberance and that of a supersonic jet impingement on a flat plat are presented. Through these studies, it is shown that hybrid RANS-LES methods are adequate at resolving the necessary frequency bandwidth relevant to these problems. Good agreement with available experimental results is obtained for both problems.

1 Introduction

It is well-known that the external skin of an aircraft, whether military or civilian, contains various protuberances. Some of these protuberances are necessary; cameras, lights and pitot tubes; and some are due to design constraints. In this paper, cylindrical protuberances of different heights are studied numerically using a hybrid RANS-LES method and the pressure field on the surrounding surfaces is computed for a supersonic boundary layer flow [1, 2]. The knowledge of these pressure fields is critical to the structures surrounding the protuberance as sonic fatigue can be detrimental to the health of these structures [3, 4]. Often times structural failures occur in areas of high loading. Similarly, the problem of supersonic jet impingement on a flat plate can result in very high sound pressure levels detrimental to neighbouring structures as well as human subjects [5–10]. An added problem for vertical take-off and landing vehicles is the lift loss experienced by the vehicle near the ground. Away from structures, the problem of jet noise itself has been studied extensively both experimentally and computationally and its components have been largely agreed upon to be due to shear and self-noise. For supersonic jets, shock noise is also an important component.

In this paper, we present computational results from two complex problems namely, supersonic flow over a surface mounted protuberance and a supersonic jet

A. Frendi (✉)
MAE Department, UAHuntsville, Huntsville, AL 35899, USA
e-mail: kader.frendi@uah.edu

S. Girimaji et al. (eds.), *Progress in Hybrid RANS-LES Modelling*,
Notes on Numerical Fluid Mechanics and Multidisciplinary Design 130,
DOI 10.1007/978-3-319-15141-0_19

impinging on a flat plate. Our previous experience using hybrid RANS-LES showed promising results for different unsteady problems [11–13] and therefore this method will be used in the present computations.

2 Mathematical Model and Method of Solution

The mathematical models are composed of the standard unsteady Reynolds Averaged Navier-Stokes equations with a two-equation eddy viscosity model based on Mentor shear stress transport model [14]. This model has been modified to accommodate the Delayed Detached Eddy Simulation of Nichols and Nelson [15]. In solving the partial differential equations subject to problem specific boundary conditions, Loci-Chem [16] and Overflow 2 [17] codes have been used. Both codes are finite volume codes having second order accuracy in space and time. Loci-Chem was used to solve the problem of a surface mounted protuberance in a supersonic turbulent boundary layer. Figure 1a shows the computational domain used for the surface mounted protuberance problem, and Fig. 1b shows a typical grid used in the vicinity of the protuberance; a hybrid unstructured grid. A two domain solution methodology was used due to computational cost; at first a large domain is used to obtain the mean flow using RANS, then the domain shown on Fig. 1a is used for hybrid RANS-LES. The shape of the smaller domain shown on Fig. 1a is chosen based on the physics of the problem and with the intent to reduce the size of the grid. The inflow boundary was set by the RANS results, while the other boundaries are farfield (vertical away from walls), outflow (downstream) and no-slip (over rigid wall). Overflow 2 was used to solve the problem of a supersonic jet impingement on a flat plate. Figure 2a

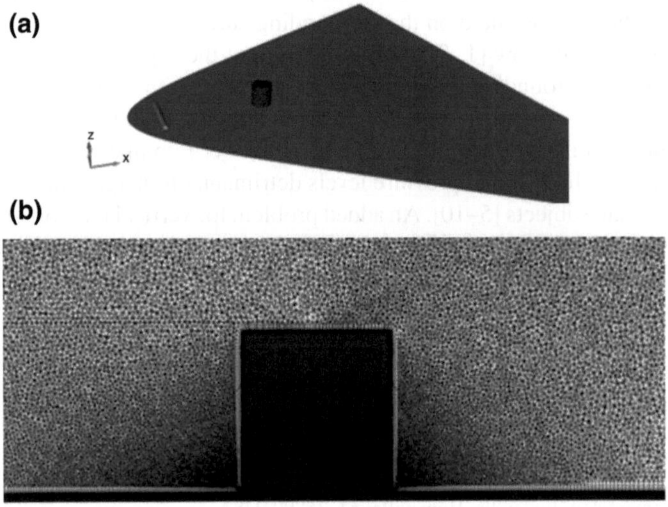

Fig. 1 **a** Computational domain used for the cylindrical protuberance problem, **b** hybrid unstructured grid near the protuberance

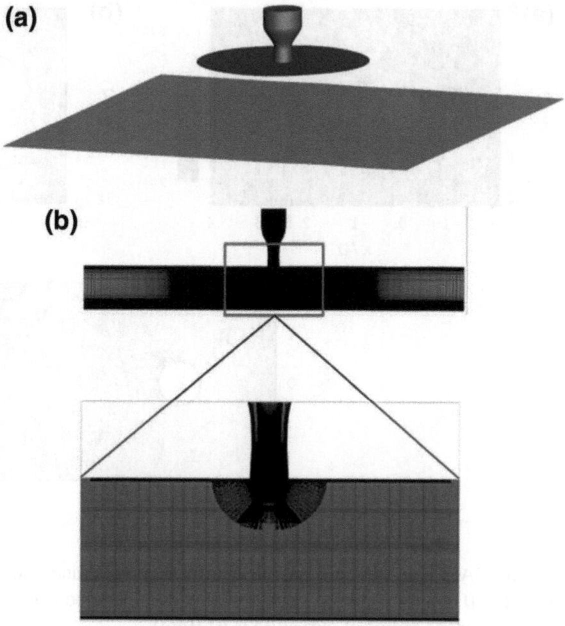

Fig. 2 **a** Computational domain used for the impinging jet problem, **b** structured overlapping grid in the vicinity of the jet

shows the three dimensional computational domain used, and Fig. 2b shows a cross section of a typical grid used for the jet problem with an added zoomed in view. In Fig. 2a, a circular plate of diameter 25.4 cm is flush-mounted to the exit of the converging-diverging nozzle to mimic the experimental setup used in [10]. In the experiment, the circular plate is used to determine lift loss. The supersonic jet was produced from a converging-diverging nozzle with an area ratio that gives ideally expanded flow at the nozzle exit. Overflow 2 supports an overset grid topology.

The grid density in both problems was such that $y^+ < 1$ in the wall-normal direction and the other two directions; x^+ and $z^+ \sim 50$. The other grid refinement criteria used is that we use at least 30 points per wavelength for the highest frequency computed to insure low dissipation and dispersion. The number of cells used to solve each problem varied between 60 and 100 million.

3 Results and Discussions

3.1 *Surface Mounted Protuberance*

In this problem, the parameters investigated are the protuberance height to boundary layer thickness and the surface curvature. The flow parameters used are; free stream Mach number 1.6, free stream air velocity 463.3 m/s, free stream total air temperature 422 K and a Reynolds number per meter of 4.92 million. Protuberance height to

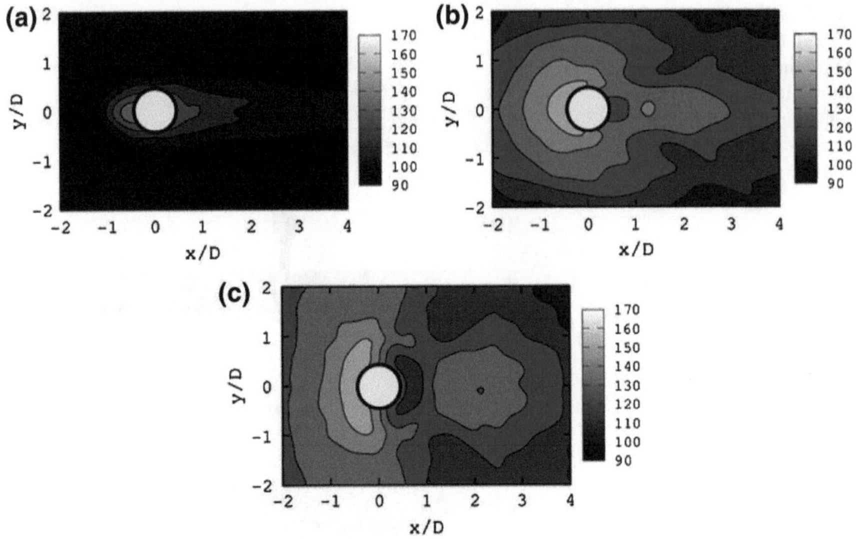

Fig. 3 OASPL in dB on a flat surface for a protuberance height-to-boundary layer thickness of **a** 0.5 **b** 2.0 and for a curved surface with a protuberance height-to-boundary layer thickness of 2.0 (**c**) (the surface radius of curvature is 61.9 cm)

boundary layer thickness ratios, h/δ, studied are 0.5, 1.0 and 2.0, the radius of curvature of the surfaces studied are; 61.9, 30.95 cm and a flat surface. The boundary layer thickness is measured at the protuberance location. The 61.9 cm curved surface corresponds to the one used in the experiments [18–20]. All protuberances were cylindrical with a diameter, D, of 1.905 cm. Figure 3 shows the OASPL map on a flat surface for a small protuberance, Fig. 3a, and a tall protuberance, Fig. 3b. It is clear from the figure that the tall protuberance, Fig. 3b, has a huge impact on the surface in terms of high acoustic loads over a large area surrounding the protuberance. Figure 3c shows the OASPL contours on a curved surface for a protuberance having a height to boundary layer thickness ratio of 2.0. The curved surface has a radius of curvature of 61.9 cm. The figure shows that the presence of the curvature, Fig. 3c, leads to the spread of the high dB-levels area in the spanwise direction, i.e. y-direction, both upstream and downstream of the protuberance. Downstream of the protuberance, Fig. 3c shows lower dB-levels over a larger area than that on a flat surface, Fig. 3b.

Figure 4a shows that the wall pressure spectra at various distances downstream of the protuberance collapse well at low frequencies using the outer scaling; i.e. using displacement thickness, δ^*, and free stream velocity, U_∞; but not so well at high frequencies. The ratio of protuberance height to boundary layer thickness in this case is 2.0 and the surface is flat. The two-point space correlation three diameters downstream of the protuberance shows the presence of large structures with stretching in the spanwise direction, Fig. 4b. The space-time correlation, Fig. 4c, shows a short lived coherence in both space and time. These results were obtained by averaging a

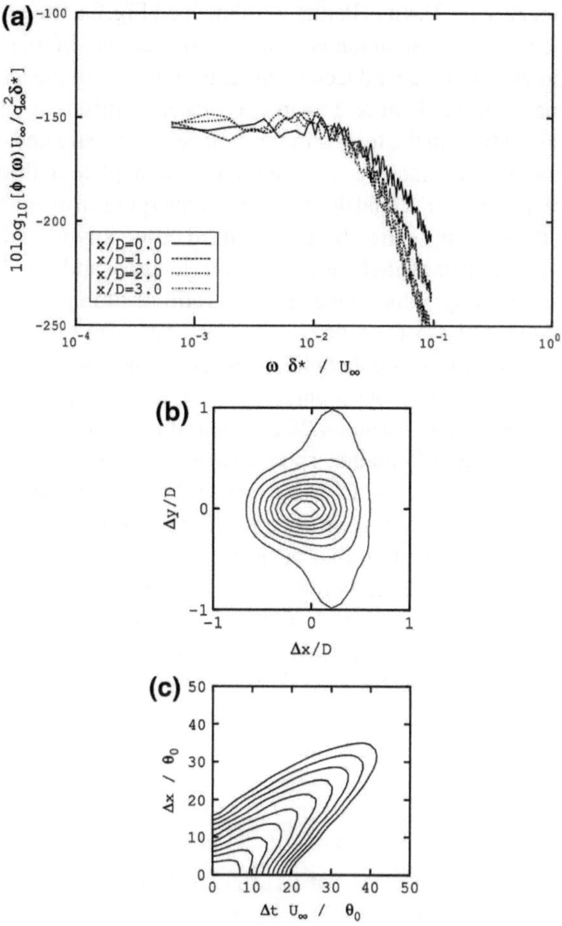

Fig. 4 **a** Wall pressure spectra at various downstream locations from the protuberance, **b** two-point space correlation three diameters downstream of the protuberance, **c** two-point space-time correlation three diameters downstream of the protuberance. The protuberance height to boundary layer thickness is 2.0 and for both the two-point and the space-time correlations the contours levels are between 0.1 and 0.9

statistically steady time sample consisting of 10,000 time steps corresponding to 50 flow through times. One can also deduce the convection velocity of the flow structures based on the slope of the curves, in this case it is found to be between 0.6 and 0.77 of the freestream.

3.2 Supersonic Impinging Jet

The geometry analysed in this study is an ideally expanded, supersonic jet impinging on a flat plate. The geometric parameters in the CFD analysis match those presented in the test setup found in Ref. [10]. In the test, the supersonic jet was produced from a converging-diverging nozzle with an area ratio that gives ideally expanded flow at the nozzle exit (Pexit = Patm). As shown on Fig. 2a, the exit of the converging-diverging nozzle was flush mounted to a 25.4 cm diameter lift plate that was used to measure the downward force induced by air entrainment into the jet. A square impingement plate measuring 2.44 m × 2.44 m was placed at different distances below the nozzle exit depending on the test configuration being investigated. The computational model did not extend to the edges of the impingement plate so the plate covers the entire bottom of the computational domain. The converging-diverging nozzle is described as having a converging section based on a third order polynomial and a conic diverging section with a constant angle of 3°. The nozzle throat and exit diameters are 2.54 and 2.75 cm respectively, which gives an area ratio resulting in a perfectly expanded exit Mach number of 1.5 for a nozzle pressure ratio (NPR) of 3.7. The parameter of interest in this study is the distance between the nozzle exit to impingement plate distance. Based on the experimental work of Krothapalli et al. [10], three distances were studied; 3.75, 4.0 and 4.25 d, where d is the nozzle throat diameter. Figure 5 shows the near field frequency spectra for a jet exit to impingement plate distance of 3.75 d. The location of the CFD observer is closer to the axis than that in the experiments, hence the difference in spectral level. Given that, the CFD computed spectrum is in fair agreement to that measured. The spectrum shows three dominant peaks, L1, L2 and L3 with L3 being more dominant than the other two. Similar agreement is obtained at the distances of 4.0 and 4.25 d as shown on Figs. 6 and 7, respectively.

Figure 8 shows instantaneous contours of the density gradients, i.e. a CFD schlieren. Acoustic waves are clearly shown to propagate from the impingement point back to the exit of the jet. This feedback loop triggers shear layer disturbances that propagate downstream to the impingement plate. In addition, there is also an

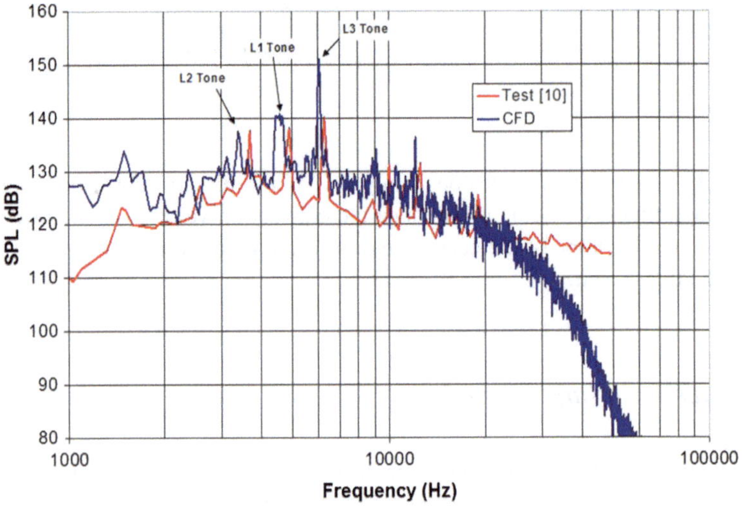

Fig. 5 Comparison of the near field frequency spectra for a nozzle exit to impingement plate distance of 3.75 d

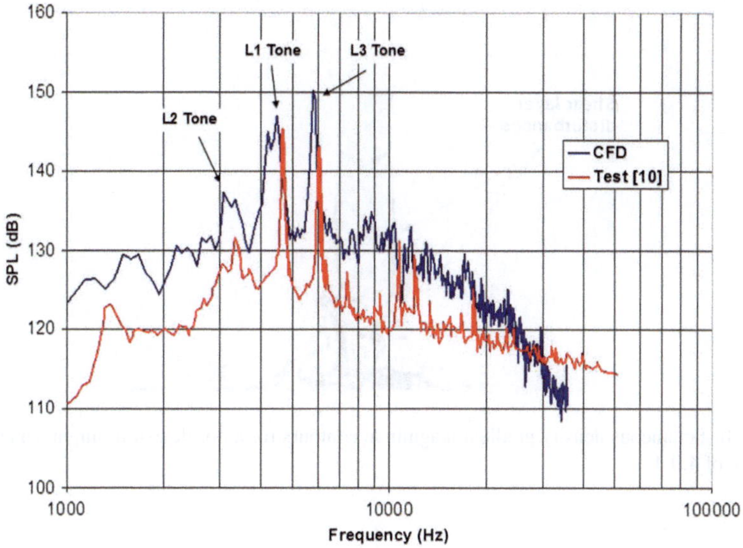

Fig. 6 Comparison of the near field frequency spectra for a nozzle exit to impingement plate distance of 4.0 d

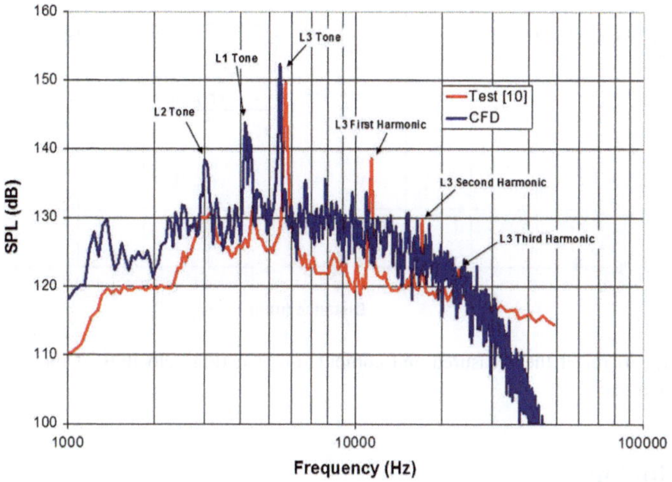

Fig. 7 Comparison of the near field frequency spectra for a nozzle exit to impingement plate distance of 4.25 d

interaction between the acoustic waves and the core of the jet itself. Figure 9 shows the convection speed of the shear layer disturbances obtained from the CFD results and those measured [10]. Good agreement is obtained between our CFD results and the experimental data.

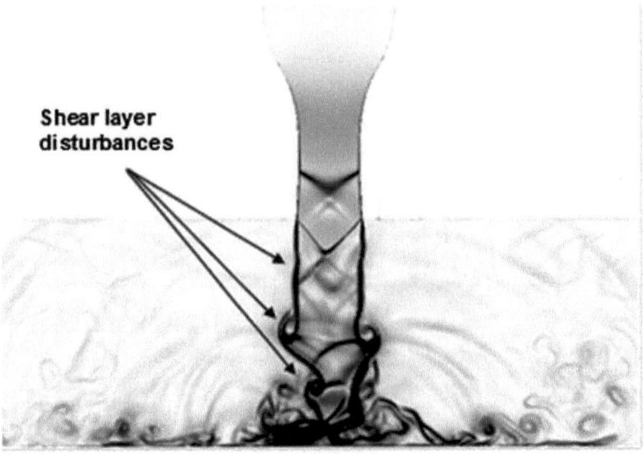

Fig. 8 Instantaneous density gradient magnitude contours for a nozzle exit to impingement plate distance of 4.0 d

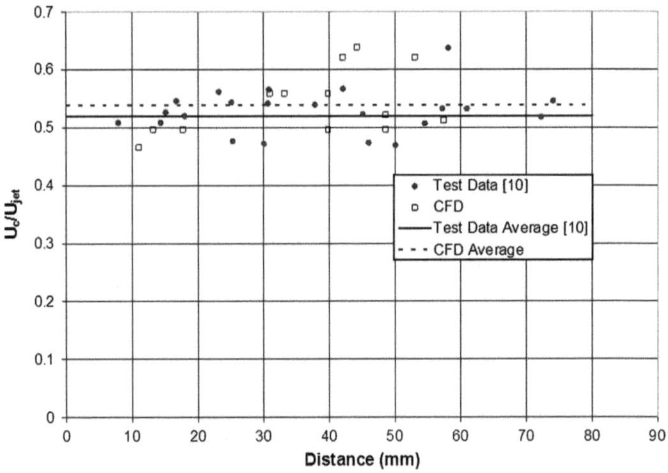

Fig. 9 Comparison of the measured and computed convection velocities of the jet shear layer vortical structures

4 Conclusions

In this paper, hybrid RANS-LES methods have been used to study two very complex acoustic problems; surface mounted protuberance in a supersonic boundary layer and a supersonic jet impinging on a flat plate. For both problems, our CFD results compared well to experimental measurements, which validates the use of hybrid RANS-LES methods for these types of problems. However, one needs to be very careful with the grids used when solving such problems in order to avoid dissipation and dispersion errors.

References

1. Hahn, P.V., Frendi, A.: Interaction of three-dimensional protuberances with a supersonic turbulent boundary layer. AIAA J. **51**(7), 1657–1666 (2013)
2. Frendi, A.: Coupling between a supersonic turbulent boundary layer and a flexible structure. AIAA J. **35**(1), 58–66 (1997)
3. Miles, J.W.: On structural fatigue under random loading. J. Aeronaut. Sci. **21**(11), 753–762 (1954)
4. Eisinger, F.L.: Acoustic fatigue of impellers of rotating machinery. J. Pressure Vessel Technol. **124**(2), 154–160 (2002)
5. Henderson, B., Powell, A.: Experiments concerning tones produced by an axisymmetric choked jet impinging on flat plates. J. Sound Vib. **168**(2), 307–326 (1993)
6. Krothapalli, A.: Discrete tones generated by an impinging underexpanded rectangular jet. AIAA J. **23**(12), 1910–1915 (1985)
7. Powel, A.: The sound-producing oscillations of round underexpanded jets impinging on normal plates. J. Acoust. Soc. Am. **83**(2), 515–533 (1988)
8. Lamont, P.J., Hunt, B.L.: The impingement of underexpanded, axisymmetric jets on perpendicular and inclined flat plates. J. Fluid Mech. **100**(3), 471–511 (1980)
9. Tam, C.K.W., Ahuja, K.K.: Theoretical model of discreet tone generation by impinging jets. J. Fluid Mech. **214**, 67–87 (1990)
10. Krothapalli, A., Rajkuperan, E., Alvi, F., Lourenco, L.: Flow field and noise characteristics of a supersonic impinging jet. J. Fluid Mech. **392**, 155–181 (1999)
11. Peugeot, J.W., Frendi, A.: Toward the understanding of flow-induced vibrations in a rocket-engine manifold. J. Propul. Power **29**(6), 1468–1477 (2013)
12. Frendi, A., Tosh, A., Girimaji, S.: Flow past a backward facing step: comparison of PANS, DES and URANS results with experiments. Int. J. Comput. Methods Eng. Sci. Mech. **8**(1), 23–38 (2006)
13. Brown, M.R., Frendi, A.: Supersonic jet impingement on a flat plate. In: 18th AIAA/CEAS Aeroacoustics Conference, AIAA-2012-2261, 2012
14. Menter, F.R.: Two-equation eddy-viscosity turbulence models for engineering applications. AIAA J. **32**(8), 1598–1605 (1994)
15. Nichols, R.H., Nelson, C.C.: Application of hybrid RANS/LES turbulence models. In: AIAA-2003-0083, 2003
16. Luke, E.A., Tong, X., Wu, J., Cinnella, P., Chamberlain, R.: Chem 3.2: A finite-rate viscous chemistry solver-the user guide (2011)
17. Nichols, R.H., Tramel, R.W., Buning, P.G.: Solver and turbulence model upgrades to OVERLOW 2 for unsteady and high-speed applications. In: AIAA-2006-2824 (2006)
18. Robertson, J.E.: Experimental program for the investigation of transonic flow around protuberances in the AEDC-16T wind tunnel facility. Wyle Laboratories Report TM 68–6, Huntsville, AL, 1968
19. Robertson, J.E.: Characteristics of the static and fluctuating pressure environments induced by three-dimensional protuberances at transonic Mach numbers. Wyle Laboratories Report WR 69–3, Huntsville, AL, 1969
20. Robertson, J.E.: Prediction of in-flight fluctuating pressure environments including protuberance induced flow. Wyle Laboratories Report WR 71–10, Huntsville, AL, 1971

On the Hybrid RANS-LES of Compressible Flows

Massimo Germano

Abstract The formulation of a RANS-LES method for compressible flows faces many fundamental problems. The usual adoption of mass weighted, Favre, averages in the RANS region, and filtered Favre averages in the LES region produces additional difficulties to the connection of the RANS and LES zones, and the dialogue between RANS and LES quantities, both as regards the post processing of data and the formulation of bridging RANS-LES closures, is not so easy. In this paper this problem is explored on the basis of the operational filtering approach that with the introduction of the generalized central moments gives a unified averaging invariant formulation both of RANS and LES. The extension of this approach to the mass weighted averages will be discussed and a mixed RANS-LES system of equations will be presented.

1 Introduction

Hybrid RANS-LES modelling is both a practical need and a theoretical challenge. From the practical point of view we would like to reduce the computational costs of the simulations, and optimal strategies could be conceived in order to apply LES where we have separated flows and wakes. On the contrary near the walls in the boundary layers we could exploit the best of RANS modeling. Many hybrid RANS-LES approaches that should connect RANS and LES simulations in different turbulent zones have been proposed and the related scenario of models is very complex and difficult to classify [1, 2]. From the fundamental point of view we can look to all that as a difficult dialogue between two very different representations of the turbulent field, the statistical and filtered ones, difficult to conciliate, both as regards the postprocessing of LES databases and the connections of different modeling approaches. The main difficulty as regards the dialogue between the RANS formulation based on the statistical average and the classical LES formulation based

M. Germano (✉)
Duke University, Durham, NC, USA
e-mail: mg234@duke.edu

S. Girimaji et al. (eds.), *Progress in Hybrid RANS-LES Modelling*,
Notes on Numerical Fluid Mechanics and Multidisciplinary Design 130,
DOI 10.1007/978-3-319-15141-0_20

on local space filters is that these two formulations are mutually inconsistent. Only in the case of homogeneous turbulent flows this dialogue is free of ambiguities, and in order to resolve this theoretical conflict a lot of studies are presently done. We have hybridization mainly based on blending or rescaling RANS or LES models with a direct dependence on the grid resolution, and some of them have been successfully applied. Near-wall models and zonal RANS/LES have been used where a great resolution is necessary, but the problem of the connection between different regions generates anomalies, in particular the so called log-layer mismatch, due to the fact that the log layers in the RANS and LES regions are not lined up and an unphysical buffer layer appears near the RANS/LES interface [3]. Hybrid models that do not contain an explicit dependance on the computational grid and try to extend RANS models to unsteady situations are the partially integrated transport model, PITM, [4] and the partially averaged Navier-Stokes, PANS, methodologies [5]. Formally they can be applied to every RANS model, and explicit damping coefficients that represent the amount of the targeted amounts of resolved kinetic energies and dissipation rates are introduced.

The extension of the hybrid RANS-LES methods to compressible flows faces additional difficulties. That is mainly due to the fact that compressible flows are usually managed in terms of the mass-weighted, Favre, averages, both as regards the statistical RANS simulations and the LES simulations, where the so called Filtered Favre averages are introduced. In this paper the problems related to the extension of the Hybrid RANS-LES modeling approaches to compressible flows are mainly examined from a basic point of view. We will first of all introduce the operational filtering approach and the generalized central moments that provide a unified and consistent formulation both of the RANS statistical approach and the LES averaging approach. We will derive some multiscale relations that could be usefully employed in modelling and finally we will apply all that to the Navier-Stokes equations for compressible flows.

2 Turbulence Without Fluctuations

It is a personal opinion of the author that a first great obstacle to a simple dialogue between RANS and LES are the fluctuations. We do not obviously refer to the fluctuations associated to the statistical average that are very useful in the applications. Their peculiar property is that their mean value is zero, and when this elementary property is not satisfied their utility becomes doubtful. That is the case of the filtered LES averages and of the mass weighted averages, both the statistical and the filtered ones. What is also very interesting is that they can be easily removed and substituted with an operational formulation that can be applied to every linear constant preserving averaging operator. Let us consider the basic problem of modeling expressed by the relation

$$\begin{aligned} \overline{ab} &= M(\overline{a}, \overline{b}) \\ \overline{abc} &= M(\overline{a}, \overline{b}, \overline{c}) \\ \overline{abcd} &= \cdots \end{aligned} \tag{1}$$

where with the overline we indicate a generic averaging operator that we only assume linear and constant preserving

$$\overline{a+b} = \overline{a} + \overline{b}; \ \overline{ka} = k\overline{a} \tag{2}$$

and where with $M(\overline{a}, \overline{b})$ we indicate a renormalization of $\overline{ab}$ in terms of the filtered quantities. In the case of the statistical average provided with the additional property

$$\overline{a\,\overline{b}} = \overline{a}\,\overline{b} \tag{3}$$

we introduce the statistical fluctuations

$$a = \overline{a} + a'; \quad b = \overline{b} + b'; \quad c = \overline{c} + c'$$

and we can write

$$\begin{aligned} \overline{ab} &= \overline{a}\,\overline{b} + \overline{a'b'} \\ \overline{abc} &= \overline{a}\,\overline{b}\,\overline{c} + \overline{a}\,\overline{b'c'} + \overline{b}\,\overline{c'a'} + \overline{c}\,\overline{a'b'} + \overline{a'b'c'} \\ \overline{abcd} &= \cdots \end{aligned} \tag{4}$$

but if the averaging operator is not provided with the property (3) that is not true and the relations (4) cannot be extended as such. A procedure that generalize the statistical relations (4) is the operational approach [6]. Let us write

$$\begin{aligned} \overline{ab} &= \lambda_1 \overline{a}\,\overline{b} + \tau(a, b) \\ \overline{abc} &= \lambda_2 \overline{a}\,\overline{b}\,\overline{c} + \lambda_3(\overline{a}\,\tau(b, c) + \overline{b}\,\tau(c, a) + \overline{c}\,\tau(a, b)) + \tau(a, b, c) \\ \overline{abc} &= \cdots \end{aligned} \tag{5}$$

where $\tau(a, b)$, $\tau(b, c)$, $\tau(c, a)$, $\tau(a, b, c)$, ... are by definition the generalized central moments that we assume provided with the Galilean invariant property

$$\tau(k, a) = \tau(k, a, b) = \tau(k, a, b, \ldots) = 0 \tag{6}$$

where k is a constant. If we impose this condition we have $\lambda_1 = \lambda_2 = \lambda_3 = 1$, and the final relations

$$
\begin{aligned}
\overline{ab} &= \overline{a}\,\overline{b} + \tau(a,b) \\
\overline{abc} &= \overline{a}\,\overline{b}\,\overline{c} + \overline{a}\;\tau(b,c) + \overline{b}\;\tau(c,a) + \overline{c}\;\tau(a,b) + \tau(a,b,c) \\
\overline{abcd} &= \cdots
\end{aligned}
\tag{7}
$$

will provide a consistent unified formulation of RANS and LES. We remark finally that in the case of the statistical average we obviously recover the usual expressions in terms of the statistical fluctuations

$$
\tau(a,b) = \overline{a'b'};\; \tau(a,b,c) = \overline{a'b'c'}; \qquad \cdots \tag{8}
$$

2.1 *Mass Weighted, Favre, Averages*

If we apply this operational formulation to compressible flows we have typically to model terms like $\overline{\rho a}$ and $\overline{\rho ab}$, where ρ is the density, and we can write

$$
\begin{aligned}
\overline{\rho a} &= \overline{\rho}\,\overline{a} + \tau(\rho,a) \\
\overline{\rho ab} &= \overline{\rho}\,\overline{a}\,\overline{b} + \overline{\rho}\;\tau(a,b) + \overline{a}\;\tau(\rho,b) + \overline{b}\;\tau(\rho,a) + \tau(\rho,a,b)
\end{aligned}
\tag{9}
$$

In order to simplify the formulation it is usual to introduce the so called mass weighted or Favre averages $\tilde{a}, \tilde{b}$ and the extension of this procedure to LES produces density weighted filtered equations [7]. Also in this case we do not use the fluctuations and we write in a common general form

$$
\begin{aligned}
\overline{\rho a} &= \overline{\rho}\,\tilde{a} \\
\overline{\rho ab} &= \overline{\rho}\,\tilde{a}\,\tilde{b} + \overline{\rho}\;\vartheta(a,b)
\end{aligned}
\tag{10}
$$

where the quantities $\vartheta(a,b)$ are by definition the mass weighted generalized central moments. It is easy to see that

$$
\begin{aligned}
\tilde{a} &= \overline{a} + \frac{\tau(\rho,a)}{\overline{\rho}} \\
\vartheta(a,b) &= \tau(a,b) - \frac{\tau(\rho,a)\;\tau(\rho,b)}{\overline{\rho}^2} + \frac{\tau(\rho,a,b)}{\overline{\rho}}
\end{aligned}
\tag{11}
$$

3 The Hybrid RANS-LES Filter

In the framework of the operational filtering formulation a simple method that naturally extends the RANS approach based on the statistical operator $\mathscr{E}$ and the LES approach based on an explicit or implicit filtering operator $\mathscr{F}$ relies on the introduction of an additive hybrid operator $\mathscr{H}$ defined as [8–10].

$$\mathscr{H} = k_F \mathscr{F} + k_E \mathscr{E}; \qquad k_F + k_E = 1 \tag{12}$$

where k_F is a blending factor possibly function of space and time. We remark that the constraint $k_F + k_E = 1$ is very important from a physical point of view: it imposes the preservation of the constants, fundamental property for an averaging operator. Given for example the hybrid velocity field $\langle u_i \rangle_H$ produced by the computation, the reconstruction of the statistical mean field $\langle u_i \rangle_E$ and of the filtered velocity field $\langle u_i \rangle_F$ are theoretically given by

$$\begin{aligned}\langle u_i \rangle_E &= \langle \langle u_i \rangle_H \rangle_E \\ \langle u_i \rangle_F &= \frac{\langle u_i \rangle_H - k_E \langle u_i \rangle_E}{k_F}\end{aligned} \tag{13}$$

where we have assumed that $\mathscr{E}\mathscr{F} = \mathscr{E}\mathscr{H} = \mathscr{E}$ and where we have introduced the notation $\langle \cdots \rangle$ to distinguish between different averaging operators. In order to derive the subgrid stress associated to this hybrid operator we apply the operational formulation of the filtering approach [6] based on the generalized central moments that provides a unified theory both for the RANS and the LES equations. Following the operational definition of the generalized central moments

$$\begin{aligned}\tau_F(u_i, u_j) &\equiv \langle u_i u_j \rangle_F - \langle u_i \rangle_F \langle u_j \rangle_F \\ \tau_E(u_i, u_j) &\equiv \langle u_i u_j \rangle_E - \langle u_i \rangle_E \langle u_j \rangle_E \\ \tau_H(u_i, u_j) &\equiv \langle u_i u_j \rangle_H - \langle u_i \rangle_H \langle u_j \rangle_H\end{aligned} \tag{14}$$

the subgrid RANS/LES stress associated to the hybrid additive operator (12) is given by

$$\begin{aligned}\tau_H(u_i, u_j) = &k_F \tau_F(u_i, u_j) + k_E \tau_E(u_i, u_j) \\ &+ k_F \langle u_i \rangle_F \langle u_j \rangle_F + k_E \langle u_i \rangle_E \langle u_j \rangle_E - \langle u_i \rangle_H \langle u_j \rangle_H\end{aligned} \tag{15}$$

We can equivalently write

$$\begin{aligned}\tau_H(u_i, u_j) = &k_F \tau_F(u_i, u_j) + k_E \tau_E(u_i, u_j) \\ &+ \frac{k_E}{k_F} \left(\langle u_i \rangle_H - \langle u_i \rangle_E \right) \left(\langle u_j \rangle_H - \langle u_j \rangle_E \right)\end{aligned} \tag{16}$$

and we notice that $\tau_H(u_i, u_j)$ is composed by a weighted average of the RANS Reynolds stresses and the LES subgrid stresses plus a second term dependent on the statistical fluctuations $\langle u_i \rangle_H - \langle u_i \rangle_E$. This term becomes more and more important with decreasing values of the blending factor k_F. We remark that the blending factor k_F is directly related to the ratio of resolved turbulent energy theoretically captured by the hybrid RANS-LES simulation. We have formally

$$\frac{\tau_E(\langle u_i\rangle_H, \langle u_j\rangle_H)}{\tau_E(u_i, u_j)} = k_F^2 \left(1 - \frac{\langle \tau_F(u_i, u_j)\rangle_E}{\tau_E(u_i, u_j)}\right) \tag{17}$$

We finally remark that the additive hybrid operator defined by (12) is linear but if k_F is space and/or time dependent it does not commute with the derivatives and we have

$$\begin{aligned} \mathscr{H}\frac{\partial}{\partial x_i} - \frac{\partial}{\partial x_i}\mathscr{H} &= \frac{\partial k_F}{\partial x_i}(\mathscr{E} - \mathscr{F}) = \frac{1}{k_F}\frac{\partial k_F}{\partial x_i}(\mathscr{E} - \mathscr{H}) \\ \mathscr{H}\frac{\partial}{\partial t} - \frac{\partial}{\partial t}\mathscr{H} &= \frac{\partial k_F}{\partial t}(\mathscr{E} - \mathscr{F}) = \frac{1}{k_F}\frac{\partial k_F}{\partial t}(\mathscr{E} - \mathscr{H}) \end{aligned} \tag{18}$$

This fact introduces a lot of new terms in the hybrid filtered Navier-Stokes equations. We remark that the quantities

$$\begin{aligned} \left(\frac{1}{k_F}\frac{\partial k_F}{\partial x_i}\right) &= \left(\frac{\partial \ln k_F}{\partial x_i}\right) = \frac{1}{l_i} \\ \left(\frac{1}{k_F}\frac{\partial k_F}{\partial t}\right) &= \left(\frac{\partial \ln k_F}{\partial t}\right) = \frac{1}{\vartheta} \end{aligned} \tag{19}$$

represent respectively the inverses of characteristic lengths l_i and time ϑ of the filter variability in space and time. If these lengths and time are large compared to the physical turbulent and geometrical lengths/time of the flow it could be justified to discard the terms produced by the non commutativity.

4 The Hybrid RANS-LES Filter Extended to Compressible Flows

The Hybrid RANS-LES filtering approach has been extended to compressible flows by Sánchez-Rocha and Menon [11] and has been applied to the hybrid RANS-LES simulation of an attached turbulent boundary layer over a flat plate. Here we will resume the main difficulties of such extension, due mainly to the fact that it requires the introduction of the associated hybrid mass weighted average. If we apply the Hybrid RANS-LES filter to the Navier-Stokes equations

$$\begin{aligned} &\partial_t \rho + \partial_j(\rho u_j) = 0 \\ &\partial_t(\rho u_i) + \partial_j(\rho u_i u_j + \pi_{ij}) = 0 \\ &\partial_t(\rho e) + \partial_j(\rho e u_j + u_i \pi_{ij} + q_j) = 0 \end{aligned} \tag{20}$$

where

$$\begin{aligned}
\pi_{ij} &= p\delta_{ij} - \sigma_{ij}; \qquad \sigma_{ij} = \mu S^d_{ij} \\
S_{ij} &= \partial_i u_j + \partial_j u_i; \qquad S^d_{ij} = S_{ij} - \frac{S_{kk}\delta_{ij}}{3} \\
q_i &= -\kappa \partial_i T; \quad \mu = \mu_0 \left(\frac{T}{T_0}\right)^\alpha; \quad Pr = \frac{\mu c_p}{\kappa} \\
e &= c_v T + \frac{u_i u_i}{2}; \qquad p = \rho R T
\end{aligned} \tag{21}$$

we produce mixed RANS-LES equations. Formally they are given by

$$\begin{aligned}
\partial_t \langle\rho\rangle_H + \partial_j \langle\rho u_j\rangle_H &= f \\
\partial_t \langle\rho u_i\rangle_H + \partial_j \langle\rho u_i u_j + \pi_{ij}\rangle_H &= \varphi_i \\
\partial_t \langle\rho e\rangle_H + \partial_j \langle\rho e u_j + u_i \pi_{ij} + q_j\rangle_H &= \psi
\end{aligned} \tag{22}$$

where the non-commutative terms are

$$\begin{aligned}
f &= \partial_t \langle\rho\rangle_H - \langle\partial_t \rho\rangle_H + \partial_j \langle\rho u_j\rangle_H - \langle\partial_j(\rho u_j)\rangle_H \\
&= \frac{\langle\rho\rangle_H - \langle\rho\rangle_E}{\tau} + \frac{\langle\rho u_j\rangle_H - \langle\rho u_j\rangle_E}{l_j}
\end{aligned}$$

$$\begin{aligned}
\varphi_i &= \partial_t \langle\rho u_i\rangle_H - \langle\partial_t(\rho u_i)\rangle_H \\
&\quad + \partial_j \langle\rho u_i u_j + \pi_{ij}\rangle_H - \langle\partial_j(\rho u_i u_j + \pi_{ij})\rangle_H \\
&= \frac{\langle\rho u_i\rangle_H - \langle\rho u_i\rangle_E}{\tau} \\
&\quad + \frac{\langle\rho u_i u_j + \pi_{ij}\rangle_H - \langle\rho u_i u_j + \pi_{ij}\rangle_E}{l_j}
\end{aligned}$$

$$\begin{aligned}
\psi &= \partial_t \langle\rho e\rangle_H - \langle\partial_t(\rho e)\rangle_H \\
&\quad + \partial_j \langle\rho e u_j + u_i \pi_{ij} + q_j\rangle_H - \langle\partial_j(\rho e u_j + u_i \pi_{ij} + q_j)\rangle_H \\
&= \frac{\langle\rho e\rangle_H - \langle\rho e\rangle_E}{\tau} \\
&\quad + \frac{\langle\rho e u_j + u_i \pi_{ij} + q_j\rangle_H - \langle\rho e u_j + u_i \pi_{ij} + q_j\rangle_E}{l_j}
\end{aligned} \tag{23}$$

We remark that the RANS and the LES average associated to the hybrid filtered density $\langle\rho\rangle_H$ continue to be recovered by the relations

$$\langle\rho\rangle_E = \langle\langle\rho\rangle_H\rangle_E; \quad \langle\rho\rangle_F = \frac{\langle\rho\rangle_H - k_E \langle\rho\rangle_E}{k_F}$$

but in the case of the velocity field we have to introduce the Favre averages $\{u_i\}_H$, $\{u_i\}_E$, and $\{u_i\}_F$ associated respectively to the hybrid filter $\mathscr{H}$ and the RANS and LES filters $\mathscr{E}$ and $\mathscr{F}$

$$\begin{aligned}\langle \rho u_i \rangle_H &= \langle \rho \rangle_H \{u_i\}_H \\ \langle \rho u_i \rangle_E &= \langle \rho \rangle_E \{u_i\}_E \\ \langle \rho u_i \rangle_F &= \langle \rho \rangle_F \{u_i\}_F\end{aligned}$$

Always by assuming that that $\mathscr{E}\mathscr{F} = \mathscr{E}\mathscr{H} = \mathscr{E}$ we have

$$\langle \rho u_i \rangle_E = \langle \langle \rho u_i \rangle_H \rangle_E; \quad \langle \rho u_i \rangle_F = \frac{\langle \rho u_i \rangle_H - k_E \langle \rho u_i \rangle_E}{k_F}$$

and we can finally derive the relations

$$\begin{aligned}\{u_i\}_E &= \frac{\langle \langle \rho \rangle_H \{u_i\}_H \rangle_E}{\langle \rho \rangle_E} \\ \{u_i\}_F &= \frac{\langle \rho \rangle_H \{u_i\}_H - k_E \langle \rho \rangle_E \{u_i\}_E}{k_F \langle \rho \rangle_F}\end{aligned}$$

More complex is the derivation of the hybrid subgrid terms. We have to introduce the Favre subgrid terms $\vartheta_H(u_i, u_j)$, $\vartheta_E(u_i, u_j)$, $\vartheta_F(u_i, u_j)$, associated respectively to the hybrid filter and to the RANS and LES filter

$$\begin{aligned}\langle \rho u_i u_j \rangle_H &= \langle \rho \rangle_H \{u_i\}_H \{u_j\}_H + \langle \rho \rangle_H \vartheta_H(u_i, u_j) \\ \langle \rho u_i u_j \rangle_E &= \langle \rho \rangle_E \{u_i\}_E \{u_j\}_E + \langle \rho \rangle_E \vartheta_E(u_i, u_j) \\ \langle \rho u_i u_j \rangle_F &= \langle \rho \rangle_F \{u_i\}_F \{u_j\}_F + \langle \rho \rangle_F \vartheta_F(u_i, u_j)\end{aligned} \tag{24}$$

and following the same procedure we can finally write

$$\begin{aligned}\vartheta_H(u_i, u_j) = & \frac{k_F \langle \rho \rangle_F}{\langle \rho \rangle_H} \vartheta_F(u_i, u_j) + \frac{k_E \langle \rho \rangle_E}{\langle \rho \rangle_H} \vartheta_E(u_i, u_j) \\ & + \frac{k_F \langle \rho \rangle_F}{\langle \rho \rangle_H} \{u_i\}_F \{u_j\}_F + \frac{k_E \langle \rho \rangle_E}{\langle \rho \rangle_H} \{u_i\}_E \{u_j\}_E \\ & - \{u_i\}_H \{u_j\}_H\end{aligned} \tag{25}$$

Different modeling strategies and different numerical implementation of the hybrid subgrid stress $\vartheta_H(u_i, u_j)$ can be conceived. If we introduce given RANS and LES models

$$\vartheta_E(u_i, u_j) \sim M_E(\{u_i\}_E, \{u_j\}_E); \quad \vartheta_F(u_i, u_j) \sim M_F(\{u_i\}_F, \{u_j\}_F) \tag{26}$$

we can write

$$\vartheta_H(u_i, u_j) \sim \frac{k_F\langle\rho\rangle_F}{\langle\rho\rangle_H} M_F(\{u_i\}_F, \{u_j\}_F) + \frac{k_E\langle\rho\rangle_E}{\langle\rho\rangle_H} M_E(\{u_i\}_E, \{u_j\}_E)$$
$$+ \frac{k_F\langle\rho\rangle_F}{\langle\rho\rangle_H}\{u_i\}_F\{u_j\}_F + \frac{k_E\langle\rho\rangle_E}{\langle\rho\rangle_H}\{u_i\}_E\{u_j\}_E \tag{27}$$
$$- \{u_i\}_H\{u_j\}_H$$

and we have a mixed RANS-LES modeling formulation. Another possible approach that is currently explored is more consistent with the data produced by the simulation. We recall that the comparison of different LES or hybrid simulation can only be done with reference to the statistical RANS values extracted by the computed database. From the RANS-LES hybrid equations we can compute the hybrid quantities $\langle\rho\rangle_H$ and $\{u_i\}_H$. Then we can recover the RANS and the filtered values associated to the LES filter $\mathscr{F}$

$$\langle\rho\rangle_E = \langle\langle\rho\rangle_H\rangle_E; \quad \{u_i\}_E = \frac{\langle\langle\rho\rangle_H\{u_i\}_H\rangle_E}{\langle\rho\rangle_E}$$
$$\langle\rho\rangle_F = \frac{\langle\rho\rangle_H - k_E\langle\rho\rangle_E}{k_F}; \quad \{u_i\}_F = \frac{\langle\rho\rangle_H\{u_i\}_H - k_E\langle\rho\rangle_E\{u_i\}_E}{k_F\langle\rho\rangle_F}$$

and finally we can recover and validate the Favre subgrid stress $\vartheta_E(u_i, u_j)$ directly by assuming that $\mathscr{E}\mathscr{F} = \mathscr{E}$

$$\vartheta_E(u_i, u_j) = \frac{\langle\langle\rho\rangle_F M_F(\{u_i\}_F, \{u_j\}_F)\rangle_E}{\langle\rho\rangle_E}$$
$$+ \frac{\langle\langle\rho\rangle_F\{u_i\}_F\{u_j\}_F\rangle_E}{\langle\rho\rangle_E} - \{u_i\}_E\{u_j\}_E$$

where $M_F(\{u_i\}_F, \{u_j\}_F)$ is the subgrid model assumed for the LES filtering operators. This procedure is both a post processing procedure of the hybrid data and a possible consistent modeling approach. In this case the hybrid stress is given by the relation

$$\vartheta_H(u_i, u_j) \sim \frac{k_F\langle\rho\rangle_F}{\langle\rho\rangle_H} M_F(\{u_i\}_F, \{u_j\}_F) + \frac{k_E}{\langle\rho\rangle_H}\langle\langle\rho\rangle_F M_F(\{u_i\}_F, \{u_j\}_F)\rangle_E$$
$$+ \frac{k_F\langle\rho\rangle_F}{\langle\rho\rangle_H}\{u_i\}_F\{u_j\}_F + \frac{k_E}{\langle\langle\rho\rangle_H}\langle\langle\rho\rangle_F\{u_i\}_F\{u_j\}_F\rangle_E \tag{28}$$
$$- \{u_i\}_H\{u_j\}_H$$

that only depends on the LES model $M_F(\{u_i\}_F, \{u_j\}_F)$.

5 Conclusions

In the paper we have examined the main difficulties of the extension of a hybrid RANS-LES procedure to compressible flows. In particular we have analyzed a mixed RANS-LES procedure recently proposed by the author [8] that is formally exact and that connects the Reynolds equations and the LES equations by means of a hybrid filter depending on a hybrid blending factor k_F. This hybrid approach has been applied to incompressible flows [9, 10] and compressible flows [11] and is now under detailed examination by a group of research in Milano composed by A. Abbà, L. Bonaventura, M. Nini and the present relator. The aim of the present research is to implement the compressible hybrid RANS-LES filtered equations in a variational formulation inside a Discontinuous Galerkin finite element numerical code. In particular the attention will be focused on the problem of the production of small scale fluctuations in the transition region from RANS to LES and on the role of the blending factor. A compressible numerical code will be used, and the preliminary study will be limited to low Mach number flow. The first basic test will be the fully developed turbulent channel flow at low Mach number in order to assess the method and its DG coupling. The no-model option and the dynamic version of the anisotropic subgrid model [12] will be carefully investigated in order to have a clear idea of the peculiarities of method in standard and well documented situations. Then more challenging and more specific tests are foreseen: the temporally developing and the spatially developing turbulent channel flows. The specific benefits that we expect from this research are to show that the proposed hybrid filtering method coupled with a DG implementation works well in standard numerical tests. Another specific benefit attended is a deeper insight in the dialogue between two very different representations of turbulence, the statistical and the filtered one.

References

1. Fröhlich, J., von Terzi, D.: Hybrid LES/RANS methods for the simulation of turbulent flows. Prog. Aerosp. Sci. **44**, 349–377 (2008)
2. Sagaut, P., Deck, S.: Large eddy simulations for aerodynamics: status and perspectives. Philos. Trans. R. Soc. A **367**, 2849–2860 (2009)
3. Hamba, F.: Analysis of filtered Navier-Stokes equation for hybrid RANS/LES simulation. Phys. Fluids **23**, 015108–015113 (2011)
4. Schiestel, R., Dejoan, A.: Towards a new partially integrated transport model for coarse grid and unsteady turbulent flow simulations. Theoret. Comput. Fluid Dyn. **18**, 443–468 (2005)
5. Girimaji, S.S.: Partially averaged Navier-Stokes model for turbulence: a Reynolds-averaged Navier-Stokes to direct numerical simulation bridging method. J. Appl. Mech. Trans. ASME **73**, 413–421 (2006)
6. Germano, M.: Turbulence: the filtering approach. J. Fluid Mech. **238**, 325–336 (1992)
7. Suman, S., Girimaji, S.S.: On the invariance of compressible Navier-Stokes and energy equations subject to density-weighted filtering. Flow Turbul. Combust. **85**, 383–396 (2010)
8. Germano, M.: Properties of the hybrid RANS/ LES filter. Theoret. Comput. Fluid Dyn. **17**, 225–231 (2004)

9. Rajamani, B., Kim, J.: A hybrid-filter approach to turbulence simulation. Flow Turbul. Combust. **85**, 421–441 (2010)
10. Abbà, A., Germano, M.: A Mixed RANS/LES Model Applied to the Channel Flow, iTi2012 Bertinoro, ITALY, Progress in Turbulence, Springer, New York (2013)
11. Sánchez-Rocha, M., Menon, S.: The compressible hybrid RANS/LES formulation using an additive operator. J. Comput. Phys. **228**, 2037–2062 (2009)
12. Abbà, A., Cercignani, C., Valdettaro, L.: Analysis of subgrid scale models. Comput. Math. Appl. **46**(4), 521–536 (2003)

Unified RANS-LES Simulations of Turbulent Swirling Jets and Channel Flows

Stefan Heinz, Michael K. Stöllinger and Harish Gopalan

Abstract The accurate and efficient simulation of both attached and separated flows represents a huge challenge. RANS methods suffer from the lack of ability to simulate instantaneous turbulence structures, and LES methods are computationally very expensive regarding the simulation of wall-bounded flows, which have to be considered very often. A promising alternative is the use of hybrid RANS-LES methods, but existing hybrid methods like DES face many questions. The paper focuses on the use of unified RANS-LES methods implied by stochastic analysis as an alternative to using existing hybrid RANS-LES methods. The theoretical basis of the approach applied and applications to turbulent channel flows and turbulent swirling jet flows will be presented. The accuracy and cost features of the unified RANS-LES model will be discussed in comparison with other (in particular DES) hybrid methods.

1 Introduction

Accurate and efficient simulations of both attached and separated flows are a requirement to properly address many relevant problems, for example, airfoil separation and stall, and shock induced separation at high angles of attack. In particular the simulation of flow separation, which is a requirement to deal with such problems, represents a huge challenge. Reynolds-Averaged Navier Stokes (RANS) equation methods were

S. Heinz (✉)
Department of Mathematics, University of Wyoming, 1000 E. University Avenue, Laramie, WY 82071, USA
e-mail: heinz@uwyo.edu

M. K. Stöllinger
Mechanical Engineering Department, University of Wyoming, 1000 E. University Avenue, Laramie, WY 82071, USA
e-mail: mstoell@uwyo.edu

H. Gopalan
Mechanical Engineering, Union College, 218 Steinmetz Hall, 807 Union Street, Schenectady, NY 12308, USA
e-mail: harish.gopalan@gmail.com

S. Girimaji et al. (eds.), *Progress in Hybrid RANS-LES Modelling*,
Notes on Numerical Fluid Mechanics and Multidisciplinary Design 130,
DOI 10.1007/978-3-319-15141-0_21

proven to have significant difficulties to deal with separated flows because of low-Reynolds number effects, transition, and the relevance of unsteady motions. The use of large eddy simulation (LES) is seen to represent a way for developing a general and accurate computational method for the prediction of turbulent flows. However, the application of LES to separated flows also faces significant problems. The biggest problem is given by the huge computational cost of such simulations, which usually require the consideration of wall effects. An attractive approach to deal with this problem is the use of hybrid RANS-LES methods to make use of the best of both approaches: the computational efficiency of RANS methods and the ability of LES to resolve unsteady motions. However, such hybrid methods like detached eddy simulation (DES) [1] and improved delayed DES (IDDES) [2] face problems regarding the simulation of both attached and separated flows.

A promising approach to address the problems described in the preceding paragraph is the use of a hybrid RANS-LES model based on a sound theory. Such a model can be obtained by using a stochastic model for turbulent velocities as basis for the development of hybrid RANS-LES methods [3–7]. The purpose of this paper is to describe the basis of this approach and recent applications of the hybrid RANS-LES model obtained. The theoretical basis will be described in Sects. 2 and 3. Recent applications to turbulent channel flows and swirling turbulent jet flows will be described in Sects. 4 and 5, respectively. Observations obtained and ongoing further developments and applications will be described in Sect. 6.

2 Realizable Stress Models

For the incompressible flow considered, the LES equations for the turbulent velocity field read

$$\frac{\partial \widetilde{U}_i}{\partial x_i} = 0, \qquad \frac{\widetilde{D}\widetilde{U}_i}{\widetilde{D}t} = -\frac{1}{\rho}\frac{\partial P}{\partial x_i} + 2\nu \frac{\partial \widetilde{S}_{ik}}{\partial x_k} - \frac{\partial \tau_{ik}^d}{\partial x_k}. \tag{1}$$

Here, $\widetilde{U}_i$ refers to the filtered velocity field (a box filter is applied), $\widetilde{D}/\widetilde{D}t = \partial/\partial t + \widetilde{U}_k\, \partial/\partial x_k$ denotes the filtered Lagrangian time derivative, $P = \widetilde{p} + 2k\rho/3$ is the modified filtered pressure that includes a contribution due to the subgrid-scale (SGS) kinetic energy k, ρ is the constant fluid mass density, and ν is the constant kinematic viscosity. The filtered rate-of-strain tensor is defined by $\widetilde{S}_{ij} = (\partial \widetilde{U}_i/\partial x_j + \partial \widetilde{U}_j/\partial x_i)/2$. The LES Eq. (1) are unclosed due to the appearance of the unknown deviatoric SGS stress τ_{ij}^d, which is defined via $\tau_{ij} = \widetilde{U_i U_j} - \widetilde{U}_i\, \widetilde{U}_j$.

An attractive approach for closing the LES Eq. (1) is to use a stochastic turbulence model that determines stochastic solutions of the LES equations [3, 5, 8]. This means, the stochastic velocity model implies the incompressibility constraint, and it exactly recovers Eq. (1) for the filtered velocity. The advantage of the stochastic model is that is also implies transport equations for all the velocity moments. In particular, it can be used to derive the following transport equation for the SGS stress τ_{ij} [3–5, 8],

$$\frac{\widetilde{D}\tau_{ij}}{\widetilde{D}t} + \frac{\partial T_{kij}}{\partial x_k} + \tau_{ik}\frac{\partial \widetilde{U}_j}{\partial x_k} + \tau_{jk}\frac{\partial \widetilde{U}_i}{\partial x_k} = -\frac{2}{\tau_L}\left(\tau_{ij} - \frac{2}{3}c_0 k\delta_{ij}\right). \quad (2)$$

Here, T_{kij} refers to the triple correlation tensor of SGS velocity fluctuations. Equation (2) involves two model parameters: the nondimensional parameter c_0, and the Lagrangian time scale τ_L. The parameter c_0 is related to the Kolmogorov constant C_0 by $c_0 = C_0/[C_0 + 2/3]$. An analysis reveals that $c_0 = 19/27 \approx 0.7$ [3–5]. An analysis of the τ_L scaling shows that $\tau_L = \ell_* \Delta k^{-1/2}$. Here, Δ denotes the filter width and the model parameter ℓ_* has a standard value $\ell_* = 1/3$ [4].

The solution of the SGS stress Eq. (2) is computationally relatively expensive. A way to reduce the computational cost is to use the stress Eq. (2) for the derivation of algebraic stress models. The quadratic stress model obtained in this way reads [4]

$$\tau_{ij} = \frac{2}{3}k\delta_{ij} - 2\nu_t\widetilde{S}_{ij} - C_n\Delta^2\left[\widetilde{S}_{ik}\widetilde{\Omega}_{kj} + \widetilde{S}_{jk}\widetilde{\Omega}_{ki} - 2\widetilde{S}_{ik}\widetilde{S}_{kj} + \frac{2}{3}\widetilde{S}_{nk}\widetilde{S}_{nk}\delta_{ij}\right]. \quad (3)$$

Here, $\widetilde{\Omega}_{ij} = (\partial\widetilde{U}_i/\partial x_j - \partial\widetilde{U}_j/\partial x_i)/2$ refers to the rate-of-rotation tensor, and $C_n = \ell_*^2/3$. The SGS viscosity is given by the expression $\nu_t = C_k\Delta k^{1/2}$, where $C_k = \ell_*/3$. This parametrization for ν_t was used in several applications [9]. The quadratic stress model (3) can be reduced to a linear stress model by setting $C_n = 0$. Hence, the stress model (3) combined with an nonequilibrium model for k (i.e., a transport equation for k) can be used as an linear or quadratic nonequilibrium model. These nonequilibrium models can be reduced to equilibrium models by using the k transport equation for deriving an equilibrium expression for k [3, 5, 8].

3 Realizable Linear Unified Stress Model

A solution for a unified formulation of computational methods was presented recently on the level of stochastic models [5]. This solution will be described here with respect to the linear stress model $\tau_{ij}^d = -2\nu_t\widetilde{S}_{ij}$, where $\nu_t = k\tau_L/3$ and $\tau_L = \ell_*\Delta k^{-1/2}$. By using the latter expressions, the velocity field is determined by the incompressibility constraint $\partial\widetilde{U}_i/\partial x_i = 0$ combined with

$$\frac{\widetilde{D}\widetilde{U}_i}{\widetilde{D}t} = -\frac{1}{\rho}\frac{\partial P}{\partial x_i} + 2\frac{\partial(\nu + \nu_t)\widetilde{S}_{ik}}{\partial x_k}. \quad (4)$$

Implied by Eq. (2), the SGS kinetic energy $k = \tau_{kk}/2$ is calculated via

$$\frac{\widetilde{D}k}{\widetilde{D}t} = \frac{\partial}{\partial x_k}\left[(\nu + \nu_t)\frac{\partial k}{\partial x_k}\right] + \nu_t|S|^2 - \frac{2(1 - c_o)k}{\tau_L}, \quad (5)$$

where $|S|^2 = 2S_{nk}S_{kn}$. Here, a diffusion model is used for the triple correlation T_{knn}. An analysis of these equations shows [5] that the LES equations correspond to a RANS model if the LES time scale $\tau_L = \ell_* \Delta k^{-1/2}$ is replaced by the RANS time scale $\tau_L = \ell_* \tau^{RANS}$. Here, τ^{RANS} represents the RANS dissipation time scale of turbulence, which is related to the turbulence frequency by the relation $\omega = 1/\tau^{RANS}$. The time scale τ^{RANS} can be calculated via a transport equation for ω that is designed in analogy to Eq. (5) [6].

The unification of RANS and LES equations can be achieved by using a transfer function that relates consistently the LES time scale $\tau_L = \ell_* \Delta k^{-1/2}$ and RANS time scale $\tau_L = \ell_* \tau^{RANS}$ [5]. However, theoretical analyses and applications to channel flow simulations suggest that this transfer function can be simplified to $\tau_L = \ell_* min(\Delta k^{-1/2}, \tau^{RANS})$. In this expression, k is provided by the unified k Eq. (5). The use of this expression for τ_L in Eqs. (4) and (5) results in a unified RANS-LES model that switches consistently from LES to RANS depending on the minimum.

This approach works well in channel flow simulations [6], but its use is hampered by the following fact. The calculation of τ^{RANS} via a transport equation for ω requires the RANS kinetic energy and velocities as input for the ω equation, i.e., this approach requires a RANS simulation prior to the unified LES simulation. This problem can be solved by using the unified SGS kinetic energy and velocity gradients as input for the ω equation, which is used to obtain τ^{RANS}. Thorough analyses and applications demonstrate the efficiency and accuracy of this dynamic coupling approach [6, 7], which was used to obtain the following results. The model obtained in this way will be referred to below as linear unified model (LUM).

4 Application to Turbulent Channel Flows

Applications to turbulent channel flows were used to study the cost and accuracy features of the unified RANS-LES model [6]. With respect to the computational cost, it was shown that the use of the unified RANS-LES model reduces the computational cost of LES by a factor of 0.07 $Re^{0.46}$, which is huge for a high Reynolds number Re, as given in most applications. The accuracy features of the unified RANS-LES model are illustrated in the following in comparison to DES for a relatively high friction Reynolds number $Re_\tau = 5{,}000$ by using the VFLES grid and defining the filter width as the large side filter width, $\Delta = \max(\Delta_x, \Delta_y, \Delta_z)$ [6]. Due to the lack of DNS data for this flow considered we apply a slightly modified version of the empirical mean streamwise velocity profile of Reichardt [10] for comparisons,

$$U^+ = \frac{1}{\kappa} \ln(1+\kappa y^+) + 7.31 \left[1 - \exp\left(-\frac{y^+}{7.87}\right) - \frac{y^+}{7.87} \exp\left(-\frac{y^+}{4.25}\right) \right]^{0.96} . \quad (6)$$

Here, $U^+ = \widetilde{U}_1/u_\tau$ is the dimensionless meanstreamwise velocity. The latter profile was shown to be in an excellent agreement with DNS data for a wide range of Reynolds numbers [6].

The location of the LES-RANS interface has been shown to affect simulation results of DES and hybrid methods [11]. Hence, it is important to have the same interface location when comparing DES and other hybrid models. The latter was accomplished in the following way. In DES, the location of the interface is determined by $C_{DES}\,\Delta = y$, which defines the switch of the length scale applied according to $\widetilde{d} = \min\,(C_{DES}\,\Delta, y)$. The filter width is calculated by using the large side filter, such that the value of Δ is constant throughout the domain and known. The unified simulation showed that the location of the RANS-LES interface was at $y^+ \approx 96$, which determines y. The use of the corresponding values of Δ and y in $C_{DES}\,\Delta = y$ then results in $C_{DES} = 0.38$.

The mean streamwise velocity profiles obtained with the LUM and DES are compared to the empirical profile (6) in Fig. 1a. Also shown in this figure are the data of the mean velocity profile obtained by using the renormalization group

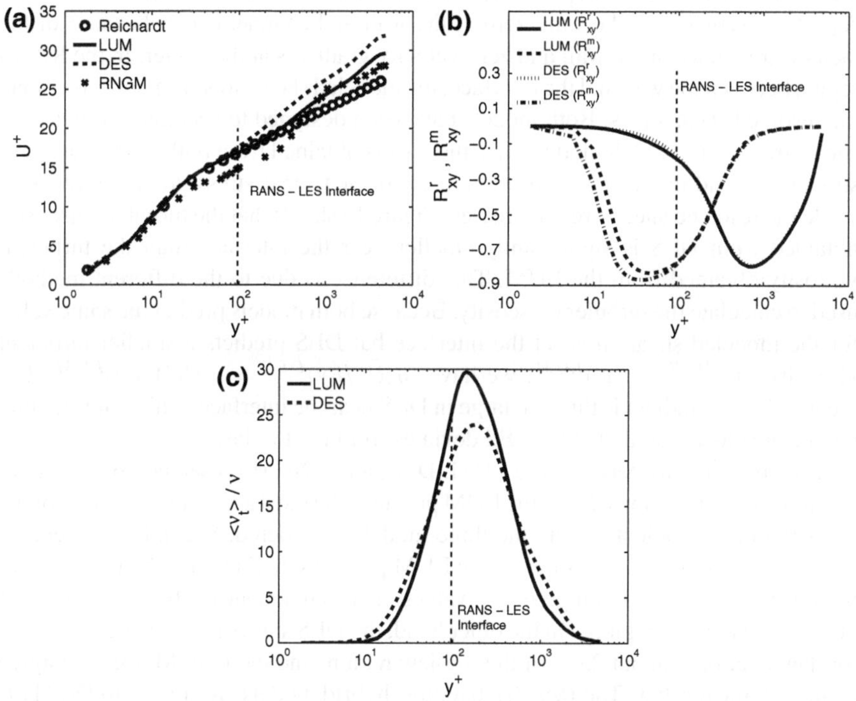

Fig. 1 Channel flow at Re_τ =5,000: **a** Comparison of the mean streamwise velocity obtained by the LUM, DES, and RNGM with the Reichardt profile (6); **b** Comparison of the modeled (denoted by m) and resolved shear stress (denoted by r) obtained by the LUM and DES; **c** Comparison of the turbulent viscosity obtained by the LUM and DES. The *vertical solid line* refers to the RANS-LES interface location

model (RNGM) [11], for which the RANS-LES interface is located at $y^+ \approx 100$. The mean velocity predicted by the RNGM only agrees with the DNS data up to $y^+ \approx 20$. For $20 < y^+ < 700$, a significant underprediction is observed, and the velocity is overpredicted for $y^+ > 700$. The LUM and DES results almost identically agree with the Reichardt profile up to the RANS-LES interface location $y^+ \approx 96$. Beyond the interface, the LUM results continue to agree well with the Reichardt profile up to $y^+ = 300$, while the DES results display a mismatch of the velocity in the log-law region, as it was also reported by [12]. This observed mismatch induces higher errors in the prediction of the skin-friction coefficient $C_f = \tau_w/(0.5\rho U_b^2)$, where U_b refers to the bulk velocity. For the simulations considered here, the error $E_{Cf} = 100\,(C_f - C_{f,\,Dean})/C_{f,\,Dean}$ in the prediction of the skin-friction coefficient (in comparison with Dean's empirical skin-friction coefficient $C_{f,\,Dean} = 0.073(2U_b\delta/\nu)^{-1/4}$ obtained from experiments [13]) was found to be -9.8, -15.5 and 16.5 % for the LUM, DES and RNGM models, respectively.

To understand the reason for the improved mean velocity profile and skin-friction coefficient obtained by the LUM, comparisons of the Reynolds shear stress (modeled and resolved) and turbulent viscosity obtained from LUM and DES simulations are shown in Fig. 1b, c respectively. Both the resolved and modeled shear stress $R_{xy} \approx -\langle \nu_t \rangle \partial \langle \tilde{U} \rangle / \partial y$ become zero at the wall and channel center because of the lack of turbulence at the wall and zero velocity gradients at the center. In the RANS region between the wall and the interface, the modeled shear stress is much larger than the resolved shear stress. Both models have been designed to accurately predict the shear stress in the RANS mode. Thus, the results obtained from both models are very similar. Beyond the interface location, the modeled shear stress gradually reduces, while the resolved shear stress increases. Figure 1c shows that the turbulent viscosity obtained from DES is significantly smaller near the interface than the turbulent viscosity obtained from the LUM. This difference is due to the different methods used to calculate the turbulent viscosity. Because both models predict the same value for the modeled shear stress at the interface but DES predicts a smaller turbulent viscosity, $\langle \nu_t \rangle^{DES} < \langle \nu_t \rangle^{LUM}$, we have $(\partial \langle \tilde{U} \rangle / \partial y)^{DES} > (\partial \langle \tilde{U} \rangle / \partial y)^{LUM}$. The mean velocity gradient is thus too large in DES near the interface, which implies that the mean velocity predicted by DES deviates from the log-law.

The comparisons between the LUM, DES, and RNGM presented above lead to two relevant conclusions. First, the LUM provides the most accurate prediction of the skin-friction coefficient among the three models considered. Second, in difference to the other two models considered, the LUM provides predictions which agree with the log-law over a wide range of y^+ values. This agreement is observed for about 50 % of the log-law region. On the other hand, the DES shows an agreement with the log-law over only about 25 % of the log-law region, and the RNGM does not agree at all with the log-law. The facts (i) that most hybrid models are similar to DES [14] and (ii) corresponding validations of the performance of other hybrid methods are unavailable for such a high Reynolds number case support the view that the LUM performs better than other (comparably simple) linear hybrid models. It has to be noted that there are ways to improve the performance of hybrid methods with regard to channel flow simulations: the addition of fluctuations or an additional filtering of the velocity field near the RANS-LES interface have been proven to overcome the log-law mismatch problem [12, 15, 16]. However, the suitability of such flow-dependent problem solutions for other applications is unclear.

5 Application to Swirling Turbulent Jet Flows

The LUM was applied to OPENFOAM simulations [17] of turbulent jets covering several swirl number regimes [7]. An illustration of the flows considered is given in Fig. 2, which shows instantaneous pressure fluctuations for a flow without vortex breakdown ($S = 0.55$) and a flow with vortex breakdown ($S = 0.6$). The flow on the left-hand side of the black disk is inside the nozzle, and the flow on the right-hand side of the black disk is the free jet. A passive scalar $0 \leq \widetilde{\phi} \leq 1$ having a value $\widetilde{\phi} = 1$ inside the nozzle was used to illustrate the flow structures. The LUM was compared to two DES methods and segregated RANS-LES model results [18]. The LUM and DES methods simulataneously simulated both the nozzle and jet flow. The segregated RANS-LES model used RANS inside the nozzle and LES outside. The geometric mean of a cell, $\Delta = (\Delta_x \Delta_y \Delta_z)^{1/3}$, was used for most simulations, but the effect of using a large side filter width, $\Delta = \max(\Delta_x, \Delta_y, \Delta_z)$, was studied, too.

The findings obtained regarding the validation of the performance of the unified RANS-LES model with respect to turbulent swirl flow simulations can be summarized in the following way. For a nonswirling $S = 0$ case and a $S = 0.23$ case of mild swirl (experimental results are unavailable for other swirl numbers [19]), it was shown that the mean velocities and turbulence intensities predicted by the unified RANS-LES model agree well with experimental results. For higher swirl numbers but in absence of vortex breakdown, the unified RANS-LES method showed that the scalar mixing efficiency increases linearly with the swirl number S, as it has to be expected [20, 21]. Unified RANS-LES simulations showed vortex breakdown for swirl numbers $S \geq 0.6$. The conclusion that $S = 0.6$ represents a critical swirl

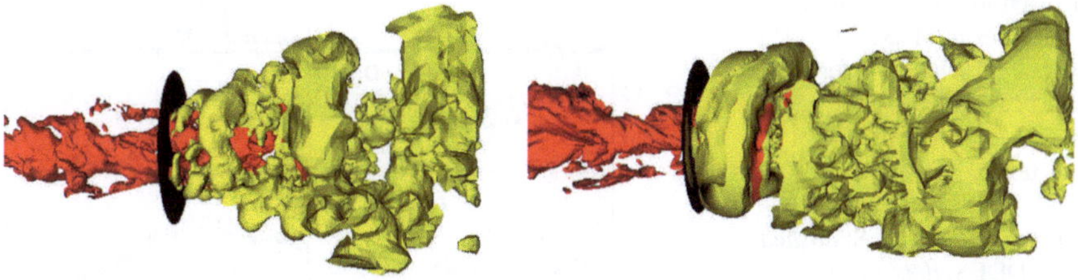

Fig. 2 Unified RANS-LES results of swirling turbulent jet flow simulations: Visualization of coherent vortex structures by means of an iso-surface $\widetilde{p} - P = -20\ Pa$ of instantaneous pressure fluctuations for swirl numbers $S = (0.55, 0.6)$ from *left* to *right*, respectively. The iso-surface is colored according to the passive scalar value: $\widetilde{\phi} \geq 0.5$ (*red*) and $\widetilde{\phi} < 0.5$ (*yellow*). The nozzle exit is illustrated by the *black* disk

number required for the onset of vortex breakdown agrees very well with observations made by theory, experiments, and other computations [7].

There are significant advantages of the unified RANS-LES method in comparison with other hybrid RANS-LES methods. (a) Compared to segregated RANS-LES models, the advantages of the unified approach is a relatively weak dependence on experimental data, which were needed here only to provide nozzle inflow data. On the other hand, the segregated RANS-LES approach requires evidence for the suitability of simulation results at the RANS-LES interface. In addition to the latter fact, the segregated RANS-LES approach may imply inaccurate flow simulations. Shortcomings of the segregated RANS-LES model for $S > 0.23$ swirl flows (the inability to correctly simulate the central vortex core, precessing vortex core (PVC), and mixing efficiency of passive scalars) are, basically, a consequence of separating the RANS and LES calculations, which leads to inaccurate RANS input for the LES region and the lack of resolved motions close to the nozzle. (b) Compared to DES, the advantage of the unified RANS-LES method presented here is its generality and better performance. In contrast to the unified RANS-LES method, both DES and IDDES suffer from the 'modeled-stress depletion' problem (see the illustration in Fig. 3), which results in inappropriate predictions of nonswirling flows and shortcomings regarding the simulation of other swirling jet flows than considered here [22]. In addition, DES results can be significantly affected by model parameter variations [22], whereas the unified RANS-LES model does not involve an adjustable parameter corresponding to C_{DES}.

The unified RANS-LES results presented here contribute to a better understanding of conditions for the onset and the mechanism of vortex breakdown. Figure 2 indicates that vortex breakdown is related to a precession of the central vortex core, which is illustrated by the red structures. Evidence for this view can be obtained by looking at the power spectral density of axial velocity fluctuations. Segregated

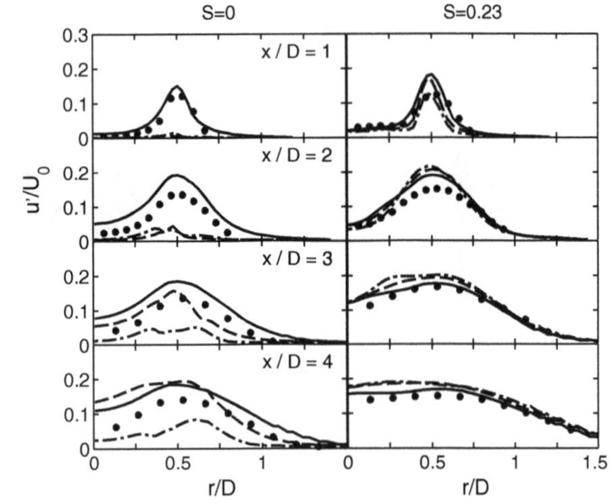

Fig. 3 Unified RANS-LES results of swirling turbulent jet flow simulations: Radial distributions of normalized intensities u' of axial velocity fluctuations for the swirl cases $S = 0$ and $S = 0.23$ at different axial positions x/D. *Solid line* unified RANS-LES model; *dot-dashed line* DES; *dashed line* IDDES; *dots* experimental data

RANS-LES simulations, which are unable to simulate the central vortex core, also show vortex breakdown for swirl numbers $S \geq 0.6$. This finding provides evidence for the correctness of the debated view [23] that vortex breakdown represents a pre-cursor to the PVC, which is only a consequence of vortex breakdown.

6 Summary and Outlook

The unified RANS-LES model described here results from a sound theory based on stochastic turbulence equations [3–8]. Applications to a variety of turbulent channel flows and swirling turbulent jet flows lead to the following conclusions. Compared to RANS equations, the significant advantage of the LUM is the incorporation of unsteady instantaneous motions. This enables, for example, accurate simulations of turbulent jet flows which cannot be accomplished with RANS methods. Compared to LES methods, the significant advantage of the LUM is the reduction of computational cost by a factor of $0.07\ Re^{0.46}$ [6], which is huge for a high Reynolds number Re, as given in most applications. In addition, the accuracy of the LUM was found to be at least as good as in resolved LES. Compared to other hybrid RANS-LES methods, it was found that the LUM also has advantages. Its performance is more universal than DES methods, which may suffer significantly from the modeled stress depletion problem (see the channel flow and non swirling jet flow discussions above). The LUM also has significant advantages compared to segregated RANS-LES methods, which suffer from their inability to simulate RANS-LES communications. With respect to channel flow simulations it was shown that the LUM performs much better than the RNGM.

Ongoing work is focused on the combination of the unified RANS-LES approach with dynamic LES derived from the same stochastic theory [8, 24]. This concept offers significant advantages compared to the combination of RANS with non-dynamic LES. First, the damping effects on LES is covered, which represents a non-trivial problem in hybrid RANS-LES methods. Second, additional unsteadiness (fluctuations) is involved, which reduces the modeled stress depletion problem seen in hybrid RANS-LES methods like DES. Third, transitional (low Reynolds number) flows, which are relevant to separated flow simulations, can be treated without modifications of the computational approach. First applications of this concept to periodic hill flows involving separation [25] (not shown) reveal the advantages.

Acknowledgments This work was partially sponsored by the Air Force Office of Scientific Research, USAF, under grant number FA9550-05-1-0485 monitored by Dr. John Schmisseur. We would also like to acknowledge support through NASA's NRA research opportunities in aeronautics program (Grant No. NNX12AJ71A) with Dr. P. Balakumar as the technical officer. The computational resources have been provided by the UW Institute for Scientific Computation.

References

1. Spalart, P., Jou, W., Strelets, M., Allmaras, S.: Comments on the feasibility of LES for wings, and on a hybrid RANS/LES approach. In: 1st AFOSR International Conference on DNS/LES, pp. 4–8, Greyden Press, Columbus, Ruston, LA, USA, 4–8 Aug 1997
2. Shur, M.L., Spalart, P.R., Strelets, M.K., Travin, A.K.: A hybrid RANS-LES approach with delayed-DES and wall-modelled LES capabilities. Int. J. Heat Fluid Flow **29**(6), 1638–1649 (2008)
3. Heinz, S.: Statistical Mechanics of Turbulent Flows, 1st edn. Springer, Berlin, Heidelberg, New York, Tokyo (2003)
4. Heinz, S.: On Fokker-Planck equations for turbulent reacting flows. Part 2. Filter density function for large eddy simulation. Flow Turbul. Combust. **70**(1–4), 153–181 (2003)
5. Heinz, S.: Unified turbulence models for LES and RANS, FDF and PDF simulations. Theoret. Comput. Fluid Dyn. **21**(2), 99–118 (2007)
6. Gopalan, H., Heinz, S., Stöllinger, M.: A unified RANS-LES model: computational development, accuracy and cost. J. Comput. Phys. **249**, 249–279 (2013)
7. Heinz, S., Zemtsop, C., Gopalan, H., Stöllinger, M.: Unified RANS-LES Simulations of Swirling Turbulent Jet Flows (2015)
8. Heinz, S.: Realizability of dynamic subgrid-scale stress models via stochastic analysis. Monte Carlo Meth. Appl. **14**(4), 311–329 (2008)
9. Lilly, D.K.: The representation of small-scale turbulence in numerical simulation of experiments. In: Goldstine, H.H. (ed.) Proceedings of the IBM Scientific Computing Symposium on Environmental Sciences, pp. 195–210, IBM, Yorktown Heights, NY, 1967
10. Reichardt, H.: Vollständige Darstellung der Turbulenten Geschwindigkeitsverteilung in Glatten Leitungen. ZAMM—J. Appl. Math. Mech. **31**(7), 208–219 (2006)
11. De Langhe, C., Merci, B., Lodefier, K., Dick, E.: Hybrid RANS/LES modelling with an approximate renormalization group. II: applications. J. Turbul. **6**(14), 1–16 (2005)
12. Keating, A., Piomelli, U.: A dynamic stochastic forcing method as a wall-layer model for large-eddy simulation. J. Turbul. **7**(1), 1–24 (2006)
13. Dean, R.B.: Reynolds number dependence of skin friction and other bulk flow variables in two-dimensional rectangular duct flow. J. Fluids Eng. **100**, 215–223 (1978)
14. Spalart, P.R.: Detached-eddy simulation. Annu. Rev. Fluid Mech. **41**, 181–202 (2009)
15. Spalart, P.R., Deck, S., Shur, M.L., Squires, K.D., Strelets, M.K., Travin, A.: A new version of detached-eddy simulation, resistant to ambiguous grid densities. Theoret. Comput. Fluid Dyn. **20**(3), 181–195 (2006)
16. Hamba, F.: Log-layer mismatch and commutation error in hybrid RANS/LES simulation of channel flow. Int. J. Heat Fluid Flow **30**(1), 20–31 (2009)
17. OPENFOAM: The Open Source CFD Tool Box, User guide, version 1.6. http://www.openfoam.org. Technical report, 2009
18. Zemtsop, C.P., Stöllinger, M.K., Heinz, S., Stanescu, D.: Large eddy simulation of swirling turbulent jet flows in absence of vortex breakdown. AIAA J. **47**(12), 3011–3021 (2009)
19. Gilchrist, R.T., Naughton, J.W.: Experimental study of incompressible jets with different initial swirl distributions: mean results. AIAA J. **43**(4), 741–751 (2005)
20. Chigier, N.A., Chervinsky, A.: Experimental investigation of swirling vortex motion in jets. J. Appl. Mech. **34**(2), 443–451 (1967)
21. Shiri, A., George, W.K., Naughton, J.W.: Experimental study of the far field of incompressible swirling jets. AIAA J. **46**(8), 2002–2009 (2008)
22. De Langhe, C., Merci, B., Dick, E.: Application of a RG hybrid RANS/LES model to swirling confined turbulent jets. J. Turbul. **7**(56), 1–19 (2006)
23. Syred, N.: A review of oscillation mechanisms and the role of the precessing vortex core (PVC) in swirl combustion systems. Prog. Energy Combust. Sci. **32**(2), 93–161 (2006)

24. Heinz, S., Gopalan, H.: Realizable versus non-realizable dynamic sub-grid scale stress models. Phys. Fluids **24**(11), 115105/1–23 (2012)
25. Balakumar, P., Rubinstein, R., Rumsey, C.L.: DNS, enstrophy balance, and the dissipation equation in a separated turbulent channel flow. In: 43rd AIAA Fluid Dynamics Conference, AIAA Paper 13-2723. San Diego, California (2013)

Hybrid RANS-LES Versus URANS Simulations of a Simplified Compressor Blades Cascade

Y. Hoarau, D. Szubert and M. Braza

Abstract Based on a previous experience from BR710 and BR725 Roll-Royce fan blade development programs, the phenomena of shock wave-boundary layer interaction (SWBLI) are known to be of critical importance regarding engine safety and performance. In the European TFAST Project (Transition Location Effect on Shock Wave Boundary Layer Interaction), experiments and numerical simulations of unsteady and transitional SWBLI phenomena in a simplified compressor blades cascade are performed. Two blades of 100 mm are set-up in a 100*100 mm channel. The grid used is 9,6 M cells with $y^+ < 1$ everywhere. The inlet turbulence intensity is set to 4 % and the Mach number has been adjusted to be 1.22 in front of the lower blade. In the present paper, unsteady transitional RANS and hybrid RANS-LES computations of the flow around two compressor blades are performed with the NSMB solver. Simulations with Spalart-Allmaras and $k - \omega$ turbulence models in their URANS and DDES formulation as well as $k - \omega$ SAS are performed.

1 Introduction

Vision-2020, whose objectives include the reduction of emissions and a more effective transport systems, puts severe demands on aircraft velocity and weight. These require an increased load on wings and aero-engine components. The greening of air transport systems means a reduction of drag and losses, which can be obtained by keeping laminar boundary layers on external and internal airplane parts. Increased

Y. Hoarau (✉)
ICUBE, CNRS - Université de Strasbourg - ENGEES - INSA, Strasbourg, France
e-mail: hoarau@unistra.fr

D. Szubert · M. Braza
IMFT, CNRS - INPT - Université de Toulouse, Toulouse, France
e-mail: dszubert@imft.fr

M. Braza
e-mail: braza@imft.fr

S. Girimaji et al. (eds.), *Progress in Hybrid RANS-LES Modelling*,
Notes on Numerical Fluid Mechanics and Multidisciplinary Design 130,
DOI 10.1007/978-3-319-15141-0_22

loads make supersonic flow velocities more prevalent and are inherently connected to the appearance of shock waves, which in turn may interact with a laminar boundary layer. Such an interaction can quickly cause flow separation, which is highly detrimental to aircraft performance, and poses a threat to safety. In order to diminish the shock induced separation, the boundary layer at the point of interaction should be turbulent, so that transition to turbulence does not occur within the SWBLI region. The main objective of this study, which is conducted within the European TFAST project, is to evaluate the effect of transition location on the SWBLI structure. The main question is how close the induced transition may be to the shock wave while still maintaining a typical turbulent character of interaction. The main study cases of this work deals with compressor blades and will help to answer open questions posed by the aeronautics industry and to tackle more complex applications. In the case of a civil turbofan engine operating at particularly high altitudes the Reynolds number can drop by a factor of 4, when compared to the see level values. The laminar boundary layer on the transonic compressor rotor blades will interact with shock waves and as a result a strong boundary-layer separation will form. This can seriously affect the aero-engine performance and operation. One way to avoid strong separation is to ensure that the boundary layer upstream of the shock wave is turbulent. Forcing transition within the boundary layer can be achieved through the application of a surface roughness or a turbulator patch. Although such passive control methods are already in use, the mechanism of the shock wave/laminar boundary layer interaction, and in particular the source of the strong shock unsteadiness are still not well understood. Furthermore, the benefits of boundary layer control obtained for low Reynolds numbers can turn into loss increase at the higher levels of Reynolds numbers. Another possibility of transition control is to use Vortex Generators driven by Air Jet (AJVG). In the compressor application the jets may be driven by the pressure difference between the suction and pressure sides of the blade. There are two effects which should be investigated. The main effect is coming from streamwise vortices generated on the blade suction side but another is the suction taking place at the blade pressure side. This study takes place in the Work Package 3 of the TFAST project whose main goal is to improve the understanding of the shock wave/laminar boundary layer interaction on the transonic compressor blade. This can potentially lead to successful new design solutions.

Shock wave/turbulent boundary layer interactions (SWBLIs) are very complex flow phenomena which can be encountered in many flows such as transonic airfoils, supersonic inlets, aircraft wings, missiles, jets, nozzles... Most of the time the shock interacts significant with the incoming boundary layer leading to separation which can give rise to unsteady flow (buffeting, inlet instability, ...). SWBLI has been extensively studied over the past 50 years. Review of such work can be found in [1–3, 5–7, 10, 14]...

The unsteadiness generated by SWBLI is characterized by a wide range of frequencies, which encompass the characteristics of the upstream boundary layer as well as motions that are typically one to two orders of magnitude lower.

The physical source of this unsteadiness is still unclear and in a recent review, [4] concluded that both fluctuations in the upstream boundary layer and large-scale instability intrinsic to the separated flow but the downstream mechanism dominates for strongly separated flows, and a combined mechanism dominates for weakly separated flows.

In this paper we compare the result of the modeling of SWBLI in a simplified compressor cascade with RANS and DDES/SAS models. This work is preliminary in the sense that no transition modeling has been introduced and the flow is fully turbulent. The next section presents details of the test configuration and of the numerical aspects of the simulations. Results and discussion are developed in Sect. 3.

2 Case Description and Numerical Modeling

The selected geometry is representative of a Low Pressure Compressor first rotor at mid-span. The Mach number in front of the blade is 1.2, the deflection 15° and the Axial Velocity Density Ratio 1.2. The original geometry has been scaled to a chord of 100 mm to fit in the transonic wind tunnel of IMP-PAN, Gdansk, Poland (Fig. 1). Fiew results of these experiments are available when this paper has been writen. Between the sea level and the cruise altitude of 51 kft, the Reynolds Number drops by a factor of 4. Preliminary simulations with Tu = 0.5 and 4.0 % show that at low Tu and altitude of 45 kft (typical cruise), the suction surface boundary layer upstream of shock is fully laminar which is the main requirement for the design of the test section. The geometry simulated in this paper is presented in the Fig. 2 and is around 2 m length and 100 mm in the spanwise direction. The inlet Mach number is 0.5 and the outlet pressure has been calibrated to have a Mach number of 1.22 in front of the first blade. The grid was generated by NUMECA/Autogrid and the y^+ is lower than 1 in all the regions of interest. The size of the 3D grid is 9.6×10^6 cells. The highest

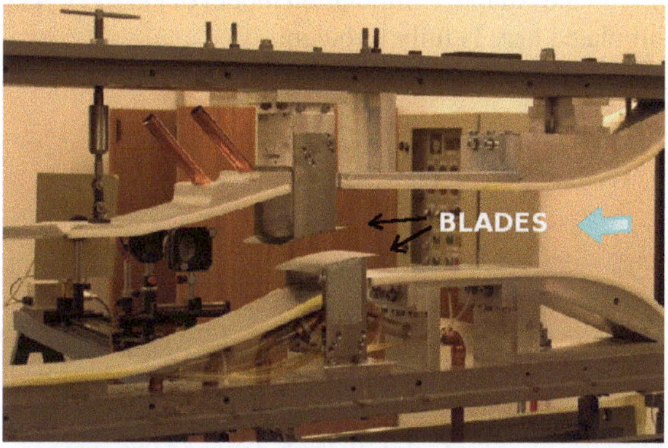

Fig. 1 Experimental setup (IMP-PAN, Gdansk, Pologne)

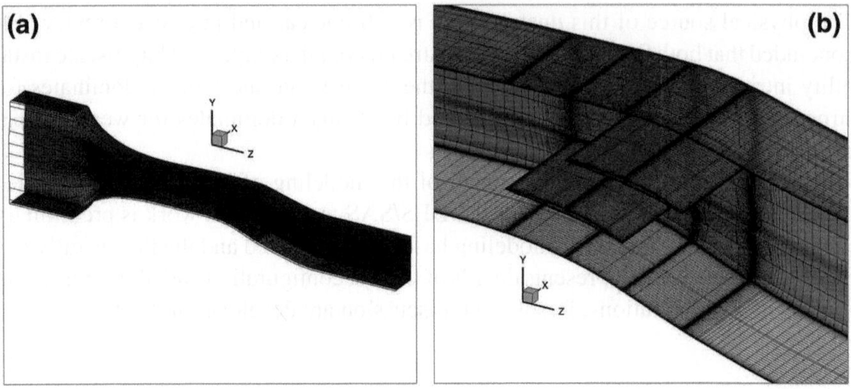

Fig. 2 **a** View of the whole grid and **b** details of the compressor cascade

Reynolds number corresponding to the sea level altitude, $Re = 4.4 \times 10^6$, and an inlet turbulence intensity of 4 % are used.

In this study the compressible NSMB flow solver [17] has been used. The NSMB code solves the compressible Navier-Stokes equations using a finite volume formulation on structured Multi-Block grids and is parallelized using the Message Passing Interface. Among the many discretization schemes available in NSMB, in this study we use the 3rd order AUSMPW+ upwind scheme [9] for the convection and a second order central scheme for the diffusion. An implicit dual-time stepping solved by Lower-Upper Symmetric Gauss-Seidel method (LS-SGS) is used for the time integration. The time-step is $\Delta t = 10^{-5}$ s and the inner loop is stopped when the convergence reaches 10^{-3} for each time-step which correspond approximatively to 60–70 iterations. Two URANS turbulence models have been used, the [8] version of the one-equation model of [15] (SA) and the two-equation $k-\omega$ SST model of [13] as well as three hybrid RANS/LES versions of these models, the Delayed Detached Eddy Simulation [16] of the SA and the $k-\omega$ SST models as well as the $k-\omega$ SST Scale Adaptive Simulation model of [11, 12]. As previously mentioned the flow simulated here is fully turbulent.

3 Results and Discussion

A Schlieren visualization of the experimental flow is provided in Fig. 3a, as well as a distribution of the Mach number over the upper side of the lower blade (b), carried out by IMP-PAN, Gdansk, in the framework of the TFAST European programme. A reduction of the Mach number is observed across the shock generated by the upper blade and impinging the lower one. The evolution of the experimental isentropic Mach number for both fully turbulent and laminar case are presented for the extrados on the lower blade and the results are in fair agreement with the numerical SA-DDES.

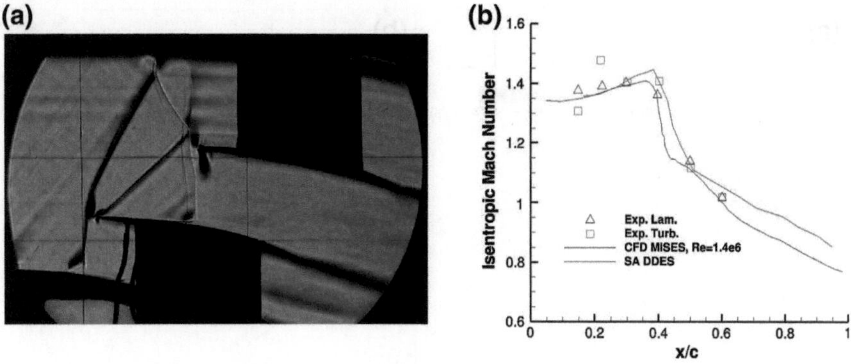

Fig. 3 Schlieren visualisation of the experimental flow (**a**) and comparison of the Isentropic-Mach number distribution over the upper side of the lower blade (**b**) (IMP-PAN, Gdansk, Pologne)

In simulations, the flow is characterised by the formation of five shocks: the first one forms in front of the lower blade, the second between the lower wall and the first blade, the third on the suction side of the first blade and in front of the second blade, the fourth one between the two blades and the last one between the second blade and the upper wall. In all the simulations the mach number before the first shock is around $M = 1.22$. The overall flow structure shown in Fig. 4 displays this shock-waves formation in good qualitative comparison with the Schlieren visualization (Fig. 3a), in particular the SA model that gave a good aspect of the first and third shocks, on front of each blade. Shocks between the upper and lower walls and the blade are not visible due to the experimental setup. This first shock is detached from the lower blade for the RANS models and oscillates on a small distance at a frequency around 160–173 Hz, whereas for the DDES models the shock moves on a bigger distance and come very closed to the lower blade with a frequency around 114–121 Hz. The frequency is lower because the extent of the movement is much higher. In the case of the SAS model, this shock is very oblique, almost attached to the lower blade and mainly it is steady as well as the third shock which is also oblique, very closed to the second blade and steady whereas for the RANS and the DDES models it oscillates around the same frequency (210–235 Hz). The second shock below the first blade oscillates on a very large distance, $\approx$25 % of the chord, at a frequency of 307–364 Hz for both RANS and DDES models. In the DDES simulations the fourth shock between the two blades remains closed to the trailing edge and moves on a short distance in a very erratic way so that it is very difficult to extract a frequency. In the RANS simulations the shock moves on $\approx$20 % of the chord up to the trailing edge but its behavior is also very erratic. Finally the opposite behavior is observed of the last shock between the upper wall and the second blade: it's amplitude is higher in the DDES simulation than with the RANS models.

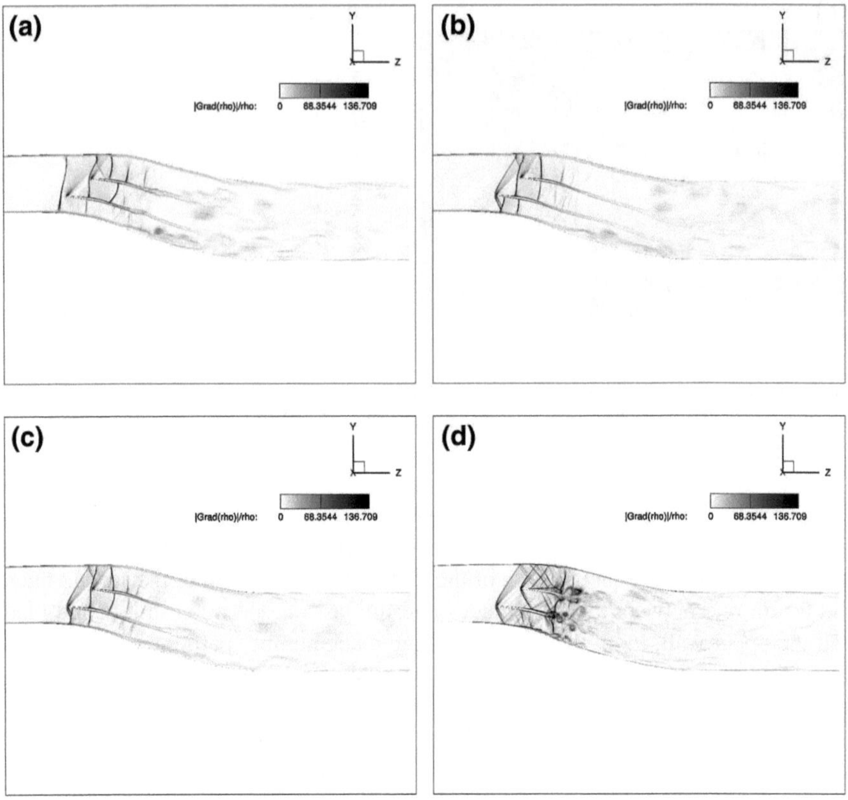

Fig. 4 Schlieren visualisation of the flow in the middle plane: **a** SA model, **b** SA DDES, **c** $k-\omega$ SST DDES and **d** $k-\omega$ SST SAS

Figure 5 features the time evolution of the lift coefficient on each blade. On the lower blade all the models except for the $k-\omega$ SST SAS give almost the same evolution, the SAS having a lower mean value whereas on the second blade all the five simulations stand in the same range. This is confirmed in the Table 1, where the mean values of the aerodynamic coefficients for the five models are presented. The SAS model gives a much lower value of the drag and lift coefficients on the lower blade and on the second blade the drag of the SAS is slightly higher. All the other models show comparable values. From the lift coefficient time evolution on the second blade of the $k-\omega$ SST model, we can extract a low frequency of 387 Hz for the SWBLI. This frequency in fact exists for all the models on each blade but there are a lot of sub-harmonics for the other models. Although the aerodynamic coefficients are quite comparable, the main flow feature is different between RANS and hybrid models, as can be seen on the Fig. 4.

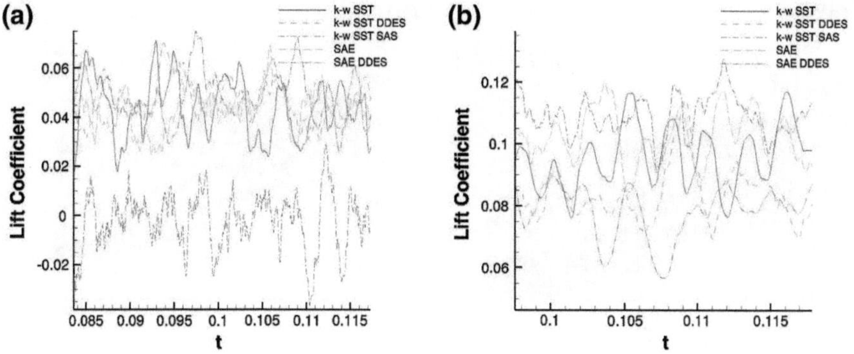

Fig. 5 Time evolution of the lift coefficient on the lower blade (**a**) and the upper blade (**b**) for all the turbulence models

Table 1 Mean values of drag and lift coefficients on each blade

		C_D	C_L
1st blade	SAE	0.0079	0.0412
	SAE DDES	0.0095	0.0506
	$k - \omega$ SST	0.0078	0.0377
	$k - \omega$ SST DDES	0.0080	0.0437
	$k - \omega$ SST SAS	0.00074	0.0082
2nd blade	SAE	0.0209	0.1045
	SAE DDES	0.0171	0.0859
	$k - \omega$ SST	0.0195	0.0982
	$k - \omega$ SST DDES	0.0183	0.0917
	$k - \omega$ SST SAS	0.0257	0.1053

The SWBLI is much more important in the case of the SAS model mainly because with this model the flow is laminar around the airfoils and on the upper and lower walls or at least the turbulence intensity is extremely small compared to the RANS ans the DDES simulations, as can be seen on the Fig. 6. The boundary layers are very sensitive to the pressure variation generated by the shocks and we can observe massive separation on each blade as well as on the upper and the lower wall. The side walls also introduce corner vortices which are captured by all the turbulence models but the SAS model gives much smaller structures than the other models.

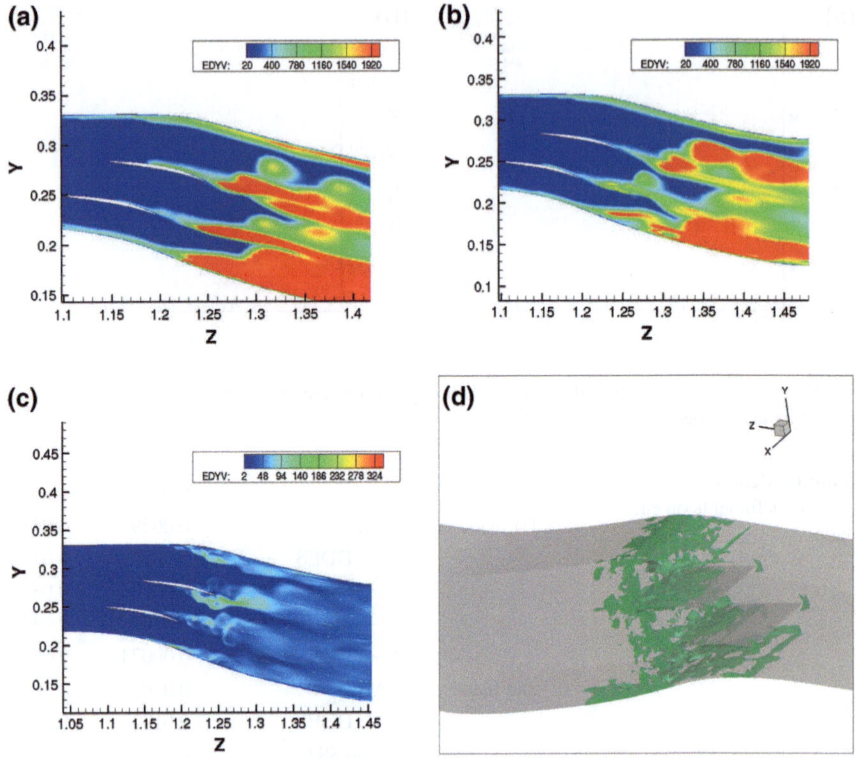

Fig. 6 Turbulent viscosity in the middle plane for the $k-\omega$ SST (**a**) $k-\omega$ SST DDES (**b**) $k-\omega$ SAS (**c**) and iso-suface of negative streamwise velocity (W $=-1$ m/s) for the $k-\omega$ SAS model

4 Conclusion

The flow around a simplified configuration of a low pressure compressor has been simulated in the preliminary configuration of the wind tunnel of IMP-PAN, Gdansk. The Reynolds number of $Re = 4.4 \times 10^6$ corresponds to flight at sea level altitude. The flow is fully turbulent with an inlet intensity of 4 %. Five simulations have been completed: two RANS and three hybrid RANS/LES. The RANS and the DDES present globally the same flow features qualitatively comparable with the experiment, the differences on the shocks location and oscillations having no tremendous effects on the mean aerodynamics coefficients whereas with the SAS model, a very small turbulence level is predicted which give rise to massive separation induced by the shock wave boundary layer interaction.

Acknowledgments This work has been carried out in the context of the European project TFAST. We acknowledge the CPU allocation on the French Supercomputing Centres IDRIS and CINES.

References

1. Babinsky, H., Harvey, J.K.: Shock WaveBoundary-Layer Interactions. Cambridge University Press, Cambridge (2011)
2. Bouhadji, A., Braza, M.: Physical analysis by numerical simulation of organised modes and shock-vortex interaction in transonic flows around an aerofoil. Part II: Reynolds number effect. J. Comput. Fluids **32**(9), 1261–1281 (2003)
3. Bourdet, S., Braza, M., Bouhadji, A., Thiele, F.: Direct numerical simulation of the three-dimensional transition to turbulence in the transonic flow around a wing. J. Flow Turbul. Combust. **71**, 203–220 (2003)
4. Clemens, N.T., Narayanaswamy, V.: Low-frequency unsteadiness of shock wave/turbulent boundary layer interactions. Annu. Rev. Fluid Mech. **46**, 469–492 (2014)
5. Doerffer, P., Hirsch, C., Dussauge, J.P., Babinsky, H., Barakos, G.N.: Unsteady Effects of Shock Wave Induced Separation. Springer, New York (2010)
6. Dolling, D.S.: Fluctuating loads in shock-wave/boundary layer interaction: tutorial and update. In: AIAA Paper, 1993-0284 (2001)
7. Dolling, D.S.: Fifty years of shock wave/boundary layer interaction: what next? AIAA J. **39**, 1517–1531 (2001)
8. Edwards, J.R., Chandra, S.: Comparison of eddy viscosity-transport turbulence models for three-dimensional, shock-separated flowfields. AIAA J. **34**(4), 756–763 (1996)
9. Kim, K.H., Kim, C., Rho, O.-H.: Methods for the accurate computations of hypersonic flows: I. AUSMPW+Scheme. J. Comput. Phys. **174**(1), 38–80 (2001)
10. Lee, B.H.K.: Self-sustained oscillations on airfoils at transonic speeds. Prog. Aerosp. Sci. **37**, 147–196 (2001)
11. Menter, F., Bender, R.: A scale-adaptive simulation model for turbulent flow prediction. In: AIAA Paper, 2003-0767, Reno, NV (2003)
12. Menter, F., Egorov, Y.: A scale-adaptive simulation model using two-equation models. In: AIAA Paper, 2005-1095, Reno, NV (2005)
13. Menter, F.: Two-equation eddy-viscosity turbulence models for engineering applications. AIAA J. **32**(8), 1598–1605 (1994)
14. Smits, A.J., Dussauge, J.P.: Turbulent Shear Layers in Supersonic Flow. AIP, Woodbury (1996)
15. Spalart, P., Allmaras, S.: A one-equation turbulence model for aerodynamic flows. In: AIAA Paper, 439 (1992)
16. Spalart, P., Deck, S., Shur, M., Squires, K., Strelets, M., Travin, A.: A new version of detached-eddy simulation, resistant to ambiguous grid densities. Theoret. Comput. Fluid Dyn. **20**, 181–195 (2006)
17. Vos, J., Chaput, E., Arlinger, B., Rizzi, A., Corjon, A.: Recent advances in aerodynamics inside the NSMB (Navier-Stokes Multi-Block) consortium. In: AIAA Paper, 0802 (1998)

References

Analysis of Scale Adaptive Approaches Based on the Rotta Transport Equation

A. Mehdizadeh, J.G. Brasseur, T. Nandi and H. Foroutan

Abstract A zonal formulation of the scale adaptive simulation (SAS) approach for wall bounded shear flows based on the Rotta's transport equation for integral length scale is contrasted with the $SST - SAS$ model of Menter and Egorov (Flow Turbul Combust 85(1):113–138, 2010) with local triggering (seamless formulation). It is known that the SAS approach does not trigger to a scale resolving mode in attached/mildly separated flows even if grid supports the transition Menter et al. (4th Symposium on Hybrid RANS-LES Methods, Beijing, China, September, 2011). This work addresses the question whether a zonal formulation of SAS ($k - \varepsilon$ formulation along with different norm for second derivative of velocity) could improve the triggering process from URANS to LES-like mode in attached/mildly separated flows. In order to study the effects of different formulations, both models were applied to different flow configurations ranging from fully attached to strongly separated, including stationary streamwise-homogeneous turbulent channel flow, flow over an S809 airfoil and swirling flow through a sudden expansion. We find that, in both formulations, even when grid is sufficiently fine to resolve the integral scale motions, the simulation only transitions to scale-resolving mode when the base URANS flow is naturally unstable.

A. Mehdizadeh (✉) · J.G. Brasseur · T. Nandi · H. Foroutan
The Pennsylvania State University, State College, USA
e-mail: aum50@psu.edu

J.G. Brasseur
e-mail: brasseur@psu.edu

T. Nandi
e-mail: tnn113@psu.edu

H. Foroutan
e-mail: hosein@psu.edu

S. Girimaji et al. (eds.), *Progress in Hybrid RANS-LES Modelling*,
Notes on Numerical Fluid Mechanics and Multidisciplinary Design 130,
DOI 10.1007/978-3-319-15141-0_23

1 Introduction

Reynolds averaged Navier-Stokes equations (RANS) models are generally used to simulate high Reynolds number turbulent flows relevant for industrial applications. However, RANS methods are not designed to capture the essential unsteady dynamics in complex flows. Large eddy simulation (LES) is an alternative approach that provides space-time accurate solutions, but LES is not feasible for many engineering applications due to its high computational cost. Hybrid URANS-LES modeling approaches can provide reasonable intermediate strategies, however, hybrid approaches present their own modeling issues.

In this investigation we analyze different formulations of hybrid turbulence modeling strategies that use models based on Rotta's equation for integral length scale to transition from URANS dynamics to a scale resolving mode through the energy-dominant integral scales similar to that predicted by LES. Our focus applications involve 4D turbulent boundary layer flow dynamics, from fully attached to fully separated flows. We focus on the essential triggering mechanisms that allow the simulation to transition between URANS mode and LES-like mode. It is known [2] that the $k - \omega - SST - SAS$ formulation fails to trigger from the URANS to LES-like mode in attached/mildly separated flows due to lack of natural transition. The straightforward goal of this work is to study if a zonal formulation in which the triggering location is predetermined could provide improvement and drive the simulation to scale resolving mode in attached/mildly separated flows.

The triggering process should make manifest appropriate equations with sub-grid models capable of transitioning the flow dynamics consistent with the resolution of the grid. To develop insight we contrast a version of a zonal hybrid approach for wall-bounded shear flows in the $k - \varepsilon$ framework using Rotta's original transport equation for turbulence integral length scale with the $SST - SAS$ model of Menter et al. [1]. In both approaches, eddy viscosity responses to the the trigger. The $k - \varepsilon$ framework changes eddy viscosity by applying a modified transport equation for the dissipation-rate ε including second derivative of the velocity away from the wall using y^+ as switch parameter. In contrast, the trigger mechanism in the $SST - SAS$ framework is local and controlled by a function which includes the "von Karman length scale" and other empirical terms.

The different triggering mechanisms have been studied by comparing the two approaches in (i) stationary streamwise-homogeneous turbulent channel flow, (ii) attached to mildly separated flows over an S809 airfoil and (iii) massively separated flow in a sudden expansion with strong swirl. We find that both triggering approaches indicate a similar characteristic, i.e. in stable attached flows the triggering mechanisms are not able to drive the simulation to a scale-resolving mode when grid support is provided. In contrast, in massively separated flows where the base flow in naturally unstable on sufficiently resolved grid (i.e. sudden change in flow geometry and/or strong swirl), transition to scale resolving mode occurs when the trigger allows eddy viscosity to drop and instability to occur.

In the following, the $SST - SAS$ model and the present approach are briefly explained and then the results obtained by comparing the two SAS models in different flows are presented and discussed.

2 Model Formulations Based on Rotta's Equation

2.1 The SST-SAS Model (Seamless Approach)

Based on Rotta's equation [3] for turbulence integral length scale, Menter et. al [1] proposed the $SST - SAS$ model by introducing an additional term into the ω transport equation in the $k - \omega - SST$ model [4] in the following form:

$$Q_{SAS} = max\left[\zeta_2 S^2 \left(\frac{L}{L_{vk}}\right)^2 - C_{SAS}\frac{2k}{\sigma_\Phi} max\left(\frac{1}{k^2}\frac{\partial k}{\partial x_j}\frac{\partial k}{\partial x_j}, \frac{1}{\omega^2}\frac{\partial \omega}{\partial x_j}\frac{\partial \omega}{\partial x_j}\right), 0\right], \quad (1)$$

In principle, Q_{SAS} should trigger the simulation to scale-resolving mode when appropriate grid support is provided.

2.2 Zonal SAS Model

Using the Rotta's equation which was originally derived based on the high Reynolds number assumptions [3], Mehdizadeh and Sadiki [5] proposed the following modified set of transport equations for turbulent kinetic energy and dissipation rate for high Reynolds number regions:

$$\frac{\partial \varepsilon}{\partial t} + \bar{U}_j \frac{\partial \varepsilon}{\partial x_j} = \underbrace{1.3}_{C_{\varepsilon_1}} \frac{\varepsilon}{k} P_k - \underbrace{2.36}_{C_{\varepsilon_2}} \frac{\varepsilon^2}{k} + 1.15\sqrt{k}\frac{|\bar{U}''|}{|\bar{U}'|} P_k + \frac{\partial}{\partial x_j}\left(\nu_t \frac{\partial \varepsilon}{\partial x_j}\right) \quad (2)$$

$$-0.33\left(\frac{\partial k}{\partial x_j}\right)^2 + 5\frac{\nu_t}{k}\left(\frac{\partial k}{\partial x_j}\right)\left(\frac{\partial \varepsilon}{\partial x_j}\right) - 2\frac{\nu_t}{\varepsilon}\left(\frac{\partial \varepsilon}{\partial x_j}\right)^2$$

$$\frac{\partial k}{\partial t} + \bar{U}_j \frac{\partial k}{\partial x_j} = P_k - \varepsilon + D_k, \quad (3)$$

where P_k, D_k and ν_t denote the production and diffusion of turbulent kinetic energy and turbulent eddy viscosity, respectively. To capture the friction dominated near-wall dynamics, Eqs. (2) and (3) are coupled with the corresponding k and ε equations designed to capture viscous effects at the wall vicinity, (in this study we apply the Launder-Sharma model [6]). This combination of the low Reynolds number model with the modified set of transport equations (Eqs. 2 and 3) has been termed as the zonal Scale-Adaptive approach. This approach allows us to describe both flow regions with the turbulence models of the same representation level using a blending function for the ε equation across the interface. For the first attempt the interface has been located 100 wall units (y^+) away from the surface. y^+ is calculated based on modeled turbulent kinetic energy using the logarithmic layer relations ($k = u_\tau^2\sqrt{c_\mu}$, where the definition of y^+ does not breakdown at separation point).

Note that Rotta's transport equation was originally based on the high Reynolds number assumptions, making it particularly suitable for flows in regions away from walls. Therefore, an appropriate combination of this equation with an appropriate low Reynolds number near wall turbulence model in a zonal framework may be a more reasonable approach. The present approach contrasts with the SST-SAS model by Menter in its zonal formulation and use of a different norm for the second derivative of the velocity vector.

3 Results and Discussions

Some representative results from fully attached stable flow to massively separated unstable flows (i.e. a stationary streamwise-homogeneous turbulent channel flow, attached to mildly separated flows over an S809 airfoil and massively separated flow in a sudden expansion with strong swirl) are presented using the open source code OpenFOAM [7]. For the momentum equations, a central differencing scheme (CDS) is used to discretise the equations in space in all simulations. The convective terms in the transport equations for turbulence quantities (k, ε, ω) used to estimate eddy viscosity are discretized using second order central differencing in channel flow (Sect. 3.1), but first order upwind is required in the attached boundary layer (Sect. 3.3) and sudden expansion (Sect. 3.2) cases to suppress numerical instabilities. For time discretization, second order method is used within an unsteady PISO-SIMPLE algorithm to solve the nonlinear coupled equations.

3.1 Channel Flow

Regarding wall bounded shear flows the present approach and the $SST - SAS$ model are applied to a stationary turbulent channel flow at $Re_\tau = 800$. For this study, the Launder-Sharma model is used as the near wall model. The computational domain has dimensions of $0.2\pi \times 0.2 \times 0.1\pi$ with $96 \times 128 \times 96$ cells, which is fine enough to support transition to scale-resolving mode to ensure that the grid will not suppress transition to scale-resolving mode. This resolution corresponds to $\Delta x^+ = 51$, $\Delta y^+ = 0.5 - 56$ (direction normal to the wall) and $\Delta z^+ = 26$.

The results are then compared to DNS results from Abe et al. [8] and also to those obtained from pure RANS models (Launder-Sharma and $k - \omega - SST$). The pure RANS models are run in non-steady mode, denoted as URANS. Figures 1 and 2 show

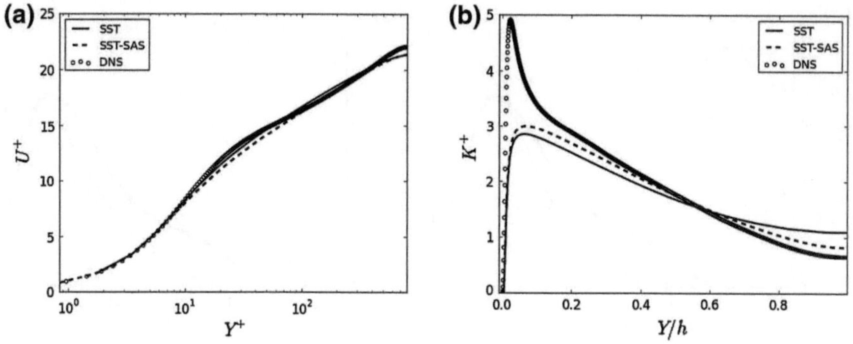

Fig. 1 **a** Streamwise mean velocity and **b** turbulent kinetic energy from DNS, $SST - SAS$ and $k - \omega - SST$ models

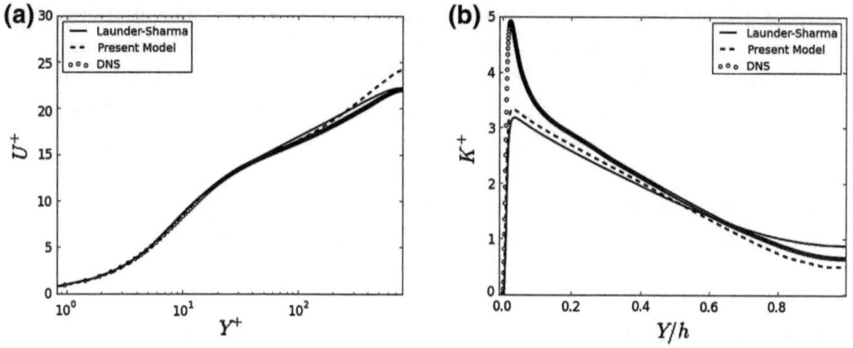

Fig. 2 **a** Streamwise mean velocity and **b** turbulent kinetic energy from DNS, Zonal SAS (present approach) and Launder-Sharma models

variations of the streamwise mean velocity and the turbulent kinetic energy from wall toward the channel center line obtained from $SST - SAS$, $k - \omega - SST$, Launder-Sharma and the present model. It seems that both SAS models modify the prediction of mean velocity around the channel center line which indicates that the SAS models differ from their underlying URANS models and can capture RANS properties better than pure RANS/URANS models. In contrast, turbulent kinetic energy in the inertial wall layer is underpredicted by all models. The is mainly because of the underlying URANS model which is dominating in this region (see Fig. 3) and is not able to capture near wall effects properly. Use of more advanced "low Reynolds number models" should improve the results in the inertial wall region.

Figure 3 indicates the behavior of the turbulent viscosity obtained from various models. The present approach shows improvement in capturing the spatial variation in turbulent kinetic energy between the wall and channel center line. The fact that none of the SAS models predict a drop in eddy viscosity of atleast an order of magnitude below RANS value indicates that the models have not transitioned from

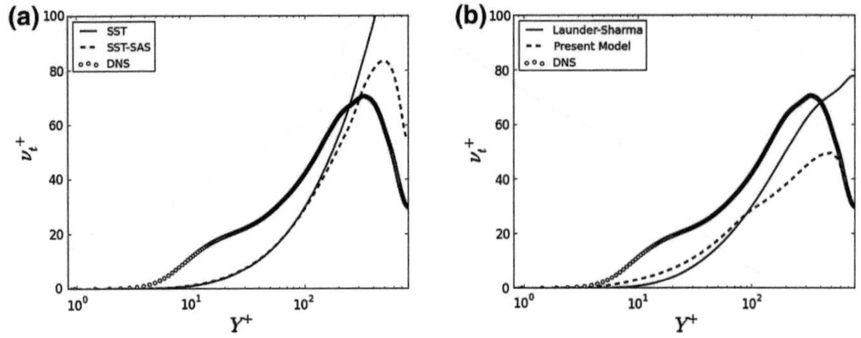

Fig. 3 Variation of turbulent viscosity from **a** $SST - SAS$, and **b** Zonal SAS (present approach)

their underlying URANS models to scale-resolving mode. However, even though the $SST - SAS$ and the present approach never trigger to scale-resolving mode, these models do not replicate precisely the underlying URANS models, indicating differences with the underlying URANS models. None of the underlying URANS models predict the proper trend of eddy viscosity around the channel center. Although the SAS models do predict the correct trend, this is not enough to trigger the instabilities required to transition to scale-resolving mode.

The results indicate that in $SST - SAS$ and the present approach, the trigger to scale-resolving mode from URANS mode does not happen even if the grid support is sufficient and even if there is an attempt to force the trigger to occur within the zonal approach.

Based on the results it may be concluded that since the baseline of URANS prediction of channel flow is sufficiently stable that even when there is opportunity to transition to a scale resolving mode, transition does not happen. Therefore, a mechanism that can generate/maintain sufficient local instability to drive the eddy viscosity to the much smaller values is required to generate a cascade of length and time scales to build an appropriate spectrum. Furthermore, both $SST - SAS$ and the present model in their steady modes have significant differences compared to their underlying models.

3.2 Swirling Flow Through a Sudden Expansion

Turbulent swirling flow through a sudden expansion is investigated using the Zonal SAS model. It is a complex flow possessing various dynamic phenomena including vortex breakdown, recirculation, detachment and reattachment, and enhanced mixing. Therefore, correctly predicting the flow behavior is quite challenging and special considerations should be paid to choice of the turbulence closure. Furthermore, the swirling flow through a sudden expansion is of high industrial interest since it resembles the flow in several technical applications including gas turbine combustor and draft tube of hydraulic turbines.

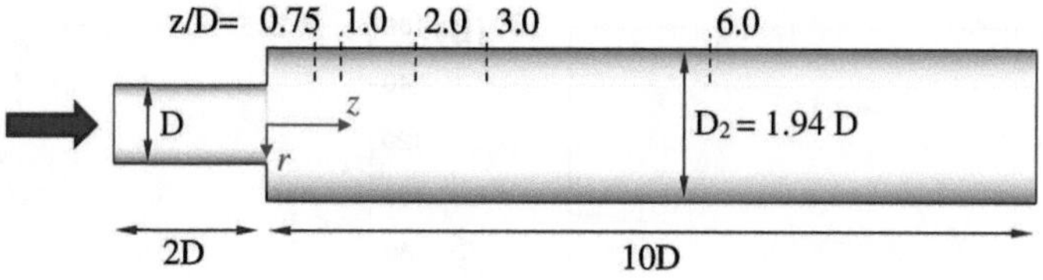

Fig. 4 Flow geometry and measurement planes

The considered test case corresponds to the experimental study of Dellenback et al. [9]. In the experiment water flows through an axisymmetric expansion (with the expansion ratio D_2/D equal to 1.94) as shown in Fig. 4. The axial and circumferential components of time-averaged and root-mean-square (RMS) velocity were measured by Dellenback et al. [9] at several cross-sections downstream of the sudden expansion. In addition, measurements were performed for a section upstream of the expansion which can be used as the inflow condition for the numerical simulations. Here, the inlet section of the computational domain is two diameters upstream of the expansion ($z/D = -2.0$) and the outlet boundary is placed at $z/D = 10$. This computational domain is shown to be sufficient for this flow problem [10]. The swirl number defined as in [11]:

$$S = \frac{\int_o^R V_{mean} U_{mean} r^2 dr}{R \int_o^R U_{mean}^2 r^2 dr} \tag{4}$$

is approximately 0.6, with $R(=D/2)$ being the inlet radius, and V_{mean} and U_{mean} are mean circumferential and axial velocities, respectively. The Reynolds number based on the inlet diameter D, and bulk velocity U_b, is 30,000. In the present simulations, the computational grid consists of 1,911,672 hexahedral cells, with the first cell center normal to the wall being placed at $r^+ \approx 1$. This resolution is sufficient to resolve the energy-dominant eddies in the flow. Figure 5a shows the change in grid aspect ratio in the main flow direction at two radial locations. Increasing aspect ratio is the result of coarsening the mesh only in the streamwise (z) direction.

Figure 5b, c indicate the variations of turbulent viscosity along the streamwise direction. In regions where vortex breakdown occurs (to plane 3), the flow is in its most unstable mode and eddy viscosity in both models is reduced indicating that transition to a scale resolving mode occurs. However, the reduction in ν_t is much larger in $SST - SAS$ model. The observed incipient numerical instability may suggest that the reduction in ν_t is too great for given grid support. On the other hand, it is possible that the present approach leads to unnecessarily large eddy viscosity. Moreover, further downstream (Fig. 6, Z/D=6) where the flow has less unsteadiness and the mesh is coarser, eddy viscosity from the present approach increases in response to the coarser grid, while ν_t in $SST - SAS$ remains of the same order. This suggests that the $SST - SAS$ model is more prone to stay in scale-resolving mode possibly explaining the numerical instability that was observed downstream.

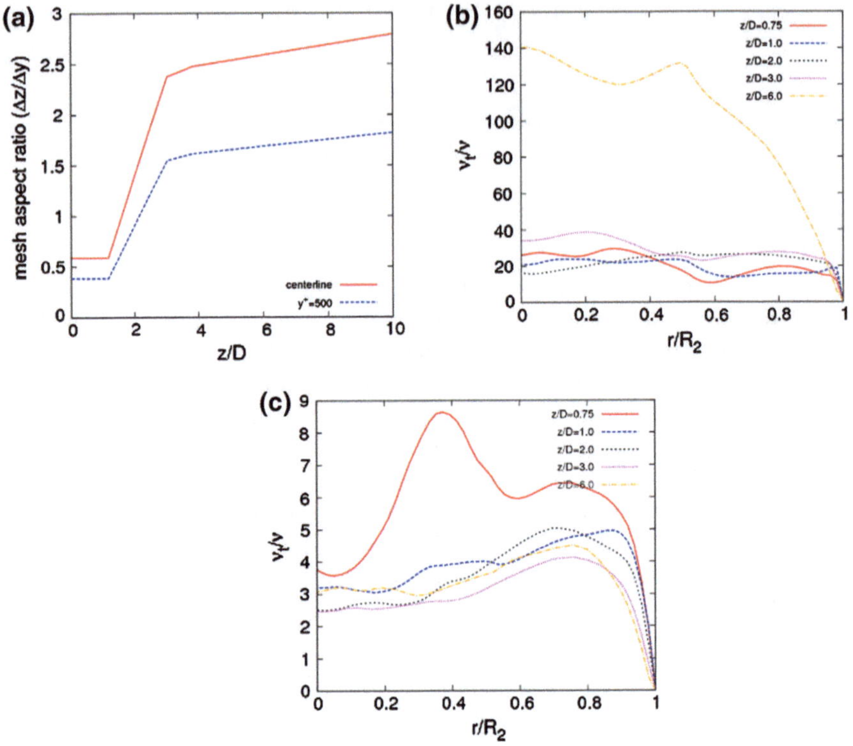

Fig. 5 **a** Mesh aspect ratio along the z direction for center line and $y^+ = 500$, variation of turbulent viscosity from **b** Zonal SAS (present approach) and **c** $SST - SAS$ model at different planes

Figure 6 shows radial distributions of the mean and RMS axial and circumferential velocities on five planes after the expansion, corresponding to the z/D values of 0.75, 1.0, 2.0, 3.0 and 6.0. Results obtained using the present model are compared to the experimental data as well as those obtained using the $SST - SAS$ model. Overall agreement between models predictions and experimental data is quite good, where the trends in the experimental data are reasonably well captured. The results indicate in contrast to stable flows like channel flow, if the base flow is inherently unstable (massively separated flows), when the SAS models provide the ability to trigger to a scale resolving mode, transition happens when appropriate grid support is provided for LES-like resolution. In this way the response to the trigger is natural and controlled so long as sufficient grid resolution is provided to allow the instability to fill in the resolved scales through a cascade process. A snapshot of the flow is presented in Fig. 7. The three-dimensional vortical structures are visualized by using isosurfaces of the λ_2 as described in [12].

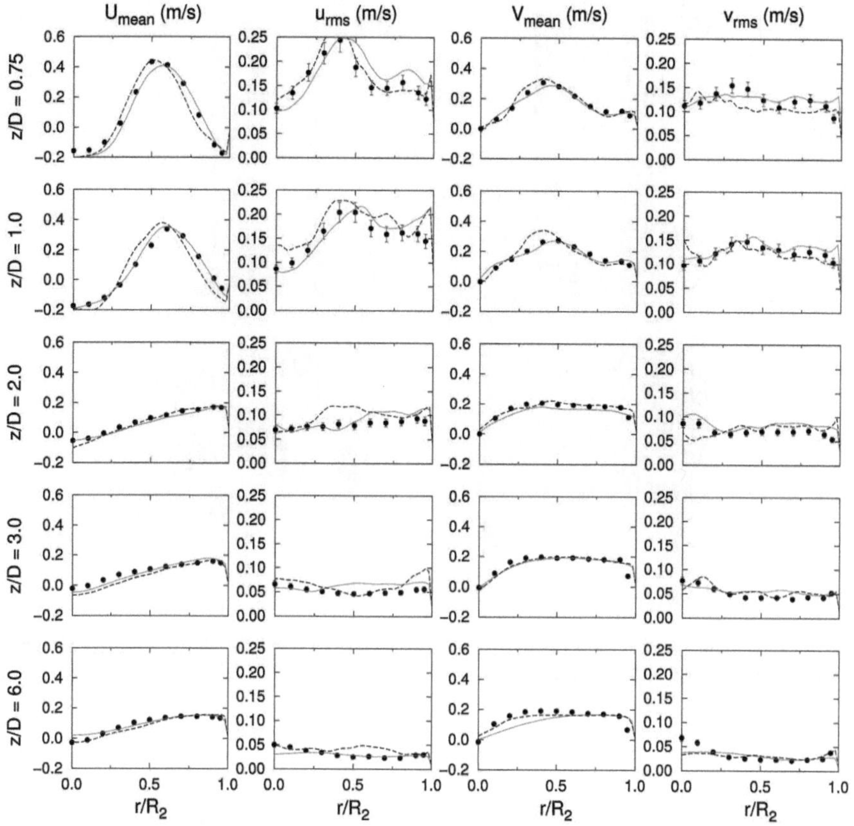

Fig. 6 Radial distributions of the mean and RMS axial and circumferential velocities on four planes after the expansion: $z/D = 6$ (*bottom*) $z/D = 0.75$ (*top*), Experiment: •, Zonal SAS (present model): —, $SST - SAS$: − − −

3.3 S809 Airfoil

The $S809$ airfoil is a 21 %-thick airfoil for horizontal-axis wind turbines. The airfoil was developed to be relatively insensitive to the leading-edge roughness and have a low-profile drag. As reference the experimental data provided by [13] are used. Figure 8 shows the computational domain which replicates the experimental test section. The uniform velocity at inlet is around 42 m/s with 1.5 % turbulence intensity. At the outlet, convective boundary condition is used. The upper and lower boundaries are treated as slip walls. The computational grid consists of approximately 2 million hexahedral cells. Figure 9 shows variation of pressure coefficient over the airfoil surface. At 5° angle of attack, where no boundary layer separation occurs, both models deliver reasonable results similar to what can be obtained using standard RANS models. For 10° AOA, Zonal SAS (present approach) shows some improvement over the first 80 % of the airfoil improving the lift coefficient

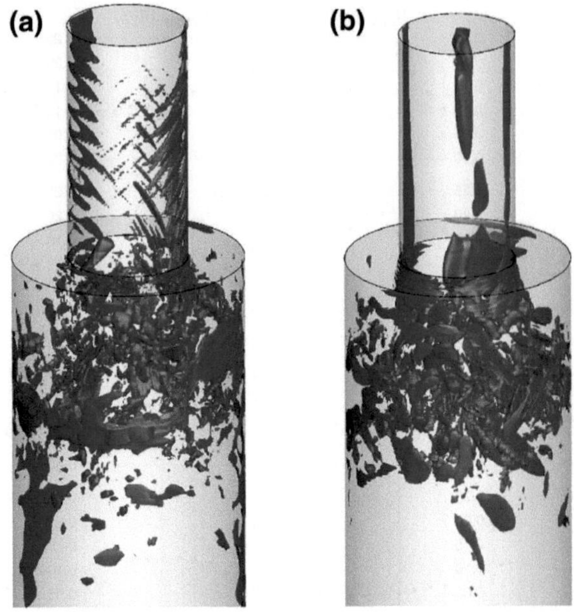

Fig. 7 Instantaneous flow field obtained from isosurfaces of $\lambda_2 = -500$, **a** the present hybrid model and **b** the $SST - SAS$ model

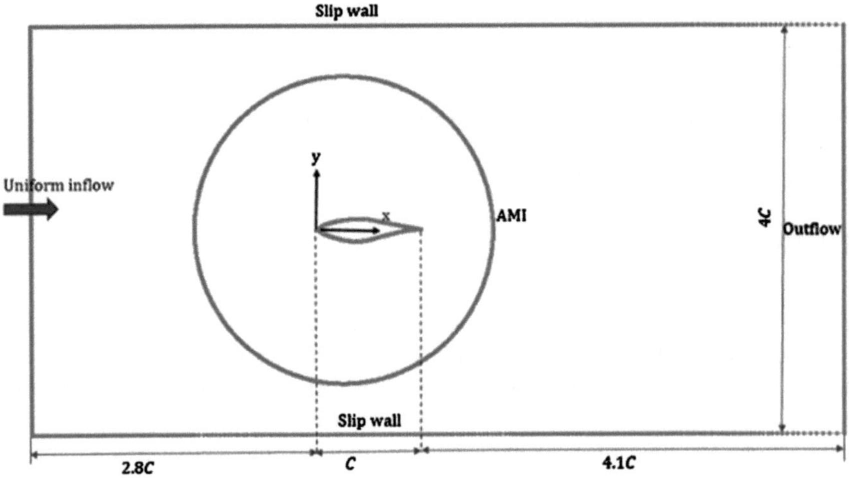

Fig. 8 Computational domain with the boundary conditions

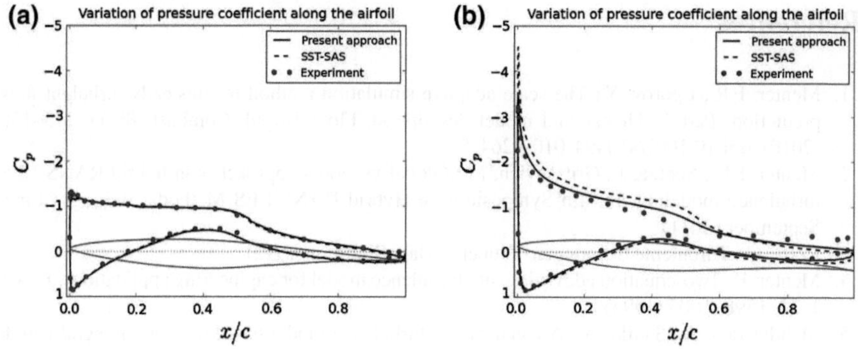

Fig. 9 Variation of pressure coefficient along the airfoil, **a** angle of attack 5° and **b** angle of attack 10°

Table 1 Lift coefficient for flow over an S809 airfoil at two angles of attack

Angle of attack (degrees)	c_l (experiment)	c_l (present approach)	c_l (SST-SAS)
5	0.6	0.62	0.62
10	0.8	0.92	1.05

(Table 1). However, both models fail to predict separation close to the trailing edge. Like channel flow both models stay in steady RANS mode as in [14]. This has been pointed out with respect to the $SST - SAS$ by Menter et al. [2] for mildly separated flows. We find a similar lack of response to the trigger with the zonal approach.

4 Conclusion

A central aim of the current study is to determine if a method that "forces" the simulation to adopt a hybrid URANS-LES solution set outside near-surface URANS region would overcome the known difficulty of the SST-SAS model in triggering to LES-like mode in attached/mildly separated boundary layer flows. We show that neither the seamless (SST-SAS) nor the zonal formulation of the SAS approach are able to trigger the simulation to LES-like mode in attached/mildly separated flows. This suggests that SAS triggering mechanisms need to be designed more intelligently to introduce dynamic forcing that triggers transition to scale resolving mode when necessary. This dynamic mechanism should generate sufficient instability to allow the cascade to develop within the resolved scales, either naturally or artificially. This should drive the model to a scale-resolving mode when grid support is provided but a self generating instability mechanism is not present.

Acknowledgments The authors acknowledge the funding of this work by Deutsche Forschungsgemeinschaft (DFG) under Grant No. ME4126/2-1.

References

1. Menter, F.R., Egorov, Y.: The scale adaptive simulation method for unsteady turbulent flow prediction, Part 1: Theory and model description. Flow Turbul. Combust. **85**(1), 113–138 (2010). doi:10.1007/s10494-010-9264-5
2. Menter, F.R., Schtze, J., Gritskevich, M.: Global vs. zonal approaches in hybrid RANS-LES turbulence modeling. In: 4th Symposium on Hybrid RANS-LES Methods, Beijing, China, September (2011)
3. Rotta, C.: Turbulente Srtmungen. Teuber Verlag, Stuttgart (1972)
4. Menter, F.: Two-equation eddy-viscosity turbulence model for engineering applications. AIAA J. **32**, 1598–1605 (1994)
5. Mehdizadeh, A., Sadiki, A.: A zonal hybrid turbulence model based on Rotta integral length scale transport equation. In: Proceeding of THMT 12, ISBN 978-1-56700-302-4, ID: E119 (2012)
6. Launder, B.E., Sharma, B.I.: Application of energy dissipation model of turbulence to the calculation of flow near a spinning disc. Lett. Heat Mass Transf. **1**(2), 131–138 (1974)
7. http://www.openfoam.com
8. Abe, H., Kawamura, H., Matsuo, Y.: Surface heat-flux fluctuations in a turbulent channel flow up to $Re_t = 1020$ with Pr = 0.025 and 0.71. Int. J. Heat Fluid Flow **25**, 404–419 (2004)
9. Dellenback, P.A., Metzger, D.E., Neitzel, P.: Measurements in turbulent swirling flow through an abrupt axisymmetric expansion. AIAA J. **26**(6), 669–681 (1988)
10. Paik, J., Sotiropoulos, F.: Numerical simulation of strongly swirling turbulent flows through an abrupt expansion. Int. J. Heat Fluid Flow **31**, 390–400 (2010)
11. Gyllenram, W., Nilsson, H.: Design and validation of a scale-adaptive filtering technique for LNR turbulence modeling of unsteady flow. ASME J. Fluid Eng. **130**(5), 051401 (2008)
12. Jeong, J., Hussain, F.: On the identification of a vortex. J. Fluid Mech. **285**, 69–94 (1995)
13. Sheng, W., Galbraith, R.A.McD., Coton, F.N.: On the S809 Airfoil‘s unsteady aerodynamic characteristics. Wind Energy **12**, 752–767 (2009)
14. Wolfe, W.P., Ochs, S.S.: CFD calculation of S809 aerodynamic characteristics, AIAA-97-0973 (1997)

Go4Hybrid: A European Initiative for Improved Hybrid RANS-LES Modelling

Charles Mockett, Werner Haase and Frank Thiele

Abstract An overview is given of a new EU-funded project targeting the mitigation of a key remaining issue with hybrid RANS-LES methods. The grey area problem arises due to inconsistencies in the transition between RANS and LES paradigms and has a detrimental impact on the early separated shear layer, particularly for e.g. shallow separating/reattaching flows. Progress on this front is seen as central to improving industrial confidence in CFD for situations typical of the limits of machine performance.

1 Introduction and Overview of Project

This paper gives an overview of the motivation and aims of a new EU-funded project targeted at further improvement in hybrid RANS-LES modelling. At the time of writing, the project is at around the 6 month stage of its 2 year planned duration. As such, a summary of project results is planned for a future publication.

The project consortium is made up of seven partners (national research laboratories, SMEs and a university), which is complemented by 11 industrial observers or associate partners drawn from the aerospace (airframe and engine), power generation and ground transportation (road and rail) sectors as well as commercial CFD software providers. The project will run for 2 years, from October 2013 to September 2015.

C. Mockett (✉) · F. Thiele
CFD Software Entwicklungs- und Forschungsgesellschaft mbH,
Berlin, Germany
e-mail: charles.mockett@cfd-berlin.com

W. Haase
Werner Haase Aeronautics Consultants, Neubiberg, Germany
e-mail: whac@haa.se

S. Girimaji et al. (eds.), *Progress in Hybrid RANS-LES Modelling*,
Notes on Numerical Fluid Mechanics and Multidisciplinary Design 130,
DOI 10.1007/978-3-319-15141-0_24

2 Technical Aspects

2.1 The Grey Area Issue and Its Impact on Applications

The aeronautical industry lacks confidence in the accuracy of CFD close to the edges of the flight envelope [1]. For many applications, hybrid RANS-LES methods are the best candidate for the next generation of CFD methods for increased fidelity at industrially-feasible expense. While these have been shown to perform considerably better than conventional (U)RANS approaches in situations with massive flow separation, they are hampered by the grey area issue when shear layer instabilities are weaker. The problem is caused by inconsistencies in the transition from RANS to LES in the early separated shear layer, generally manifesting itself through insufficient levels of overall (resolved + modelled) turbulence. Unfortunately, precisely such flows are of high importance for aircraft performance (e.g. flight near maximum lift, turbo-machinery at maximum loading). As a result, future acceptance of hybrid RANS-LES methods will strongly depend on the ability to reduce the extent of the grey area.

Results from the recently-completed EU project ATAAC [4] can be seen to confirm the importance of the grey area as the key remaining problem in hybrid RANS-LES. An example for a shallow separating/reattaching flow is shown in Fig. 1. The RANS result exhibits a strongly exaggerated recirculation region. A non-zonal hybrid RANS-LES method ("SST-IDDES" in the legend) performs somewhat better than RANS in this respect, however the skin friction inside the recirculation bubble is significantly less accurate than the RANS solution. This is caused by the grey area

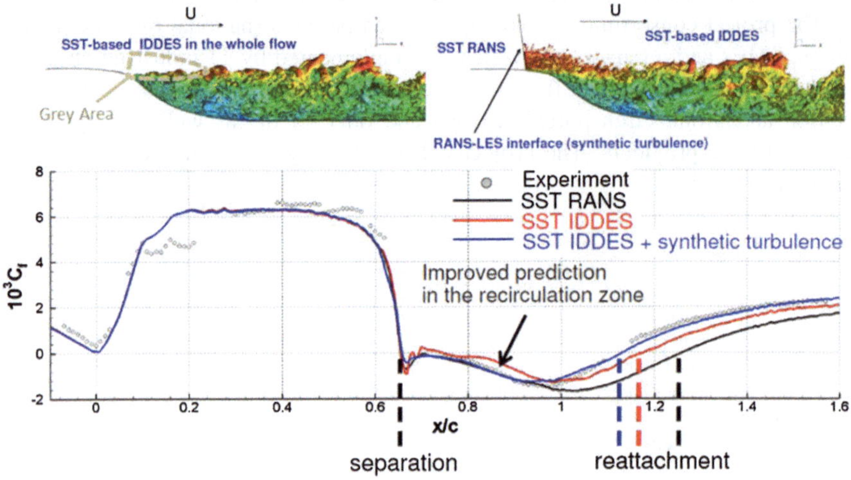

Fig. 1 Example of *grey* area impact for shallow separating/reattaching flow over a 2D wall-mounted hump (simulations by NTS [2], conducted within the ATAAC project)

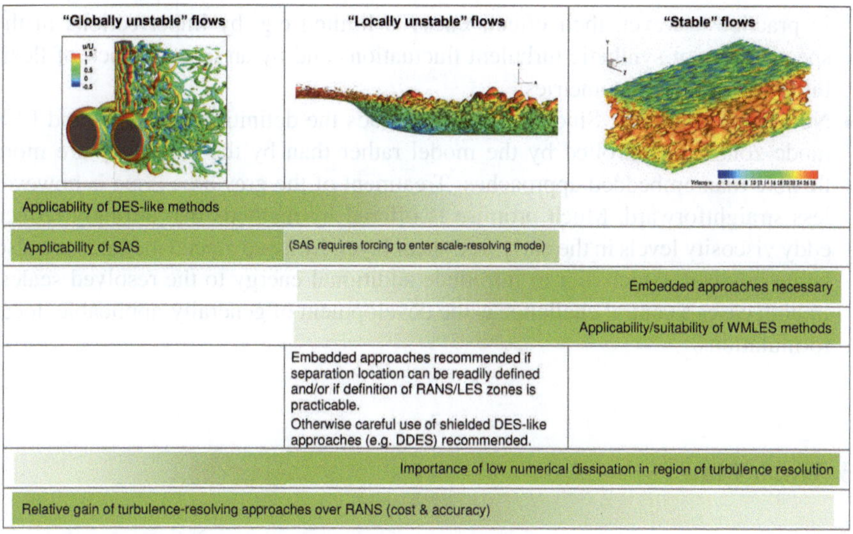

Fig. 2 Summary of method applicability prior to the Go4Hybrid project: Best practice output from the previous ATAAC project [4] showing the suitability (indicated by intensity of the *green* shading) of different classes of hybrid RANS-LES methods for different flow types (categorisation proposed by Menter et al. [3])

problem, in the form of delayed development of resolved turbulence in the initial separated shear layer. When synthetic turbulent fluctuations are imposed upstream of separation using a zonal method, the prediction matches the experimental data very well throughout the recirculation, reattachment and recovery regions.

This issue was furthermore represented in the best practice guidelines formulated in the ATAAC project, which sought to describe the applicability of different classes of hybrid model to different types of flow. Figure 2 provides a summary, whereby the approach to flow classification according to Menter et al. [3] is adopted. It is clearly apparent that the intermediate class of "locally unstable" flows is poorly served by currently available methods.

2.2 *Addressing the Grey Area Issue*

The goal of the Go4Hybrid project is essentially to improve the situation encapsulated in Fig. 2 by improving the performance of hybrid methods with respect to grey-area dominated flows. Improvements will be pursued for two classes of approach:

- **Embedded approaches**: These introduce explicit RANS-to-LES (and vice-versa) coupling at user-specified interfaces between the RANS and LES zones. In principle these have the potential to completely eliminate the grey area problem.

In practice however, their effectiveness is limited e.g. by imperfections in the specification of synthetic turbulent fluctuations and by an inherent lack of flexibility for complex geometries.

- **Non-zonal methods**: Since for these methods the definition of RANS and LES mode zones is controlled by the model rather than by the user, they are more flexible than embedded approaches. Treatment of the grey area issue is however less straightforward. Much promise is offered by methods that strongly reduce eddy viscosity levels in the early free shear layer. More advanced methods are also being pursued, which aim to introduce additional energy to the resolved scales. In both cases a central challenge is the development of generally-applicable, local formulations.

3 Work Plan

New method formulations will be assessed through direct comparison on two fundamental test cases, namely a spatially-developing shear layer (for non-zonal methods) and a flat plate boundary layer (for embedded approaches). Additionally, validation of the real-world robustness and effectiveness of the developed strategies will be carried out on a suite of complex demonstration test cases, namely a complex helicopter fuselage, a delta wing, a three-element high lift system, a 2D wall-mounted hump and the prediction of flow and noise from a round jet. These cover a wide range of applications and flow topologies.

An innovative element to the project will be an attempt to carry out a fully-direct comparison of the developed improvements. The "common assessment platform" concept goes beyond the usual approach of common grids to include also the underlying numerical methods. To this end, a selection of the most promising developments will be implemented into one solver and compared directly for the two fundamental cases. A central outcome of the project will finally be given in the form of best practice guidelines on the applicability of the developed approaches for different applications.

4 Conclusion

Go4Hybrid is a highly-focussed R&D effort aiming at improvement for one of the major remaining issues in hybrid RANS-LES. Progress is underway towards reduced grey area severity for non-zonal methods on the one hand, and improved accuracy and flexibility of embedded approaches on the other. Industrial relevance and applicability are high priorities, and the inclusion of 11 industrial observers or associate partners is one key vehicle to assure this. Although the project is in its early stages, promising initial results have already been achieved (some of which are reported in these proceedings). A further paper summarising the project outcomes will be published shortly after the conclusion of the project.

Acknowledgments The Go4Hybrid project ("Grey Area Mitigation for Hybrid RANS-LES Methods") is funded by the European Union Seventh Framework Programme *FP7/2007–2013* under grant agreement no. 605361. The Partners of the Go4Hybrid project are CFD Software E+F GmbH (DE—coordinators), German Aerospace Center (DE), National Aerospace Laboratory (NL), New Technologies and Services LLC (RU), ONERA (FR), Swedish Defence Research Agency (SE) and the University of Manchester (GB). The ATAAC project ("Advanced Turbulence Simulation for Aerodynamic Application Challenges") was funded by the European Seventh Framework Programme *FP7/2007–2013* under grant agreement no. 233710.

References

1. Abbas, A., Becker, K.: Numerical simulation "Airbus vision and strategy". In: Fu, S. et al. (eds.) Progress in Hybrid RANS-LES Modelling. NNFM, vol. 117, pp. 1–13. Springer, New York (2012)
2. Adamian, D., Travin, A.: An efficient generator of synthetic turbulence at RANS-LES interface in embedded LES of wall-bounded and free shear flows. In: Kuzmin, A. (eds.) Computational Fluid Dynamics 2010, pp. 739–744. Springer, New york (2011)
3. Menter, F., Schütze J., Gritskevich, M.: Global vs. zonal approaches in hybrid RANS-LES turbulence modelling. In: Fu, S. et al. (eds.) Progress in Hybrid RANS-LES Modelling. NNFM, vol. 117, pp. 15–28. Springer, New York (2012)
4. Schwamborn, D., Strelets, M.: ATAAC—an EU-project dedicated to hybrid RANS/LES methods. In: Fu, S. et al. (eds.) Progress in Hybrid RANS-LES Modelling. NNFM, vol. 117, pp. 59–75. Springer, New York (2012)

Automatic Hybrid RANS/LES Strategy for Industrial CFD

Grégoire Pont, Paola Cinnella, J.C. Robinet and Pierre Brenner

Abstract An automatic HRL (Hybrid RANS/LES) strategy is investigated in FLUSEPA, a finite-volume solver developed by Airbus Defense and Space. A HRL turbulence model is coupled to a high-order hybrid numerical approximation method. Concerning the turbulence model, the well-known $k - \varepsilon$ two equations RANS turbulence model is sensitized to the grid as suggested by Perot and Gadebusch (Phy Fluids 19:1–11, 2007). Concerning the numerical strategy, a third-order accurate upwind approximation method is locally re-centered in vortex dominated regions to achieve non-dissipative fourth-order accuracy. Results are presented for a 2D backward facing step and an an axisymmetry backward facing step, which represent good prototypes of after body flows.

1 Introduction

There are three kinds of viscosity in a numerical simulation of turbulent flow : laminar viscosity ν, eddy viscosity ν_t introduced by RANS (Boussinesq hypothesis) or SGS model in use , and numerical viscosity ν_n intrinsic to the numerical scheme. The second kind of viscosity is equal to zero for so called "implicit" modelling approaches (see e.g. Ref. [2]). The main idea of a hybrid turbulence model is to reduce the turbulent viscosity ν_t to locally solve unsteady turbulent structures. In the resolved part of the turbulent spectrum, the numerical viscosity must be negligible compared to eddy viscosity, otherwise, turbulent structures will be dissipated by the numerical scheme. In this work, the numerical solver in use is FLUSEPA, the unstructured finite-volume solver developed by Airbus Defense and Space company to calculate

G. Pont (✉)
Dynfluid Laboratory, Airbus Space and Defense, Les Mureaux, France
e-mail: gregoire.pont@ensam.eu

P. Cinella · J.C. Robinet
Dynfluid Laboratory, Les Mureaux, France

P. Brenner
Airbus Space and Defense, Les Mureaux, France

S. Girimaji et al. (eds.), *Progress in Hybrid RANS-LES Modelling*,
Notes on Numerical Fluid Mechanics and Multidisciplinary Design 130,
DOI 10.1007/978-3-319-15141-0_25

compressible, multidimensional, unsteady, viscous and reactive flow over bodies in relative motion. The numerical scheme used in this solver is designed for highly compressible flows and has good shock capturing capabilities. The solver is based on a high order Godunov type method along with a MUSCL-like reconstruction. All the required derivatives are calculated by a successive corrections algorithm. The Godunov methods are well known to be suitable for compressible flows with shocks because they introduce a numerical dissipation that damps non-physical oscillations and ensures the stability of the method, but they are much too dissipative for HRL calculations. In the quest for a compromise between computational cost and resovability properties of the numerical method, we retain a hybrid scheme, combining a 3rd-order accurate version of the present scheme in inviscid regions with a fourth-order non dissipative scheme in regions dominated by turbulent flow structures.

2 Governing Equations

We look for the numerical solutions of the compressible Navier-Stokes equations within Reynolds-Averaged form or in filtered form, and supplemented by a turbulence or subgrid model, respectively. In the following, the Reynolds/subgrid stress tensor is described by using an eddy viscosity model, and is supposed to be related to the average/filtered velocity gradient via an eddy viscosity coefficient ν_t. To compute ν_t, automatics HRL models are considered, based on a modification of well known RANS models. Precisely, in this work we consider the hybrid $k-\varepsilon$ model developed by Perot and Gadebusch [1] (referred-to as PG hereafter). Grid sensitization of the underlying RANS models is achieved by introducing an energy transfer parameter α:

$$\alpha = 1.5\left(1 - C^*\left(\frac{k}{k+k_r}\right)^2\left[\left(\frac{\Delta x_i}{\sqrt{k_r}}\frac{\partial\sqrt{k_r}}{\partial x_i}\right)^2 + 0.11\right]^{-1}\right)$$

where k_r is the resolved turbulent kinetic energy, k the modelled kinetic energy and $C^* = 0.28$. The α parameter, comprised between -1 and 1, detects well-resolved flow regions and turns the baseline two equations models into a subgrid model while keeping a classical $k - \varepsilon$ model in under-resolved regions. The α parameter acts through two mechanisms: it pre-multiplies the Reynolds stress tensor, contributing to lower the modelled Reynolds stresses in well-resolved regions; it pre-multiplies the production term in the turbulent kinetic energy equation, contributing to lower the amount of modelled kinetic energy produced by the model and, indirectly, the eddy viscosity. The eddy viscosity formulation is also modified with respect to the standard model, by introducing a weighting factor equal to the ratio of the modelled to the total kinetic energy. Note that α may become negative in highly resolved regions characterized by a too large amount of modelled kinetic energy. This amplifies flow instabilities by converting turbulent diffusion in an antidiffusion, and enables effective model transition from RANS to LES mode by enriching the frequency

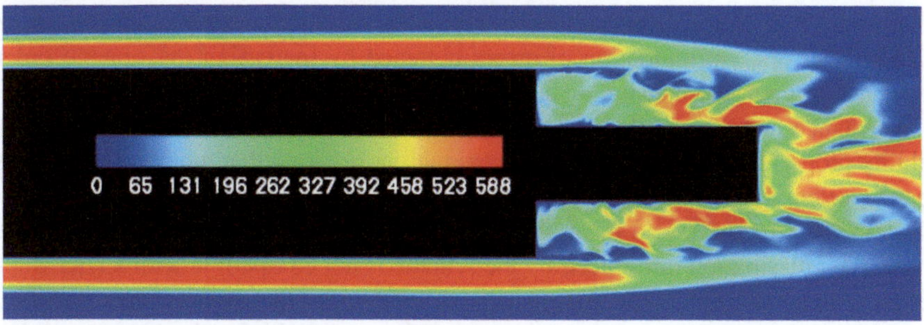

Fig. 1 Snapshot of the eddy viscosity ratio ν_t/ν

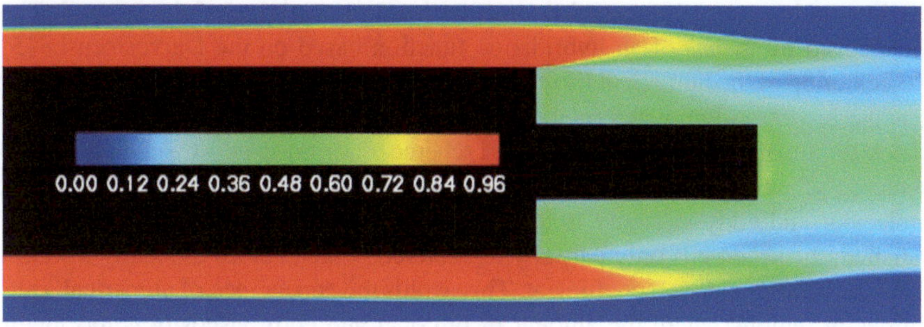

Fig. 2 Time-averaged modeled to total kinetic energy ratio

spectrum. The model is implemented along with a shielding function that enforces a RANS mode in attached boundary layers. Figure 1 shows a snapshot of eddy viscosity and the time-averaged distribution of modelled to total kinetic energy ratio for an axisymmetric backward facing step flow discussed in the following. The eddy viscosity is found to decrease in the shear layers and in the recirculation bubble, which allows the resolved kinetic energy to increase. The figure shows that in attached boundary layers, turbulent kinetic energy is fully modelled, i.e. they are treated in a RANS mode (Fig. 2).

3 Numerical Method

For the numerical approximation of the Navier-Stokes equations, we consider an unstructured finite volume methodology. Fluxes are integrated on each interfaces of control volumes using a high-order reconstruction formula. The reconstruction involves the derivatives of the primitive variables, approximated through a successive correction method [3]. This kind of method allows to ensure high-order accuracy on general meshes. The truncation error and the spectral properties of the preceding numerical scheme have been studied in details for finite volume operators ranging from the second to the fifth order of accuracy in [4]. Due to the use of an upwind numerical flux at cell interfaces, all schemes in the family exhibit relatively high damping errors, and reduction of errors obtained by increasing the scheme order does not compensate the significant increase in computational complexity and cost. For instance, we showed that even for schemes of 4th and 5th accuracy, the resolvability properties of the numerical approximations are such that more than 10 points per wavelength are required to keep damping errors to within a reasonable limit. This means that, with grid resolutions typically used in industrial applications, the numerical dissipation introduced by the scheme is not compatible with a HRL simulation. In an attempt to reach the best possible compromise between resolvabiliity and computational cost, we restrict our attention to the third-order scheme, which provides sufficiently low dispersion errors with a moderate increase in computational complexity with respect to the baseline second-order scheme, and reduce numerical dissipation in vortex-dominated regions by local re-centering of the method. This is achieved by means of an hybridation function based on the Ducros sensor [5]. Re-centering of the numerical fluxes leads to a fourth-order accurate, non dissipative scheme in vortex-dominated regions, while keeping a 3rd-order upwind scheme in shock-dominated regions. The proposed scheme, referred-to as VC scheme (vortex-centered scheme), is stable if the grid Reynolds number, based on the sum of the molecular and eddy viscosity is found to be below 2. When this condition is not satisfied, the scheme is only partially re-centered, which comes to locally lowering the numerical dissipation coefficient. This avoids the appearance of numerical instabilities in regions where the amount of physical and eddy viscosity is too low to dam ill-resolved solution modes. Temporal integration is carried out by means of a second-order method, namely, the Heun scheme. Each cell is advanced with nearly its maximum allowable time step according to the CFL condition, and time consistency is ensured by sub-iterating over the cells with local time step lower than the maximum value over the domain. This time integration is well suited for unsteady problems characterized by small physical time scales.

4 Backward Facing Step Flows

Backward facing step flows are a good prototype for after-body aerodynamics, and exhibit several unsteady phenomena. First, a Kelvin-Helmholtz vortex shedding appear at the separation point followed by a pairing process highlighted by the hot-wire techniques experiment of Troutt et al. [6]. Hairpin vortices have been observed in the recirculation bubble by Kiya and Sasaki [7]. On the other side, several researcher agree in indicating the presence of another unsteady phenomenon, called flapping motion, which is a low frequency instability. The backward facing step flow is a very good testing bench for self-adaptive HRL models because the Kelvin-Helmholtz instability is difficult to capture without delay due to the convective character of this instability and of the turbulent viscosity coming from the boundary layer before the separation point.

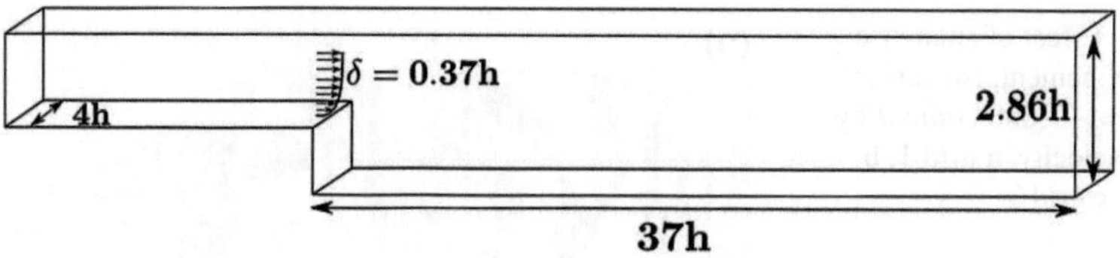

Fig. 3 Sketch of the computational domain

Table 1 Details of calculations and experiment

	Present calculation	ZDES ONERA, [9]	Experiment
Reynolds number	40,000	40,000	40,000
Number of elements	Grid 1: 7×10^5, $z^+ = 300$, $x^+ = y^+ = 15$ Grid 2: 1.4×10^6, $z^+ = 150$, $x^+ = y^+ = 15$ Grid 3: 2.8×10^6, $z^+ = 75$, $x^+ = y^+ = 15$	4×10^6	Moreau et al. [8]
Turbulence	Perot and Gadebusch	ZDES	

4.1 2D Backward Facing Step

The first test case is a geometrically 2D backward facing step with an upper wall experimentally studied by Moreau et al. [8]. The geometrical features and measures of the computational domain used in this study are detailed in Fig. 3. The length before the step is chosen so as to obtain a boundary layer thickness of 0.37 h just before the separation point. Hereafter, we show results obtained with the PG model. Numerical simulations using ZDES[1] modelling on a fine mesh of 3.9 million points were provided by Deck [9], which are also displayed for comparison. Details of the simulations and references are presented in the Table 1. Grids 1, 2 and 3 differ only by refinement in the spanwise direction, where the number of grid point is doubled. The y^+ and x^+(at the beginning of the step) for grids 1, 2 and 3 is equal to 15 (Fig. 4). Figure 3 shows the isosurfaces of the λ_2 criterion colored by mean velocity. Refinement in the spanwise direction leads to the appearence of finer and finer structures, indicating some form of grid convergence of the model as precedingly observed by Perot and Gadebusch for isotropic decaying turbulence [1]. Also note the formation of hairpin vortices in the recirculation bubble after a vortex pairing mecanism, and the generation of longitudinal vortices downstream of the reattachment point. Figure 5 shows the average streamwise velocity field and streamlines, as well as velocity profiles taken at different streamwise locations in the recirculation bubble. Present results are in good agreement with the experiments. The location of the reattachment point for grids 1, 2 and 3 ($X_r/H \simeq 6.2$) is close to the experiment of Hall et al. [10], of $X_r/H \simeq 6.8$, Driver et al. [11], $X_r/H \simeq 6.1$ and the direct simulation of Le et al. [12] with $X_r/H \simeq 6.28$. On the other hand, the shape of the secondary

[1] Zonal Detached Eddy Simulation [9].

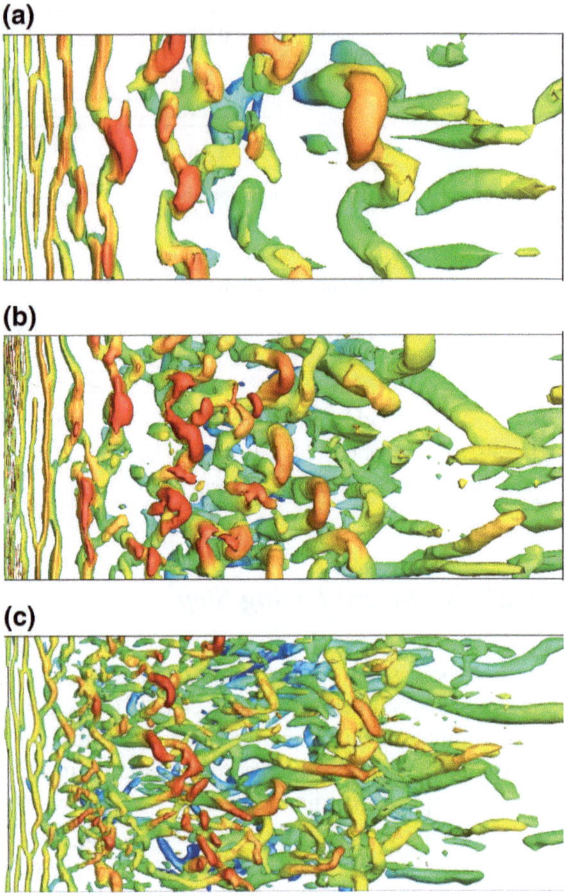

Fig. 4 Effect of spanwise grid refinement, iso surface of $\lambda_2 = -7{,}000$ *colored* by mean velocity, **a** grid 1, **b** grid 2, **c** grid 3

recirculation bubble in the corner is in good agreement with the PIV streamlines of the experiment of Hall et al. [10]. ZDES calculations by Deck [9], obtained on a grid of 4×10^6 cells are also reported for reference. Present results reproduce accurately both the reference calculation and the experimental data, in spite of substantially coarser grids. Of course, simulation result for grid 3 better predict the mean flow in the recirculation bubble, especially for the first three profiles of Fig. 5, but results on the two coarser grids are also satisfactory. RMS of longitudinal velocity fluctuations are shown in Fig. 6. They are in good agreement with the reference ZDES calculation and reasonably close to the experiments, despite the coarseness of grid 1. Thanks to the non-dissipative numerical scheme, there is no delay in the appearance of Kelvin Helmholtz instabilities in the shear layer: physical perturbations are not damped, and backscatter of energy is fostered.

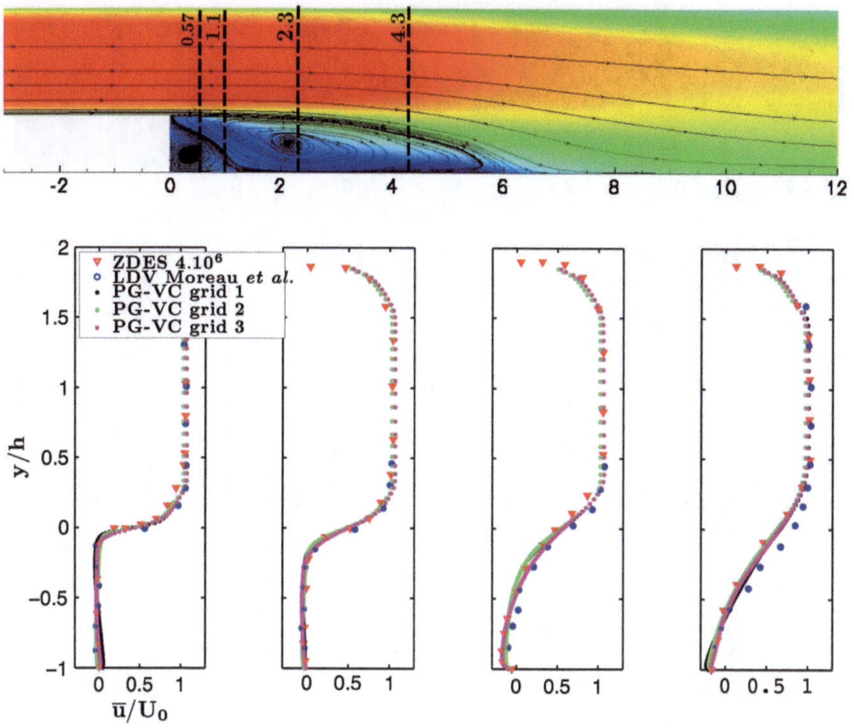

Fig. 5 Mean longitudinal velocity at different locations in the recirculation bubble

4.2 Axisymmetric Backward Facing Step

The second test case is an axisymmetric backward facing step experimentally studied by Deprés et al. [13] and Meliga et al. [14]. The geometrical features and measures are provided on Fig. 7. The upstream length before the step is chosen so as to obtain a boundary layer thickness of 0.2D just before the separation point. Present calculation was carried out by using the PG turbulence model coupled with the VC scheme on a grid of 5.7×10^6 cells with average y^+ of the first cell close to the wall of about 50. The x^+ at the beginning of the step is equal to 50. This is rather coarse, but allows keeping computational costs to within an industrially acceptable level. Results are compared to the available experimental data [13, 14] and to ZDES calculations [15, 16] with differents grids listed in the Table 2. Figure 8 represents a snapshot of the instantaneous flow field, showing the coherent structures in the recirculation area downstream of the backward facing step. We can see a three-dimensionalization process very similar so that of the 2D backward facing step flow. Hairpin vortices appear during the pairing mechanism after the separation point and longitudinal vortices are generated downstream of the reattachment point.

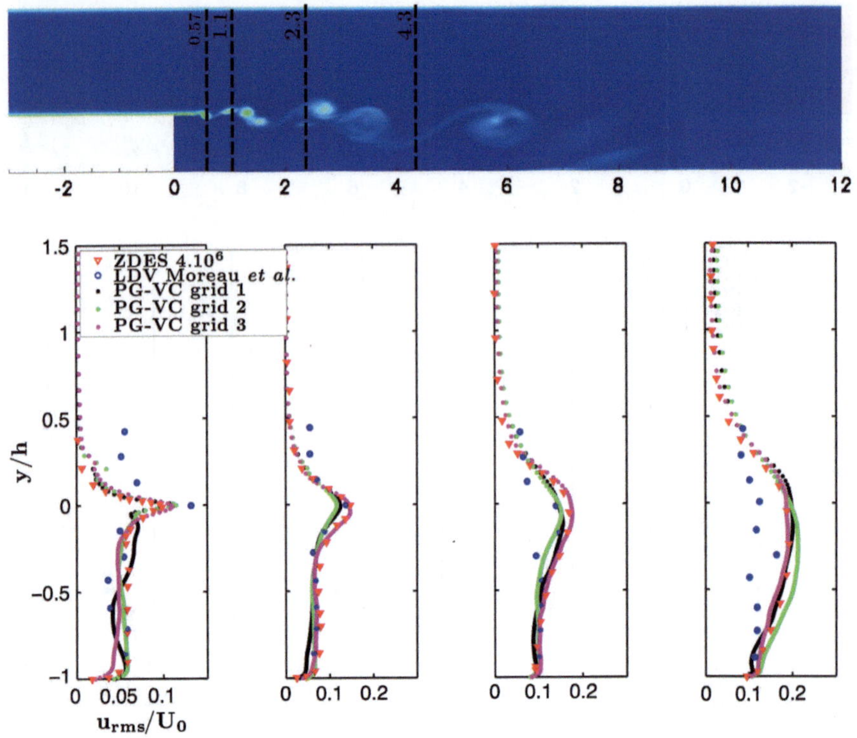

Fig. 6 Longitudinal velocity fluctuations at different locations in the recirculation bubble

Fig. 7 Schematic of the axisymmetric backward facing step

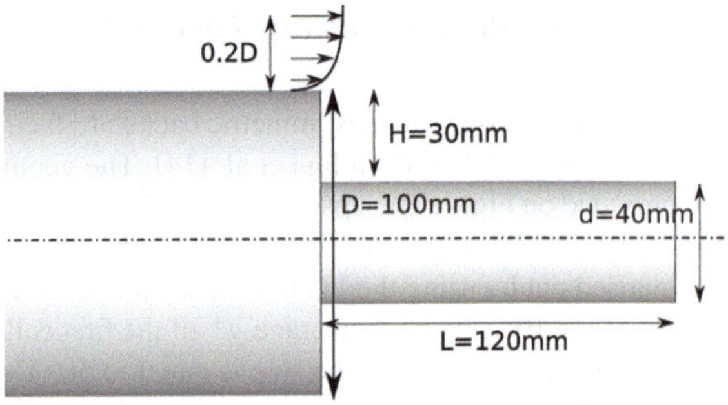

Figure 9 shows the streamlines of the computed mean flow, which is characterized by a recirculation bubble with reattachment point located at $x/D = 1.15$. This value is close to the reattachment point predicted by the ZDES calculation ($x/D = 1.1$) and by the measurements of Lê [18] ($x/D = 1.11$) and Deprés et al. ($x/D = 1.3$). The mean flow is also characterized by a secondary corner vortex. Figure 10 shows mean velocity profiles taken at the same location of the 2D case. The shape of

Table 2 Detail on calculations and experiments

	ZDES [9]	Present calculation	Experiments
Re	1.2×10^6	1.2×10^6	1.2×10^6
Number of cells	Grid 1: 5×10^6, $N_z = 97$ [15] Grid 2: 8×10^6, $N_z = 147$ [15] Grid 3: 12×10^6, $N_z = 240$ [16, 17]	5.7×10^6 $N_z = 120$ $x^+ = y^+ = 50$	Deprés et al. [13] Meliga et al. [14]

N_z is the azimuthal resolution

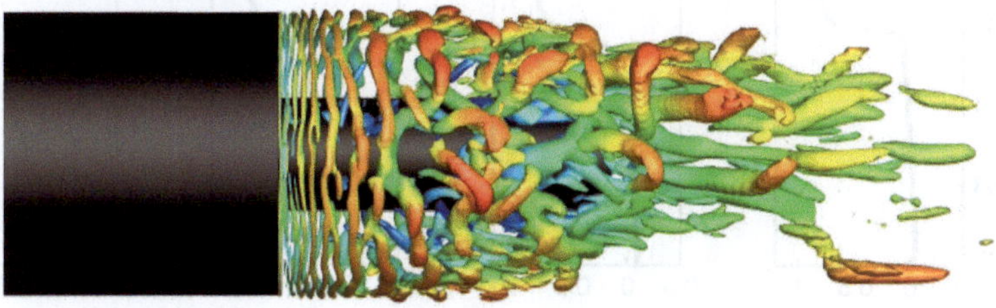

Fig. 8 Coherent structures = isosurface of $\lambda_2 D/u_\infty = 10$ *colored* by instantaneous longitudinal velocity

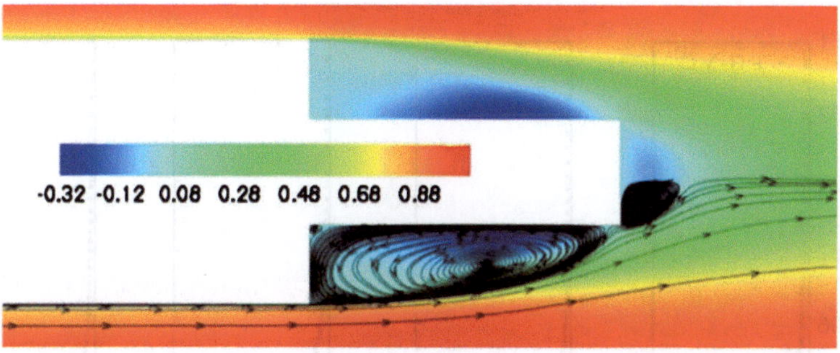

Fig. 9 $\overline{u}/U_\infty$ ratio

velocity profiles is very similar to those shown in Fig. 5 for the 2D case. This is also true for longitudinal velocity fluctuations, shown in Fig. 11. Figure 12a displays the pressure coefficient distribution along the extrusion: present results provide an accurate prediction of the average pressure coefficient, but the point of minimum C_p is located downstream with respect to the ZDES calculations. The dowstream shift of the reattachment point and of the minimum of pressure coefficient is due to the fact that there is a little delay in the triggering of Kelvin Helmholtz instabilities. This likely to be due to the high value of y^+ used in the present calculation, which does not allow a good resolution of the shear layer just before the separation point. The distribution of the root mean square of pressure fluctuations (Cp_{rms}) is presented on Fig. 12b. Present results (5.7 million cells) are more accurate than ZDES calculations on grid

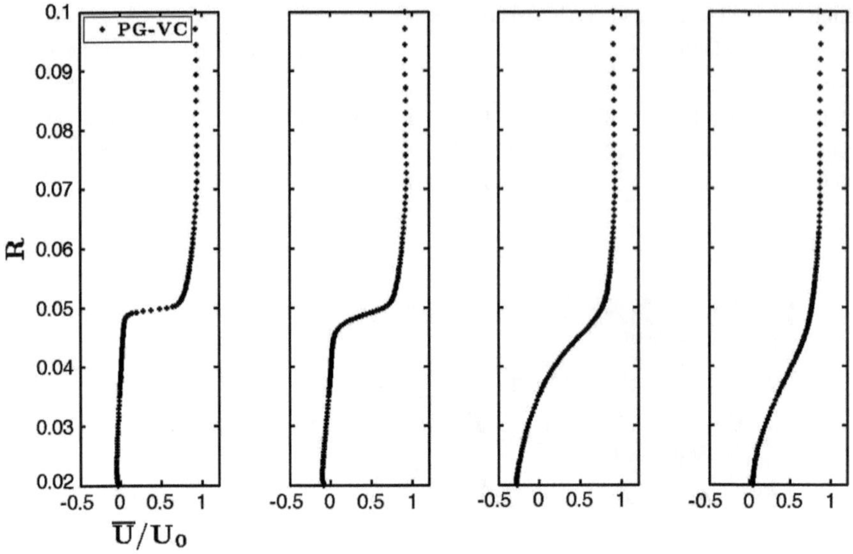

Fig. 10 Mean longitudinal velocity at different locations in the recirculation bubble

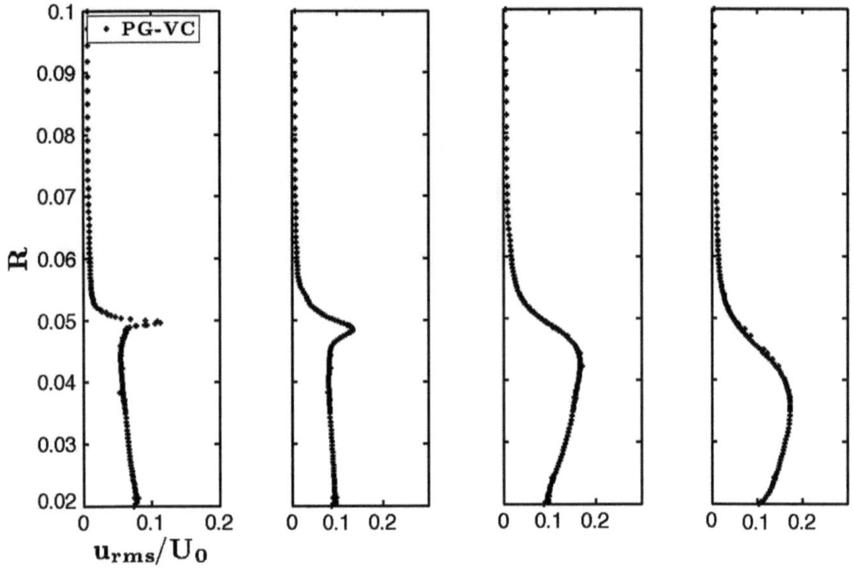

Fig. 11 Longitudinal velocity fluctuations at different locations in the recirculation bubble

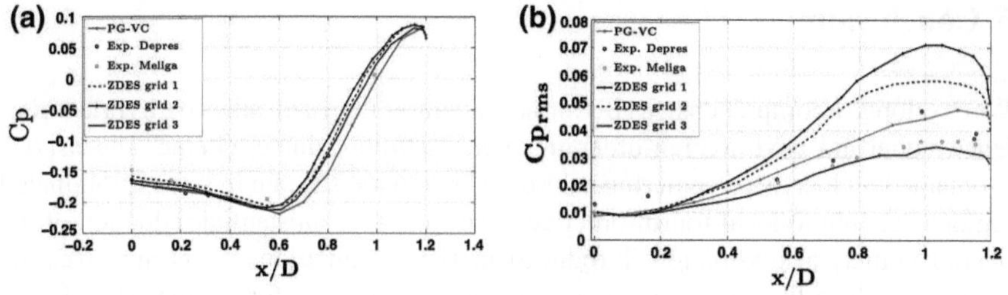

Fig. 12 Pressure coefficient on the wall in the recirculation area

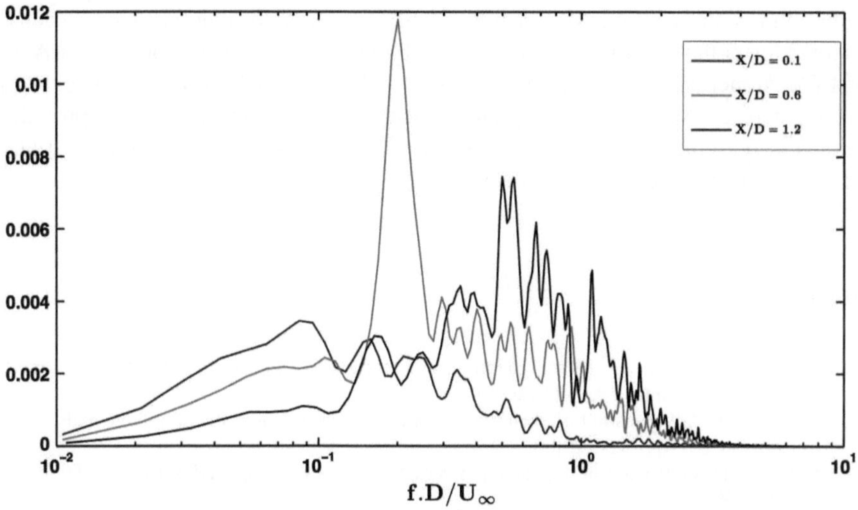

Fig. 13 PSD of pressure on three sensors along the emergency

2 (8 million cells) and below ZDES on grid 3 (12 million cells), which demonstrates the good resolvability of the proposed approach. The level of pressure fluctuations is clearly related to the numerical dissipation, since the latter determine the size of vortices generated in the separated region. Using a low dissipative numerical scheme allows an accurate prediction of fluctuations using a coarser grid with respect to other methods. Figure 13 shows the spectra of wall pressure fluctuations for several stations along the model. At the beginning of recirculation area $x/D = 0.1$ the dominating frequency is $Str = 0.08$; for $x/D = 0.6$, the dominating frequency is $Str = 0.2$ and near the reattachment point, the dominating frequency corresponds to $Str = 0.58$. Str is the Strouhal number based on the diameter D and the freestream velocity. The computed dominating frequencies and their location are in good agreement with the spatial Fourier analysis of Weiss [17] and with the study of Deck and Torigny [15]. The frequency $Str = 0.2$ is observed after the calculation of loads on the body by integrating the pressure along the small cylinder since is a shedding like instability.

5 Conclusion

We developed a comprehensive HRL strategy combining a self-adaptive hybrid turbulence model and a hybrid high-order unstructured finite volume scheme. The last one is obtained by locally re-centering a third-order upwind scheme in vortex-dominated regions, leading to local fourth-order accuracy and to a substantial reduction of the overall numerical dissipation introduced in HRL simulations. This improves the scheme resolvability while ensuring a very good robustness for compressible flows. The HRL model in use is an extension to wall-bounded flows of the one proposed by Perot and Gadebush. This model includes automatic mechanisms for switching from a classical $k - \varepsilon$ model in under-resolved regions and attached boundary layers, to a subgrid model in well-resolved region. It also includes a local antidiffusion mechanism to foster the development of flow instabilities in transition regions from RANS to LES. Numerical results shown for both 2D and axisymmetric flows over backward facing steps are very encouraging and indicate that the proposed method provides results of reasonable accuracy on relatively coarse grids, leading to an industrially acceptable overall computational cost.

References

1. Perot, B.J., Gadebush, J.: A self-adapting turbulence model for flow simulation at any mesh resolution. Phy. Fluids **19**, 1–11 (2007)
2. Drikakis, D.: Advances in turbulent flow computations using hight-resolution methods. Prog. Aerosp. Sci. **39**, 412–424 (2003)
3. Haider,F.: Discrétisation en maillage non structuré général et application LES. Ph.D. thesis, UPMC (2009)
4. Pont, G., Cinnella, P., Robinet, J.C., Brenner, P.: Development of numerical schemes for hybrid rans/les modelling in an industrial cfd solver. In: AIAA Conference (2013)
5. Ducros, F., Ferrand, V., Nicoud, F., Weber, C., Darracq, D.: Large eddy simulation of the shock/turbulence interaction. J. Comput. Phys. **152**, 517–549 (1999)
6. Troutt, T.R., Scheelke, B., Norman, T.R.: Organizes structures in a reattaching separated flow filed. J. Fluid Mech. **143**, 413–427 (1984)
7. Sasaki, K., Kiya, M.: Three-dimensional vortex structure in a leading-edge separation bubble at moderate reynolds numbers. J. Fluid Eng. **113**, 405–410 (1991)
8. Moreau, P., Labb, J., Dupoirieux, F., Borghi, R.: Experimental and numerical study of a turbulent recirculation zone with combustion. Turbul. Shear Flows **5**, 337–346 (1987)
9. Deck, S.: Recent improvement in the Zonal Detached Eddy Simulation (ZDES) formulation. Theor. Comput. Fluid Dyn. **26**, 523–550 (2011)
10. Hall, S.D., Behinia, M., Fetcher, C.A.J., Morrison, G.L.: Investigation of the secondary corner vortex in a benchmark turbulent backward-facing step using cross-correlation particle imaging velocimetry. Exp. Fluids **35**, 139–151 (2003)
11. Driver, D.M., Seegmiller H.L., Marvin, J.G.: Time-dependant behavior of a reattaching shear layer. AIAA J. **25**, 345–356 (1985)
12. Le, H., Moin, P., Kim, J.: Direct numerical simulation of turbulent flow over a backward-facing step. J. Fluid Mech. **330**, 349–2374 (1997)
13. Deprés, D., Reijasse, P., Dussauge, P.: Analysis of unsteadiness in after-body transonic flows. AIA J. **42**, 2541 (2004)

14. Meliga, P., Reijasse, P.: Unsteady transonic flow behind an axisymmetric afterbody with two boosters. In: Proceedings of the 25th AIAA Applied Aerodynamics Conference, Paper No. 2007-4564, Miami (2007)
15. Deck, S., Torigny, P.: Unsteadyness of an axisymmetric separating-reattaching flow. Phys. Fluids **19**, 065103 (2007)
16. Weiss, P.E., Deck, S., Robinet, J.C., Sagaut, P.: On the dynamics of axisymmetric turbulent separating/reattaching flows. Phys. Fluids, **21**, 075–103 (2009)
17. Weiss, P.E.: Simulation numrique et analyse physique d'un coulement d'arrire-corps axisymtrique et application au contrle des charges latrales. Ph.D. thesis, Universit Pierre et Marie Curie (2010)
18. Lê, T.H.H.: Etude expérimentale du couplage entre lécoulement transonique darrière-corps et les charges latérales dans les tuyères propulsives. Ph.D. thesis, University of Poitiers (2005)

Reynolds Stress Closure in Hybrid RANS-LES Methods

Michael Stoellinger, Stefan Heinz and Pankaj Saha

Abstract The feasibility of using the elliptic blending Reynolds stress model (EB-RSM) in hybrid RANS-LES methods is investigated in this paper. The advantage of the EB-RSM is that it does not use any geometrical wall distance or wall normal vector information which makes it well suited for application in flows with complex wall geometries. A slight modification to the original EB-RSM is proposed to improve the performance for flows with separation. The model is also extended to a sub-grid scale model for fully resolved LES and several possibilities for use as a hybrid RANS-LES model are presented. The RANS EB-RSM model performed overall well in plane channel flows, the periodic hill flow and the flow over a NACA 4412 airfoil with trailing edge separation. In LES, the EB-RSM model provided very good results in a plane channel flow at low Reynolds number. When used as a zonal hybrid RANS-LES model, the EB-RSM displayed a significant log-layer mismatch although the relevance of the modeled and resolved stresses switched right at the prescribed interface.

1 Introduction

The development of so called hybrid RANS-LES models that attempt at combining the advantages of the two modeling approaches has been a research focus for more than decade. Different models have emerged such as the Detached Eddy Simulation (DES) method of Spalart [1], the Scale Adaptive Simulation (SAS) method of Menter and Egorov [2], the Partially Averaged Navier-Stokes (PANS) method of Girimaji

M. Stoellinger (✉)
Department of Mechanical Engineering, University of Wyoming, 1000 E. University Ave, Laramie, USA
e-mail: mstoell@uwyo.edu

S. Heinz
e-mail: heinz@uwyo.edu

P. Saha
e-mail: psaha@uwyo.edu

S. Girimaji et al. (eds.), *Progress in Hybrid RANS-LES Modelling*,
Notes on Numerical Fluid Mechanics and Multidisciplinary Design 130,
DOI 10.1007/978-3-319-15141-0_26

[3], the Partially Integrated Transport Model PITM [4] or the unified RANS-LES model Heinz [5] and Gopalan et al. [6] to name a few. Most of these methods are based on one-equation models (e.g. DES) or two-equation models that invoke the eddy-viscosity assumption for the modeled turbulence stress. When applied in wall bounded flows, these models are usually integrated to the wall but still need a damping function to yield the correct shear stress in the log-layer. A notable exception here is the PANS approach based on the $k-\varepsilon-\zeta-f$ model of Basara et al. [7]. In flows with stagnation points (e.g. aerodynamics applications) the eddy viscosity model leads to an excessive production of turbulence kinetic energy and hence limiters have to be introduced [8].

Turbulence models based on second-moment closure directly solve modeled transport equations for the Reynolds stress and thus do not need to invoke the eddy-viscosity assumption. However, the main modeling issue related to the pressure redistribution term is usually addressed for nearly homogeneous flows and hence the redistribution models need to be modified to be applicable to wall bounded flows. The effect of kinematic blocking on the redistribution term and its elliptic nature was successfully modeled by Durbin [9] using the so called elliptic relaxation model. The elliptic relaxation model is based on the solution of six elliptic equations to adjust any homogeneous redistribution model to yield the correct near wall behavior. More recently, a simpler albeit slightly less accurate model that solves only one additional elliptic equation was proposed by Manceau and Hanajalić [10] and Thielen et al. [11]. The model is based on a blending between any homogeneous redistribution model like the LRR model [12] and a near wall redistribution model that has the desired asymptotic behavior. The blending variable is based on the solution of a elliptic equation. No geometric wall distance variable is needed which makes the elliptic blending (EB) model particular suitable for flows with complex geometries. It should be noted that the EB model was adopted in the PITM approach of Fadai-Ghotbi et al. [4] in a channel flow with low Reynolds number. The aim of this paper is to further investigate the potential of the EB-RSM in hybrid RANS-LES simulations of flows with separation.

2 RANS EB-RSM Model

We consider only incompressible flows in this work. The EB-RSM used in this paper is a modification of the original EB model [10] and the improved model presented by Thielen et al. [11]. The modeled transport equation for the Reynolds stress tensor is given by

$$\frac{\partial \overline{u_i u_j}}{\partial t} + U_j \frac{\partial \overline{u_i u_j}}{\partial x_j} = P_{ij} + \Phi_{ij}^* - \varepsilon_{ij} + \frac{\partial}{\partial x_k} \left[\left(\nu \delta_{kl} + C_k \tau \overline{u_k u_l} \right) \frac{\partial \overline{u_i u_j}}{\partial x_l} \right], \quad (1)$$

where P_{ij} is the production, Φ^*_{ij} is the pressure redistribution term, ε_{ij} is the dissipation rate tensor and the last term represents molecular diffusion and turbulent transport according to the Daly-Harlow model [13]. The production term is given by

$$P_{ij} = -\overline{u_i u_k}\frac{\partial U_j}{\partial x_k} - \overline{u_j u_k}\frac{\partial U_i}{\partial x_k}. \tag{2}$$

In the EB model the redistribution term is given by a "linear blending" of a near wall model Φ^w_{ij} and a homogeneous model Φ^h_{ij} that is appropriate away from the wall:

$$\Phi^*_{ij} = (1 - f_\alpha)\Phi^w_{ij} + f_\alpha \Phi^h_{ij}, \tag{3}$$

where $f_\alpha = \alpha^2$ is the blending function which is based on the variable α that defines the "closeness" to a solid wall and that satisfies an elliptic equation:

$$\alpha - L_d^2 \nabla^2 \alpha = 1. \tag{4}$$

The boundary conditions are such that at solid walls $\alpha = 0$ and in the free stream $\alpha = 1$. The Durbin-limited [8] length scale L_d is given by

$$L_d = C_L max\left(\frac{k^{3/2}}{\varepsilon}, C_\eta \frac{\nu^{3/4}}{\varepsilon^{1/4}}\right), \tag{5}$$

with constants $C_L = 0.161$ and $C_\eta = 80$ [11]. The dissipation rate tensor is also given by a blending between the near wall anisotrop form and the common isotropic form far away from the wall:

$$\varepsilon_{ij} = (1 - f_\alpha)\frac{\overline{u_i u_j}}{k}\varepsilon + f_\alpha \frac{2}{3}\varepsilon \delta_{ij}. \tag{6}$$

The homogeneous part of the redistribution term is modeled according to the LRR model [12]

$$\Phi^h_{ij} = -C_1 \varepsilon a_{ij} - C_2\left(P_{ij} - \frac{1}{3}P_{ll}\delta_{ij}\right), \quad a_{ij} = \frac{\overline{u_i u_j}}{k} - \frac{2}{3}\delta_{ij}. \tag{7}$$

The near wall form of the redistribution model was obtained in [10] such that the correct asymptotic behavior is obtained

$$\Phi^w_{ij} = -5\frac{\varepsilon}{k}\left(\overline{u_i u_k} n_j n_k + \overline{u_j u_k} n_i n_k - \frac{1}{2}\overline{u_k u_l} n_k n_l \left(n_i n_j + \delta_{ij}\right)\right), \tag{8}$$

where the wall normal vector $\mathbf{n}$ is also obtained from the elliptic variable α by

$$\mathbf{n} = \frac{\nabla\alpha}{||\nabla\alpha||}. \tag{9}$$

Finally, the modeled transport equation for the dissipation rate of the turbulence kinetic energy is given by [14]

$$\frac{\partial\varepsilon}{\partial t} + U_j\frac{\partial\varepsilon}{\partial x_j} = C_{\varepsilon 1}P_{kk}\frac{\varepsilon}{2k} - C_{\varepsilon 2}f_\varepsilon\frac{\tilde{\varepsilon}\varepsilon}{k} + E_\varepsilon + \frac{\partial}{\partial x_k}\left[\left(\nu\delta_{kl} + C_\varepsilon\tau\overline{u_k u_l}\right)\frac{\partial\varepsilon}{\partial x_l}\right] \tag{10}$$

with

$$E_\varepsilon = C_{\varepsilon 3}\nu\tau\overline{u_k u_l}\frac{\partial^2 U_i}{\partial x_k\partial x_j}\frac{\partial^2 U_i}{\partial x_l\partial x_j}, \tag{11}$$

and

$$\tilde{\varepsilon} = \varepsilon - 2\nu\left(\frac{\partial\sqrt{k}}{\partial n}\right)^2. \tag{12}$$

The function f_ε is modified from the Re_t dependence to be a function of the elliptic near wall variable α and is given by

$$f_\varepsilon = 1 - \frac{C_{\varepsilon 2} - 1.4}{C_{\varepsilon 2}}\exp\left[-(6\alpha)^5\right]. \tag{13}$$

The model coefficients are given by:

$$C_{\varepsilon 1} = 1.44,\quad C_{\varepsilon 2} = 1.92,\quad C_{\varepsilon 3} = 0.1,\quad C_\varepsilon = 0.18. \tag{14}$$

The boundary conditions at the wall are

$$U_i = 0,\quad \overline{u_i u_j} = 0,\quad \varepsilon = 2\nu\frac{k}{y_1^2},\quad \alpha = 0. \tag{15}$$

It should be noted that the above model does not dependent on a geometrically defined wall distance and wall normal direction nor does it depend on a turbulence Reynolds number. The former attribute will be of great importance in complex flow and the latter will be useful when using the model in a hybrid RANS-LES approach.

3 LES EB-RSM Model

In LES a spatial filter is used instead of the Reynolds average and by assuming that the filter function with filter size Δ is homogeneous the filter operation commutes with differentiation in space and time. Therefore, the filtered Navier-Stokes equations are formally equivalent to the RANS equations. The transport equation for sub-grid scale (SGS) stress is also formally identical to the Reynolds stress equation (1). The stress equation can thus be seen as an extension of SGS-stress model of Deardorff [15] to flows with solid walls. The main difference to the RANS model is given by the model for the SGS dissipation rate which is modeled according to [15, 16],

$$\varepsilon = C_{LES}\frac{k^{3/2}}{\Delta} + 2\nu\frac{\partial\sqrt{k}}{x_n}\frac{\partial\sqrt{k}}{x_n}, \tag{16}$$

where the second term is added to provide a non-zero dissipation rate at the wall [16]. The model coefficient is set to $C_{LES} = 2$ and the filter size is related to the grid spacing by $\Delta = \left(\delta_x\delta_y\delta_z\right)^{1/3}$. Since the length scale of the unresolved motion is given by Δ the relevant length scale in the elliptic equation (4) is now modeled as $L_d = C_L\Delta$ with $C_L = 0.7$. This means that the wall-blocking effects only the unresolved scales of size Δ, further away from the wall the blocking effect of the wall is treated explicitly through the pressure boundary condition [4]. All other model parameters are identical to their RANS values.

4 Hybrid RANS-LES EB-RSM Model

To use the EB-RSM model in a hybrid RANS-LES context mainly requires to modify the dissipation rate which we will call ε_{hb}. The EB-RSM model parameters are not altered from their respective RANS values (see Sect. 2). Three different strategies can be adopted: a zonal approach [17], a unified approach [6], or a PANS [3] approach. In the zonal approach we prescribe a fixed RANS-LES interface according to

$$\varepsilon_{hb} = \begin{cases} \varepsilon & \text{if} y^+ < y^+_{lim}, \\ C_{LES}\frac{k^{3/2}}{\Delta} & \text{otherwise}, \end{cases} \tag{17}$$

where y^+_{lim} is the desired location of the interface and ε is the solution of Eq. (10). This zonal model is useful to study the behavior of the hybrid EB-RSM model for different locations of the RANS-LES interface as shown in Sect. 5.

The second approach to model the dissipation rate for the hybrid model is based on the unified time-scale approach suggested by Heinz [5] and Gopalan et al. [6]. In this approach, a "unified" time-scale is calculated based on the RANS time scale $\tau_R = k/\varepsilon$ and the LES time scale $\tau_L = \Delta/(C_{LES}k^{1/2})$ according to

$$\tau_u = min(\tau_R, \tau_L). \tag{18}$$

The hybrid dissipation rate is then given by

$$\varepsilon_{hb} = \frac{k}{\tau_u}, \tag{19}$$

which provides the desired limits RANS limit $\varepsilon_{hb} = \varepsilon$ and LES limit $\varepsilon_{hb} = C_{LES}\frac{k^{3/2}}{\Delta}$.

Finally, the EB-RSM could be operated using the PANS concept [3] which suggests to specify a constant resolution parameter f_k and then to modify the model coefficient for the destruction term in the dissipation rate equation according to

$$C_{\varepsilon 2} = C_{\varepsilon 1} + f_k \left(C_{\varepsilon 2} - C_{\varepsilon 1}\right). \tag{20}$$

In the PANS approach the grid resolution does not explicitly appear and hence the user has to ensure a sufficient mesh resolution for a given choice of f_k.

5 Simulation Results

The EB-RSM model is applied to flows with and without flow separation: plane channel flow at $Re_\tau = 395, 590, 2003$, the periodic hill case [18], and the NACA 4412 airfoil at $Re = 1.52 \times 10^6$ with an angle of attack of 13.87°.

5.1 RANS EB-RSM Results

The EB-RSM model given in Sect. 2 was implemented in the open source CFD software OpenFOAM. The solver is based on the PISO pressure-velocity coupling and second order accurate schemes are used for spatial and time discretization. The model parameters were calibrated using the plane channel flow DNS data of Hoyas and Jiménez [19] at $Re_\tau = 2003$. A 2-d grid was adopted consisting of 4×80 cells stretched in wall normal direction such that $y^+ < 1$. A periodic boundary condition (BC) was applied in stream-wise direction and a uniform pressure gradient was added to the momentum equation to keep the prescribed mean flow rate constant. Figure 1 shows results from the $Re_\tau = 2003$ case for the mean velocity (left) and Reynolds stresses (left). The mean velocity is in very good agreement with the DNS data. The Reynolds stress components are predicted very well near the wall but the normal stresses decay somewhat too fast in the channel center. It should be noted that for the lower Reynolds number cases $Re_\tau = 395, 590$ the agreement is almost perfect (not shown). Further tuning of the model parameters could possibly improve the results for the $Re_\tau = 2003$ but was not attempted.

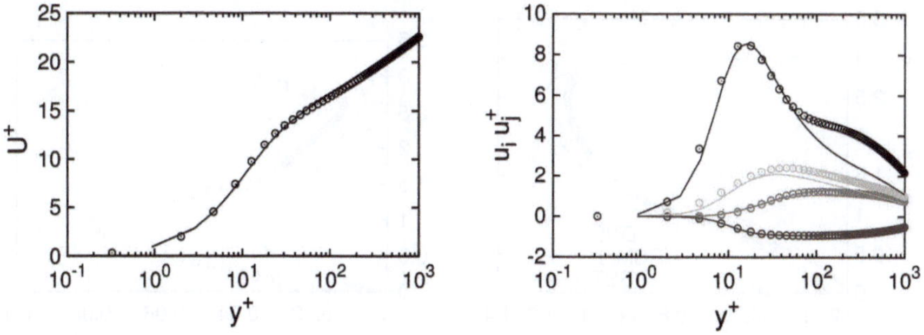

Fig. 1 Results of the RANS EB-RSM (*solid-line*) for plane channel flow at $Re_\tau = 2003$ compared to DNS data (*open circles*)

Fig. 2 Contour plot of the elliptic variable α in the $Re_b = 2800$ periodic hill flow case

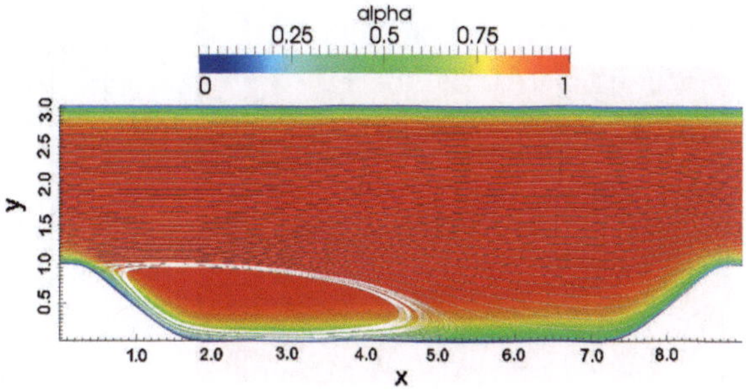

Results for the periodic hill flow at $Re_b = 2800$ are considered next. Figure 2 shows a contour plot of the elliptic variable α and the stream lines. The variable α clearly follows smoothly the wall geometry and one can see a thickening of the near wall region in the trough of the hill. A comparison of the EB-RSM results with DNS data [18] and with results from Menter's SST model are shown in Fig. 3 at $x/h = 4$. The mean velocity predicted by the EB-RSM model shows a slightly better agreement with the DNS data than the SST model. The prediction for the stream wise Reynolds stress from the EB-RSM model agrees much better with the DNS data than the SST model demonstrating that the EB-RSM model can indeed capture the strong stress anisotropies in this challenging flow.

Figure 4 shows a contour plot of the elliptic variable in the NACA 4412 test case and corresponding EB-RSM and SST results for the pressure coefficient. The simulation was performed on a 449×129 point C-grid provided at NASA's turbulence modeling web page. The Reynolds number based on the chord length is $Re = 1.52 \times 10^6$ with an angle of attack of $13.87°$. The predictions for the pressure coefficient are comparable between the two models and no significant improvement is achieved with the EB-RSM model over the simpler SST model. The simulations did not account for the tripped boundary layer created in the experiments. The effect of the tripped boundary layer might be relevant and will be investigated in a future paper.

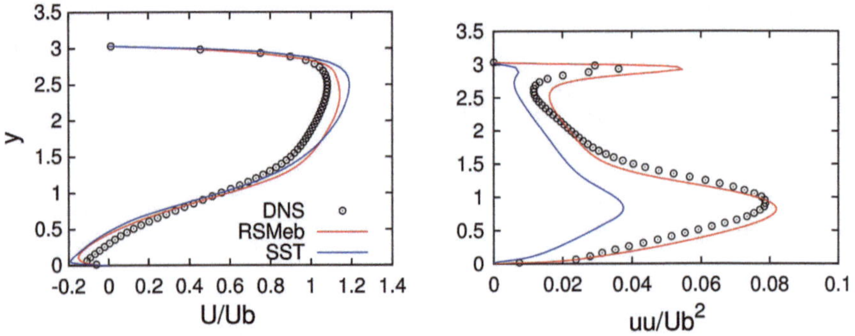

Fig. 3 Results of the RANS EB-RSM (*red-line*) and the SST model (*blue line*) for the periodic hill flow at $Re_b = 2800$ compared to DNS data (*open circles*) at $x/h = 4$

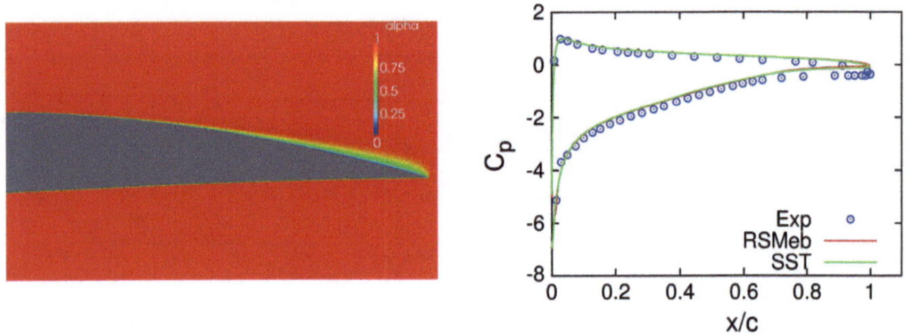

Fig. 4 Contour plot of the elliptic varibale α (*left*) and pressure coefficient (*right*) for the NACA 4412 airfoil test case

5.2 LES EB-RSM Results

The LES EB-RSM model was validated with simulations of the plane channel flow at $Re_\tau = 395$. A $64 \times 64 \times 64$ mesh was used on a $2\pi \times 2 \times \pi$ domain leading to $y_1^+ \approx 1.5$, $\Delta x^+ = 40$ and $\Delta z^+ = 20$. The mesh is thus on the coarse side for a fully resolved LES. After the initial transient time averages were taken over a time of $T_a = 10L_x/U$. Figure 5 shows results for the mean velocity (left) and the shear stress (right). Both, the mean velocity profile and the total (modeled + resolved) shear stress are in very close agreement with the DNS data. The resolution of the mesh is indeed sufficient for a fully resolved LES as can be seen by the small fraction of the modeled shear stress component compared to the resolved component.

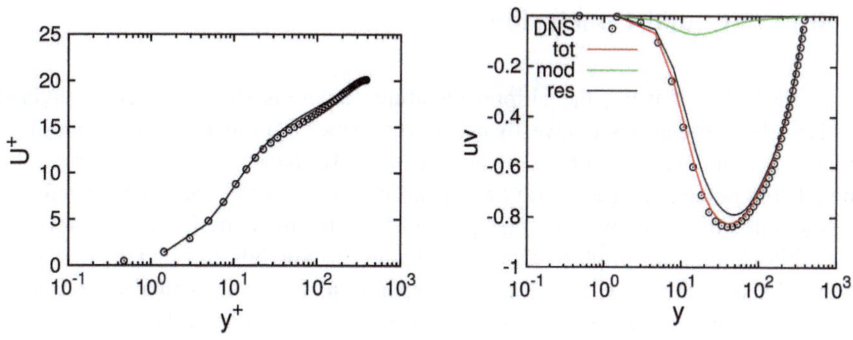

Fig. 5 LES EB-RSM results for the plane channel flow at $Re_\tau = 395$ compared to DNS data (*open circles*)

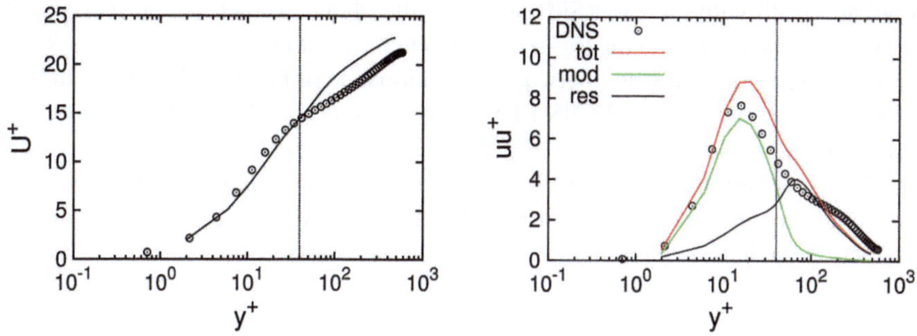

Fig. 6 Hybrid EB-RSM results for the plane channel flow at $Re_\tau = 590$ using the zonal approach with an interface location $y^+_{lim} = 40$ compared to DNS data (*open circles*)

5.3 Hybrid EB-RSM Results

The performance of the hybrid EB-RSM model was investigated using simulations of the plane channel flow at $Re_\tau = 590$ using the same $64 \times 64 \times 64$ mesh as in the LES case. For this higher Reynolds number the grid spacing in wall units is $y_1^+ \approx 3$, $\Delta x^+ = 80$ and $\Delta z^+ = 40$ which is not sufficient for a fully resolved LES. To study the influence of the interface location, only the zonal approach described in Sect. 4 was adopted. Figure 6 show simulation results for an interface location $y^+_{lim} = 40$ which is slightly after the tke peak. The mean velocity is predicted reasonable well up to the interface but a significant log-layer mismatch can be observed afterwards. The stream wise component of the modeled stress is larger in the RANS region an becomes smaller than the resolved stress right at the interface location. This is encouraging since not all hybrid models provide this essential behavior. The total Reynolds stress (red line) is predicted quite well. The computational cost increased by about 80 % for the EB-RSM model when compared to a simple two-equation hybrid model.

6 Summary and Conclusions

The feasibility of using the elliptic blending Reynolds stress model for hybrid RANS-LES simulations is investigated in this paper. To this end, some modifications to the model are presented to improve the performance in separated flows. The model does not use any geometric wall distance or wall normal vector which makes it very suitable for flows over complex geometries. The model performed overall well in RANS simulations of plane channel flow, the periodic hill flow and the flow over an NACA 4412 airfoil with trailing edge separation. The model can also be used as a sub-grid stress model in fully resolved LES and simulations of the plane channel flow at $Re_\tau = 395$ showed very good agreement with the DNS data. Three different approaches to use the EB-RSM model as hybrid RANS-LES model are presented: the zonal approach, the unified approach and the PANS approach. Initial tests of the zonal approach with the EB-RSM model for the plane channel flow at $Re_\tau = 590$ showed that the issue with the log-layer mismatch was not resolved by using a RSM. However, a smooth transition between the dominance of the modeled stress and the resolved stress right at the RANS-LES interface could be observed. The computational cost increased by about 80 % for the EB-RSM model when compared to a simple two-equation hybrid model. To improve the mean velocity predictions will most likely require a smoother transition from a RANS dissipation model to an LES dissipation model. This will be investigate in a future paper.

References

1. Spalart, P.R.: Ann. Rev. Fluid Mech. **41**, 181–202 (2009)
2. Menter, F., Egorov, Y.: 43rd AIAA Aerospace Sciences Meeting and Exhibit, Reno, Nevada. AIAA 2005–1095 (2005)
3. Girimaji, S.: J. Appl. Mech. **73**, 413 (2006)
4. Fadai-Ghotbi, A., Friess, C., Manceau, R., Borée, J.: Phys. Fluids **22**, 055104 (2010)
5. Heinz, S.: Theor. Comput. Fluid Dyn. **21**, 99–118 (2007)
6. Gopalan, H., Heinz, S., Stöllinger, M.: J. Comput. Phys. **249**, 249–274 (2013)
7. Basara, B., Krajnovic, S., Girimaji, S., Pavlovic, Z.: AIAA J. **49**, 2627–2637 (2011)
8. Durbin, P.A.: Fluid Dyn. Res. **41**, 012203 (2009)
9. Durbin, P.A.: J. Fluid Mech. **249**, 465–498 (1993)
10. Manceau, R., Hanajalić, K.: Phys. Fluids **14**, 744–753 (2002)
11. Thielen, L., Hanjalic, K., Jonker, H., Manceau, R.: Int. J. Heat Mass Transf. **48**, 1583–1598 (2005)
12. Launder, B.E., Reece, G.J., Rodi, W.: J. Fluid Mech. **68**, 537–566 (1975)
13. Daly, B.J., Harlow, F.H.: Phys. Fluids **13**, 2634–2649 (1970)
14. Chaouat, B.: AIAA J. **44**, 2390–2403 (2006)
15. Deardorff, J.W.: ASME J. Fluids Eng. **95**, 429–438 (1973)
16. Inagaki, M.: Int. J. Heat Fluid Flow **32**, 26–40 (2011)
17. Fröhlich, J., Terzi, D.V.: Progr. Aerosp. Sci. **44**, 349–377 (2008)
18. Breuer, M., Peller, N., Rapp, C., Manhart, M.: Comput. Fluids **38**, 433–457 (2009)
19. Hoyas, S., Jiménez, J.: Phys. Fluids **18**(011702), 1–4 (2006)

Hybrid RANS-LES and URANS Simulations of a Laminar Transonic Airfoil

D. Szubert, F. Grossi, Y. Hoarau and M. Braza

Abstract Laminar flow is a potential way of minimizing drag and reducing aircraft emissions. However, the interaction of laminar boundary layers with shock waves at transonic speeds can cause severe detrimental aerodynamic effects and remains an opened question. In this way, in the framework of the TFAST European project, a laminar transonic airfoil has been developed for both numerical and experimental studies on such laminar interactions. The so-called V2C profile has been studied to provide natural laminar flow from the leading edge to the shock wave for a wide range of freestream Mach numbers and angles of attack. In the present paper, a numerical investigation of the transonic flow around the V2C airfoil is conducted by means of URANS and hybrid RANS-LES computations. At sufficiently-high angles of attack and moderate freestream Mach number (0.70), the transonic interaction develops buffet. In the paper, special attention is paid to the differences between the URANS and hybrid RANS-LES predictions of the shock-induced separation.

1 Introduction

Research for more effective transport systems and the reduction of emissions, which put severe demands on aircraft velocity and drag reduction, is intense. This can be accomplished by keeping laminar boundary layer on external and internal airplane parts. However supersonic flow velocities are more frequent due to the increased loads on aerodynamic surfaces and aero-engine components, generating shock waves that interact with boundary layers. Laminar shock-wave/boundary-layer interaction can quickly cause flow separation, which is highly detrimental to aircraft performance and poses a threat to safety. This issue can be fixed by forcing the laminar-turbulent

D. Szubert (✉) · F. Grossi · M. Braza
Institut de Mécanique des Fluides de Toulouse, UMR 5502 Toulouse, France
e-mail: damien.szubert@imft.fr

Y. Hoarau
ICube,, UMR 7357 Strasbourg, France
e-mail: hoarau@unistra.fr

S. Girimaji et al. (eds.), *Progress in Hybrid RANS-LES Modelling*,
Notes on Numerical Fluid Mechanics and Multidisciplinary Design 130,
DOI 10.1007/978-3-319-15141-0_27

transition upstream of the interaction, but that would induce an increase of drag and losses, which is contrary to the objectives of the greening of air transport.

In this context, the European research program TFAST (Transition Location Effect on Shock Wave Boundary Layer Interaction) was hosted, with the aim of studying the effects of transition location on the structure of the shock-wave/boundary-layer interaction, and several ways of controlling the position of the transition, by the collaboration of European industrials as well as laboratories and universities. To this end, in the context of external flows, a supercritical laminar wing (the V2C) has been designed by Dassault Aviation. This profile allows the boundary layer to remain laminar up to the shock foot, even in the environment of transonic wind tunnels of the laboratories involved in the project, and up to the angle of attack of 7°. The transonic buffet has been studied experimentally in detail since the 70s [11] on circular-arc airfoils, and most recently on supercritical airfoil. The physics governing the transonic buffet is complex and still remains to be clarified, though several theories have been proposed, like the effect of the feedback mechanism of wave propagating from the trailing edge, or the onset of a global instability.

Navier-Stokes simulations of transonic buffet as well as of the shock-vortex interaction at moderate Reynolds numbers were reported by Bouhadji and Braza [2], as well as DNS by Bourdet et al. [3]. While Computational Fluid Dynamics (CFD) techniques may provide considerably accurate results around the cruising design point of an airplane, for example, they become less reliable as one approaches the limits of the flight envelope, where nonlinear effects such as flow separation and shock waves get pronounced. Furthermore, the typical high Reynolds numbers of aerodynamic applications require the use of an appropriate closure for the turbulent stresses and time-resolved computations are frequently necessary. Concerning transonic buffet, the unsteady shock-wave/boundary-layer interaction represents a major challenge for turbulence models and the low frequencies associated with the shock-wave motion can make the simulations very expensive. Since the pioneering simulations by Levy [8] and Seegmiller et al. [11] for a circular-arc airfoil, Unsteady Reynolds-Averaged Navier-Stokes (URANS) computations using eddy-viscosity turbulence models have been largely used to predict the phenomenon over two-dimensional airfoils. Hybrid RANS-LES methods combine the robustness and near-wall physics offered by URANS in the near region, as well as LES capabilities of vortices and instabilities development in the detached flow regions, as for example in [5] capturing the buffet phenomenon around the OAT15A airfoil. Regarding their application to the transonic buffet problem, Grossi et al. [7] performed a Delayed Detached-Eddy Simulation (*DDES*) of the unsteady flow over the OAT15A supercritical airfoil that succeeded in predicting the self-sustained motion of the shock wave near the experimental buffet onset boundary. In the present paper, a similar strategy is applied on the V2C airfoil within the TFAST program, at 7°, the maximum angle of attack allowed by the design. A 2D study is first carried out to investigate the main characteristics of the airfoil regarding the effects of the angle of attack as well as the influence of the turbulence model. The transition location effects are also studied in the buffeting regime, by imposing the laminarity at several positions.

These results are developped in section 3.3. The results of the 3D computations, URANS and DDES, are presented in the last section of this paper.

2 Numerical Method and Turbulence Modelling

2.1 Flow Configuration

The technique employed for laminarity and an initial design in respect of transition prediction was based on the e^N method (Ref. [4] for instance) and the airfoil surface was generated in such a way that the N-factor remains small for low-to-moderate turbulence intensity levels, such as in the wind tunnels used for that test case for the experimental study currently in progress. The design was validated numerically by Dassault on a 0.25 m-chord length (c) profile by means of RANS computations for various angles of attack (α) at freestream Mach numbers of 0.70 and 0.75, yielding chord-based Reynolds numbers of approximately 3.245×10^6 and 3.378×10^6 respectively. The study was performed using a compressible Navier-Stokes code adopting a two-layer $k - \varepsilon$ model, with the transition location being determined from the fully-turbulent flowfield using a three-dimensional compressible boundary-layer code by means of the N-factor amplification with a parabola method. At Mach 0.70, the flow separated between $\alpha = 6°$ and $7°$. The amplification factor N was shown to be smaller than 3 up to the shock wave, thus guaranteeing laminar flow. At Mach 0.75, the value of N remained smaller than 2 up to $\alpha = 7°$. For this Mach number, there were not buffeting phenomenon, whatever the angle of attack. Moreover, for incidence higher than $1°$, the shock induces a separation of the boundary layer until the trailing edge.

2.2 Numerical Method

The simulations of the V2C configuration have been performed with the Navier-Stokes Multi-Block (*NSMB*) solver. The NSMB solver is the fruit of a european consortium that included Airbus from the beginning of 90s, as well as main European aeronautics research Institutes like KTH, EPFL, IMFT, ICUBE, CERFACS, Univ. of Karlsruhe, ETH-Ecole Polytechnique de Zurich, among other. This consortium is coordinated by CFS Engineering in Lausanne, Switzerland. NSMB solves the compressible Navier-Stokes equations using a finite volume formulation on multi-block structured grids. It includes a variety of efficient high-order numerical schemes and of turbulence modelling closures in the context of LES, URANS and of hybrid turbulence modelling. NSMB includes an ensemble of the most efficient CFD methods, as well as efficient fluid-structure coupling for moving and deformable structures. For the study presented here, the third-order of accuracy Roe upwind scheme [10] associated with the MUSCL van Leer flux limiter scheme [16] is used to discretize the

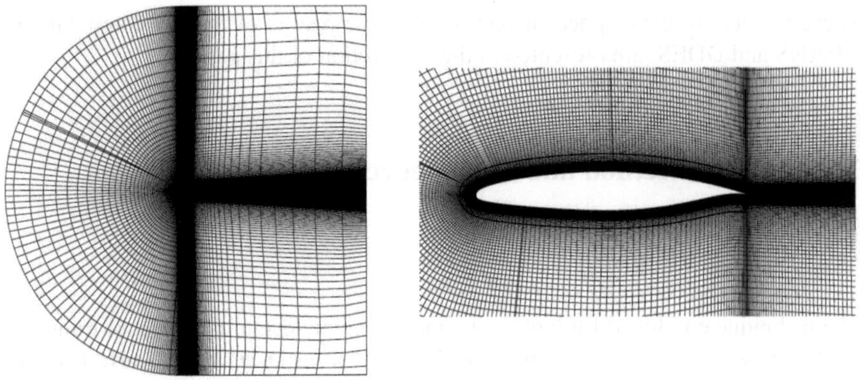

Fig. 1 Multiblock domain

convective fluxes. For the unsteady RANS, implicit time integration using the dual time stepping technique has been performed. A physical time step of 5 μs has been adopted for 2D simulations. For 3D calculation, the time step has been reduced to 0.1 μs. A typical number of inner iterations of 30 was necessary for the convergence in each time step.

The 2D planar grid has a C-H topology, and is of size 163,584 cells. The limit of the domain is located at a mean distance of 80 chords from the obstacle. A convergence study on the mesh refinement has been carried out, by means of steady computations (local time stepping) for the flow at $M_\infty = 0.70$ and $\alpha = 4.0°$ using the Menter's SST model and assuming fully-turbulent behavior, with two other grids: one 50 % coarser, and another 30 % finer. Detailed results of this convergence study can be found in [7]. The grid retained for the present study gave a maximum value of non-dimensional wall distance y^+ of about 0.55 with respect to the turbulence modelling. Figure 1 shows the grid and the computational domain.

For the 3D computations, the planar grid has been extruded to 59 cells uniformly distributed in the spanwise direction over a distance of $0.33 \times c$. The 3D grid contains about 9.65 M cells.

2.3 Boundary Conditions

On the solid wall, impermeability and no-slip conditions are employed. The far-field conditions are the characteristic variables extrapolated in time: the total pressure ($P_0 = 10^5$ Pa) and total temperature ($T_0 = 290$ K), as well as the upstream Reynolds number of 3.245 million and Mach number of 0.70. The upstream turbulence intensity is $Tu = 0.08$ %.

2.4 Turbulence Modelling

The two-equation $k-\omega$ SST model of Menter [9] with turbulence-sustaining ambiant terms to prevent the free decay of the transported turbulence variables [15] has been used for the (U)RANS and the DDES computations.

3 Results

3.1 Two-Dimensional Study: Effects of the Angle of Attack

In order to characterize the aerodynamics of the V2C airfoil at $M_\infty = 0.70$, the angle of attack is varied from 1° up to 7°, which is the maximum angle of attack for which the boundary layer is supposed to remain laminar from the leading edge to the shock wave. Initially, the computations adopt local time stepping. If convergence is not reached (i.e., a relative reduction of 10^{-6} in the residual), time-accurate simulations with a time step of 5×10^{-6} s are then performed. Near the critical angle regarding the buffet, the angle of attack is varied by an increment of 0.5° in order to refine the buffet boundary.

Figure 2 shows the final distributions of the pressure coefficient for the full range of incidences and the skin-friction coefficient for the steady cases. For angles of attack up to 5°, the flow is steady and rear separation is always present. The shock wave can already be distinguished at 2°. As the angle of attack is further increased, the shock initially moves downstream, then it goes upstream for $\alpha > 3°$. From $\alpha = 4°$, a bubble separation appears and develops, and the amount of rear separation steadily increases with the angle of attack (Fig. 2b).

Flow unsteadiness, characterized by an oscillating shock wave, has been detected from 5.5°. The main frequency increases with incidence in the range of 80–82 Hz.

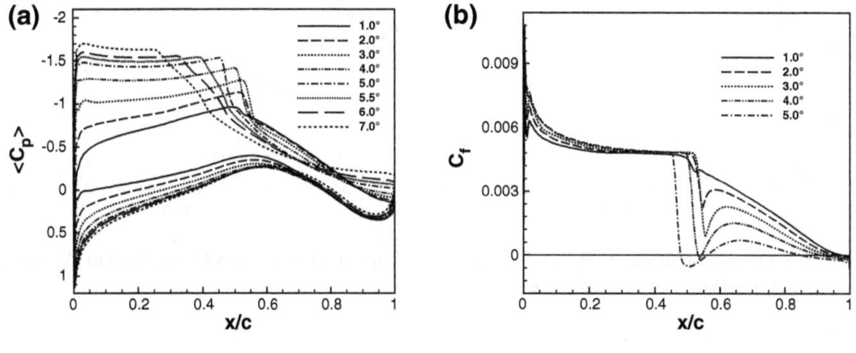

Fig. 2 Steady and mean surface distributions of the pressure (**a**) and friction (**b**) coefficients

At 5.5° the amplitude of the shock-wave motion is still small, resulting in a slight slope in the C_p curve.

It has to be mentioned that the experimental study of this flow configuration was just started within TFAST. For this reason, comparison with experiments are not yet available. A detailed comparison of the results obtained in the present study by the NSMB code has been carried out instead by using the Edge code, an unstructured compressible finite volume CFD code developed by the FOI since 1997 in collaboration with industrial and academic partners. This comparison showed small differences closed to the critical angle, but the results were very similar at lower and higher angles of attack.

In the next section, these results are compared to the same calculations performed in (U)RANS based on the Spalart-Allmaras turbulence model.

3.2 *Influence of the Turbulence Model*

The sensitivity of the critical angle to the buffet phenomenon at Mach number 0.70 has been studied in terms of the turbulence models used in the 2D URANS calculations. Results from the two-transport equations $k-\omega$ SST turbulence model presented in the previous section, are compared to those obtained by the Edwards and Chandra variant [6] of the one-equation model of Spalart-Allmaras [13], using the compressibility correction of Secundov [12] (SAE+CC). The mean surface pressure coefficient is plotted in Fig. 3 at the angles of attack of 4° and 5°. At the lowest incidence (Fig. 3a), below the critical angle, the modified Spalart-Allmaras gives a steady position of the shock wave downstream than with the $k-\omega$ SST model.

The modified Spalart-Allmaras model reaches the critical angle of $\alpha = 5.0°$ whereas, as seen in the previous section, the onset of the buffeting phenomenon appeared at an angle of $\alpha = 5.5°$ for the $k-\omega$ SST model.

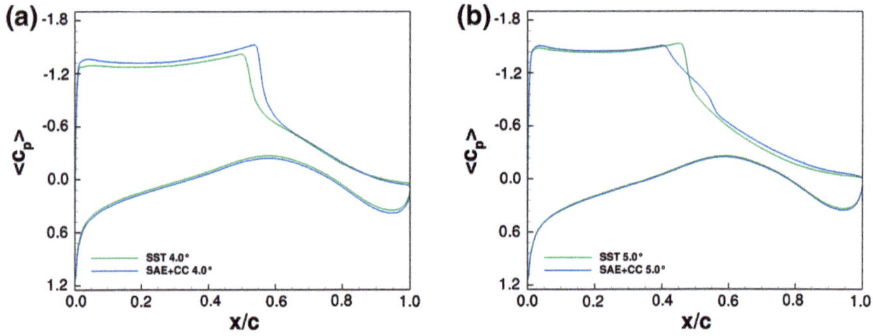

Fig. 3 Comparison of mean pressure coefficient distribution for $k-\omega$ SST and Spalart-Allmaras models at incidences 4° (**a**) and 5° (**b**)

3.3 Transition Study

Two flow conditions have been selected for a numerical investigation of the transition location effect on the shock-wave/boundary-layer interaction, due to their interesting flow physics. First, the steady interaction arising at $\alpha = 4.0°$ is addressed, featuring a reasonably strong shock just below the critical angle of attack for buffet onset. The second flow condition is the fully-established buffet regime at $\alpha = 7.0°$, which presents a large shock-wave motion region.

The transition is forced at the position x_t by imposing the turbulent viscosity $\nu_t = 0$ for $x < x_t$. Its location x_t is varied from the leading edge up to as close as possible to the shock wave. The influence of the tripping point over the selected steady and unsteady transonic flowfields is presented in the following two subsections.

This study has been carried out to assess the influence of the transition point on the properties of the well-developed buffeting flow at 7°. Besides the fully-turbulent case, three tripping locations have been considered: $x_t/c = 0.09, 0.16$ and 0.24. For the latter, the most upstream position of the shock wave during buffet has been of about $x_t/c = 0.25$. This limits the displacement of the tripping point since imposing $\nu_t = 0$ inside the shock-motion region would not be an acceptable approximation.

Figure 4a presents the statistical pressure distributions obtained for each boundary layer tripping position. While the most upstream limit of the shock-motion range is not much sensitive to the transition location, its most downstream limit is strongly affected by the boundary layer state. As seen for the case $\alpha = 4°$, a larger extent of laminar boundary layer tends to move the shock wave further downstream by altering the displacement thickness distribution around the airfoil. In fact, this effect can also be observed in the unsteady case regarding the mean shock-wave position, which roughly corresponds to the point of maximum pressure unsteadiness in Fig. 4b. As the tripping point is placed downstream, the amplitude of shock motion becomes wider, increasing the fluctuation levels in the shock-wave region as well as the trailing edge unsteadiness. This can be observed in the series presented in Fig. 5, in terms of statistical pressure fluctuation fields. Comparing to the fully-turbulent simulation with the most laminar case ($x/c = 0.24$), the pressure unsteadiness increases by approximately 20 % in the shock region and gets nearly two times larger near the trailing edge. The development of the shock-motion area as a function of the transition location is clearly visible in Fig. 5.

Table 1 gives the average lift, drag and pitching moment coefficients for the three transition cases as well as for the fully-turbulent computation. The standard deviation σ of the aerodynamic forces is also presented. As for the steady flow at 4°, the values of the mean lift and of the moment magnitude increase as the triggering location moves towards the trailing edge. A slight augmentation in the mean drag is also noticed. As a result of the increasing shock-motion amplitude and of the overall flow unsteadiness, the standard deviations of the lift and drag coefficients also become larger as the extent of laminar boundary layer gets longer. Therefore the mean lift over mean drag ration doesn't show much improvement whereas the laminar region is increased. Indeed, as the transition is located closer to the

Table 1 Transition location effect on the unsteady global coefficients

x_t/c	Fully Turbulent	0.09	0.16	0.24
$\overline{C_D} \times 10^2$	6.163	6.501	6.604	6.715
$\sigma(C_D) \times 10^2$	0.9419	1.250	1.384	1.533
$\overline{C_L}$	0.9423	0.9718	0.9927	1.018
$\sigma(C_L)$	0.0854	0.1047	0.1132	0.1204
$\overline{C_m} \times 10^2$	−4.223	−4.932	−5.267	−5.676
$\overline{C_L}/\overline{C_D}$	15.3	14.9	15.0	15.2

shock-wave/boundary-layer interaction, the boundary layer downstream detaches more easily than the fully-turbulent case, which gave here the higher lift-to-drag ratio. Moreover, due to the high angle of attack, the most upstream location of the shock is near 25 % of the chord, which limits the flexibility on the position of the transition.

3.4 Three-Dimensional Simulation of the Fully-Turbulent Case

The Delayed Detached-Eddy Simulation approach was also used in [7] to simulate the transonic flow over the OAT15A supercritical airfoil at conditions of angle of attack just above the buffet onset boundary. In that case, the predicted shock-wave motion region was in good agreement with the experiments and the solution exhibited a rich content of resolved flow structures. Nevertheless, probably due to a long delay in the formation of resolved structures in LES regions, the DDES was shown to produce too-intense pressure and velocity fluctuations in the region downstream the oscillating shock wave.

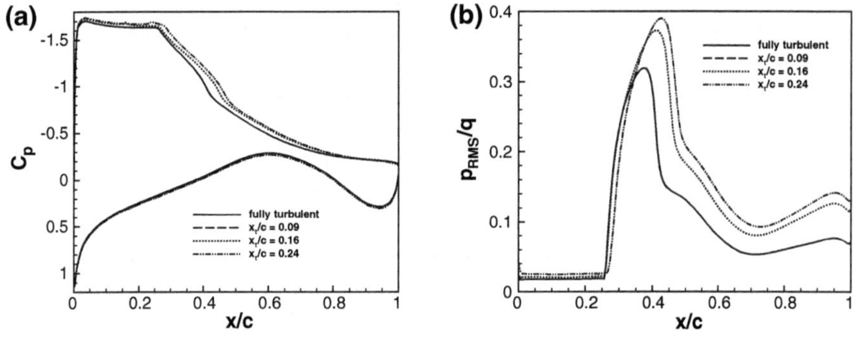

Fig. 4 Statistical wall pressure at $\alpha = 7.0°$. **a** Mean pressure coefficient, **b** RMS value of pressure on the upper surface

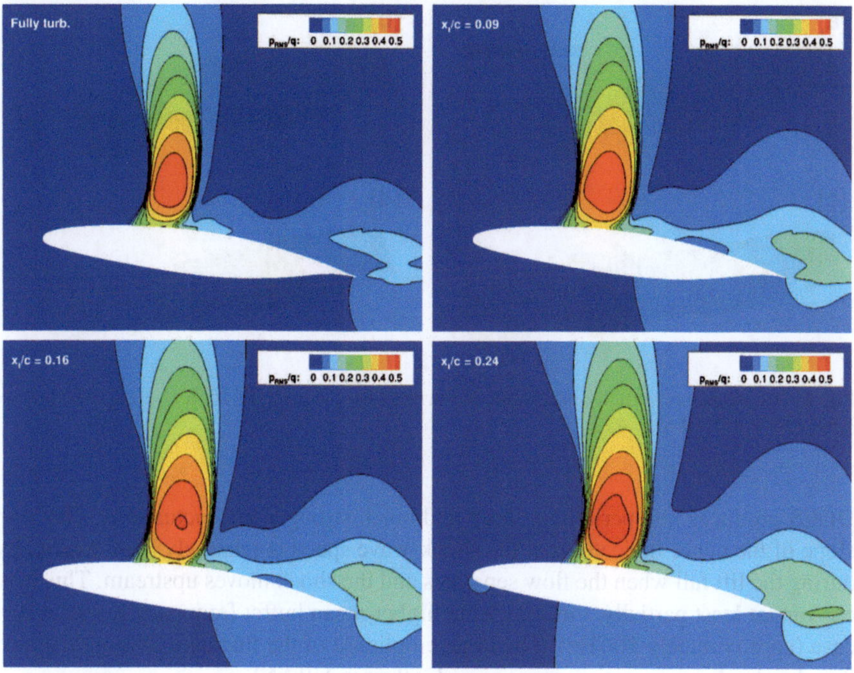

Fig. 5 RMS pressure fields for different transition locations at $\alpha = 7.0°$

In this study, the DDES method is applied, based on similar numerical configuration (same numerical scheme and time step $\Delta t = 10^{-7}$ s), to the well-developed transonic buffet occurring over the V2C airfoil at $M_\infty = 0.70$ and $\alpha = 7.0°$. The objective is to investigate whether the same issues detected in the OAT15A test case exist for a stronger shock-induced separation by comparing the DDES results with those of the previous URANS simulation. Due to the high computational cost of the simulation and for simplicity, only the fully-turbulent configuration is considered. The three-dimensional grid has been obtained by copying the planar grid used in the URANS simulations in the spanwise direction over a distance $L_z/c = 0.33$. To obtain $\Delta i \approx \Delta k$, 59 cells have been distributed along the span keeping a constant spacing, resulting in a final grid of about 9.65 M cells.

In the SST-based DDES, the turbulence length scale provided by the RANS part is computed using local turbulence properties and is given by $\sqrt{k}/(\beta^* \omega)$.

3.4.1 Flowfield Dynamics

The time history of lift after the transient period is presented in Fig. 10 for both the DDES and the fully-turbulent 2D URANS computation. While in URANS the lift coefficient oscillates quasi-harmonically at a frequency of approximately 82 Hz, the

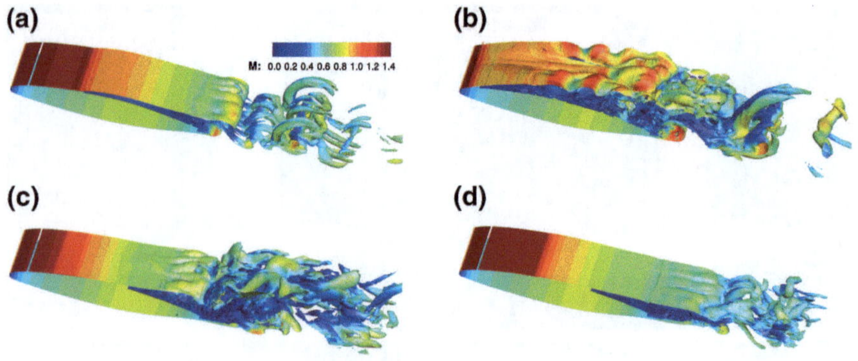

Fig. 6 Instantaneous vorticity magnitude isosurfaces for $|\overrightarrow{\omega}|c/U = 10$. **a** $t^* = -1.60$, **b** $t^* = 1.59$, **c** $t^* = 4.01$, **d** $t^* = 4.83$

DDES produces non periodical sharp and much stronger lift fluctuations. The high slope of the curve indicates that the shock-wave speed is relatively high, especially during the lift fall when the flow separates and the shock moves upstream. This may explain, at least partially, the somewhat higher mean buffet frequency found in that case (approximately 106 Hz). The large amplitude of the fluctuations suggests also that the shock-wave motion range is wider than in URANS.

A series of flow snapshots is presented in Fig. 6 for one period of buffet. It helps understanding the dynamics of the flow predicted by the DDES. The figures illustrate instantaneous isosurfaces of non-dimensional vorticity magnitude for $|\overrightarrow{\omega}|c/U = 10$ as a function of the non-dimensional time $t^* = tU/c$, where $t^* = 0$ is an instant of maximum lift. Surfaces are colored with the Mach number. During the upstream travel of the shock (Fig. 6a), alternate vortex shedding can be observed at the trailing edge. The primary structures are always three-dimensional. As the shock approaches the leading edge, the flow over the upper surface gets fully separated and the shear layer becomes unstable (Fig. 6b). Such intense separation generates a large wake combining the eddies produced in the shear layer and the trailing edge structures. As the shock and the separation point move downstream, the height and streamwise extension of the separation region decrease and the amount of resolved flow structures reduces as seen in the Fig. 6c. Unlike in URANS, a considerable amount of separation always exists on the rear part of the airfoil. While the shear layer becomes stable as the shock wave approaches its most downstream position, the alternate vortex shedding at the trailing edge is always present during buffet (Fig. 6d).

The distribution of the RANS and LES has been monitored in order to evaluate the ability of the present SST-based DDES to switch between the two modes during buffet. This analysis shows the existence of a RANS-mode layer covering the near-wall region around the V2C airfoil. The overall height of this layer seems to be relatively small. This might cause some degree of modeled-stress depletion (*MSD*) [14] due to the erroneous penetration of the LES mode into attached boundary layers, which facilitates separation. The instantaneous distributions of the function $1 - f_d$

at four phases of buffet are given in Fig. 7. The irregular black areas over the upper surface indicate large regions of separation, even when the shock is at its most downstream position (Fig. 7d) where a large amount of rear separation exists on the upper surface. It should be remembered that the hybrid method used here is originally intended to massively separated flows so that the present application should be regarded as an extended use of DDES as the height of the separation region remains small during buffet.

3.4.2 Statistical Flow Properties

The differences between DDES and URANS simulations in terms of mean pressure distributions, plotted in Fig. 8a, are important. The shock wave and thus the separation point reach the leading edge in their travel upstream, as indicated by the lack of a supersonic plateau in the DDES. The shock-motion range is much wider than in URANS, covering about 40 % of the chord, and the flattened aspect downstream the shock-motion region suggests the occurrence of a large amount of separation. The RMS pressure distributions show in Fig. 8b that the maximum

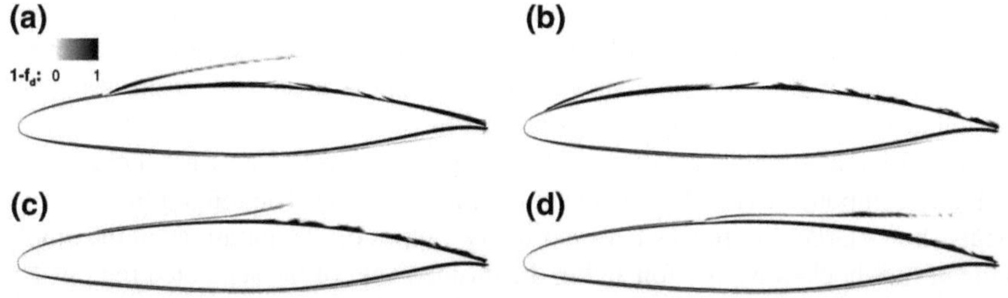

Fig. 7 RANS and LES regions around the V2C airfoil. **a** Maximum lift ($t^* = 0$), **b** shock upstream ($t^* = 1.46$), **c** minimum lift ($t^* = 2.96$), **d** shock downstream ($t^* = 6.02$)

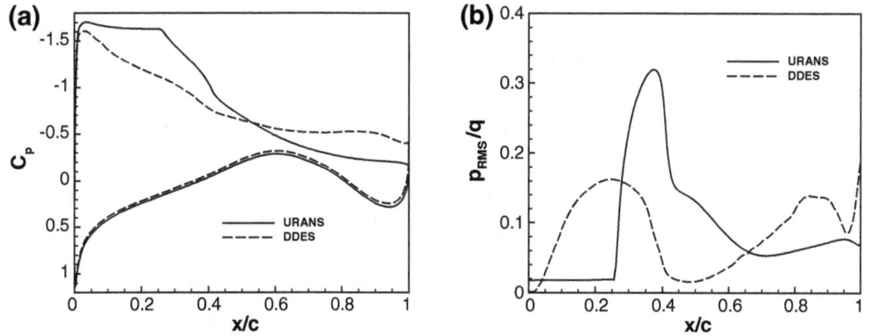

Fig. 8 URANS and DDES statistical pressure distributions at $M_\infty = 0.70$ and $\alpha = 7.0°$. **a** Mean pressure coefficient, **b** RMS pressure on the upper surface

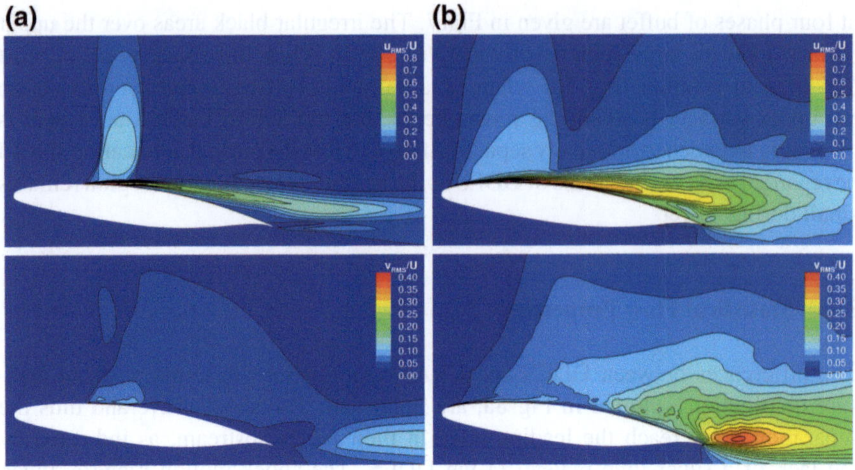

Fig. 9 RMS value of longitudinal (*top*) and vertical (*bottom*) velocity fields around the V2C airfoil. **a** URANS, **b** DDES

unsteadiness levels in the shock region are lower than in the URANS computations, whereas over the rear part of the airfoil, the pressure fluctuations predicted by the DDES are much higher. This result suggests that, the delay in the development of resolved flow structures in the shear layer and in the separated region arising from the 'gray area' is critical in DDES, even in this case of shock-induced separated flows at relatively high angles of attack. The RMS fields of the velocity components (Fig. 9) lead to similar interpretation. The longitudinal components shows large differences between the two types of computations in the amplitude of the shock-wave motion and in the unsteadiness of the separated region. The discrepancies at the trailing edge are large, being primarily caused by alternate vortex structures, which can been seen by the transversal component of the velocity. Although such phenomenon might indeed exist, its strength and stability is probably overestimated in the simulation.

3.4.3 Three-Dimensional URANS

Based on the results of the DDES simulations detailed in the previous section, a new set of computations have been carried out on a grid which has 30 % more cells in all directions, leading to grid of about 23 M cells, with the hope to evaluate more accurately the limit between the RANS and LES. Two 3D URANS computations are performed. One is based the Edwards and Chandra's variant [6] of the Spalart-Allmaras turbulence model [13] with the Secundov's compressibility correction [12] to improve the behavior of the one-equation model in compressible mixing layers. The other uses the $k - \omega$ SST model [9] with ambiant terms [15], as in all the

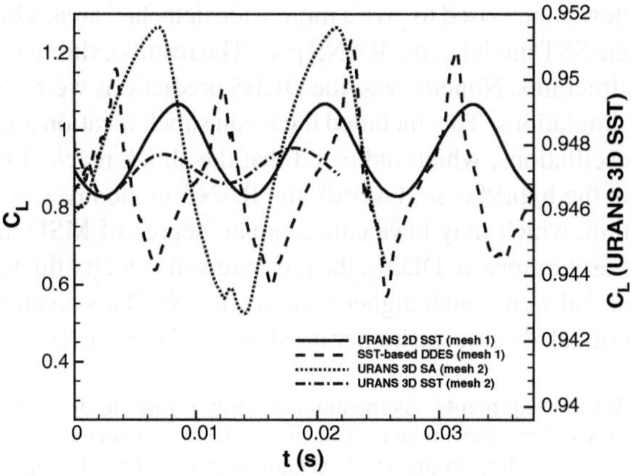

Fig. 10 Comparison of lift coefficient between 2D and 3D URANS and DDES computations

results presented until now (except for the effects of the turbulence model studied in Sect. 3.2).

Lift coefficient is plotted in the Fig. 10 of the transient phase, in addition to the results obtained with the previous grid. On the one hand, amplitudes of the lift coefficient from the Spalart-Allmaras simulation are large and have a low frequency compared to the full established buffet regime in 2D. On the other hand, the variations of the lift for the $k - \omega$ SST (values on the right vertical axis) are extremely low.

These computations need to be continued to confirm the tendency of the results, and the hybrid method needs to be applied and compared with those presented in this study.

4 Conclusion

The transonic V2C profile has been first investigated in terms of 2D (U)RANS computations. The different flow phenomena occurring around the airfoil for various angles of attack at Mach number 0.70 have been analysed. The pressure and skin friction distributions have shown the effects of the angle of attack on the shock wave position, as well as on the state of the boundary layer and its interaction with the foot of the shock. The critical angle regarding the transonic buffet onset has been found to be dependent on the turbulence closure model: while the Spalart-Allmaras model gave a critical angle of 5°, the onset of the buffeting phenomenon appeared at a higher angle, 5.5°, for the $k-\omega$ SST model. The transition study has been carried out in the unsteady case. The buffeting conditions (angle of attack of 7°) has been analysed with the same method, for a maximum downstream position of the transition at $x/c = 0.24$, due to the shock-motion amplitude. A DDES simulation has been performed for the transonic buffet over the V2C airfoil at a relatively high angle of attack, where the flow is supposed to give a more wide detached area. This approach has been used with the SST model in the RANS part. The results exhibited a rich content of resolved flow structures. Nonetheless, the DDES predictions were very different from the URANS simulations. This included the mean shock location and amplitude of the shock-wave oscillations, which indicated that the shock reached the leading edge during buffet in the hybrid case. Overall, the RANS-mode layer around the airfoil was relatively thin, which may have caused some degree of MSD and thus facilitated separation. Furthermore, in DDES, the pressure and velocity fluctuations over the rear part of the airfoil were much higher than in URANS. This result suggests the gray area issue as critical, despite the higher incidence and the existence of more eddies in the flowfield.

Acknowledgments As mentionned several times in this paper, this work was supported by the TFAST European project—Transition Location Effect on Shock Wave Boundary Layer Interaction. The research team thankfully acknowledges the French computing centers CINES and CALMIP for the HPC resources allocated as well as for their availability.

References

1. Barbut, G., Braza, M., Hoarau, Y., Barakos, G., Sévrain, A., Vos, J.B.: Prediction of transonic buffet around a wing with flap. In: Progress in Hybrid RANS-LES Modelling, Notes on Numerical Fluid Mechanics and Multidisciplinary Design, Vol. 111, pp. 191–204. Springer (2010)
2. Bouhadji, A., Braza, M.: Organised modes and shock-vortex interaction in unsteady viscous transonic flows around an aerofoil: Part I: Mach number effect. J. Comput. Fluids **32**(9), 1233–1260 (2003)
3. Bourdet, S., Bouhadji, A., Braza, M., Thiele, F.: Direct numerical simulation of the three-dimensional transition to turbulence in the transonic flow around a wing. Flow Turbul. Combust. **71**(1–4), 203–220 (2003)
4. Crouch, J.D., Ng, L.L.: Variable N-factor method for transition prediction in three-dimensional boundary layers. AIAA J. **38**(2), 211–216 (2000)
5. Deck, S.: Numerical computation of transonic buffet over a supercritical airfoil. AIAA J. **43**(7), 1556–1566 (2005)
6. Edwards, J.R., Chandra, S.: Comparison of eddy viscosity-transport turbulence models for three-dimensional, shock-separated flowfields. AIAA J. **34**(9), 756–763 (1996)
7. Grossi, F., Braza, M., Hoarau, Y.: Prediction of transonic buffet by delayed detached-eddy simulation. AIAA J. **52**(10), 2300–2312 (2014)
8. Levy Jr, L.L.: Experimental and computational steady and unsteady transonic flows about a thick airfoil. AIAA J. **16**(6), 564–572 (1978)
9. Menter, F.R.: Two-equation eddy-viscosity turbulence models for engineering applications. AIAA J. **32**(8), 1598–1605 (1994)
10. Roe, P.L.: Approximate riemann solvers, parameter vectors, and difference schemes. J. Comput. Phys. **43**(2), 357–372 (1981)
11. Seegmiller, H.L., Marvin, J.G., Levy Jr, L.L.: Steady and unsteady transonic flow. AIAA J. **16**(12), 1262–1270 (1978)
12. Shur, M., Strelets, M., Zaikov, L., Gulyaev, A., Kozlovand, V., Secundov, A.: Comparative numerical testing of one- and two-equation turbulence models for flows with separation and reattachment. In: AIAA Paper (95–0863) (1995)
13. Spalart, P.R., Allmaras, S.R.: A one-equation turbulence model for aerodynamic flows. La Recherche Aérospatiale **1**, 5–21 (1994)

14. Spalart, P.R., Deck, S., Shur, M.L., Squires, K.D., Strelets, M.K., Travin, A.: A new version of detached-eddy simulation, resistant to ambiguous grid densities. Theor. Comput. Fluid Dyn. **20**, 181–195 (2006)
15. Spalart, P.R., Rumsey, C.L.: Effective inflow conditions for turbulence models in aerodynamic calculations. AIAA J. **45**(10), 2544–2553 (2007)
16. van Leer, B.: Towards the ultimate conservative difference scheme. V. A second-order sequel to Godunov's method. J. Comput. Phys. **32**(1), 101–136 (1979)

Model-Invariant Hybrid LES-RANS Computation of Separated Flow Past Periodic Hills

Stephen Woodruff

Abstract The requirement that physical quantities not vary with a hybrid LES-RANS model's blending parameter imposes conditions on the computation that lead to better results across LES-RANS transitions. This promises to allow placement of those transitions so that LES is performed only where required by the physics, improving computational efficiency. The approach is applied to separated flow past periodic hills, where good predictions of separation-bubble size are seen due to the gradual, controlled, LES-RANS transition and the resulting enhanced near-wall eddy viscosity.

1 Introduction

Full realization of the potential of hybrid LES-RANS computations requires that LES regions be placed only where necessary to capture the physics the RANS model cannot, and RANS be used everywhere else for computational efficiency. That current hybrid methods lack this flexibility, even in a simple case like plane channel flow, was demonstrated by Nikitin et al. [13] in 2001. They found that placing the LES-RANS transition in the log layer led to what is now commonly referred to as "log-layer mismatch": the LES log layer is shifted upwards and the mean velocity in the central portion of the channel is too large. Many authors have reported similar results. As a consequence, if a RANS model fails in certain parts of a flow (say near the wall-mounted injector of a scramjet or in a smooth-body separation bubble), one's only alternatives are to accept the errors resulting from the LES-RANS transition or to render the errors neglegible by defining unnecessarily large LES regions with transitions far enough away from the region of interest.

The model-invariant hybrid computation [21, 22] addresses this problem by establishing a basis for interpreting the results of a hybrid computation in those parts of the flow where the model is somewhere between a pure RANS model and a pure LES

S. Woodruff (✉)
NASA Langley Research Center, Hampton, VA 23681, USA
e-mail: stephen.l.woodruff@nasa.gov

S. Girimaji et al. (eds.), *Progress in Hybrid RANS-LES Modelling*,
Notes on Numerical Fluid Mechanics and Multidisciplinary Design 130,
DOI 10.1007/978-3-319-15141-0_28

model. A blending parameter characterizes this transition, controlling the mixture of RANS and LES at a given point in the flow. As a non-physical artifact of the turbulence model, the value of the blending parameter should not affect physically meaningful results in the transition region if both RANS and LES are valid there. Model-invariant computations ensure this is the case.

Combinations of flow variables that do not change when the blending parameter is varied are model invariants; physical variables must be expressible in terms of these model invariants. The total turbulent kinetic energy should be a model invariant, for example. Expressing it as the sum of the resolved turbulent kinetic energy and the modeled turbulent kinetic energy yields an approximate model invariant, whose accuracy is limited by the fidelity of the underlying RANS and LES models and by the manner of blending them.

Results of model-invariant computations were given for decaying, isotropic, homogeneous turbulence in Ref. [21] and for plane channel flow in Ref. [22]. The channel-flow results are particularly compelling: the model-invariant computation has no log-law shift and the mean velocity in the channel center is accurately predicted. Additionally, the model-invariant computation attenuates the modeled shear stress and amplifies the resolved shear stresses across the LES-RANS transition, so the RANS zone is closer to pure RANS and the LES zone is closer to pure LES than for a conventional hybrid computation. Predictions of physical quantities are largely unchanged by varying the height and thickness of the LES-RANS transition or the shape of the blending function.

A few examples of other approaches to facilitating LES-RANS transitions involve stochastic forcing [9], definition of a hybrid filter [6], modifying the RANS eddy viscosity to account for resolved Reynolds stresses [2, 10] and modeling commutation error [20]. A discussion of the relation between the present and other approaches may be found in Ref. [22].

In the present work, a model-invariant computation is performed for the flow past streamwise-periodic hills in a channel (Fig. 1). The flow separates in the lee of each hill and the length of the separation bubble has been particularly

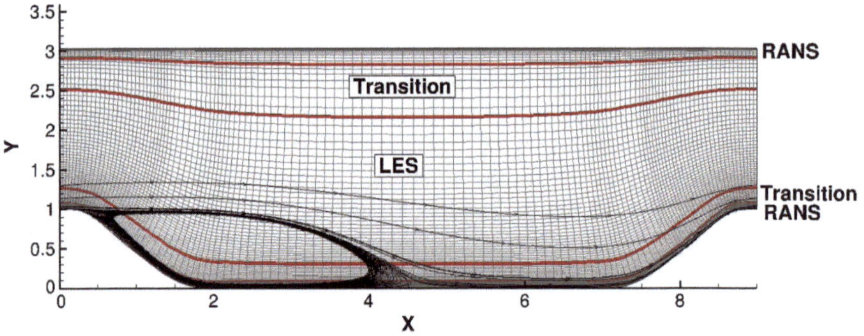

Fig. 1 Flow configuration, showing mesh, edges of RANS-LES transition zones (*thick red lines*) and streamlines from model-invariant computation

difficult to predict accurately with most RANS or hybrid computations. Lengths are non-dimensionalized by the hill height and velocities by the bulk velocity over the hill. An artificial body force provides an imposed pressure gradient and is controlled to maintain a constant mass flux through the channel. The flow is incompressible and the Reynolds number (Re), based on bulk velocity and hill height, is 10,595. The thick solid red lines in Fig. 1 represent the edges of the LES-RANS transitions.

LES computations [3, 5, 19, 23] provide a largely consistent reference for comparisons. They yield, for $Re = 10{,}595$, separation at $x_s \approx 0.19 - 0.22$ and reattachment at $x_r \approx 4.68 - 4.72$, though one of the computations of Ref. [5] gives $x_r \approx 4.56$. Hybrid results [2, 4, 7, 18] are much more dispersed, with most giving separation in the range 0.23–0.6 and reattachment in the range 4.6 – 5.8. The best results include a one-equation hybrid model [2] ($x_r \approx 4.75$) and one based on an explicit algebraic stress model [7] (several computations with $x_r \approx 4.65 - 4.75$). Most hybrid computations, however, like the Spalart-Almaras DES [15] reported in Ref. [2] ($x_r \approx 5.20$), yield separation bubbles that are noticeably too large. RANS computations (as discussed in, for example, Refs. [1, 8, 12, 14]) generally lead to an overly long separation bubble, with reattachment delayed to as far as $x_r \approx 6 - 8$. Nevertheless, several RANS model improvements [8, 12, 14] have succeeded in enhancing the level of turbulence in the bulk of the separation bubble and advancing the reattachment point to a more realistic location.

The thin tail of the separation bubble means that large variations in the reattachment location can result from very small changes in the flow field near the wall: existing hybrid techniques in fact predict mean velocities and correlations in the bulk of the flow fairly well. The work of Temmerman et al. [19] shows that wall conditions can significantly affect the reattachment location. It will be seen in what follows that the model-invariant computation intensifies the turbulence in this critical region near the wall ahead of separation and reattachment to reduce the size of the separation bubble.

2 Model-Invariant Computations

A model-invariant computation is one in which model invariants are preserved as the computation evolves. It turns out that this is possible only if terms involving derivatives of the hybrid LES-RANS model blending parameter are added to the equations of motion. This is because the flow variables are affected both by flow dynamics and by changes in the blending parameter across the LES-RANS transition; the model-invariant terms are required to cancel the non-physical effects of the changing blending parameter and restore the physical balances of the equations.

These new terms necessarily also involve quantities expressing the sensitivity of the flow variables to changes in the blending parameter. The framework following from the model-invariance concept yields a number of methods for determining these model sensitivities. They may be computed, expensively, by finite differences using multiple, simultaneous flow simulations [21]. They may also be determined, more cheaply, from approximations to the model invariants [22]; the only additional computational cost of this approach is the evaluation of the new terms.

The model-invariant computation is set up with gradual transitions between RANS and LES regions, to permit better control. A continuous model [16], valid "in between" RANS and LES, is employed within the transition regions. The continuous model has a blending parameter λ that is 0 when the model is in RANS mode and 1 when the model is in LES mode. The RANS and LES models are arbitrary, as is the manner of blending them together. The decomposition of the flow variables into resolved and modelled components is defined by the blended model; no explicit decomposition, such as a filter, is used.

If useful flow information is to be derived from the computation in the transition regions, the computed variables must be connected with physical quantities. Model invariant quantities provide the means for doing so, but model invariance is lost as the equations evolve when the blending parameter varies in space and time. A simple thought experiment shows why: a hybrid computation of a homogeneous flow would show variations of the mean flow variables as the transition from RANS to LES is traversed; the modeled kinetic energy would decrease and the resolved stresses would increase, for example. These variations lead to unphysical gradients in the equations of motion and erroneous results.

To see how to incorporate the variable blending parameter into a hybrid computation without also introducing unphysical gradients, let the blending parameter vary in space and time and connect each point in the computation with an alternative computation conducted with a constant blending parameter of the same value. This second computation has no unphysical gradients, of course, because the blending parameter does not vary. The two computations are connected by the coordinate transformation $t = t'$, $\mathbf{x} = \mathbf{x}'$, $s = \xi - \lambda(t', \mathbf{x}')$, where $(t, \mathbf{x}, s)$ are the coordinates of the variable-λ case and $(t', \mathbf{x}', \xi)$ are the coordinates of the constant-λ case.

Connecting the variable-λ and constant-λ computations through the coordinate transformation guarantees that properties of the computations will be connected similarly. In particular, the property of model invariance is unaffected by variations in λ provided the governing equations conform to the coordinate transformation. This means that time and space derivatives transform according to (c.f. Ref. [6])

$$\frac{\partial}{\partial t'} = \tilde{\partial}_t \equiv \frac{\partial}{\partial t} - \frac{\partial \lambda}{\partial t}\frac{\partial}{\partial s} \quad \text{and} \quad \nabla' = \tilde{\nabla} \equiv \nabla - (\nabla\lambda)\frac{\partial}{\partial s} \tag{1}$$

(∇' is the gradient operator on the primed variables); the new terms in these expressions cancel out the unphysical gradients caused by the variation of λ.

Employing (1), the continuity and momentum equations become

$$\tilde{\nabla} \cdot \mathbf{v} = 0, \quad \tilde{\partial}_t \mathbf{v} + (\mathbf{v} \cdot \tilde{\nabla})\mathbf{v} = -\tilde{\nabla} p + \tilde{\nabla} \cdot \left[\left(\frac{1}{Re} + \nu_t \right) \tilde{\nabla} \mathbf{v} \right], \tag{2}$$

where $\mathbf{v}$ is the resolved-scale velocity, p is the resolved-scale pressure and ν_t is the eddy viscosity.

The basis for the hybrid model employed in this paper is Strelets' Detached-Eddy Simulation (DES) hybrid model [17], which is in turn based on Menter's SST model [11],

$$\tilde{\partial}_t k + (\mathbf{v} \cdot \tilde{\nabla})k = P - \left[(1-\lambda)\beta^* \omega k + \lambda k^{3/2}/(C_{DES}\Delta)\right] + \tilde{\nabla} \cdot \left[\left(\frac{1}{Re} + \sigma_k \nu_t\right) \tilde{\nabla} k\right]$$

$$\tilde{\partial}_t \omega + (\mathbf{v} \cdot \tilde{\nabla})\omega = \frac{\gamma}{\nu_t} P - \beta \omega^2 + \tilde{\nabla} \cdot \left[\left(\frac{1}{Re} + \sigma_\omega \nu_t\right) \tilde{\nabla} \omega\right]. \quad (3)$$

The production P is expressed as $P = \nu_t \tilde{S}_{ij} \tilde{S}_{ij}$, where $\tilde{S}_{ij}$ is the symmetric part of the tensor $\tilde{\nabla}\mathbf{v}$. The Strelets DES expression for the dissipation term in the kinetic energy equation has been replaced by a blended combination of the RANS and LES dissipation terms to make it a blended, continuous, model. The transition to LES occurs below the outer part of the boundary layer where the $k - \varepsilon$ branch would be active, so it is not included. Standard $k - \omega$ constants [11] are employed.

The derivatives of the flow variables with respect to s, the model sensitivities appearing in (1), are determined by means of the crude approximation employed previously for plane channel-flow turbulence [22]. The destruction terms in the k and ω equations are both considered to be model invariants, because they are sinks for the total amounts (resolved and modelled) of these quantities. Differentiating these terms with respect to s and setting the results to zero yields expressions for k_s and ω_s. If the total kinetic energy $k + |\mathbf{v}|^2/2$ (modeled plus resolved kinetic energies) is also a model invariant, differentiating it yields $|\mathbf{v}|_s$ in terms of k_s. It remains to split $|\mathbf{v}|_s$ into components: this is properly done through model invariants based on components of the Reynolds stress tensor, but in this crude approximation, the total $|\mathbf{v}|_s$ is simply split into components according to the empirically determined proportions 30 % spanwise and 70 % normal to the transition layer.

Computations are performed with a modification of the code employed in Refs. [21, 22]. Streamwise and normal directions are discretized by fourth-order finite differences and the spanwise direction is discretized spectrally. Second-order time advancement is performed via Newton iteration. Continuity is imposed at each time step; this also requires iteration, due to the presence of the model-invariant terms. The grid of Fröhlich et al. [5] was adapted to yield a 128 × 84 streamwise-normal two-dimensional grid that retains the original near-wall normal spacing. Twenty spectral modes are employed in the spanwise direction. A Spalart-Almaras DES [15] computation was performed for validation and yielded results similar to those found previously [2], with reattachment at $x_r \approx 5.2$.

3 Results

The results of the model-invariant computation just outlined are now compared with a non-model-invariant computation which is the same in all respects except that the model-invariant terms are not included. All quantities presented are averaged over

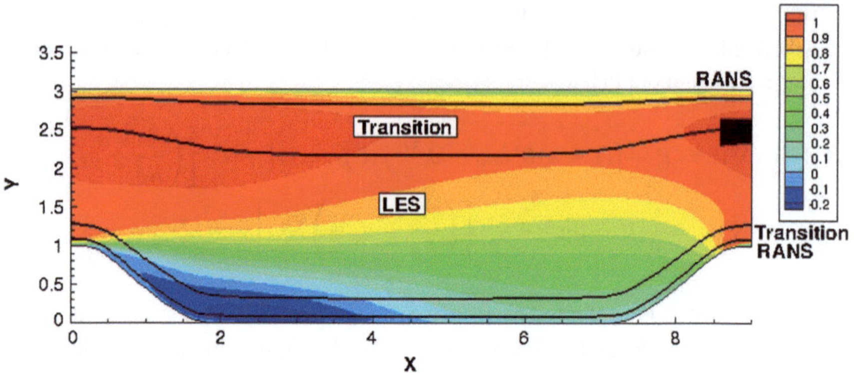

Fig. 2 Streamwise velocity contours for non-model-invariant hybrid computation

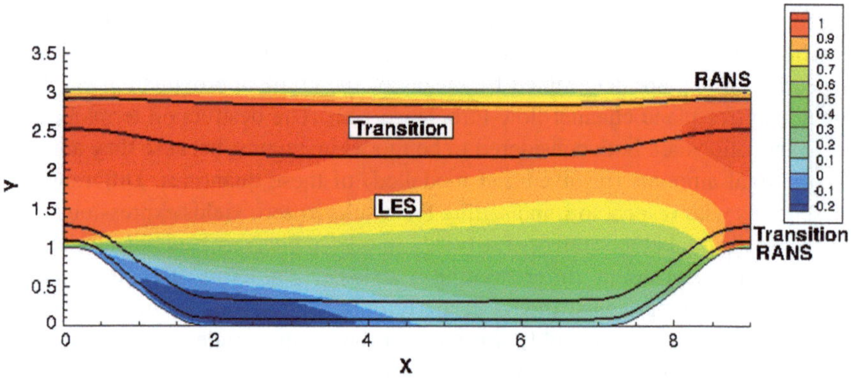

Fig. 3 Streamwise velocity contours for model-invariant hybrid computation

the spanwise direction and over a period of 50 flow-through times (based on bulk velocity and channel length), beginning after time averages of the velocity field have settled to fixed values.

Plots of the streamwise velocity contours for the non-model-invariant (Fig. 2) and the model-invariant (Fig. 3) computations show the flow separating at approximately the same point ($x_r \approx 0.24$ and $x_r \approx 0.23$, respectively) but reattaching at $x_r \approx 4.78$ in the non-model-invariant computation and $x_r \approx 4.67$ in the model-invariant computation. As noted in the Introduction, the bulk of existing hybrid techniques do noticeably worse.

Both the non-model-invariant (Fig. 4) and model-invariant (Fig. 5) modelled kinetic-energy contour plots show spots of high kinetic energy at the inception of the free shear layer defining the separation bubble; the spot is more localized to the transition and RANS layers in the model-invariant computation. The model-invariant computation also shows increased kinetic energy in the RANS layer in regions ahead of the separation and reattachment points. (It should be noted that the peak kinetic

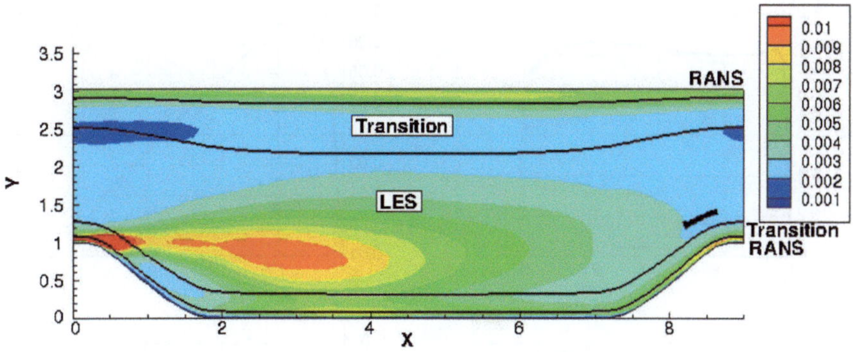

Fig. 4 Modelled kinetic-energy contours for non-model-invariant hybrid computation

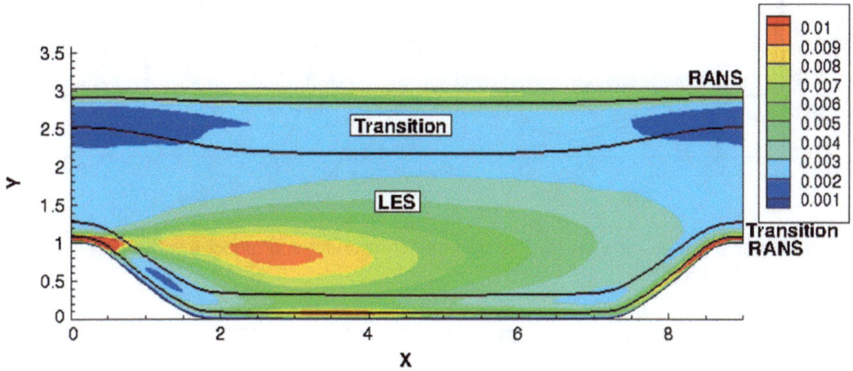

Fig. 5 Modelled kinetic-energy contours for model-invariant hybrid computation

energy, at the inception of the free shear layer, is approximately 0.02 in both cases; the contour levels have been chosen to bring out differences in the two plots elsewhere in the flow.)

Profiles of flow quantities further indicate that the model-invariant terms have a minimal effect in the LES region, but increase the RANS eddy viscosity upstream of the separation and reattachment points. First, cross-channel profiles of the mean streamwise velocity, mean normal velocity, streamwise normal Reynolds stresses and the Reynolds shear stress are given at $x = 0.05$, just upstream of the separation point (Fig. 6). The computations with and without the model-invariant terms yield very similar results, except for minor differences in the RANS region near the wall. Both computations agree fairly well with the LES of Fröhlich et al. [5], though the magnitude of the second-order correlations are somewhat low. Profile comparisons are similar at other streamwise locations.

Profiles of the mean contribution of the model-invariant terms to the streamwise momentum, normal momentum, k and ω equations at $x = 0.05$ are shown in Fig. 7. The mean eddy viscosity is also shown to illustrate how the model-invariant

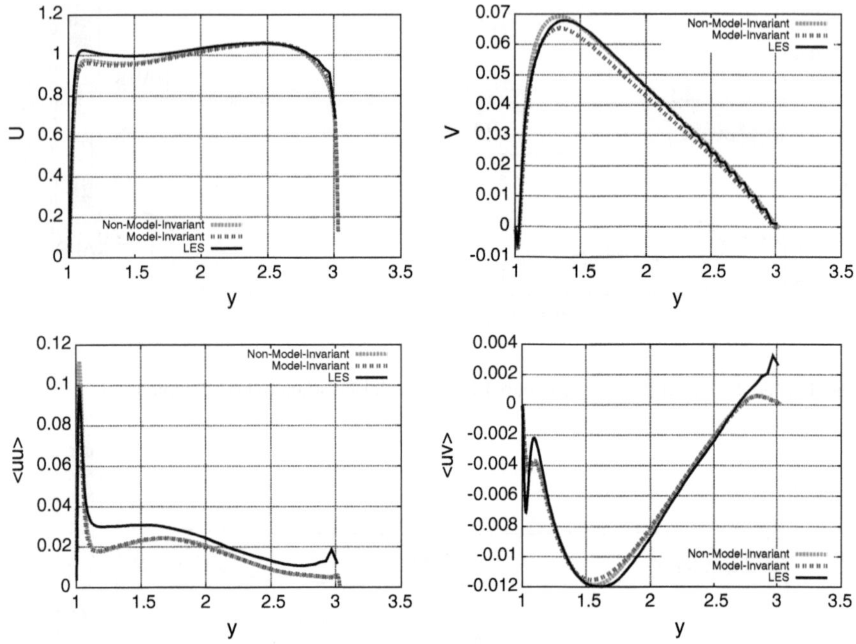

Fig. 6 Moment profiles at $x = 0.05$ of non-model-invariant and model-invariant computations compared with LES [5]

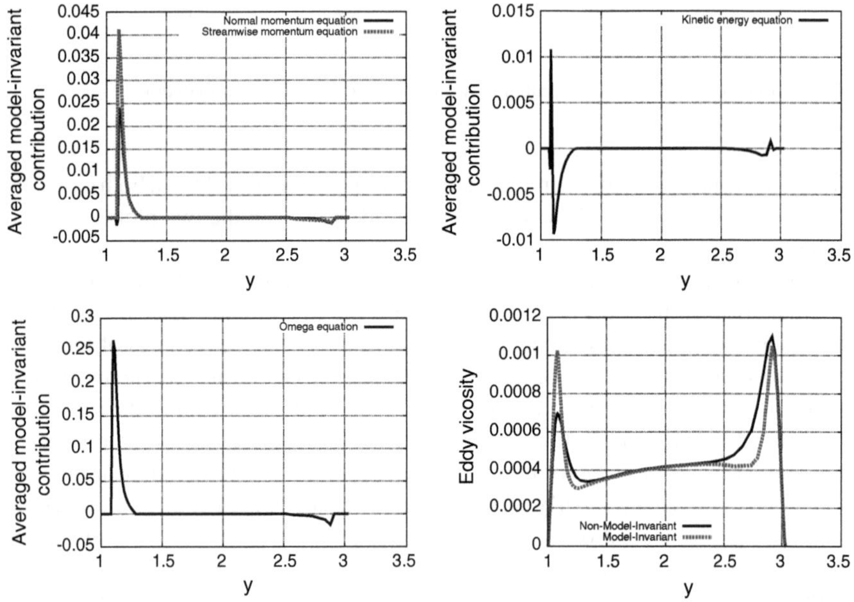

Fig. 7 Mean contribution of model-invariant terms to streamwise and normal momentum equations, kinetic-energy equation and ω equation and mean eddy viscosity at $x = 0.05$

contributions to the k and ω equations affect separation and reattachment. For the most part, the model-invariant terms act as one would expect to speed the transition from RANS to LES: the terms serve to increase the resolved velocity and decrease the modelled kinetic energy. They also act to reduce the rate of decrease of ω and the modelled kinetic energy is increased at the RANS edge of the transition region. The mean eddy viscosity profile shows that the latter actions help to localize the eddy-viscosity to the RANS layers in the model-invariant computation and to significantly increase, by 50 %, the peak eddy viscosity in the model-invariant computation at this location just upstream of the separation point. The increased modelled kinetic energy upstream of the reattachment point (Fig. 5) indicates a similar enhancement of the eddy viscosity occurs there.

In general, higher turbulence levels delay separation and advance reattachment. For Boussinesq-approximation RANS models like that used here, higher turbulence levels correspond to a higher eddy viscosity, and enhancing the eddy viscosity was the mechanism for improving RANS separated-flow predictions in Refs. [12, 14]. The present model-invariant computation succeeds in part by enhancing the eddy viscosity in the RANS layer upstream of the reattachment point, moving it forward. It also does so upstream of the separation point, moving it slightly downstream. Unlike the channel-flow computations of Ref. [22], the turbulence levels in the LES region were predicted fairly well by the non-model-invariant computation (as they are by most computations in the literature) and are largely unaffected by the addition of the model-invariant terms.

The actions of the model-invariant terms on the flow are, overall, relatively minor in the present computation. This is largely due to the wide LES-RANS transition zones (Fig. 1) which allow the blending-parameter gradient $\nabla\lambda$ in (1) to be small, reducing the effect of the model-invariant terms. In situations where the flow configuration prevents such a gradual transition, these terms would play a larger role and the effects discussed above would be more significant.

4 Conclusion

The observation that physically meaningful quantities cannot depend on the blending parameter in a hybrid LES-RANS computation led to the need to accommodate variations in the blending parameter in the derivatives of the governing equations. The concept of model invariance provides a framework for understanding this and other aspects of hybrid computations, as well as a means for computing the model sensitivities that arise in the new terms.

The significant improvements in flow predictions of the non-model-invariant computation compared to SA DES and many other hybrid computations in the literature indicated the very gradual LES-RANS transition kept the model-invariant terms fairly small; including these terms further reduced the size of the separation bubble. Examination of resolved and modeled turbulence quantities showed how the introduction of the model-invariant terms led to enhanced eddy viscosity in the RANS layer upstream of the separation and reattachment points, delaying and advancing them, respectively.

References

1. Abe, K., Jang, Y.-J., Leschziner, M.A.: An investigation of wall-anisotropy expressions and length-scale equations for non-linear eddy-viscosity models. Heat Fluid Flow **24**, 181–198 (2003)
2. Breuer, M., Jaffrezic, B., Arora, K.: Hybrid LES-RANS technique based on a one-equation near-wall model. Theoret. Comput. Fluid Dyn. **22**, 157–187 (2008)
3. Breuer, M., Peller, N., Rapp, C., Manhart, M.: Flow over periodic hills—numerical and experimental study in a wide range of Reynolds numbers. Comput. Fluids **38**, 433–457 (2009)
4. Fröhlich, J., von Terzi, D.: Hybrid LES/RANS methods for the simulation of turbulent flows. Progr. Aerosp. Sci. **44**, 349–377 (2008)
5. Fröhlich, J., Mellen, C.P., Rodi, W., Temmerman, L., Leschziner, M.A.: Highly resolved large-eddy simulation of separated flow in a channel with streamwise periodic constrictions. J. Fluid Mech. **526**, 19–66 (2005)
6. Germano, M.: Properties of the hybrid RANS/LES filter. Theoret. Comput. Fluid Dyn. **17**, 225–231 (2004)
7. Jaffrezic, B., Breuer, M.: Application of an explicit algebraic Reynolds stress model within a hybrid LES-RANS method. Flow Turbul. Combust. **81**, 415–448 (2008)
8. Jakirlic, S., Maduta, R.: On "Steady" RANS modeling for improved prediction of wall-bounded separation. AIAA Paper 2014–0586 (2014)
9. Keating, A., De Prisco, G., Piomelli, U.: Interface conditions for hybrid RANS/LES calculations. Heat Fluid Flow **27**, 777–788 (2006)
10. Medic, G., Templeton, J.A., Kalitzin, G.: A formulation for near-wall RANS/LES coupling. Int. J. Eng. Sci. **44**, 1099–1112 (2006)
11. Menter, F.R.: Two-equation eddy-viscosity turbulence models for engineering applications. AIAA J. **32**, 1598–1605 (1994)
12. Menter, F.R., Egorov, Y.: The scale-adaptive simulation method for unsteady turbulent flow predictions. Part 1: theory and model description. Flow Turbul. Combust. **85**, 113–138 (2010)
13. Nikitin, N.V., Nicoud, F., Wasistho, B., Squires, K.D., Spalart, P.R.: An approach to wall modelling in large-eddy simulations. Phys. Fluids **12**, 1629–1632 (2000)
14. Rumsey, C.: Exploring a method for improving turbulent separated-flow predictions with k-ω models. NASA TM-2009-215952 (2009)
15. Spalart, P.R.: Detached-eddy simulation. Annu. Rev. Fluid Mech. **41**, 181–202 (2009)
16. Speziale, C.G.: Turbulence modeling for time-dependent RANS and VLES: a review. AIAA J. **36**, 173–184 (1998)
17. Strelets, M.: Detached eddy simulation of massively separated flows. In: AIAA 2001–0879 (2001)
18. Temmerman, L., Hadziabdic, M., Leschziner, M.A., Hanjalic, K.: A hybrid two-layer URANS-LES approach for large eddy simulation at high Reynolds numbers. Heat Fluid Flow **26**, 173–190 (2005)
19. Temmerman, L., Leschziner, M.A., Mellen, C.P., Fröhlich, J.: Investigation of wall-function approximations and subgrid-scale models in large-eddy simulation of separated flow in a channel with streamwise periodic constrictions. Heat Fluid Flow **24**, 157–180 (2003)
20. Wallin, S., Girimaji, S.: Commutation error mitigation in variable-resolution PANS closure: proof of concept in decaying isotropic turbulence. AIAA Paper 2011–3105 (2009)
21. Woodruff, S.L.: Coupling turbulence in hybrid LES-RANS techniques. Seventh International Symposium on Turbulence and Shear-Flow Phenomena, Ottawa, Canada (2010)

22. Woodruff, S.L.: A new formulation for hybrid LES-RANS computations. AIAA Paper 2013–2722 (2013)
23. Ziefle, J., Stolz, S., Kleiser, L.: Large-eddy simulation of separated flow in a channel with streamwise-periodic constrictions. AIAA J. **46**, 1705–1718 (2008)

Part V
Bridging Methods

Partially-Averaged Navier-Stokes (PANS) Simulations of Lid-Driven Cavity Flow—Part 1: Comparison with URANS and LES

Bhanesh Akula, Pratanu Roy, Pooyan Razi, Steven Anderson and Sharath Girimaji

Abstract Multiple-resolution simulations of lid-driven cavity flows at Reynolds number of 10,000 are performed at two cavity aspect ratios (SAR) of 3:1:1 and 1:1:1. The objective is to assess the LES (Large eddy simulations), URANS (unsteady Reynolds-averaged Navier-Stokes) and PANS (partially-averaged Navier-Stokes) results against available experimental data. All of the approaches reasonably capture the mean flow velocity field for this low Reynolds number case. It is also shown that the second moments are rather small. Therefore, turbulence, while present, does not profoundly affect the mean flow statistics. It is shown that much of the flow unsteadiness is due to large scale structures that are reasonably resolved even in the URANS computation. It is expected that the distinction between the computations of different degrees of fidelity will be evident only at much higher Reynolds numbers.

1 Introduction

The lid driven cavity flow is widely considered as one of the benchmark problems for closure model development and code validation in computational fluid dynamics. The relatively complex flow phenomena that appear in a seemingly simple geometry have attracted much work on this problem in the past few decades. The flow geometry for the lid driven cavity is shown in Fig. 1. The lid moves in the X direction, Y is in the vertical direction and the spanwise direction is Z. The spanwise aspect ratio of the cavity (SAR) is defined as the ratio of spanwise length (L) to the height (H) of the cavity. Koseff and Street [1] performed experiments for SAR 3:1:1 at Re $=$ 10,000 and measured mean velocity profiles close to the symmetry plane. They also performed measurements for turbulent shear stress at selected locations. Prasad and Koseff [2] measured second moment of velocities at the symmetry plane for different SAR and Reynolds numbers. As most of the experimental data and numerical simulations in literature are available at the SAR of 3:1:1 and 1:1:1 and for Re $=$ 10,000, these flow conditions are considered for current investigation.

B. Akula · P. Roy · P. Razi · S. Anderson · S. Girimaji (✉)
Texas A&M University, College Station, TX, USA
e-mail: girimaji@tamu.edu

S. Girimaji et al. (eds.), *Progress in Hybrid RANS-LES Modelling*,
Notes on Numerical Fluid Mechanics and Multidisciplinary Design 130,
DOI 10.1007/978-3-319-15141-0_29

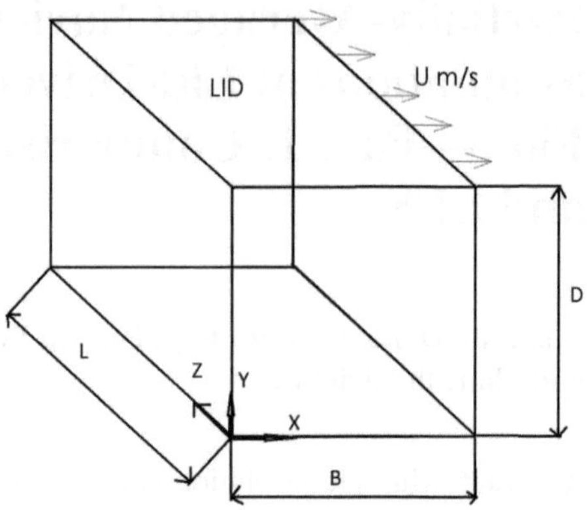

Fig. 1 Schematic diagram of the lid driven cavity with main dimensions and axes directions. SAR is defined as the ratio $\frac{L}{D}$ and the lid moves in the X direction

The objective of this work is to compare the PANS method against URANS and LES for lid driven cavity flow simulation. PANS is a hybrid model that is intended to bridge smoothly between URANS and LES/DNS. In PANS, the accuracy of results can be optimized based on available computational resources. In the URANS method, only the unsteady mean flow i.e. scales that are comparable to the geometry of the flow are resolved, whereas all other scales are modeled. In LES, all the large scale motions or energy carrying eddies are computed, and the small scale motions, that are more universal in nature, are modeled [3]. To illustrate the operative regions of PANS, a typical spectrum of energy as a function of wavelength for turbulent flow is shown in Fig. 2. The relative cut-off for unresolved flow scales is shown for RANS, PANS, and LES. The cut off parameters f_k and f_ε for PANS are defined as the ratio of the unresolved to resolved turbulent kinetic energy and dissipation, respectively. Value of f_k close to zero indicates the limit of DNS (Direct Numerical Simulation) and the value of unity is essentially a URANS simulation. As an important part of the closure development process, it is essential to perform PANS simulations on benchmark flow problems in order to verify its ability to resolve wider range of scales than URANS, and closely reproduce experimental data at higher degree of fidelity on much coarser grids than LES/DNS.

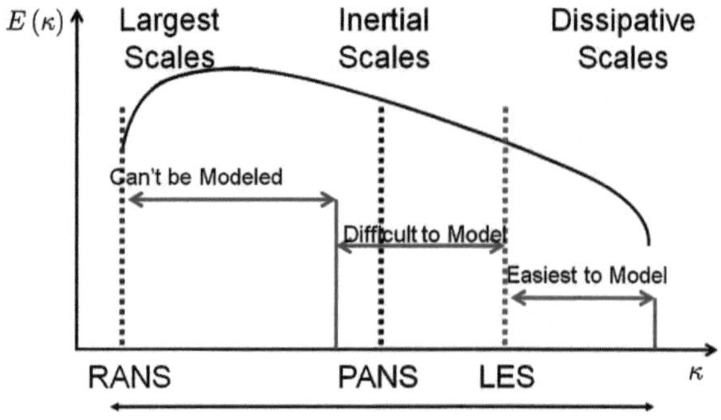

Fig. 2 Energy spectrum of turbulent flow with relative URANS, PANS, and LES spectrum cut-off

In this paper, we simulate lid driven cavity problem using PANS $k - \omega$ turbulence model. The parameter f_k is varied between simulations and its effect on mean flow and turbulence quantities is examined. Section 2 presents the PANS formulation and equations. Numerical simulation setup is presented in Sect. 3. The results are presented in Sect. 4 and we conclude in Sect. 5 with a brief discussion. In the companion paper [4], we examine the vortical structures within the cavity flow as computed by PANS.

2 Hybrid Models

The development of all the hybrid models commences from the incompressible Navier-Stokes equations and continuity equation:

$$\frac{\partial V_i}{\partial t} + V_j \frac{\partial V_i}{\partial x_j} = -\frac{\partial p}{\partial x_i} + \nu \frac{\partial^2 V_i}{\partial x_j \partial x_j}, \tag{1}$$

$$\frac{\partial V_i}{\partial x_i} = 0. \tag{2}$$

The difference between URANS and PANS/ LES lies in how the averaged or filtered velocity equations are obtained from Eq. 1. URANS uses an averaging operator leading to equations that describe the mean velocity field. On the other hand, PANS/ LES uses a generalized homogeneous filter to decompose the velocity into resolved and unresolved part [5]:

$$V_i = U_i + u_i, \quad U_i = \langle V_i \rangle, \quad \langle u_i \rangle \neq 0 \tag{3}$$

By applying the filter to the Navier-Stokes equations, the momentum equations for the resolved field are given as:

$$\frac{\partial U_i}{\partial t} + U_j \frac{\partial U_i}{\partial x_j} = -\frac{\partial \tau(V_i, V_j)}{\partial x_j} - \frac{\partial \langle p \rangle}{\partial x_i} + \nu \frac{\partial^2 U_i}{\partial x_j \partial x_j}, \tag{4}$$

$$\tau(V_i, V_j) = \langle V_i V_j \rangle - \langle V_i \rangle \langle V_j \rangle. \tag{5}$$

The term $\tau(V_i, V_j)$ in Eq. 4 represents the "sub-filter stress", given by Eq. 5 [6]. The sub filter stress term is modeled differently in various turbulence models.
LES: The generalized sub-filter stress in LES is modeled via Boussinesq-type approximation [6–8]:

$$\tau_{ij} = \tau(V_i, V_j) = \frac{2}{3} k \delta_{ij} - \nu_T S_{ij} \tag{6}$$

where ν_T, and k are the "unresolved eddy viscosity" and the "unresolved kinetic energy" respectively. S_{ij} is the resolved strain-rate tensor defined as:

$$S_{ij} = \frac{1}{2}\left(\frac{\partial U_i}{\partial x_j} + \frac{\partial U_j}{\partial x_i}\right) \tag{7}$$

In LES Smagorinsky model, the unresolved eddy viscosity is modeled algebraically by assuming that the energy production and dissipation are in equilibrium, resulting in the following relationship [9]:

$$\nu_T = C\Delta^2|S|, \ where\ |S| = (2S_{ij}S_{ij})^{1/2} \tag{8}$$

Here, Δ is the grid spacing and C is the model coefficient. If the filter cut-off is in the inertial range, then the values of the Smagorinsky constant ($C_s = \sqrt{C}$) usually lie between 0.18 and 0.23. In the dynamic Smagorinsky model, the model parameter C is not constant, rather it is calculated from the energy content of the smallest resolved scales [10]. In order to make the model self-consistent, an additional test filter ($\widehat{\Delta} > \Delta$) is introduced and $C = C_d(x, y, z, t)$ is dynamically adjusted based on the following identity:

$$L_{ij} = T_{ij} - \widehat{\tau_{ij}} \tag{9}$$

where $L_{ij} = \widehat{\overline{u_i}\,\overline{u_j}} - \widehat{\overline{u_i}}\,\widehat{\overline{u_j}}$ is the Loenard stress, and $T_{ij} = \widehat{\overline{u_i u_j}} - \widehat{\overline{u_i}}\,\widehat{\overline{u_j}}$ is the 'test-level' subgrid scale stress or subtest stress. It is assumed that the subtest stress can also be expressed with eddy viscosity model:

$$T_{ij} - \frac{1}{3}\delta_{ij}T_{kk} = -2C\widehat{\Delta}^2|\widehat{\overline{S}}|\widehat{\overline{S_{ij}}} \tag{10}$$

Incorporating Eqs. 6, 8 and 10 into 9, we obtain an equation for determining C [9]:

$$L_{ij} - \frac{1}{3}\delta_{ij}L_{kk} = \alpha_{ij}C - \widehat{\beta_{ij}C} \tag{11}$$

where,

$$\alpha_{ij} = -2\widehat{\Delta}^2|\widehat{\overline{S}}|\widehat{\overline{S_{ij}}} \tag{12}$$

$$\beta_{ij} = -2\Delta^2|\overline{S}|\overline{S_{ij}} \tag{13}$$

PANS: The filtering procedure in PANS modeling is similar to LES, which separates the flow into resolved and unresolved features. However, while LES uses explicit filtering, PANS uses implicit filtering via the parameters f_k and f_ω (or f_k and f_ε) in such a way that the governing equations resolve a selected portion of the unsteady scales of motion. Thus, the degree of PANS resolution is dependent upon the parameters $f_{k,\omega,\varepsilon}$. These parameters designate the ratio of unresolved to resolved kinetic energy (k), dissipation (ε), or turbulence frequency (ω), which are defined as:

$$f_k = \frac{k_u}{k} \quad f_\omega = \frac{\omega_u}{\omega} \quad f_\varepsilon = \frac{\varepsilon_u}{\varepsilon} \tag{14}$$

In PANS, the sub-filter stress term in Eq. 5 is also modeled with Boussinesq approximation:

$$\tau_{ij} = \tau(V_i, V_j) = \frac{2}{3}k_u\delta_{ij} - \nu_u\left(\frac{\partial U_i}{\partial x_j} + \frac{\partial U_j}{\partial x_i}\right). \tag{15}$$

Where ν_u is the "unresolved eddy viscosity" defined as $\nu_u = k_u/\omega_u = C_\mu k_u^2/\varepsilon_u$. Based on the cut-off parameters defined in Eq. 14, the governing equations for PANS $k-\omega$ model are given by

$$\frac{\partial k_u}{\partial t} + U_j\frac{\partial k_u}{\partial x_j} = P_u - \beta^* k_u\omega_u + \frac{\partial}{\partial x_j}\left[(\nu + \nu_u/\sigma_{ku})\frac{\partial k_u}{\partial x_j}\right], \tag{16a}$$

$$\frac{\partial \omega_u}{\partial t} + U_j\frac{\partial \omega_u}{\partial x_j} = \alpha\frac{\omega_u}{k_u}P_u - \beta'\omega_u^2 + \frac{\partial}{\partial x_j}\left[\left(\nu + \nu_u/\sigma_{\omega_u}\right)\frac{\partial \omega_u}{\partial x_j}\right], \tag{16b}$$

$$P_u = f_k\left(P - \beta^*\frac{k_u\omega_u}{f_\omega f_k}\right) + \beta^* k_u\omega_u, \tag{16c}$$

$$\beta' = \alpha\beta^* - \alpha\frac{\beta^*}{f_\omega} + \frac{\beta}{f_\omega}. \tag{16d}$$

Note that σ_{ku} and $\sigma_{\omega u}$ are given by Eq. 17.

$$\sigma_{ku} = \sigma_k\frac{f_k}{f_\omega}, \quad \sigma_{\omega u} = \sigma_\omega\frac{f_k}{f_\omega} \tag{17}$$

The values of the closure coefficients are as follows: $\beta^* = 0.09, \alpha = 5/9, \beta = 0.075$, $\sigma_k = 2.0$, and $\sigma_\omega = 2.0$.

URANS: In URANS, Eq. 5 becomes the familiar Reynolds stress term [11–13], where the filter $\langle\rangle$ denotes an ensemble average. With two-equation URANS models, the Reynolds stress term is given by:

$$\overline{u_i u_j} = \frac{2}{3}k\delta_{ij} - \nu_T\left(\frac{\partial U_i}{\partial x_j} + \frac{\partial U_j}{\partial x_i}\right) \tag{18}$$

where ν_T is the eddy viscosity, defined as $\nu_T = C_\mu k^2/\varepsilon = k/\omega$. When $f_k = f_\varepsilon = 1.0$ in PANS formulation, we recover the evolution equations for URANS.

3 Numerical Simulation Setup

OpenFoam [14], which is is an open source finite volume code written in C++, is used for numerical simulation of lid driven cavity flow. This code is open source, easily parallelizable and it can be easily modified to include new turbulence models. Out of the many available solvers in OpenFOAM, an incompressible transient solver named PisoFOAM was used with a second order accurate spatial discretization scheme. A backward Euler time stepping scheme, which is first order accurate, was applied for time integration.

The simulations for two cavity aspect ratios of 3:1:1 and 1:1:1 are performed on the TAMU EOS and TACC Lonestar supercomputers. For SAR = 3:1:1, the domain size is 1.0 m × 1.0 m × 1.5 m in x, y and z directions, respectively. Since the flow is symmetrical at z-midplane, the domain length in z-direction is half of the actual cavity length. The validity of using a symmetry plane instead of simulating the entire domain depends on the symmetry of the physical problem. Jordan and Ragab [15] suggest that symmetry assumption may be reasonable for this Reynolds number. They concluded that utilization of the symmetry plane for Re =10,000 at SAR of 3:1:1 introduces averaged error of 2 % compared to the experiment. Therefore, in the current investigation, half of the cavity is simulated for SAR = 3:1:1. However, for SAR = 1:1:1, the entire domain is simulated and symmetry boundary is not invoked.

No slip boundary conditions are specified at all the stationary walls. For the case of SAR = 3:1:1, the symmetry plane is specified with symmetry boundary condition. The moving lid is given a fixed velocity of 1.0 m/s. The eddy turnover time (τ) is thus $\tau = H/U_{lid} = 1$ s in the streamwise direction. It was observed that the primary flow reached a statistically steady state around $t = 60\tau$, thus the time averaged statistics were taken from $t = 60\tau$ to $t = 150\tau$.

4 Results and Discussions

Numerical simulations are performed using both URANS and PANS $k - \omega$ model for SAR 3:1:1 and 1:1:1 cavity at $Re = 10{,}000$ with two different grid sizes of 65^3 and 85^3. For SAR 3:1:1 cavity, PANS simulations with f_k values of 0.6 and 0.8 are performed on the 65^3 grid. For the finer grid of 85^3, PANS calculations with f_k values of 0.25, 0.35, and 0.6 are carried out. For SAR 1:1:1 cavity, two PANS simulations are performed for 65^3 and 85^3 grids at f_k value of 0.5. URANS and LES simulations are also performed for comparison purposes for both cavity aspect ratios.

Figure 3 shows the total mean turbulent kinetic energy along the vertical centerline at the symmetry plane for SAR = 3:1:1. The total kinetic energy is the sum of the resolved and unresolved turbulent kinetic energy. It can be observed that for both the RANS and PANS simulations with different f_k values and grid sizes, the mean turbulent kinetic energy profiles reasonably collapse onto each other. More importantly, Fig. 3 shows that the total turbulent kinetic energy of the system is very small (less than 1 %) compared to the total energy imparted by the lid, indicating the low Reynolds number nature of the flow. The low intensity level implies weakly developed turbulence with a narrow range of turbulence scales.

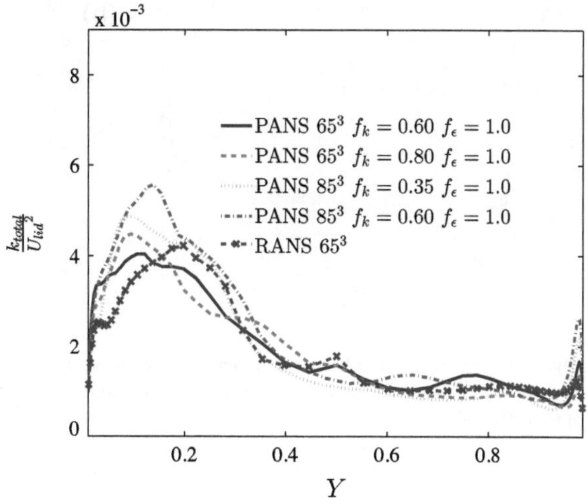

Fig. 3 Total energy $k_{total} = < u'^2 > + < v'^2 > + < w'^2 > + k_u$ variation across the *vertical center line* on the symmetry plane for different sets of simulations performed for 3:1 cavity problem at Re = 10,000

4.1 Mean Velocity Profiles

The time averaged U and V velocities obtained from the simulations are plotted at vertical and horizontal cavity centerlines on the symmetry plane. Figure 4a, b compare the U and V velocity profiles obtained from PANS simulations with URANS, LES and experimental results [1] for SAR 3:1:1 cavity. As seen in Fig. 4, all the simulations captured the main flow features relatively well except for LES Smagorinsky model. For this reason, LES Smagorinsky model is excluded for comparison purposes of the remaining investigations on mean flow and turbulent statistics. Hereafter, all the references to LES dynamic model is denoted as LES.

The results from Fig. 4 indicate that URANS and PANS performs adequately well in predicting the velocity gradients and peak velocities close to the walls. In addition, close agreement with the experimental measurement is observed in the core region for URANS and PANS calculations. For the vertical mean velocity profile, PANS with 85^3 resolved the profile better compared to other models close to the upstream wall. For the U velocity, LES deviates more from experiments close to the bottom wall compared to other models. Similar behavior is seen for the V velocity profile close to the upstream wall for this model. This shows that LES is not very effective compared to PANS and URANS at these coarser grid sizes. Clearly, on finer grids both constant and dynamic LES can be expected to perform much better than URANS, especially for higher Reynolds number flows.

Mean velocity profiles obtained along vertical and horizontal centerlines on the symmetric plane for SAR 1:1:1 cavity are shown in Fig. 5. The mean velocity profiles obtained for this case are very similar to SAR 3:1:1 cavity. For SAR 1:1:1, it was observed that the eddy viscosity ratio $\frac{\nu_t}{\nu}$ in the core region of the cavity is an order of magnitude smaller that the corresponding value for cavity with SAR = 3:1:1 at the same Reynolds number. This is an indication of low turbulence generated

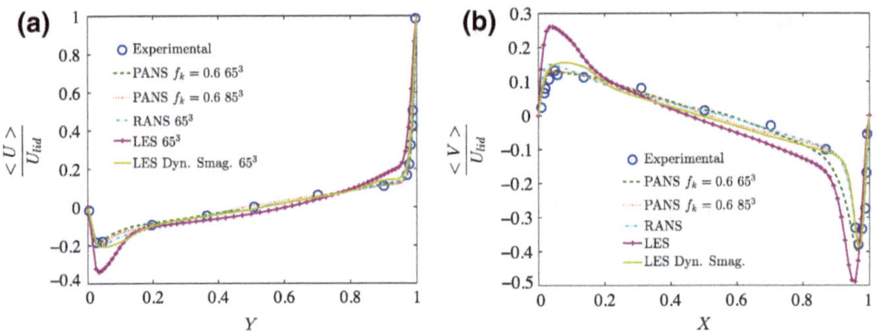

Fig. 4 Comparison of **a** U mean velocity along the *vertical center line* of the symmetry plane, **b** V mean velocity profile across *horizontal center line* of the symmetry plane for SAR 3:1:1 cavity at Re = 10,000

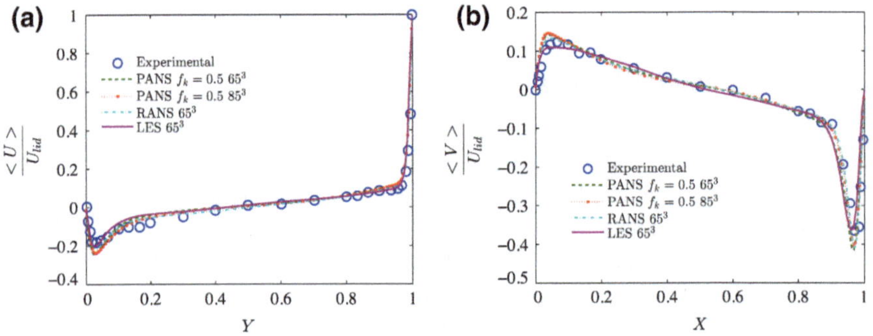

Fig. 5 Comparison of **a** U mean velocity along the *vertical center line* of the symmetry plane **b** V mean velocity profile across *horizontal center line* of the symmetry plane for SAR 1:1:1 cavity at Re = 10,000

for SAR 1:1:1 cavity. This can be attributed to the 'breaking effect' [16] caused by the closer wall distances at lower aspect ratio of cavity. In contrast to the SAR = 3:1:1 case, LES performs well in predicting the peaks and gradients of the mean velocity profiles for the SAR 1:1:1. In fact, it matches well with the experimental measurements closer to the upstream wall and bottom wall for the mean vertical and horizontal velocity profiles, respectively. This may be due to the lower grid resolution needed to resolve the low turbulent flow features at SAR 1:1:1. In addition, PANS and RANS calculations predict the mean velocity profiles and their gradients accurately at this low Reynolds number simulation. This is consistent with the conclusions of SAR = 3:1:1 at Re = 10,000. Figures 4 and 5 demonstrate that for the current aspect ratios studied at Re = 10,000, URANS is able to predict the mean velocity profiles adequately well. The results indicate that PANS simulations yield a slight improvement in the mean velocity profiles over URANS calculation. The difference between URANS and PANS calculations might be more apparent at high Reynolds numbers simulations where a wider range of turbulence length scales are present in the flow. Unfortunately, experimental data for lid driven cavity flow is not available at higher Reynolds numbers and limits the study to the low Reynolds number of 10,000.

4.2 Turbulent Shear Stress

The total turbulent shear stress is computed by adding the unresolved and resolved components. The resolved value is calculated based on the following equation

$$< u'v' >_{resolved} = < UV > - < U >< V > \tag{19}$$

where $< U >$ and $< V >$ are the time averaged resolved velocities. The unresolved part $< u'v' >$ is calculated from Boussinesq's approximation and is time averaged during the simulation. Only the case of SAR = 1:1:1 is considered here as experimental data is not available for the other case.

Figure 6 shows the total shear stress profiles along the cavity centerlines for different models. As seen from the experimental data, the shear stress peaks close to the walls. Variation of shear stress along the vertical line is shown in Fig. 6a. The results show that none of the simulations capture the peak precisely. However, the trends for shear stress variation along the centerline are well captured by both the PANS cases shown. For the variation along the horizontal centerline, shown in Fig. 6b, the peak near the upstream wall is negative and small compared to the peak near the downstream wall. PANS model $f_k = 0.5$ for 65^3 grid predicts the peaks well compared to other models. In addition, URANS calculation is not able to capture the peak values of shear stress close to the walls. It is interesting to note that, URANS does capture all the peaks and valleys in the interior of the domain reasonably well.

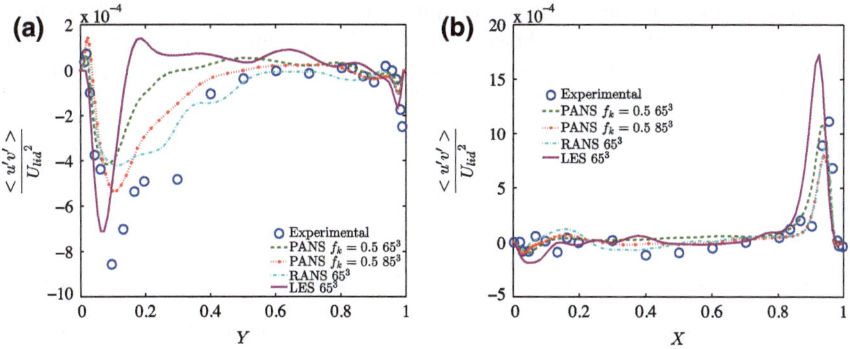

Fig. 6 Comparison of total turbulent shear stress $< u'v' > = < u'v' >_{resolved} + < u'v' >_u$ **a** along the *vertical center line* of the symmetry plane **b** along the *horizontal center line* of the symmetry plane, for SAR 1:1:1 cavity at Re = 10,000

The results presented here for shear stress profiles indicate that overall, PANS performs adequately in predicting the trends and peak values throughout the domain. Additionally, URANS calculation is able to resolve the shear stress components in the interior of the domain faily well. As discussed in the previous section, this performance of URANS simulation is attributed to the low turbulence levels presented in this flow. On the other hand, LES fails to reproduce the right behavior of shear stress in both near wall regions and interior of the domain for the grid resolutions studied here.

5 Conclusion

Numerical simulations of lid driven cavity flow for two aspect ratios of SAR = 3:1:1 and SAR = 1:1:1 at Re=10,000 are performed using PANS $k-\omega$ model. The results from PANS simulations are compared against URANS, LES Smagorinsky model and LES dynamic Smagorinsky model. It is important to reiterate that, the near wall resolution in the present simulations may be quite inadequate for LES computations. Nonetheless, it is the objective of the paper to compare PANS and LES on identical grids. LES Smagorinsky model fails to capture the mean velocity profiles accurately for the current grid resolutions and only LES dynamic Smagorinsky model is primarily used for comparison purposes. Overall, PANS simulations adequately capture the behavior of mean velocity profiles and peak values of shear stress along the cavity centerlines compared to other models. The total turbulent kinetic energy obtained by different simulations is shown to be very small compared to the total energy imparted, indicating low Reynolds number nature of the flow. Under the current conditions, the flow is highly unsteady and complex, but does not exhibit a wide range of turbulence scales. Therefore, URANS appears to be nearly as good as PANS or LES simulations in predicting the mean velocity and shear stress profiles.

Acknowledgments The authors would like to thank Kyle Knox for his assistance in the initial phase of this project. The Texas A&M Supercomputing Facility (http://sc.tamu.edu/) is gratefully acknowledged for providing computing resources useful in conducting the research reported in this paper. The work was supported by NASA NRA #NNXA161A.

References

1. Koseff, J.R., Street, R.L.: The lid-driven cavity flow: a synthesis of qualitative and quantitative comparisons. Trans. ASME **106**, 390–398 (1984)
2. Prasad, A.K., Koseff, J.R.: Reynolds number and end-wall effects on a lid-driven cavity flow. Phys. Fluids **1**(2), 208–218 (1989)
3. Piomelli, U.: Large-eddy simulation: achievements and challenges. Prog. Aerosp. Sci. **35**(4), 335–362 (1999)

4. Razi P., Venugopal, V., Jagannathan, S., Girimaji, S.: Partially-averaged Navier-Stokes (PANS) simulations of lid-driven cavity flow—part ii: flow structures. In: 5th Symposium on Hybrid RANS-LES Methods (2014)
5. Lakshmipathy, S., Girimaji, S.S.: Partially-averaged Navier-Stokes method for turbulent flows: k-ω model implementation. AIAA Paper 119:2006 (2006)
6. Germano, M.: Turbulence: the filtering approach. J. Fluid Mech. **238**, 325–336 (1992)
7. Murthi, A., Reyes, D., Girimaji, S., Basara, B: Turbulent transport modelling for pans and other bridging closure approaches. In: Proceedings of V European Conference on CFD, ECCOMAS CFD (2010)
8. Lakshmipathy, S.: Partially averaged Navier-Stokes method for turbulence closures: characterization of fluctuations and extension to wall bounded flows. Ph.D. dissertation, Texas A&M University (2009)
9. Germano, M., Piomelli, U., Moin, P., Cabot, W.H: A dynamic subgrid-scale eddy viscosity model. Phys. Fluids A: Fluid Dyn. (1989–1993), **3**(7), 1760–1765 (1991)
10. Ghosal, S., Lund, T.S., Moin, P., Akselvoll, K.: A dynamic localization model for large-eddy simulation of turbulent flows. J. Fluid Mech. **286**, 229–255 (1995)
11. Girimaji, S.S., Abdol-Hamid, K.S.: Partially-averaged Navier-Stokes model for turbulence: implementation and validation. Number 0502. AIAA (2005)
12. Girimaji, S.S.: Partially-averaged Navier-Stokes model for turbulence: a Reynolds-averaged Navier-Stokes to direct numerical simulation bridging method. J. Appl. Mech. **73**(3), 413–421 (2006)
13. Girimaji, S.S., Suman, S.: Partially averaged Navier Stokes (PANS) method for turbulence simulations: theory and practice. In: Progress in Hybrid RANS-LES Modelling, pp. 29–43. Springer, New York (2012)
14. OpenCFD: OpenFOAM—The Open Source CFD Toolbox—User's Guide, 1.4 edn. OpenCFD Ltd., Bracknell (2007)
15. Jordan, S.A., Ragab, S.A.: On the unsteady and turbulent characteristics of the three-dimensional shear-driven cavity flow. J. Fluids Eng. **106**, 386–389 (1984)
16. Shankar, P.N., Deshpande, M.D.: Fluid mechanics in the driven cavity. Ann. Rev. Fluid Mech. **32**, 93–136 (2000)

Predictive Capability Assessment of the PANS-ζ-f Model of Turbulence. Part I: Physical Rationale by Reference to Wall-Bounded Flows Including Separation

C.-Y. Chang, S. Jakirlic, B. Basara and C. Tropea

Abstract The present work deals with validation of the PANS (Partially-Averaged Navier-Stokes) computational methodology coupled with the $\zeta - f$ RANS model of turbulence [10], formulated and implemented into the CFD software package AVL-FIRE by Basara et al. [2], in a broad range of wall-bounded flows featured by separation, swirling and tumbling motion. The configurations considered in the part I include a fully-developed flow in a plane channel (DNS by Moser et al. [14]) and flow separating from a continuous curved surface (reference database is provided by LES, [7] and [3]). The PANS approach, whose validation represents the prime objective of the present work, provides a seamless transition from the fully-modelled Unsteady RANS to the fully-resolved direct numerical solution (DNS) as the unresolved-to-total ratios of kinetic energy and its dissipation are appropriately varied. In addition, the complementary RANS (by using the $\zeta - f$ model) computations of the flow configurations considered are also performed. The results obtained illustrate the PANS model capability to capture the turbulence unsteadiness leading consequently to a correct prediction of time-averaged flow quantities, unlike its RANS counterpart. The companion part II [4] is concerned with the PANS model application to more complex flow configurations including a tumbling vortex generation and compression and a swirling flow in a tube, the outlet of which is designed as an orifice with eccentric opening.

C.-Y. Chang · S. Jakirlic (✉) · C. Tropea
Institute of Fluid Mechanics and Aerodynamics/Center of Smart Interfaces
Technische Universität Darmstadt, Alarich-Weiss-Straße 10,
64287 Darmstadt, Germany
e-mail: s.jakirlic@sla.tu-darmstadt.de

B. Basara
Advanced Simulation Technologies—AST, AVL List GmbH,
Hans-List-Platz 1, A-8020 Graz, Austria

S. Girimaji et al. (eds.), *Progress in Hybrid RANS-LES Modelling*,
Notes on Numerical Fluid Mechanics and Multidisciplinary Design 130,
DOI 10.1007/978-3-319-15141-0_30

1 Introduction

The employment of an appropriately adjusted RANS (Reynolds-Averaged Navier-Stokes) model describing the dynamics of the residual turbulence in an LES-relevant (Large-Eddy Simulation) computational scheme has experienced growing popularity in the CFD (Computational Fluid Dynamics) community. These so-called hybrid LES/RANS methods aim primarily at a reduction of spatial and temporal resolution as well as remedying the difficulties a conventional LES has with respect to the near-wall treatment. Their goal is to combine the advantages of both particular methods in order to provide a computational procedure that is capable of capturing the organized, large-scale coherent structures and the bulk flow unsteadiness, as encountered in flows involving separation (all the features being beyond the reach of the conventional RANS method), but at affordable costs—it relates especially to flows at higher Reynolds numbers. Following this basic idea, a number of proposals for coupling RANS and LES approaches has been reported: the (seamless) hybrid LES/RANS methods (with an appropriately modified RANS model mimicking a sub-grid-scale (SGS) model in the entire flow domain) and novel Unsteady RANS methods (RANS model plays here the role of a subscale model). The relevant methods have been proposed by Menter and Egorov [13] and Maduta and Jakirlić [12] (see also [11]); both works deal with the so-called SAS strategy (Scale Adaptive Simulations) employing the $k-\omega$-SST and a differential Reynolds stress RANS models as background models respectively, Girimaji [9] and Basara et al. [2] (PANS—Partially-Averaged Navier Stokes), Schiestel and Dejoan [15] and Chaouat and Schiestel [5] (PITM—Partially-Integrated Transport Model). The common feature of all these models is an appropriate modification of the scale-determining equation resulting in a dissipation rate level which suppresses the turbulence intensity towards the residual level and consequent enhancement of the turbulence activity originating from the resolved motion. Herewith, the evolution of the structural features of the turbulence associated with the regions where large coherent structures with a broader spectrum dominate the flow is enabled. Whereas an appropriate dissipation level enhancement in both PANS and PITM methods is achieved by reducing selectively (e.g. in the separated shear layer region) the destruction term in the model dissipation equation (i.e. its coefficient), an additional production term was introduced into the ω-equation (with $\omega \propto \varepsilon/k$ representing inverse turbulent time scale) in the SAS framework. This term is modelled in terms of the von Karman length scale comprising the second derivative of the velocity field ($\nabla^2\mathbf{U}$), which is capable of capturing the vortex size variability.

The work reported here aims at an in-depth validation of the variable-resolution PANS method coupled with the near-wall RANS- $\zeta-f$ model [2] illustrating its physical rationale in wall-bounded flow configurations characterized by boundary layer separation, for which a detailed reference database (obtained by means of Direct Numerical Simulation and highly-resolved LES) including mean flow and turbulence quantities exists. Admittedly, both presently considered configurations, fully-developed flow in a plane channel and flow over a two-dimensional, axisymmetric hill, considered as particularly suitable for studying near-wall turbulence and separation at a continuous curved surface (the latter differing substantially from the fixed-point separation at a sharp-edged configuration), are characterized by lower bulk Reynolds numbers, $Re_m = 13{,}750$ ($Re_\tau = 395$) and $Re_H = 10{,}600$ respectively. However, it doesn't affect the interpretation of the physical rationale of the PANS model. The companion paper demonstrates the PANS method capability in capturing the flow configurations affected by swirl, tumbling motion, its generation and destruction, [4].

2 Computational Method

The continuity and momentum equations governing the velocity field in a three-dimensional, incompressible ($\rho \neq f(x_i)$) and unsteady flow read

$$\frac{\partial \rho}{\partial t} + \rho \frac{\partial \left(\bar{U}_j - U_{bj}\right)}{\partial x_j} = 0 \tag{1}$$

$$\frac{\partial \left(\rho \bar{U}_i\right)}{\partial t} + \rho \frac{\partial \left[\bar{U}_i \left(\bar{U}_j - U_{bj}\right)\right]}{\partial x_j} = -\frac{\partial \bar{p}}{\partial x_i} + \frac{\partial}{\partial x_j}(\overline{\tau}^{\mu}_{ij} + \overline{\tau}^{t}_{ij}) \tag{2}$$

where U_{bi} stands for the velocity of moving boundaries of the computational domain. This velocity and the temporal variation of the fluid density and viscosity relate to the tumbling vortex compression case only (considered in the companion article). All other cases considered presently are computed in a non-moving coordinate system accounting for $U_{bi} = 0$ and ρ, $\mu = const$. In Eq. (2)

$$\overline{\tau}^{\mu}_{ij} = 2\mu \bar{S}_{ij} - \frac{2}{3}\mu \bar{S}_{kk}\delta_{ij}; \quad \bar{S}_{ij} = \frac{1}{2}\left(\frac{\partial \overline{U}_i}{\partial x_j} + \frac{\partial \overline{U}_j}{\partial x_i}\right)$$

represents the viscous stress tensor formulated in terms of the mean rate of strain $\bar{S}_{ij}$. The turbulent stress tensor representing either subgrid-stress tensor $\bar{\tau}_{ij}$ in the LES framework or the Reynolds stress tensor $\overline{u_i u_j}$ (mimicking the subgrid-scale stress tensor) in the PANS-framework ($\overline{u_i u_j}$ represent the fully-modelled Reynolds stress tensor in the conventional (U)RANS framework) is expressed in terms of the mean strain tensor via the Boussinesq relationship:

$$\overline{\tau}^{t}_{ij} \equiv -\rho \overline{\tau}_{ij} = \mu_{SGS}\left(\frac{\partial \overline{U}_i}{\partial x_j} + \frac{\partial \overline{U}_j}{\partial x_i} - \frac{2}{3}\frac{\partial \overline{U}_k}{\partial x_k}\delta_{ij}\right) - \frac{2}{3}\rho \delta_{ij} \overline{\tau}_{kk} \tag{3}$$

$$\overline{\tau}^{t}_{ij} \equiv -\rho \overline{u_i u_j} = \mu_t\left(\frac{\partial \overline{U}_i}{\partial x_j} + \frac{\partial \overline{U}_j}{\partial x_i} - \frac{2}{3}\frac{\partial \overline{U}_k}{\partial x_k}\delta_{ij}\right) - \frac{2}{3}\rho k \delta_{ij} \tag{4}$$

For the determination of the turbulent viscosity within the conventional LES framework the most widely used subgrid-scale model of Smagorinsky, representing a zero-equation model, utilizes the eddy-viscosity of the residual motion:

$$\mu_{SGS} = \rho\,(C_S\Delta)^2\,|\overline{S}|\,;\;\; \Delta = \left(\Delta_x\Delta_y\Delta_z\right)^{1/3};\;\; |\bar{S}| = \left(2\overline{S}_{ij}\overline{S}_{ij}\right)^{1/2} \tag{5}$$

modeled in terms of the representative mesh size (filter width) Δ and the strain rate modulus $|\bar{S}|$; the Smagorinsky constant C_S takes the value of 0.1.

PANS method. The PANS method represents a seamless RANS-LES coupling method, where a RANS-type model mimics a sub-grid-scale model. This so-called bridging method provides smooth and seamless transition from RANS to LES in terms of a "filter-width control parameter" variation, representing the unresolved-to-total ratios of kinetic energy ($f_k = k_u/k$) and its dissipation ($f_\varepsilon = \varepsilon_u/\varepsilon$) which varies within the flow domain, enabling the transition from a fully-averaged computation to a completely resolved simulation. The afore-mentioned filter-width control parameter is pertinent to the spectrum cut-off, which is to be moved continuously from the RANS limit towards the full LES limit by partially integrating the spectrum. In the Basara's et al. [2] PANS model the parameter f_k is implemented in the computational procedure as a dynamic variable, changing at each computational node at the end of corresponding time step, being afterwards used as a fixed value at the same location during the next time step. The equations governing the unresolved turbulence quantities, such as kinetic energy k_u, corresponding dissipation rate ε_u, as well as the ratio $\zeta_u = (\overline{v^2})_u/k_u$ and associated elliptic-relaxation function f are systematically derived from the Hanjalic's et al. [10] RANS- $\zeta - f$ model. This model represents an appropriate upgrade of the Durbin's $\overline{v^2} - f$ model [6] towards increased numerical robustness. Introduction of the $\overline{v^2}/k$ ratio instead of $\overline{v^2}$ itself led to much more suitable wall boundary conditions for the elliptic function f. Furthermore, the production rate of the kinetic energy of turbulence P_k appears in the ζ-equation; it is more convenient to model than the dissipation rate ε_{22} appearing in the $\overline{v^2}$-equation. Interested reader is referred to original references by Hanjalic et al. [10] and Basara et al. [2] for more details. Presently, only the unresolved dissipation rate equation whose destruction term is appropriately modified will be explained more extensively

$$\frac{\mathbf{D}\varepsilon_u}{\mathbf{D}t} = \frac{C_{\varepsilon 1}P_u - C^*_{\varepsilon 2}\varepsilon_u}{T_u} + \frac{\partial}{\partial x_j}\left[\left(\nu + \frac{\nu_u}{\sigma_{\varepsilon u}}\right)\frac{\partial \varepsilon_u}{\partial x_j}\right] \tag{6}$$

Here, the eddy viscosity of the unresolved scales and the corresponding time scale switch take the following forms: $\nu_u = C_\mu \zeta_u k_u T_u$ and $T_u = max[6\tau_{Ku}, k_u/\varepsilon_u]$, with $\tau_{Ku} = (\nu/\varepsilon_u)^{1/2}$ representing the Kolmogorov time scale. The form of the functional dependency in the model coefficients is of decisive importance. This is especially the case with the coefficient multiplying the destruction term in the dissipation equation (as well as the diffusion coefficient $\sigma_{\varepsilon u}$). The appropriate formulation is given by

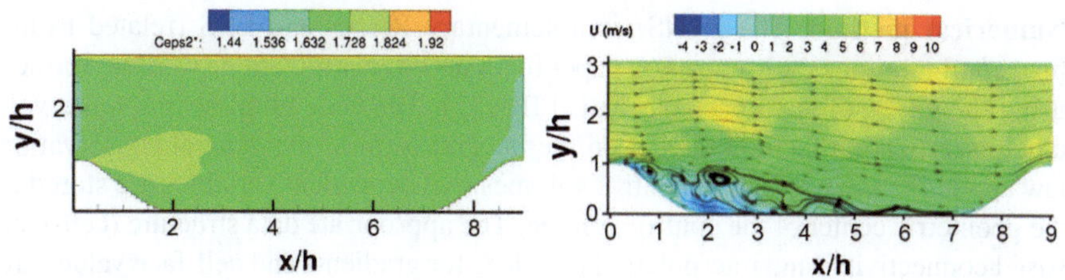

Fig. 1 2D hill flow: iso-contours of the $C^*_{\varepsilon 2}$ coefficient (Eqs. 7 and 8) and instantaneous streamlines

$$C^*_{\varepsilon 2} = C_{\varepsilon 1} + \frac{f_k}{f_\varepsilon}\left(C_{\varepsilon 2} - C_{\varepsilon 1}\right); \quad \sigma_{\varepsilon u} = \sigma_\varepsilon \frac{f_k^2}{f_\varepsilon} \tag{7}$$

with the model constants taking their standard values: $C_{\varepsilon 1} = 1.44$ and $C_{\varepsilon 2} = 1.92$. Such a formulated model coefficient (note its field in Fig. 1, left illustrating the decrease of the value $C^*_{\varepsilon 2} = 1.92$ (corresponding to $f_k = 1$), prevailing in the near-wall region, towards the value $C^*_{\varepsilon 2} \approx 1.6 - 1.7$ in the reminder of the periodic 2D hill flow domain) provides a dissipation rate level which suppresses the turbulence intensity towards the subgrid (i.e. subscale) level in the region where large coherent structures with a broader spectrum dominate the flow, as for instance in the separated shear layer region, allowing in such a way evolution of structural features of the associated turbulence, Fig. 1, right. Herewith, a seamless coupling, i.e. a smooth transition from LES to RANS and opposite is enabled. The parameter f_k is formulated in terms of the grid spacing following Basara et al. [2]:

$$f_k = \frac{1}{\sqrt{C_\mu}}\left(\frac{\Delta}{\Lambda}\right)^{2/3} \tag{8}$$

where Δ represents the grid cell size (defined by Eq. 5) and Λ ($= k^{3/2}/\varepsilon$) is the turbulent length scale. It was assumed in this derivation that the length scale of the smallest resolved eddies, defined in accordance with the Kolmogorov-scale-like formulation as $(\nu_u^3/\varepsilon_u)^{1/4}$, corresponds to the grid spacing Δ. Furthermore, the equality $\varepsilon_u = \varepsilon$ resulting in $f_\varepsilon = 1$ was assumed. This is justified by the fact that a PANS-related grid size is still too coarse to be capable of even approximately approaching the resolution being necessary to resolve the dissipation range (some attempts to propose a variable f_ε parameter have been recently undertaken; see e.g. [8]). The PANS asymptotic behaviour goes smoothly from RANS to DNS with f_k decreasing from the value 1 towards zero value. Accordingly, in the computational procedure used here, the lowest value of the parameter f_k is adjusted to the given grid as it was implemented as a dynamic parameter, changing at each grid node. The values obtained at the end of a time step are used in the following time step; see the field of the model parameter f_k obtained by using the formulation given in Eq. (8) compared with the finally evaluated ratio $f_k (= k_u/k)$ in Fig. 5 and associated discussion.

Numerical method. The PANS, complementary RANS and LES (related to the flow configurations the companion paper focusses on) computations were performed using the CFD software package AVL FIRE [1]. The code employs the finite volume discretization method resting on the integral form of the general conservation law applied to the polyhedral control volumes. All dependent variables are stored at the geometric center of the control volume. The appropriate data structure (cell-face based connectivity) and interpolation practices for gradients and cell-face values are introduced to accommodate an arbitrary number of cell faces. The convection can be approximated by a variety of differencing schemes; the present channel and 2D hill flow configurations were computed by using the 2nd order central differencing scheme (CDS) employed in a differed correction manner. The diffusion is approximated using central differencing. Temporal discretization is accomplished by the 2nd order accurate Crank-Nicolson scheme. The overall solution procedure is iterative and is based on the SIMPLE-like segregated algorithm, which ensures coupling between the velocity and pressure fields.

3 Results and Discussion

The predictive capabilities of the PANS model in capturing the near-wall turbulence and the flow separation from a curved continuous surface are tested by computing a fully-developed channel flow and the flow over a periodical arrangement of the smoothly contoured 2-D hills. In both cases periodic inlet/outlet boundary conditions have been applied with the streamwise pressure gradient imposed in accordance with the relevant Reynolds number. The governing equations are solved in conjunction with the exact wall boundary conditions (integration to the wall itself) implying the wall-adjacent computational node situated in the viscous sublayer. Selected results shown in following figures illustrate the unsteady nature of these flow configurations returned by the present PANS eddy-resolving model, being beyond the reach of its RANS- $\zeta - f$ counterpart. The model capability to account for the large-scale structures and bulk unsteadiness led consequently to a correct prediction of time-averaged flow quantities. Some integral flow characteristics (as e.g., friction coefficient), mean velocity and turbulence quantities are illustrated.

3.1 Fully-Developed Channel Flow

The fully-developed flow in a plane channel represents the most important flow configuration for studying the near-wall turbulence. The solution domain adopted for the presently studied channel flow at $Re_\tau = 395 (L_x \times L_y \times L_z = 4h \times 2h \times 2h$; with h representing the half channel width) was meshed by $(N_x, N_y, N_x) = (64, 100, 64)$ grid cells, implying the resolution corresponding to $\Delta x^+ = 25$ and $\Delta z^+ = 12.5$; the height of the wall-next grid cell is $\Delta y^+ = 1.4$. Maximal value of the ratio

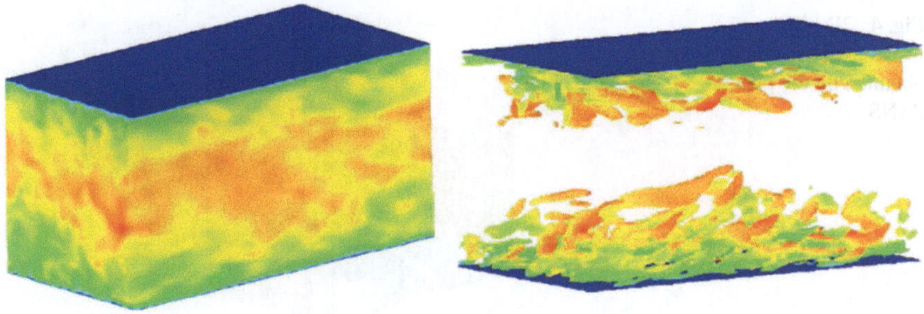

Fig. 2 Channel flow: structural flow properties visualized by instantaneous velocity field (*left*) and Q-criterion (*right*; $Q = 0.005\,s^{-2}$)

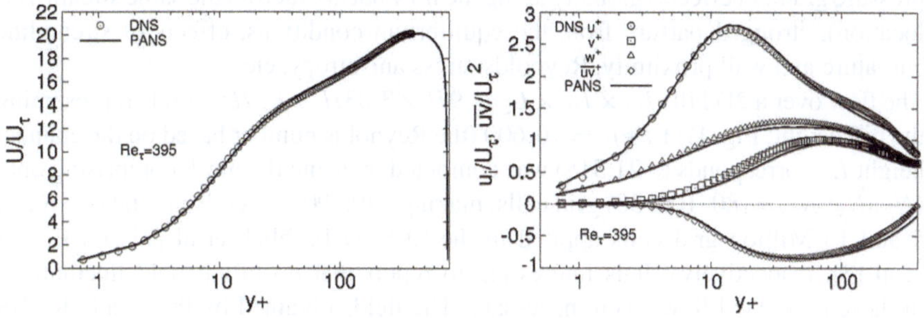

Fig. 3 Channel flow: mean velocity and Reynolds-stress components

Δ/η_k corresponds to 16 in the near-wall region. Figure 2 illustrates the instantaneous flow field obtained by the present PANS method started from the mean flow and turbulence fields obtained by the $\zeta - f$ model within the steady RANS framework. The profiles of the mean velocity and all four Reynolds stress components exhibit very good agreement with the DNS results of Moser et al. [14], Fig. 3. Herewith, good predictive capabilities of the present PANS model in capturing the fluctuating turbulence also in such a globally stable flow with no inherent forcing are illustrated. It is furthermore demonstrated that even an eddy-viscosity model, if appropriately adjusted as the constituent of a hybrid LES/RANS modelling strategy, can correctly return all Reynolds stress components.

3.2 Periodic Flow Over a 2D Hill

Beside the streamwise periodicity this flow configuration (Fig. 4) exhibits a number of features typically associated with a separating flow: boundary layer separation from a continuous curved surface, reattachment, highly-unsteady shear layer separating the main stream from the recirculation flow, relaxation in the post-reattachment

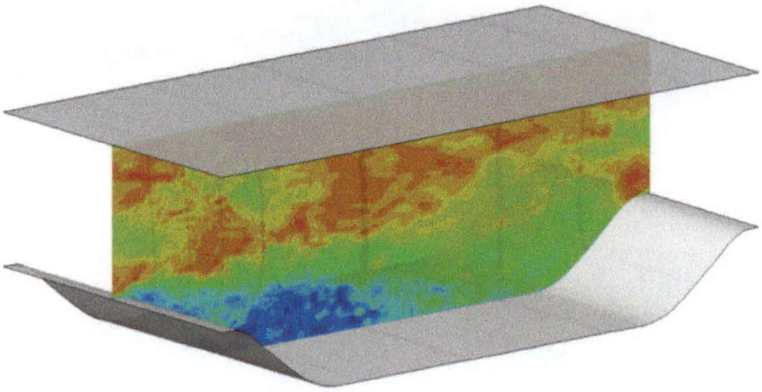

Fig. 4 2D hill: instantaneous velocity field obtained by the present PANS

region, alternating adverse (flow deceleration) and favourable (flow acceleration) pressure gradient effects (globally along the flow but also across the same streamwise location), strong departure from the equilibrium conditions, effects of streamline curvature and wall proximity, Reynolds-stress anisotropy, etc.

The flow over a 2D hill ($L_x \times L_y \times L_z = 9H \times 3.03H \times 4.5H$; with H representing the hill height; Fig. 4) at $Re_H = 10{,}600$ (the Reynolds number based on the channel height L_y corresponds to 21,518) was computed by using the mesh comprising only $(N_x, N_y, N_z) = (80, 100, 30)$ grid cells, making 240,000 grid cells in total (we recall 5 and 13 Million grid cells applied in the LES by Fröhlich et al. [7] and Breuer et al. [3], respectively). It is interesting to report that no initial field fluctuations in these periodical flows were necessary. The fields obtained by the steady RANS computations using the $\zeta - f$ model served for the initialization of the computations with the present PANS- $\zeta - f$ formulation.

The 2D hill flow represents the configuration featured by the flow separation at a curved continuous surface characterized by a high level of natural instability, originating primarily from the highly intermittent separation region which oscillates over a wider wall area. Consequently, a typical outcome of any RANS model (we recall that the application of a conventional RANS model applied in the Unsteady RANS framework results in a steady solution), almost independent of the modelling level, is a too low intensity of the turbulence activity in the separated shear layer resulting in a much larger recirculation zone, see e.g. Wang et al. [16]. Important prerequisite for capturing the flow topology correctly is the reproduction (at least to a certain extent) of the fluctuating turbulence pertinent to the highly unsteady separated shear layer, the feature being beyond the conventional RANS framework. Figure 4 displays the instantaneous velocity field illustrating the capability of the present PANS model to return the fluctuating turbulent flow field.

Figure 5 illustrates the spatial variation of the coefficient f_k denoting basically the resolution parameter defined as the ratio of the unresolved kinetic energy of turbulence to its total counterpart ($f_k = k_u/k$). This represents the key parameter in the PANS computational methodology indicating the potential of the numerical mesh, in conjunction with the turbulence model applied, to resolve the structural properties of the associated turbulence to an appropriate extent. Accordingly, the model has to be appropriately modified to be "grid-dependent", as in the present PANS method,

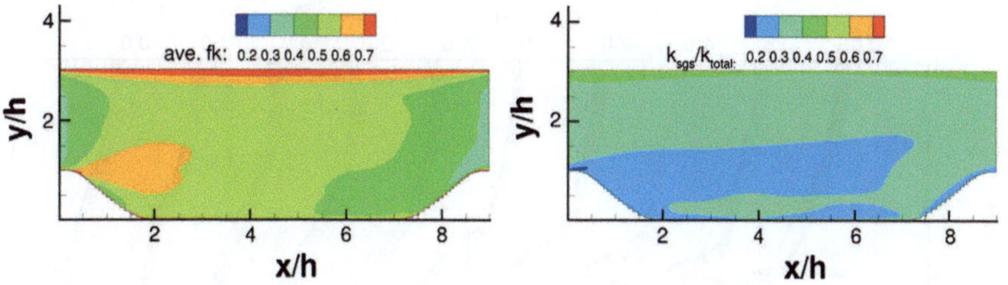

Fig. 5 2D hill: modelled-to-total ratios of the kinetic energy of turbulence—*(left)* f_k obtained by using the formulation (Eq. 8) and *(right)* $f_k(= k_u/k)$ obtained by evaluating the final results

to enable its "interaction" with the grid resolution. Figure 5, left illustrates the f_k-field determined by using the formulation given by Eq. (8). It represents a modelling input, actually the PANS model coefficient expressed by a relevant functional dependency, formulated in terms of grid spacing Δ (Eq. 5; pertinent to the given grid resolution). The value $f_k \approx 1$, characteristic for the very immediate wall vicinity (implying the RANS operating mode of the PANS method), decreases towards $f_k \approx 0.6$ pertinent to the separated shear layer (see also discussion associated with Fig. 1). Such a formulated f_k is directly adjusted to the present mesh size. Figure 5, right represents the parameter f_k evaluated from the finally computed flow field. Also here the characteristic decrease of the f_k value by approaching the separated flow region is obvious. It is important to note appropriately lower level of the finally evaluated f_k-parameter compared to the modelled f_k (Eq. 8). This result represents the desired outcome of the PANS method which implies that the appropriately smaller values of the f_k-parameter are supported by the numerical mesh applied, or with other words, the finally evaluated f_k-parameter is consistent with the underlying grid resolution. The f_k-parameter takes values between 0.2 and 0.5 in the largest portion of the flow field (see Fig. 5, right), implying that about (20–50) % of the turbulence energy spectrum has been modelled; this would certainly be too coarse for an LES (one cannot even talk about LES when only 50 % of the turbulence energy spectrum is to be resolved), but not for the PANS method. Due to application of a sophisticated RANS-based, sub-scale modelling, coarser grid resolutions, commensurate with the spectral cut-off position at the very beginning of the sub-inertial region or even in the productive region, can also be applied. Finally, the denotation PANS staying for "Partially-Averaged Navier-Stokes" is equivalent to "partially-resolved" turbulence. The fine vortex structures cannot certainly be captured if the modelled turbulence fraction would be as high as 50 %, but the "partially-resolved" vortex structures, representing the outcome of the PANS method under given conditions, can certainly be sufficient to enable correct reproduction of the time-averaged flow field and associate turbulence quantities. This represents finally the prime objective of any hybrid LES/RANS method. The lowest f_k values (about 0.2) found in the lower wall region (Fig. 5, right) are in accordance with a highly intensified turbulence activity in the immediate near-wall region within the recirculation bubble and the post-reattachment region (see corresponding discussion in [7]).

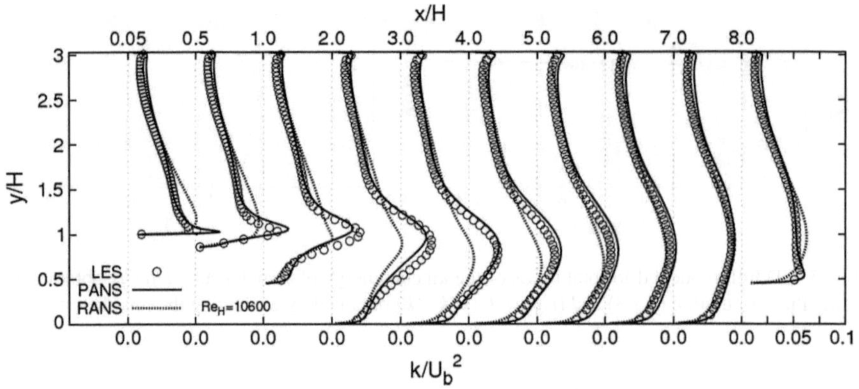

Fig. 6 2D hill: evolution of the turbulent kinetic energy profile

The afore-described situation concerning a significantly underpredicted level of the turbulence kinetic energy and shear stress component representing the outcome of the application of the background $\zeta - f$ RANS model is clearly illustrated in Figs. 6 and 7. The turbulence activity in the separated shear layer related especially to the narrow region around the separation point is appropriately intensified after application of the presently investigated PANS model resulting finally in the correct reproduction of the kinetic energy of turbulence and shear stress magnitudes and of the recirculation zone size.

The latter is indicated by the velocity field development (Fig. 8) and the friction factor evolution, Fig. 9. The computationally obtained position of the reattachment point (length of the separation bubble) defined by the zero C_f-value $(x/H)_{RP} = 4.98$ compares well with the reference LES result: $(x/H)_{RP} \approx 4.72$; we recall the RANS outcome corresponding to $(x/H)_{RP} = 5.70$.

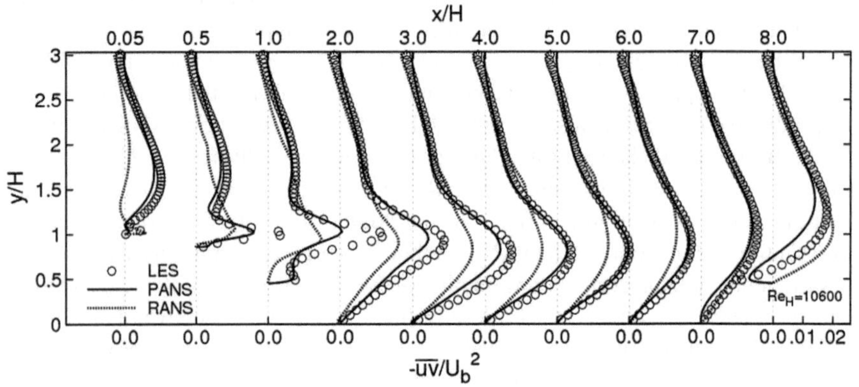

Fig. 7 2D hill: evolution of the shear stress component profile

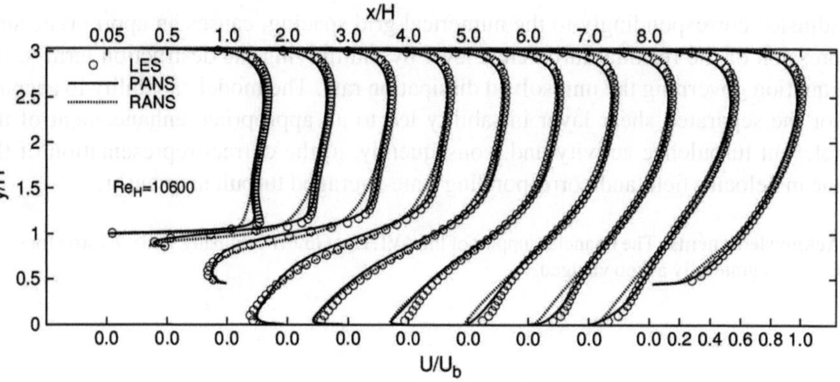

Fig. 8 2D hill: evolution of the mean velocity profile

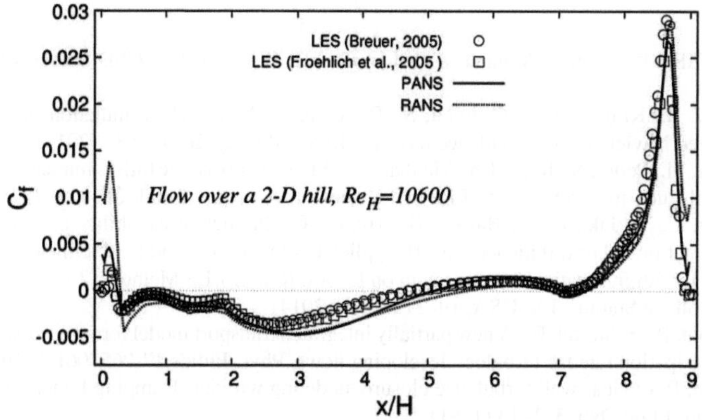

Fig. 9 2D hill: friction factor evolution

4 Conclusions

The functional principle of the PANS (Partially-Averaged Navier Stokes; Basara et al. [2]) method is computationally illustrated by reference to fully-developed channel flows with flat walls and a series, in a periodical manner, of axisymmetric smoothly-contoured hills, implying the boundary layer separation from a curved continuous surface. The presently applied PANS method employs a RANS model relying on the transport equation governing the ratio of the normal-to-the-wall Reynolds stress component $\overline{v^2}$ to the kinetic energy of turbulence k: $\zeta = \overline{v^2}/k$, Hanjalic et al. [10], being capable of capturing the near-wall Reynolds-stress-anisotropy. In both configurations considered presently, the initial mean flow and turbulence fields correspond to the results obtained by the complementary steady $\zeta - f$ RANS model with no fluctuations imposed. The resolution parameter f_k (taking the values between 0 and 1), adjusted correspondingly to the numerical grid spacing, causes an appropriate suppression of the residual turbulence level by multiplying the destruction term in the equation governing the unresolved dissipation rate. The model capability to account for the separated shear layer instability led to an appropriate enhancement of the relevant turbulence activity and, consequently, to the correct representation of the mean velocity field and corresponding time-averaged turbulent quantities.

Acknowledgments The financial support of the AVL List GmbH Company, Graz, Austria for C.-Y. Chang is gratefully acknowledged.

References

1. AVL FIRE Programme Manual: AVL List GmbH, Graz, Austria. CFD Solver Version 2011 (2011)
2. Basara, B., Krajnovic, S., Girimaji, S., Pavlovic, Z.: Near-wall formulation of the partially averaged Navier-Stokes turbulence model. AIAA J. **49**(12), 2627–2636 (2011)
3. Breuer, M., Peller, N., Rapp, Ch., Manhart, M.: Flow over periodic hills—numerical and experimental study in a wide range of Reynolds numbers. Comput. Fluids **38**, 433–457 (2009)
4. Chang, C.-Y., Jakirlic, S., Basara, B., Tropea, C.: Predictive capability assessment of the PANS/ζ-f model of turbulence. Part II: application to swirling and tumble/mean-compression flows. In: 5th International Symposium on Hybrid RANS-LES Methods, Texas A&M University, College Station, TX, USA, 19–21 March (2014)
5. Chaouat, B., Schiestel, R.: A new partially integrated transport model for subgrid-scale stresses and dissipation rate for turbulent developing flows. Phys. Fluids **17**(065106), 1–19 (2005)
6. Durbin, P.A.: Near-wall turbulence closure modeling without 'Damping Functions'. Theoret. Comput. Fluid Dyn. **3**, 1–13 (1991)
7. Fröhlich, J., Mellen, C.P., Rodi, W., Temmerman, L., Leschziner, M.A.: Highly resolved large-eddy simulation of separated flow in a channel with streamwise periodic constrictions. J. Fluid Mech. **526**, 19–66 (2005)
8. Girimaji, S.S.: Variable resolution modelling of turbulence: paradigms of closure. In: 11th World Congress on Computational Mechanics (WCCM XI), 6th European Conference on Computational Fluid Dynamics (ECFD VI), Barcelona, 20–25 July 2014
9. Girimaji, S.S.: Partially-averaged Navier-Stokes model for turbulence: a Reynolds-averaged Navier-Stokes to direct numerical simulation bridging method. J. Appl. Mech. **73**, 413–421 (2006)
10. Hanjalic, K., Popovac, M., Hadziabdic, M.: A robust near-wall elliptic-relaxation eddy-viscosity turbulence model for CFD. Int. J. Heat Fluid Flow **25**(6), 1047–1051 (2004)
11. Jakirlić, S., Maduta, R.: Extending the bounds of "steady" RANS closures: towards an instability-sensitive Reynolds stress model. Int. J. Heat Fluid Flow **51**, 175–194 (2015)
12. Maduta, R., Jakirlić, S.: An eddy-resolving Reynolds stress transport model for unsteady flow computations. In: Fu, S., Haase, W., Peng, S.-H., Schwamborn, D. (eds.) Advances in Hybrid RANS-LES Modelling 4. Notes on Numerical Fluid Mechanics and Multidisciplinary Design, vol. 117, pp. 77–89. Springer, Verlag (ISBN 978-3-642-31817-4) (2012)
13. Menter, F., Egorov, Y.: The scale-adaptive simulation method for unsteady turbulent flow predictions. Part 1: Theory and model description. Flow Turbul. Combust. **85**, 113–138 (2010)
14. Moser, R.D., Kim, J., Mansour, N.N.: Direct numerical simulation of turbulent channel flow up to $Re_\tau = 590$. Phys. Fluids **11**(4), 943–945 (1999)

15. Schiestel, R., Dejoan, A.: Towards a new partially integrated transport model for coarse grid and unsteady turbulent flow simulations. Theoret. Comput. Fluid Dyn. **18**(6), 443–468 (2005)
16. Wang, C., Jang, Y.J., Leschziner, M.A.: Modelling two- and three-dimensional separation from curved surfaces with anisotropy-resolving turbulence closures. Int. J. Heat Fluid Flow **25**, 499–512 (2004)

Predictive Capability Assessment of the PANS-ζ-f Model of Turbulence. Part II: Application to Swirling and Tumble/Mean-Compression Flows

C.-Y. Chang, S. Jakirlic, B. Basara and C. Tropea

Abstract The present work is concerned with the application of the PANS-ζ-f (Partially-Averaged Navier-Stokes) variable resolution model by Basara et al. (AIAA J 49:2627–2636, 2011), formulated in conjunction with the universal wall treatment, to the process of generation and destruction of a tumbling vortex in a square-piston compression machine (investigated experimentally by Borée et al. Phys Fluids 14:2543–2556, 2002) and the swirling flow in a tube generated by two tangential inlets with the outlet geometry resembling an orifice with an eccentrically positioned opening (reference experiment is by Grundmann et al. Int J Heat Fluid Flow 37:51–63, 2012). In addition, the complementary RANS computations (by using the $\zeta - f$ model of Hanjalic et al. Int. J. Heat Fluid Flow 25:1047–1051, 2004), representing also the background RANS formulation in the present PANS model) and LES (in conjunction with the Standard Smagorinsky subgrid-scale model) of both configurations are performed. The PANS-ζ-f model description and its preliminary validation by simulating a fully-developed channel flow and a separating flow over a series of axisymmetric hill-shaped constrictions are given in a companion article by Chang et al. (5th International Symposium on Hybrid RANS-LES Methods, 2014)

1 Introduction

The flows subjected to mean compression and to a strong swirl represent the configurations of high practical relevance [8]. The most prominent examples are piston-cylinder assemblies relevant to Internal-Combustion engines, aircraft gas combustors, turbomachinery, cyclone separators, etc. The related flow structure is

C.-Y. Chang · S. Jakirlic (✉) · C. Tropea
Institute of Fluid Mechanics and Aerodynamics/Center of Smart Interfaces
Technische Universität Darmstadt, Alarich-Weiss-Straße 10, 64287 Darmstadt, Germany
e-mail: s.jakirlic@sla.tu-darmstadt.de

B. Basara
Advanced Simulation Technologies—AST, AVL List GmbH, Hans-List-Platz 1, 8020 Graz, Austria

S. Girimaji et al. (eds.), *Progress in Hybrid RANS-LES Modelling*,
Notes on Numerical Fluid Mechanics and Multidisciplinary Design 130,
DOI 10.1007/978-3-319-15141-0_31

strongly influenced by joint action of different phenomena such as injecting jets, tumbling and swirling motion, wall shear and confinement, expansion and compression. Accordingly, a complex unsteady recirculating flow pattern and cyclic large-scale motion are generated. Because of their practical and theoretical importance, numerous experimental investigations, computational modeling studies and, more recently, direct numerical simulations, (DNS) and large-eddy simulations (LES) of these flows have been reported in the literature. Relevant reviews can be found in Jakirlic et al. [9, 10]. The common practice in predicting respective engineering applications is to use the Reynolds-averaged Navier-Stokes (RANS) models. However, the RANS models are single point closures relying on the assumption of self-similarity of the turbulence spectrum, the fact leading to only one characteristic turbulence length scale, defining the entire spectrum. Consequently, the physics of the flows dominated by the organized, large-scale coherent structures could not be captured satisfactorily in such a way. Therefore, an LES-related, eddy-resolving scheme should be employed in order to correctly capture flow effects mentioned above. Accordingly, the present work focusses on application of a scale-resolving, seamless hybrid LES/RANS method denoted as PANS [3, 5] in conjunction with the universal wall treatment in relevant complex flow configurations featured by swirl and mean compression phenomena.

2 Computational Method

The mathematical description of the PANS method and its physical rationale, demonstrated exemplarily by capturing the near-wall turbulence in a plane channel and flow separation from a curved continuous surface, were the topics of the companion article [5]. Here, only some specific numerical details are given pertinent to the numerical discretization of the computational model equations and especially to the treatment of the near-wall region.

Numerical Method. The PANS, associated RANS and LES computations were performed using the CFD software package AVL FIRE [1] employing the finite volume discretization method, which rests on the integral form of the general conservation law applied to the polyhedral control volumes. The convection can be approximated by a variety of differencing schemes; the presently considered swirling pipe and compression chamber cases are computed by applying the 2nd order MINMOD convective scheme, representing a TVD (Total Variation Diminishing) scheme which combines the CDS and LUDS (Linear Upwind Differencing Scheme). The diffusion is approximated using central differencing. Temporal discretization is accomplished by the 2nd order accurate Crank-Nicolson scheme.

All computational models used presently are applied in conjunction with the so-called universal wall treatment. This method blends the integration up to the wall (exact boundary conditions) with the wall functions, enabling well-defined boundary conditions irrespective of the position of the wall-closest computational node. This method is especially attractive for computations of industrial flows in complex domains where higher grid flexibility, i.e. weaker sensitivity against grid nonuniformities in the near wall regions, featured by different mean flow and turbulence phenomena (flow acceleration/deceleration, streamline curvature effects, separation, etc.), is desirable. Popovac and Hanjalic [11] proposed the so-called compound wall treatment with a blending formula for the quantities specified at the central node P of the wall-closest grid cell as $f_P = f_n e^{-G} + f_t e^{-1/G}$, where '$\nu$' denotes the viscous and 't' the fully turbulent value. The variables f apply here to the wall shear stress, production and dissipation of the turbulence kinetic energy. A somewhat simplified approach was introduced under the name "Hybrid Wall Treatment" in the numerical code AVL FIRE. Whereas the original compound wall treatment of Popovac and Hanjalic [11] includes the tangential pressure gradient and convection, a simpler approach utilizing the standard wall functions as the "upper" bound is used presently, Basara et al. [2].

3 Results and Discussion

The feasibility of the PANS method in capturing the turbulent flow structure modulation due to mean compression and strong swirl imposed is illustrated against experiments of a tumbling vortex generation and destruction in a square-piston compression chamber (Fig. 1) and of a swirling flow in a circular tube whose exit geometry represents an abrupt constriction in the form of an eccentric opening (Fig. 8). The experimental results are made available by Borée et al. [4] and Grundmann et at. [7]. In addition, the PANS results are comparatively assessed with those of RANS and LES computations performed in parallel.

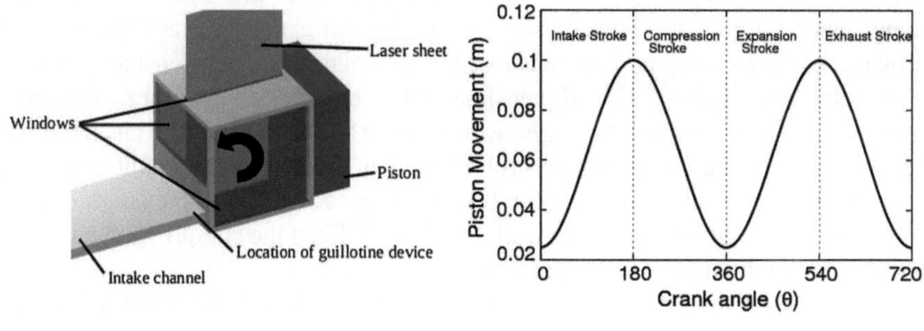

Fig. 1 Schematic of the compression chamber (adopted from Borée et al. [4]) and piston movement kinematics

3.1 Tumbling Vortex Compression in a Square-Piston Compression Machine

Reference database was provided by Borée et al. [4] who designed a square-cylinder (with the cross-section $b \times b = 100 \times 100\,\text{mm}^2$) compression machine equipped with a flat head piston (Fig. 1, left) to investigate the generation and breakdown of a tumbling motion experimentally (a relevant computational study using LES method was performed by Toledo et al. [13]). This simple geometry, in comparison with a realistic Internal Combustion (IC) engine, provides well defined boundary conditions and good optical access. Detailed PIV (Particle Image Velocimetry) data corresponding to the evolution of the vortex induced during the intake process and its consequent compression were generated. The data set comprises fluctuating and phase-averaged (over 120 cycles) velocity and turbulence fields measured in the central plane of the compression chamber. Kinematics of the piston movement is described by a sinusoidal function, Toledo et al. [13]: $a(t) = b - (V_p/\omega)[1 + \cos(\omega t)]$ (Fig. 1, right), with ωt representing the crank angle (CA) and the maximum piston velocity $V_p = 0.809$ m/s and engine speed $\omega = 206\,\text{rpm}$. Accordingly, a and b are the piston positions at a time t and at a time instant corresponding to the Bottom-Dead-Center (BDC). The length of the square cylinder volume at the Top-Dead-Center (TDC)—pertinent to the compression stroke—corresponds to $a_{min} = 25\,\text{mm}$. The piston stroke—the way the piston has to cover until reaching the Bottom-Dead-Center (BDC) at the end of the intake stroke, i.e. at the beginning of the compression stroke—amounts 75 mm. Accordingly, the compression ratio (CR), representing the ratio of the maximum chamber volume ($b^2 a_{max}$; with $a_{max} = 100\,\text{mm}$) to the current one, takes the values between 4 (TDC) and 1 (BDC). The inflow system represents a channel functioning as an "intake/outtake valve" in a four-stroke engine, being opened during the intake stroke (every uneven expansion) and the exhaust stroke (every even compression) and closed in all remaining expansion and compression cycles, see e.g. Fig. 2. The dimensions of this eccentrically positioned channel (see Figs. 1 and 2) are (*length*, *height*, *width*) = (300, 10, 96 mm). The channel flow Reynolds number during the intake stroke, based on the relevant hydraulic diameter, corresponds approximately to 12,000. It is assumed, according to the *length/height* ratio – $300/10 = 30$, that the near-wall flow at the inlet of the compression chamber corresponds to the fully-developed turbulence.

In the case of the present 1-D compression, $U_{bi} = (V_p, 0, 0)$ (see Eq. 2 in the companion article, Chang et al. [5]) with V_p representing the piston velocity. The velocity U_{bi} is to be determined by solving an additional equation describing conservation of space—in accordance to the space conservation law, see e.g. Demirdzic and Peric [6]—simultaneously with the continuity and momentum equations. By assuming that acoustic waves have an insignificant effect on the turbulence (e.g., Reynolds [12]), the fluctuating field can be viewed as being incompressible (divergence free), interacting with a compressed mean flow. Accordingly, the mean gas density change is approximated as being a function of time.

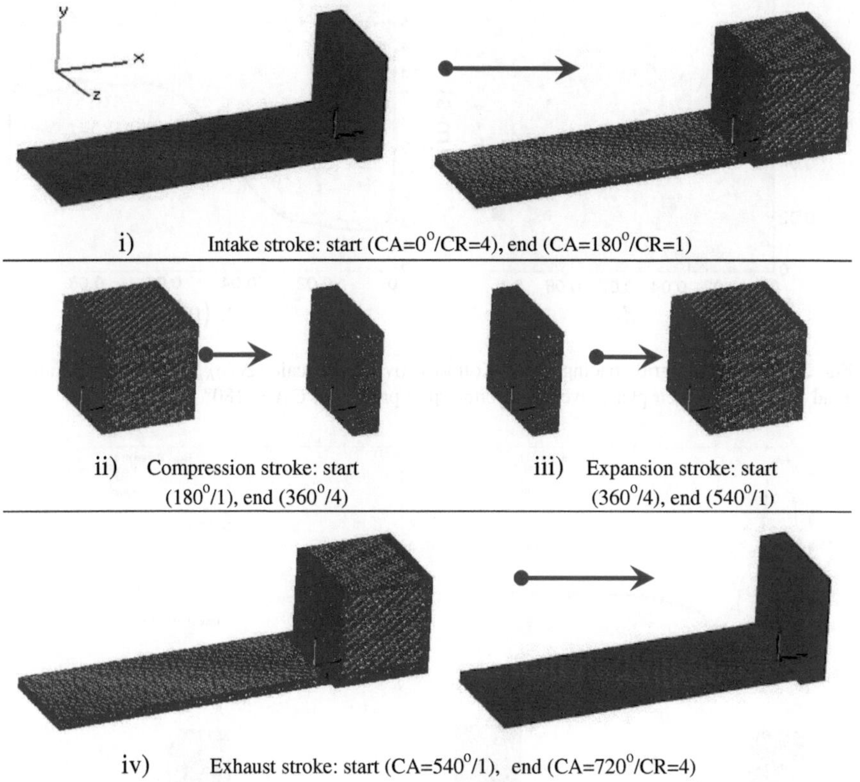

Fig. 2 Temporal variation of the flow domain corresponding to different operating modes of the compression machine

The size of the solution domain (comprising the intake channel and compression chamber) corresponds closely to the experimental configuration. The channel was meshed by 253,080 grid cells in total; the grid size of the compression chamber during the intake/expansion and exhaust/compression strokes corresponds to 544,600 cells ($N_x, N_y, N_z = 80, 85, 80$). The volume of the chamber part of the solution domain accommodating the piston is variable in accordance with the piston movement, see Fig. 2. The maximum of the non-dimensional wall distance values at the wall-next node along the chamber walls are between $y^+ = 0.7$–0.9 (corresponding to $CA = 30°$ and 180°) and 1.3–1.6 (at $CA = 180° - 360°$). Figure 3 displays the field of the ratio of the characteristic grid spacing to the Kolmogorov length scale (Δ/η_K) representing one important grid quality assessment measure. This parameter takes the values well under 10 in the largest part of the flow domain. The initial velocity field was generated by computing five full four-stroke cycles (atmospheric pressure was assumed at the intake channel inlet plane): intake, compression, expansion and exhaust strokes. The phase-averaged results obtained by both LES and PANS methods correspond to ten further cycles. Figure 3 illustrates the mean axial velocity obtained after phase-

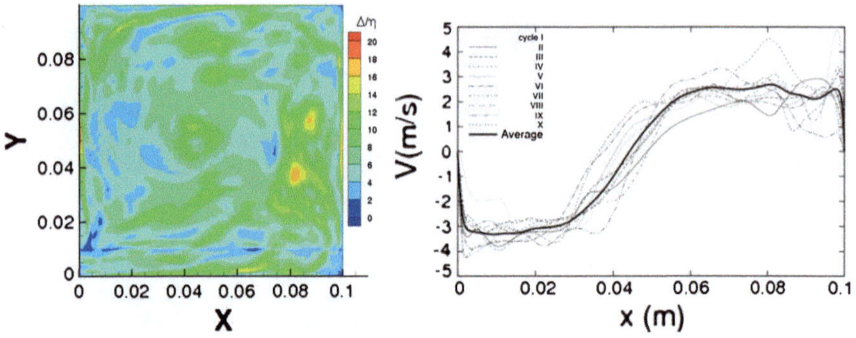

Fig. 3 Ratio of the grid spacing to the Kolmogorov length scale (Δ/η_K) and the instantaneous axial velocity and their phase-averaged counterpart profiles at $CA = 180°$

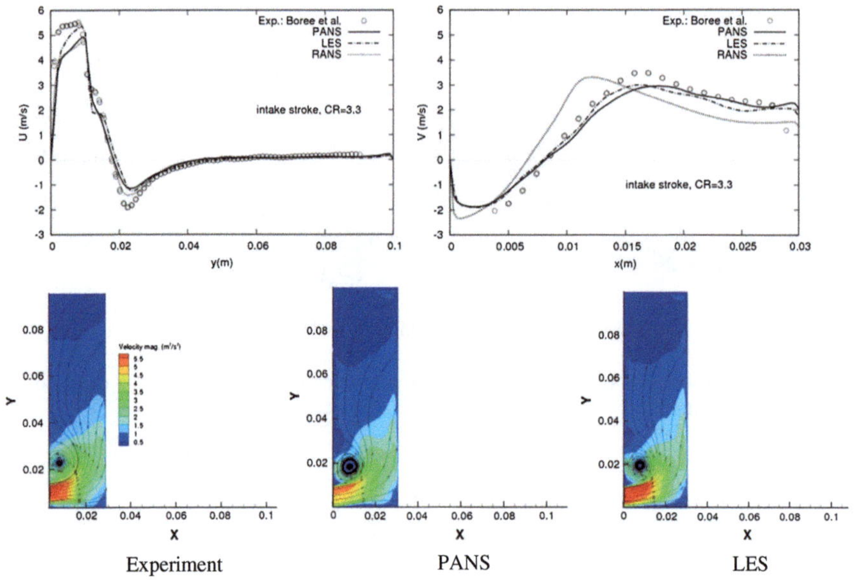

Fig. 4 Intake stroke, CA = 30°, CR = 3.33: mean velocity profiles across the tumble core and phase-averaged mean flow streamlines and iso-contours of the corresponding velocity field coloured by its magnitude

averaging of the instantaneous velocity field after ten cycles in the central vertical plane (z = 0).

Some selected results obtained by applying both LES and PANS at different time instants during the intake and compression strokes are depicted in Figs. 4, 5, 6 and 7. The figures reveal a number of features typically associated with the highly-unsteady jet discharging from the inflow channel, separating from its sharp corners and transforming into a tumbling vortex being characterized by high velocity values.

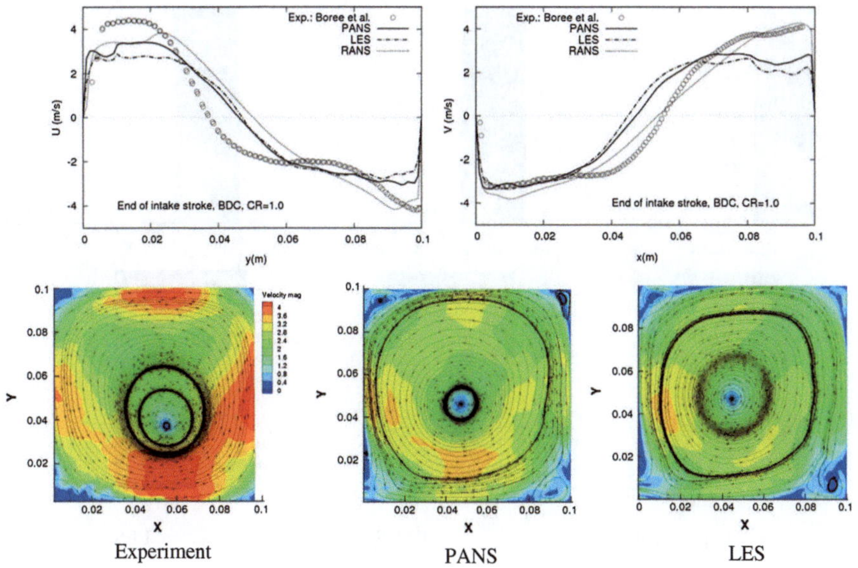

Fig. 5 Bottom-Dead-Center, CA = 180°, CR = 1.0: mean velocity profiles across the tumble core and phase-averaged mean flow streamlines and iso-contours of the corresponding velocity field coloured by its magnitude

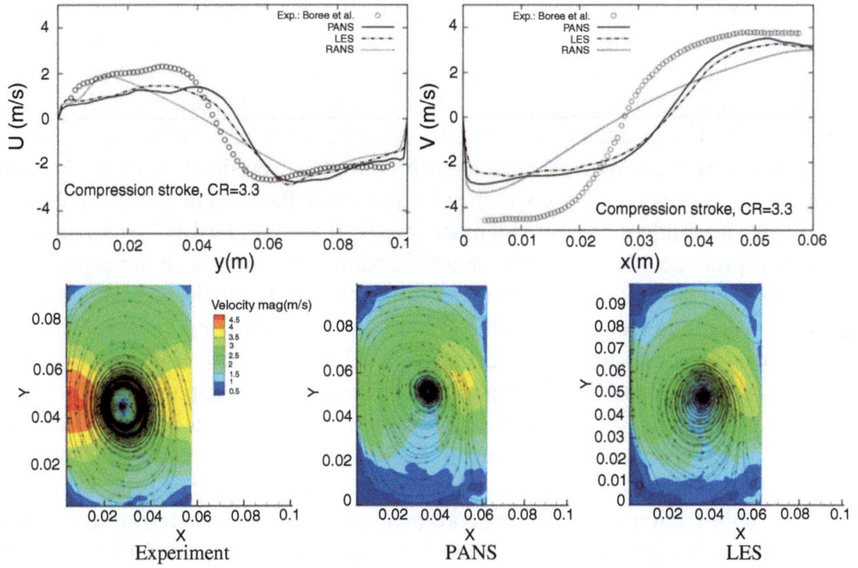

Fig. 6 Compression stroke, CA = 273.8°, CR = 1.67: mean velocity profiles across the tumble core and phase-averaged mean flow streamlines and iso-contours of the corresponding velocity field coloured by its magnitude

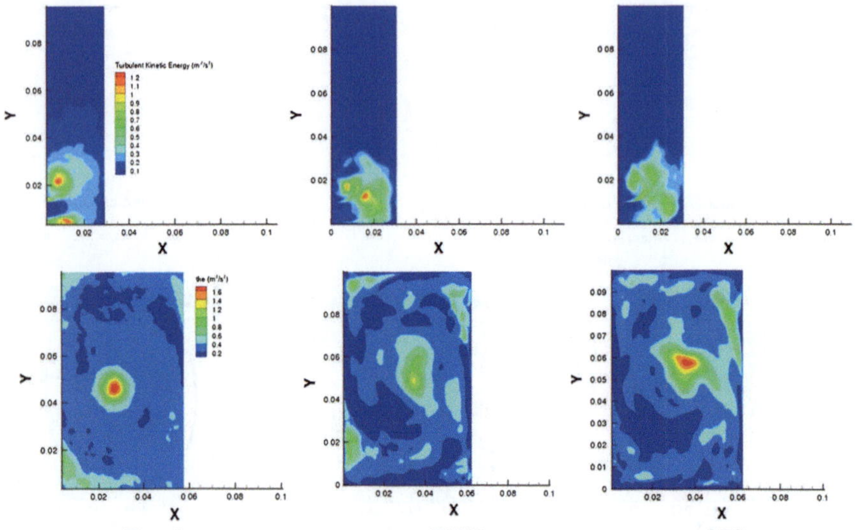

Fig. 7 Intake stroke, CA = 30°, CR = 3.33 (*upper*) and compression stroke, CA = 273.8°, CR = 1.67 (*lower*): iso-contours of the turbulent kinetic energy field

This tumbling motion occupies gradually the entire compression chamber. After onset of the compression stroke its systematic retardation takes place; the most intensive deceleration occurs along the chamber/piston walls propagating up to the vortex core; one notes the flattening of the velocity profiles in the largest portion of the cross-section. The profiles of all variables are depicted across the tumbling vortex core being characterized by the most intensive turbulence production. The maximum of the kinetic energy coincides with the position where the velocity components take zero value. It could be said in summary that both PANS and LES computational methods reproduced the mean flow in a reasonable agreement with the experimental results with respect to both vortex core position and velocity magnitude. This relates especially to the process of the generation of the tumbling motion. Here, the kinetic energy profiles are characterized by dual peaks, originating also from the vortex core fluctuations (Fig. 7, upper). This phenomenon is captured qualitatively by both simulations. Agreement weakens during the compression stroke. Velocity magnitude corresponding to the "annular" region of the tumbling vortex is somewhat underpredicted (Fig. 6). The kinetic energy profiles reveal only one peak indicating a certain stabilization of the vortex core precession, Fig. 7, lower. The turbulence enhancement concentrated to the core region is qualitatively captured although with a substantial underestimation.

3.2 Swirling Flow in a Tube with Eccentric Outlet Orifice

The presently considered case is investigated experimentally by Grundmann et al. [7] using Magnetic Resonance Velocimetry (MRV), providing three-dimensional and three-component velocity distributions within the swirl tube. The swirl tube configuration has important practical relevance as it represents an internal cyclone-cooling flow configuration being the part of the blade cooling system of a gas turbine. The present generic vortex tube with a double-tangential inlet (each having the cross sectional area of 29.33 mm × 8.8 mm swirl generator following a ring-shaped annular diffuser is depicted in Fig. 8. Deionized water, utilized as working medium, enters the inlet pipe (having a constant diameter of 25.4 mm and a length of 157 mm) upstream of the diffuser. The annular diffuser supplies the swirl generator with a constant flow rate. The swirl tube has a length-to-diameter ratio (*L/D*) of 10. The tube's inner diameter (D) is 44 mm. All axial positions are denoted as the dimensionless parameter *z/D*, with $z/D = 0$ denoting the end of the swirl generator and $z/D = 10$ the outlet orifice location. The experimental results are available up to the location $z/D = 3.5$. At the end of the swirl tube, an eccentric outlet orifice geometry, illustrated in Fig. 8, is applied. The experiments are carried out at Reynolds number of 15,000, based on the axial bulk velocity in the swirl tube, the tube diameter as characteristic length and kinematic viscosity of water at 20 °C. This corresponds to the flow rate provided by the pump of 31.5 l/min. The initial swirl intensity entering the pipe corresponds to S = 1.6 (see experimental reference for the definition of the swirl intensity). This flow configuration, belonging to the large length-to-diameter class of the King's classification, is characterized by a strong 'circumferential' interaction between tube wall and vortex flow. Interested readers should consult Grundmann et

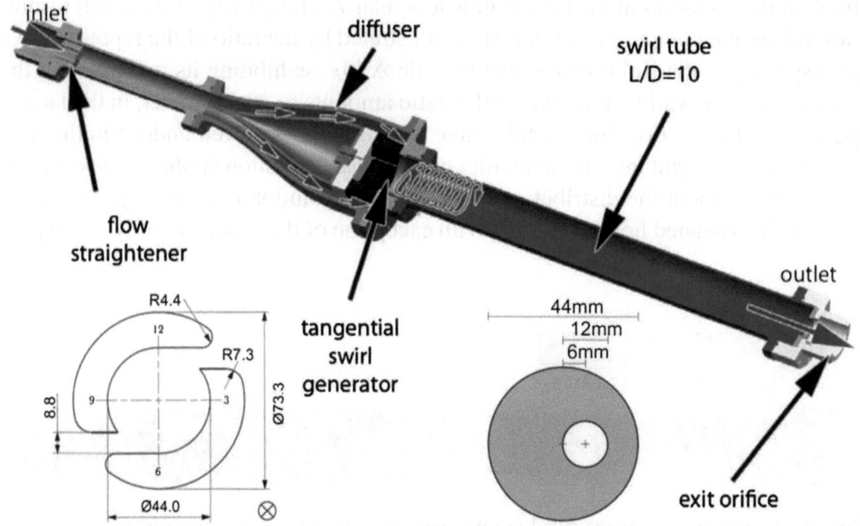

Fig. 8 Flow configuration investigated presently; adopted from [7]

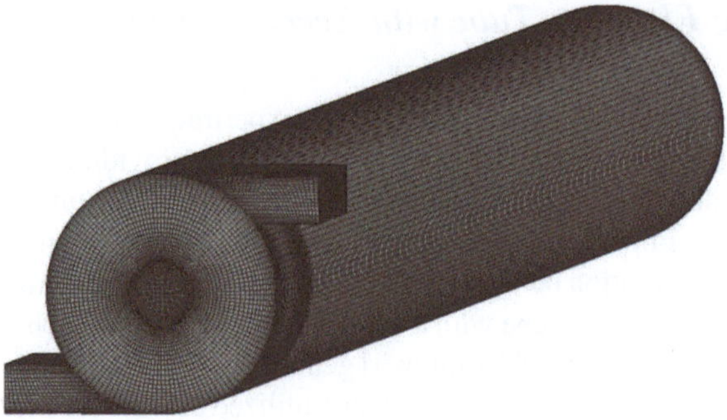

Fig. 9 Solution domain and numerical mesh

al. [7] for more details about the flow configuration considered as well as about the MRV measurement technique.

The computational domain considered presently is displayed in Fig. 9. It is designed as a somewhat simplified geometry compared to the experimental setup with two tangential duct inlets connected to the swirl tube. Fixed mass flow rate 0.258 kg/s is prescribed at each inlet implying a uniform velocity profile of a corresponding magnitude; no fluctuations are imposed at the duct inflow planes. Pressure boundary condition for the velocity field and zero-gradient boundary condition for the equations governing the turbulent quantities are applied at the swirl pipe's outlet plane. Each inlet duct is meshed by $(25 \times 20 \times 40)$ cells in streamwise, normal and spanwise directions. The whole domain is meshed by around 1.2 million hexahedral cells with 160 cells in axial direction, Fig. 9. Between the inlet duct and cylinder, appropriate interpolation is applied to avoid too high aspect ratio of the cells. The maximum y^+-values at the tube wall is less than 7. The quality of the mesh resolution within the computational domain is illustrated by the ratio of the representative grid spacing to the Kolmogorov length scale Δ/η_K, exhibiting its maximum in the immediate tube wall vicinity where this ratio amounts to 40. However, in the largest portion of the solution domain this value is well under 15, even under 5 in the flow core. Further insight into the suitability of the mesh resolution applied can be gained by reporting about the distribution of the model resolution parameter f_k (based on the finally evaluated fields), Fig. 10. With exception of the wall proximity of the pipe

Fig. 10 Distribution of the modelled-to-total-ratio of the turbulent kinetic energy f_k

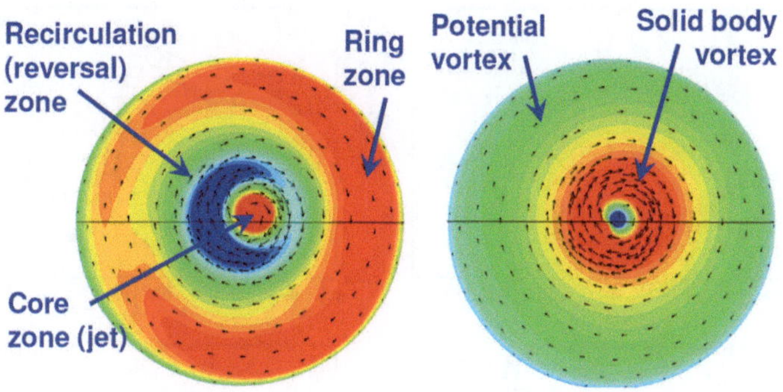

Fig. 11 Iso-contours of the axial (*left*) and tangential (*right*) velocity fields in a cross-section of the swirling pipe illustrating different flow regions

inflow region, where $f_k \approx 0.3–1$, this parameter takes the values around 0.1 in the largest portion of the flow domain. The mean flow and turbulent statistics is obtained after averaging for 5 s of physical flow time after reaching fully-developed turbulent state.

Selected results, relating primarily to the axial and tangential velocity fields, obtained by applying all three presently considered computational methods (LES, RANS and PANS) are shown in Figs. 11, 12 and 13. The computational results are analysed along with the experimental results, being available only up to the streamwise location $z/D = 3.5$. The alternation of the flow direction across the pipe indicating a three-layered structure in the axial velocity distribution induced by swirl in combination with the outlet cross-section contraction is obvious. A fluid jet with high axial velocity develops along the tube's centerline, denoted as core jet. The core jet is the consequence of the annular flow-reversal stream induced by the swirl imposed. It flows back impinging eventually onto the swirl-generator floor (repre-

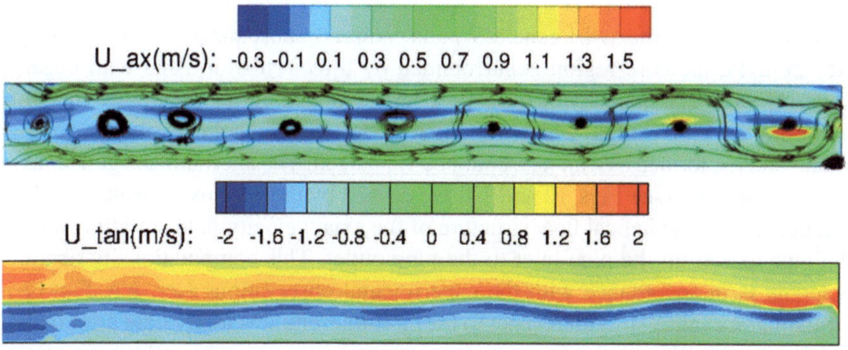

Fig. 12 Mean flow streamlines and axial (*upper*) and tangential (*lower*) velocity contours in the central vertical plane z-r obtained by PANS; pipe length is 10D

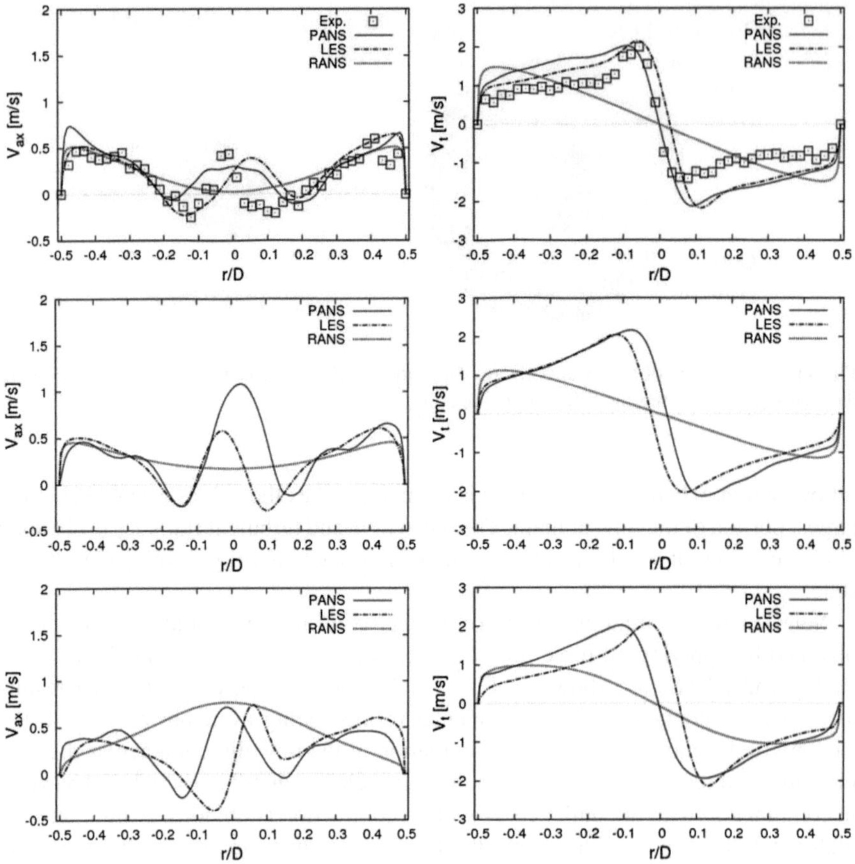

Fig. 13 Axial (*left*) and tangential (*right*) velocity profiles at z/D = 1.75 (*upper*), 5 (*middle*) and 9 (*lower*)

senting the end of the swirl generator) and redirects subsequently into the core jet. Accordingly, the core region is surrounded by a recirculation zone with negative axial velocity. The majority of mass flow rate is moved through the outer ring zone bordering the pipe wall, in which low-intensity left-handed helical structures of high axial velocity, termed as wall jets, could be visualized. The alternating negative 'axial velocity packets' (see Figs. 12 and 13) are caused by the high axial, wall-jet velocities underlying the helical flow structure of the axial velocity field spiralling around the pipe axis from the pipe inlet to the pipe outlet. This computationally obtained outcome is in accordance with the experimental observations (see, e.g. Fig. 17 in Grundmann et al. [7]). With respect to the circumferential velocity field a solid body vortex occurs in the core region, whereas a potential vortex exists in the recirculation and ring zones, Figs. 12 and 13. The experimentally determined velocity component profiles are available only in the entrance region of the swirl pipe at the streamwise position $z/D = 1.75$; some conclusions about their consequent development can be extracted from the three-dimensional representation of the velocity profile evolution up to the position $z/D = 3.5$–4 (we recall here the pipe outlet location corresponding to $z/D = 10$) displayed in the work of Grundmann et al. [7] (not shown here).

Such a cyclone-type flow structure is very different from that encountered in a typical combustion-chamber flow configuration characterized by a swirl-induced recirculation zone situated in the flow core. Accordingly, it represents a great challenge for computational methods. The results obtained by applying LES and PANS methods are in a good qualitative and quantitative agreement with the experimentally available profiles of both axial and tangential velocity implying their correct shape and magnitude. The position coinciding with the pipe radius where a transition from the solid-body-rotation profile shape to the potential-vortex one takes place is correctly reproduced. In the axial flow direction, the intensity of core jet increases towards the outlet orifice, being in accordance with the flow acceleration in the central pipe region caused by the cross section contraction. The flow reversal intensity immediately around the core region weakens after some distance, the occurrence being in line with the core jet magnitude increase. As expected, the circumferential velocity field computed by the RANS method employing the $\zeta - f$ EVM model exhibits a solid body rotation profile shape being typical of the eddy-viscosity modeling group due to the scalar nature of the turbulent viscosity, see e.g. Jakirlic et al. [10].

4 Conclusions

The PANS method, providing a smooth and seamless transition from RANS to LES in terms of a variable model resolution parameter f_k, representing the unresolved-to-total kinetic energy ratio is applied to a flow in a swirling pipe with an eccentrically designed exit orifices and in a square piston-cylinder compression chamber. A complex three-layered structure (core jet, recirculation and ring zone) characterizing the axial velocity distribution and a two-layered structure (solid-body vortex and potential vortex) of the tangential velocity field in a swirling tube and a specific alternation of the axial velocity sign indicated by a series of distinct flow-reversal regions is returned by PANS in good agreement with the experimentally obtained results. PANS computations of a four-stroke rapid compression machine captured the generation of a tumbling vortex and its consequent destruction by a mean compression process with respect to the structural characteristics of the instantaneous and phase-averaged flow field demonstrating its potential in computing flow configurations characterized by a broader frequency range.

Acknowledgments The financial support of the AVL List GmbH, Graz, Austria is gratefully acknowledged. Our special thanks go to S. Grundmann and F. Wassermann (swirling tube) and J. Borée (tumbling vortex compression) for making their experimental data available.

References

1. AVL FIRE Programme Manual: AVL List GmbH, Graz, Austria. CFD Solver Version 2011 (2011)
2. Basara, B., Aldudak, F., Schrefl, M., Jakirlic, S., Hanjalic, K., Tropea, C., Mayer, J.: Experimental investigations and computations of unsteady flow past a real car using a robust elliptic relaxation closure with a universal wall treatment. SAE Technical Paper Series, Paper No. 2007–01-0104 (also in SAE 2007 Transactions, Journal of Passenger Cars—Mechanical Systems, March 2008; Vol. V116–6, ISBN 978-0-7680-1985-8) (2007)
3. Basara, B., Krajnovic, S., Girimaji, S., Pavlovic, Z.: Near-Wall formulation of the partially averaged Navier-Stokes turbulence model. AIAA J. **49**(12), 2627–2636 (2011)
4. Borée, J., Maurel, S., Bazile, R.: Disruption of a compressed vortex. Phys. Fluids **14**(7), 2543–2556 (2002)
5. Chang, C.-Y., Jakirlic, S., Basara, B., Tropea, C.: Predictive capability assessment of the PANS/ζ-f model of turbulence. Part I: physical rationale by reference to wall-bounded flows including separation. In: 5th International Symposium on Hybrid RANS-LES Methods, Texas A&M University, College Station, TX, USA, 19–21 Mar 2014
6. Demirdzic, I., Peric, M.: Finite volume method for prediction of fluid flow in arbitrarily shaped domains with moving boundaries. Int. J. Numer. Methods Fluids **10**, 771–790 (1990)
7. Grundmann, S., Wassermann, F., Lorenz, R., Jung, B., Tropea, C.: Experimental investigation of helical structures in swirling flows. Int. J. Heat Fluid Flow **37**(5), 51–63 (2012)
8. Hanjalic, K., Popovac, M., Hadziabdic, M.: A robust near-wall elliptic-relaxation eddy-viscosity turbulence model for CFD. Int. J. Heat Fluid Flow **25**(6), 1047–1051 (2004)
9. Jakirlić, S., Hadžić, I., Pascal, H., Hanjalić, K., Tropea, C.: Computational study of joint effects of shear, compression and swirl on flow and turbulence in a valveless piston-cylinder assembly. SAE Technical Paper Series, Paper No. 2001–01-1236, pp. 1–40 (also in SAE 2001 Transactions, Journal of Engines, September 2002; ISBN 0-7680-0875-1) (2001)
10. Jakirlić, S., Hanjalić, K., Tropea, C.: Modelling rotating and swirling flows: a perpetual challenge. AIAA J. **40**(10), 1984–1996 (2002)
11. Popovac, M., Hanjalic, K.: Compound wall treatment for RANS computation of complex turbulent flows and heat transfer. Flow Turbul. Combust. **78**(2), 177–202 (2007)
12. Reynolds, W.C.: Modelling of fluid motion in engines—an introductory overview. In: Mattavi, J.N., Amman, C.A. (eds.) Combustion Modelling in Reciprocating Engines, pp. 41–68. Plenum Press, New York (1980)
13. Toledo, M.S., Le Penven, L., Buffat, M., Cadiou, A., Padilla, J.: Large eddy simulation of the generation and breakdown of a tumbling flow. Int. J. Heat Fluid Flow **28**, 113–126 (2007)

Status of PANS for Bluff Body Aerodynamics of Engineering Relevance

Siniša Krajnović and Guglielmo Minelli

Abstract This paper is a review of the use of Partially-Averaged Navier-Stokes (PANS) simulations for various bluff body flows of engineering relevance. The application range from flows around generic bluff bodies such as a finite cylinder to flows around simplified road vehicles, trains and landing gears. Active flow control of bluff body flow is one of important application of PANS discussed in this paper. Judgment of the PANS performance is made in comparison with LES and other numerical techniques such as DES and RANS. The review illustrates the success of PANS and its advantage over LES in several bluff body flows, such as the flow around a rudimentary landing gear and the flow around finite cylinder. It also shows that there is a need of improvement in existing PANS methodology with variable f_k in flows with varying boundary conditions.

1 Introduction

Bluff body flows are characterized by regions of separated flows containing wide spectra of turbulent scales. The only affordable technique to predict bluff body flows is RANS, which relies heavily on turbulence modeling. Despite more than 40 years of development, the RANS turbulence models are not capable of accurate prediction of flows in wakes behind bluff bodies. This is perhaps understandable considering that RANS models are calibrated using flows that are very different from the wake behind bluff bodies. They normally have only one turbulent scale, and it is hard to believe that such a model can represent the turbulent flow behind a bluff body. Large Eddy Simulations (LES) of bluff body flows during the 1990s have boosted optimism among researchers (including the first author) that LES can be part of a solution for

S. Krajnović (✉) and G. Minelli
Department of Applied Mechanics, Chalmers University of Technology, 41296 Gothenburg, Sweden
e-mail: sinisa@chalmers.se

G. Minelli
e-mail: guglielmo.minelli@chalmers.se

S. Girimaji et al. (eds.), *Progress in Hybrid RANS-LES Modelling*,
Notes on Numerical Fluid Mechanics and Multidisciplinary Design 130,
DOI 10.1007/978-3-319-15141-0_32

prediction of bluff body flows. Unfortunately, it was rather soon clear that only flows at relatively low Reynolds numbers could be predicted with LES. In particular, flows with sharp edge separations were amenable for LES, while prediction of pressure induced separation was found to require very fine computational grids. Detached Eddy Simulation (DES) [17] was constructed as a technique that could overcome the poor predictability of the RANS in the wake and the large computational effort of LES in the boundary layer. Although DES has been successful in the prediction of several bluff body flows, it was less successful in other bluff body flows. For example, DES predictions of the wake flow around a simplified vehicle body at strong yaw [6] or the flow around an Ahmed body with a rear slanted surface at an angle of 25° [16] have shown poor agreement with the experimental data. As the DES is a zonal method with RANS in the boundary layers and LES away from the body, it is constructed as a "compromise technique". An example of a "compromise" is in the boundary conditions for the LES region (coming from the RANS solution without physical oscillations). Early DES of bluff body flows often suffered from grid induced separation, which was later improved by modification of the length scale in so called Delayed Detached Eddy Simulation (DDES). This, on the other hand, could cause problems of different kinds, leading to the dependency of the results on the initial solutions [3]. Despite problems with the DES formulation, the technique has become the most common hybrid RANS-LES technique and, as such, has received great attention by the research community which produced substantial improvement of DES. There are several competing flavors of DES (many of them are described in this book by other authors) and there will, certainly, be more research activity in the DES improvement. The authors' main objection for DES in bluff body flows is that DES is a zonal technique. The flow around bluff bodies is not so simple that we can partition it in one RANS region near the wall and one LES part away from the wall. The authors are thinking of a general bluff body flow of engineering relevance for which we do not know the resulting flow. The boundary layer (predicted with RANS in DES) separates from the body, forming a shear layer that, in DES, is predicted with LES rather soon (when the distance from the wall leads to the LES switch). The question is how important the prediction of the dynamics of the boundary layers is for the accurate prediction of the separating shear layers. This is, in particular, interesting in prediction of flow shear layer instability behind bluff bodies, which is the mechanism of formation of the wake region behind bluff bodies. Another interesting question in DES approach is: What happens in the LES region when the computational grid in that region is not sufficient for LES resolution? Could such a situation be helped if RANS eddy viscosity is used in these cells instead of much smaller LES subgrid-scale viscosity? The Partially-Averaged Navier-Stokes (PANS) technique is a bridging technique between RANS and DNS. In the PANS version proposed by Basara et al. [1], the turbulent eddy viscosity is varied gradually from cell to cell and from time step to time step. Our hope is that such an approach in blending the turbulence modeling and scale resolving simulation will be more suitable for predictions of bluff body flows. The aim of the present paper is to present an overview of bluff body predictions with PANS by the first author and his co-workers and to compare predictions with simulations using other techniques (when available).

1.1 Partially-Averaged Navier-Stokes

Partially-Averaged Navier-Stokes (PANS) is a method originally proposed by Girimaji [4] as a bridging technique between RANS and DNS. The switch in PANS is continuous and based on the ratios of unresolved to total kinetic energy and dissipation.

The PANS equations [5] read:

$$\frac{\partial U_i}{\partial t} + U_j \frac{\partial U_i}{\partial x_j} + \frac{\partial \tau(V_i, V_j)}{\partial x_j} = -\frac{1}{\rho}\frac{\partial p}{\partial x_i} + \nu \frac{\partial^2 U_i}{\partial x_j \partial x_j} \tag{1}$$

Here a turbulent velocity V_i field is decomposed into two parts by an arbitrary filter as $V_i = U_i + u_i$. $\tau(V_i, V_j) = -2\nu_u S_{ij} + 2/3 k_u \delta_{ij}$ where k_u and ε_u are the unresolved turbulent kinetic energy, and the dissipation and the eddy viscosity of unresolved scales is given as

$$\nu_u = C_\mu \frac{k_u^2}{\varepsilon_u} \tag{2}$$

in the original PANS $k - \varepsilon$ model by [4]. The resolved stress tensor is given by

$$S_{ij} = \frac{1}{2}\left(\frac{\partial U_i}{\partial x_j} + \frac{\partial U_j}{\partial x_i}\right) \tag{3}$$

The model equations for the unresolved kinetic energy k_u and the unresolved dissipation ε_u are required to close the system of equation given previously.

The PANS $\zeta - f$ model [1] that is used for most of the PANS predictions of the bluff body flows discussed in this review is given by the following equations:

$$\nu_u = C_\mu \zeta_u \frac{k_u^2}{\varepsilon_u} \tag{4}$$

$$\frac{\partial k_u}{\partial t} + U_j \frac{\partial k_u}{\partial x_j} = P_u - \varepsilon_u + \frac{\partial}{\partial x_j}\left[\left(\nu + \frac{\nu_u}{\sigma_{ku}}\right)\frac{\partial k_u}{\partial x_j}\right] \tag{5}$$

$$\frac{\partial \varepsilon_u}{\partial t} + U_j \frac{\partial \varepsilon_u}{\partial x_j} = C_{\varepsilon 1} P_u \frac{\varepsilon_u}{k_u} - C^*_{\varepsilon 2}\frac{\varepsilon_u^2}{k_u} + \frac{\partial}{\partial x_j}\left[\left(\nu + \frac{\nu_u}{\sigma_{\varepsilon_u}}\right)\frac{\partial \varepsilon_u}{\partial x_j}\right] \tag{6}$$

$$C^*_{\varepsilon 2} = C_{\varepsilon 1} + f_k (C_{\varepsilon 2} - C_{\varepsilon 1}) \tag{7}$$

$$\frac{\partial \zeta_u}{\partial t} + U_j \frac{\partial \zeta_u}{\partial x_j} = f_u - \frac{\zeta_u}{k_u} P_u + \frac{\zeta_u}{k_u}\varepsilon_u (1 - f_k) + \frac{\partial}{\partial x_j}\left[\left(\nu + \frac{\nu_u}{\sigma_{\zeta u}}\right)\frac{\partial \zeta_u}{\partial x_j}\right] \tag{8}$$

$$L_u^2 \nabla^2 f_u - f_u = \frac{1}{T_u}\left(c_1 + c_2 \frac{P_u}{\varepsilon_u}\right)\left(\zeta_u - \frac{2}{3}\right) \tag{9}$$

where constants C_μ, c_1, c_2 and $C_{\varepsilon 2}$ are equal to 0.22, 0.4, 0.65 and 1.9, respectively. L_u is the length scale and T_u is the time scale defined by using unresolved kinetic energy:

$$T_u = max\left[\frac{k_u}{\varepsilon}, C_\tau\left(\frac{\nu}{\varepsilon}\right)^{1/2}\right]; \quad L_u = C_L max\left[\frac{k_u^{3/2}}{\varepsilon}, C_\eta\left(\frac{\nu^3}{\varepsilon}\right)^{1/4}\right] \tag{10}$$

where constants C_L and C_τ are equal to 0.36 and 6.0, respectively. We assume that our numerical meshes support the cutoff in the energy containing scales and inertial range and choose $f_\varepsilon = 1$ (i.e. $\varepsilon_u = \varepsilon$). $C_{\varepsilon 1}$ is computed as $C_{\varepsilon 1} = 1.4(1 + 0.045/\sqrt{\zeta_u})$. This represents the four-equation PANS $\zeta - f$ model with enhanced near-wall attributes.

The unresolved-to-total-kinetic-energy ratio f_k in the original PANS model was prescribed as a constant. f_k in the PANS $\zeta - f$ model [1] is a variable parameter that depends upon the grid spacing as follows

$$f_k \geq \frac{1}{\sqrt{c_\mu}}\left(\frac{\Delta}{\Lambda}\right)^{2/3}; \quad \Lambda = \frac{k^{3/2}}{\varepsilon}. \tag{11}$$

k in Eq. 11 is computed as $k = k_r + k_u$ and $\Delta = (\Delta x \Delta y \Delta z)^{1/3}$.

2 Sharp Edge Bluff Body Flows

The first group of bluff body flows that we discuss here is that of the bodies with sharp edge separations. Examples include flow around a cuboid [12] surface mounted cube in Fig. 1 at a Reynolds number of 4×10^4 and flow around a pyramid in Fig. 3 at a Reynolds number of 3×10^4. These flows are dominated by large scale structures originating from separation at the sharp edges. One would expect that such flows are made for LES. However, even with such a relatively simple prediction challenge, an LES requires sufficient resolution of the scales. The prediction of the flow around a surface mounted cube at a Reynolds number of 4×10^4 is shown in Fig. 2. The PANS results using a very coarse grid are clearly better than LES results on the

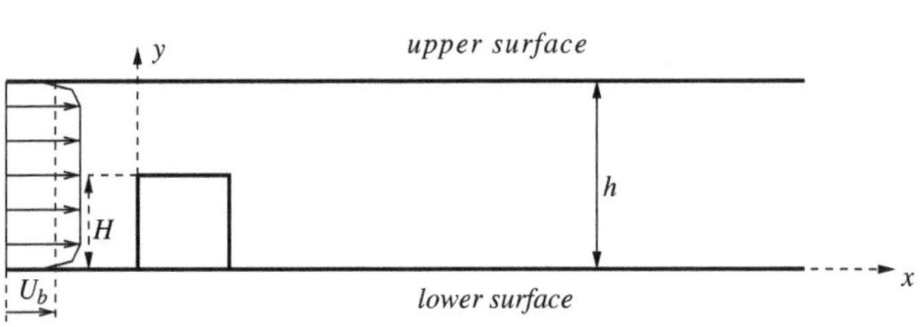

Fig. 1 Geometry of the surface-mounted cube

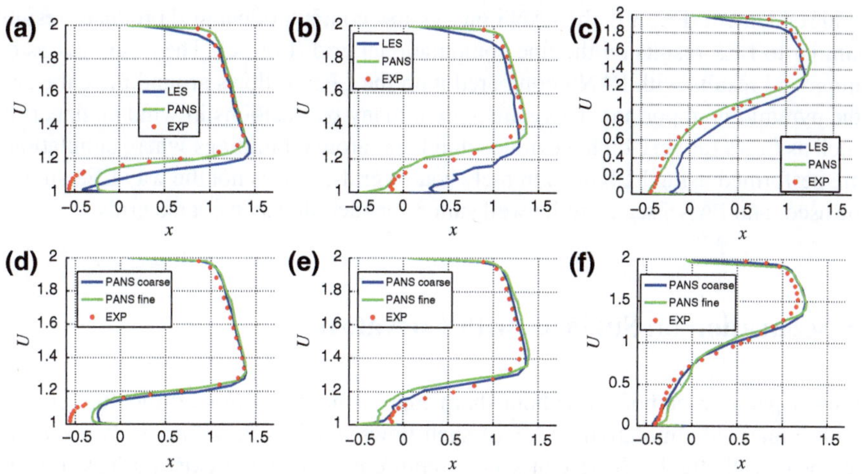

Fig. 2 Comparison of the PANS and LES of the flow around a surface mounted cube. Figure **a–c** Show comparisons of PANS and LES using a coarse grid with 3×10^5 computational nodes. Figure **d–f** show comparisons of PANS using the coarse and the fine (6×10^5 nodes) grids. Locations of the profiles are in the symmetry plane of the cube at positions in the middle of the cube (**a**, **d**), at the trailing edge of the cube (**b**, **d**) and in the wake of the cube (**c**, **f**)

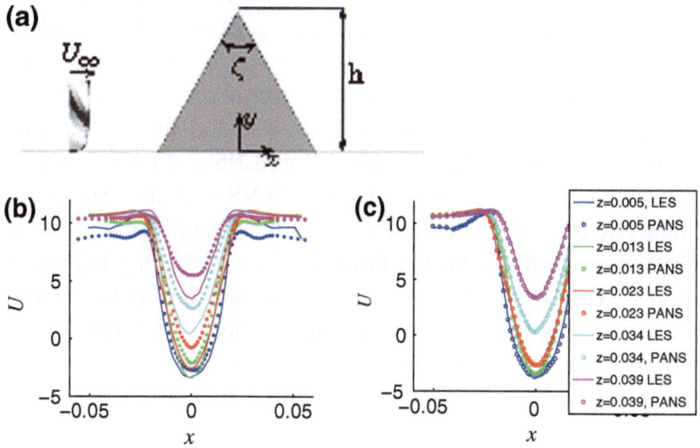

Fig. 3 **a** Geometry of the set-up with the surface-mounted pyramid. Comparison of velocity profiles behind the pyramid between: **b** LES on coarse and fine computational grids. **c** PANS using the coarse grid and the LES using the fine grid

same grid. Figure 2d–f show that the results of PANS do not change a great deal with the grid refinement. Figure 3 shows some results for the velocity profiles behind the pyramid mounted on the ground. The difficulty in predicting this flow is for the region just behind the pyramid tip, as found in [15]. While LES requires a very fine computational grid for accurate prediction and the results vary strongly with grid refinement (Fig. 3b), the PANS on a coarse computational grid produces results similar to LES results on the fine computational grid (Fig. 3c). The observations of improved results with PANS compared with the LES of flows around the cube and the pyramid is important. There are a great many publications in wind engineering such as flows (at very high Reynolds numbers) around buildings where simulations are performed using LES on extremely coarse grids. This is not the way LES should be used, and PANS seems to be well suited for such flows on coarse grids.

3 Separation on Surfaces with Curvature

Flow separation on bodies without sharp edges is normally more difficult to predict with numerical simulations. LES of such flows requires very fine grids resolving the near wall streaks. Note that such a requirement does not exist (at least for the prediction of the position of the separation) in LES with sharp edges. It is the accurate prediction of the boundary layer that governs the prediction of the separation. The first PANS simulations of bluff body flow with separation at the curved surface was that around a finite circular cylinder in [14] using PANS $k - \varepsilon$. Although relatively good results were obtained using PANS $k - \varepsilon$, the improved PANS $\zeta - f$ produced even better results. Flow around a rudimentary landing gear studied in [11] is an example of a flow with both sharp edge separations and smooth surface separations. The Reynolds number of 10^6 based on the diameter of the wheel is very high for wall resolved LES and would require a grid with several hundred million nodes. The computational grids used in [11] had spanwise and streamwise lengths of the computational cells of up to 500 wall units. Both LES and PANS simulations were made. However, the flow predicted by LES and PANS were different. While the LES produced unphysical small scale structures on the lateral sides of the wheels, the PANS prediction resulted in attached flows on the wheels, in good agreement with the experiments. When the grid was made even coarser, the mean PANS results were almost unchanged. This shows that PANS was capable of adjusting to the existing grid, using more turbulence modeling where it was needed.

4 Bluff Bodies in the Wakes of Other Bluff Bodies

Prediction of the flow around one bluff body in the wake of another bluff body is challenging because the prediction of the surrounding flow is strongly influenced by the accuracy of the prediction of the upstream flow. A large number of engineering applications includes combinations of bluff bodies behind each other, such as buildings, convoys of vehicles (platooning) or some flow control devices. A comparison of prediction of two such flows is presented in [7]. Comparisons were made between URANS and PANS, both using $k - \varepsilon - \zeta - f$ turbulence model, and LES using a coherent structure subgrid-scale model. All methods used identical grids created

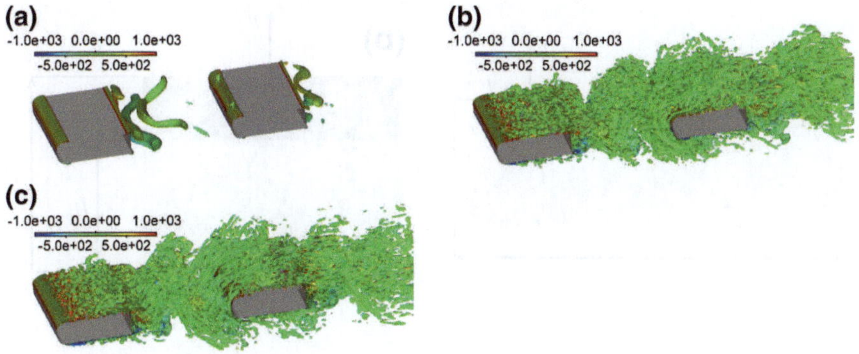

Fig. 4 Iso-surface of the second invariant of the velocity gradient ($Q = 1.0 \times 10^4$) for two D-shaped bodies in platoon: **a** 3D URANS; **b** PANS; and **c** LES, *colored* by pressure

initially for LES. The PANS and the LES simulations produced similar flows in relatively good agreement with the experimental observations. Figure 4 shows that PANS resolves the small flow scales when the grid allows it. The URANS using an identical computational grid predicted only the large flow scales.

5 The Ahmed Body with Slanted Surface at 25°

The flow around the so called Ahmed body with a slanted surface at the back at 25° has been shown in the past to be impossible to predict with methods that rely on turbulence modeling in boundary layers. This is not only true for RANS and URANS but also for DES which uses RANS near walls. The difficulty with this flow is in the prediction of the reattachment of the flow on the slanted surface. PANS is also expected to use RANS in the region near the wall but in a less zonal way than DES. Besides, PANS uses turbulent viscosity that is very different from the corresponding LES subgrid-scale viscosity in the wake region of a DES. Thus, the PANS prediction of this flow is expected to differ from a DES prediction. Two PANS predictions of this flow were made using PANS $k - \varepsilon$ model [2] and PANS $\zeta - f$ model [10]. PANS prediction of passive flow control of the Ahmed body with the rear slant of 25° was explored in [10], where impinging devices in the form of short cylinders shown in Fig. 5 were used for flow control.

Figure 6 shows a comparison of LES and PANS for flow around the rear end of the Ahmed body from [10]. The Reynolds number based on the model length was $Re_L = 1.35 \times 10^6$. Figure 6a, c show a comparison of predictions of the natural flow while Fig. 6b, d show a comparison of predictions of the controlled flow. The separation bubble on the slanted surface of the body in the natural flow predicted using PANS in Fig. 6c is much larger than the one predicted using LES in Fig. 6a (which is in agreement with the experimental observations). Note, however, that the

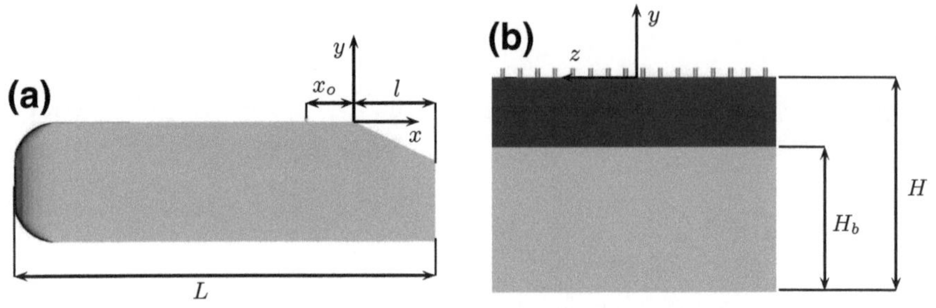

Fig. 5 **a** Geometry of the Ahmed body with impinging devices. **b** Ahmed body seen from behind

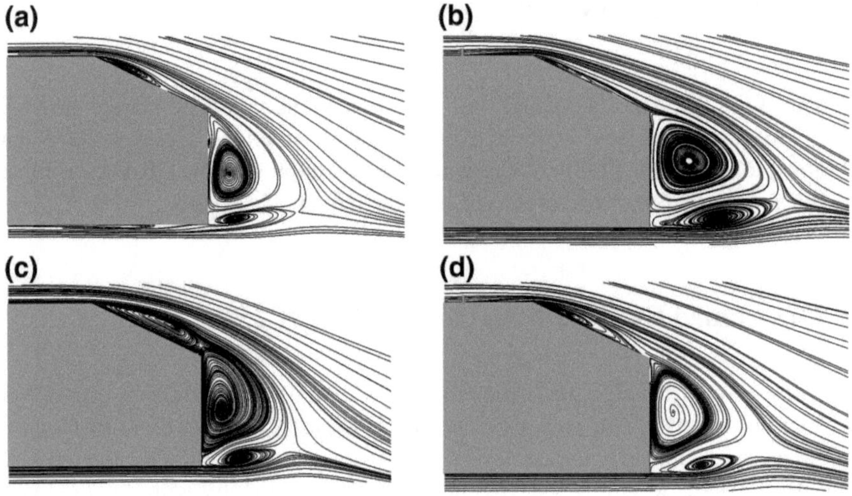

Fig. 6 Streamlines projected onto the symmetry plane: **a** LES natural flow, **b** LES controlled flow, **c** PANS natural flow and **d** PANS controlled flow

influence of the cylindrical impinging devices to decrease the size of the separation bubble was also observed in PANS (compare Fig. 6c, d).

PANS of the flow around the Ahmed body in [2] was found to produce different results depending on the computational grid. In general the results were in better agreement with the experimental data than DES predictions known to the authors. The reattachment of the flow at the slanted surface was predicted when the grid was sufficiently fine. However, the separation bubble on the slanted surface was found to be larger than the one predicted using LES and an identical computational grid. The change of the prediction from the one similar to the URANS prediction to more LES-like results with grid refinement is in agreement with our expectations. However, the difference in the PANS predictions using fine grid (suitable for LES) with experimental observations indicates that PANS has problems with this flow. The existing investigations are not sufficient to draw conclusions about the under-lying reasons for the discrepancies. A possible explanation for the wrong prediction could be that RANS modeling is used in such flow regions as shear layers, where small scale dynamics are important and must be resolved.

6 Flow Control

Numerical prediction of the flow control of bluff body flows is one of the driving forces behind our efforts to develop hybrid RANS-LES methods such as PANS. Bluff body flows are known to have strong Reynolds number dependency, and this is also the case for bluff body flows actuated with different flow control methods. A typical example is the above mentioned Ahmed body flow, for which it was found that optimal parameters for active flow control using periodic blowing and suction change with Reynolds number. LES predictions of flow control at decreased Reynolds numbers are useful for our understanding of flow control processes. However, at operational Reynolds numbers, we will need to rely on a method that includes significant regions of turbulence modelling. The first PANS prediction of active flow control was presented in [8], where flow control around a D-shaped body (2D Ahmed body with square back), with periodic blowing and suction at the rear, was simulated. The key feature of this flow is the shear layer instability leading to interaction of the two shear layers behind the body and the von-Kárman-like vortex shedding. Accurate prediction of the shear layer instability requires prediction of small scale flow in the shear layer. Actuation with blowing and suction complicates flow prediction even further with new flow physics but also possibly decreases accuracy and convergence because of a local increase in the CFL (Courant-Friedrichs-Lewy) number. PANS simulations performed in [8] show a prediction accuracy comparable to the one from previous LES [9]. Despite the relatively good prediction of the present formulation of PANS in [8], with f_k computed using Taylor scale, it is not appropriate for flow with changing boundary conditions such as the one with periodic blowing and suction. The reason for that is found in the need for total turbulent kinetic energy in the expression for the Taylor scale Λ in Eq. 11 and thus f_k. The resolved part of the turbulent kinetic energy in the expression for Λ is computed using time-averaging, and it is not clear how such averaging should be done if the boundary conditions are changing.

7 Simplified Vehicle at Yaw

Flow around a simplified car model (a so called Willy model) at yaw is another bluff body flow that is challenging for numerical prediction. The difficulty here is in the prediction of the separation on a curved surface and in the prediction of the wake flow. Previous simulations using DES, RANS and LES [13] showed difficulties in

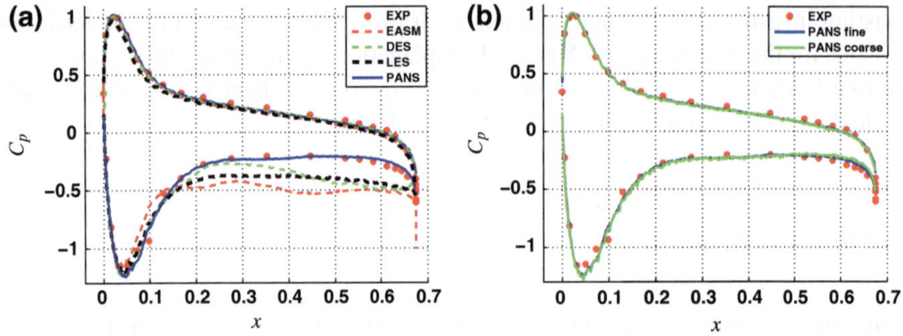

Fig. 7 **a** Comparison of the C_p between PANS, LES, DES and RANS and the experimental data on the surface of the Willy vehicle model at 30° yaw. **b** Comparison of two PANS using fine and coarse grids

predicting the flow on the lee side of the vehicle at a large yaw angle of 30°. Figure 7 shows a comparison of PANS with results of previous simulations for the vehicle at a yaw angle of 30°. The PANS prediction is in very good agreement with the experimental data and in much better agreement than, for example, LES with an identical computational grid. One interesting observation is that the PANS prediction does not change very much when the computational grid is coarsened (Fig. 7). This is a desirable feature, indicating that the grid resolution can be decreased substantially compared with an LES resolution.

8 Conclusions and Discussion

Applications of PANS for various bluff body flows during recent years have shown the potential of this hybrid technique. The review presented here shows several successful PANS predictions of flows where the high Reynolds number of the flow does not allow wall resolved LES. The ability of the method to adjust to the given computational grid is a particularly desirable feature as it allows an engineering approach to the simulation. The idea is that the computational grid will most of the time allow resolution of some turbulent scales. However, at its extreme of a grid resolution typical for RANS, an URANS solution is obtained. Thus the prediction is never worse than an URANS prediction but most of the time better. This should be compared with a LES prediction which, if the grid is too coarse, can result in predictions that are worse than those of even steady RANS simulations. Variable f_k, being a function of the computational grid, is a desirable feature in engineering applications as it allows the method to adapt to the resolution. The formulation of variable f_k from [1] used by the authors has been successful in most applications in bluff body flows. However, its present implementation, with averaging of resolved Reynolds stresses during the calculation, is not suitable for flows with changing boundary conditions such as in active flow control with periodic blowing/suctions, gusty inlet velocity profiles or solid body rotations in the flow. PANS was also shown to have difficulty predicting the flow over an idealized car body (Ahmed body with slanted rear end at 25°), for unknown reasons.

Future research in PANS of bluff body flows should include other applications of bluff bodies and, in particular, the application of combinations of bluff bodies with interacting flows. It is also desirable that we explore more systematically the behavior of PANS with grid refinement/coarsening for bluff body flow. It would be interesting to see whether there is some golden ratio of computer effort (in terms of resolution) and desirable accuracy.

References

1. Basara, B., Krajnović, S., Girimaji, S., Pavlović, Z.: Near-wall formulation of the partially averaged Navier stokes turbulence model. AIAA J. **49**(12) (2011)
2. Basara, B., Krajnović, S., Girimaji, S.: PANS vs. LES for computing of the flow around a 3D bluff body. In: 7th International ERCOFTAC Symposium on Engineering Turbulence Modelling and Measurements, Limassol, Cyprus, 4–6 June 2008
3. Frölich, J., von Terzi, D.: Hybrid LES/RANS methods for the simulation of turbulent flows. Progr. Aerosp. Sci. **44**, 349–377 (2008)
4. Girimaji, S., Srinivasan, R., Jeong, E.: PANS turbulence models for seamless transition between RANS and LES: fixed point analyses and preliminary results. In: ASME Paper FEDSM2003-4336 (2003)
5. Girimaji, S.: Partially-averaged Navier-Stokes model for turbulence: a Reynolds-averaged Navier-Stokes to direct numerical simulation bridging method. J. Appl. Mech. **73**(3), 413–421 (2006)
6. Guilmineau, E., Chikhaoui, O., Deng, G., Visonneau, M.: Cross wind effects on a simplified car model by a DES approach. Comput. Fluids **78**, 29–40 (2013)
7. Han, X., Krajnović, S., Bruneau, C.-H., Mortazavi, I.: Comparison of URANS, PANS, LES and DNS of flows around simplified ground vehicles with passive flow manipulation. In: DLES9 ERCOFTAC WORKSHOP Direct and Large-Eddy Simulation 9, Dresden, Germany, 3–5 April 2013
8. Han, X., Krajnović, S.: Study of active flow control for a simplified vehicle model using the PANS method. Int. J. Heat Fluid Flow **42**, 139–150 (2013)
9. Krajnović, S., Fernandes, J.: Numerical simulation of the flow around a simplified vehicle model with active flow control. Int. J. Heat Fluid Flow **32**(1), 192–200 (2011)
10. Krajnović, S., Han, X.: LES and PANS of passive and active control of flows around generic vehicle bodies. In: Seventh International Conference on Computational Fluid Dynamics (ICCFD7), Big Island, Hawaii (2012)
11. Krajnović, S., Lárusson, R., Basara, B.: Superiority of PANS compared to LES in predicting a rudimentary landing gear flow with affordable meshes. Int. J. Heat Fluid Flow (2012). http://dx.doi.org/10.1016/j.ijheatfluidflow.2012.04.013
12. Krajnović, S., Ringqvist, P., Basara, B.: Comparison of partially averaged Navier stokes and large-eddy simulations of the flow around a cuboid influenced by crosswind. ASME: J. Fluids Eng. **134** (2012)
13. Krajnović, S., Sarmast, S.: Numerical investigation of the influence of side winds on a simplified car at various yaw angles. In: 3rd Joint US-European Fluids Engineering Summer Meeting, Symposium on Issues and Perspectives in Ground Vehicle Flows, FEDSM-ICNMM2010-30766, Montreal, Canada, 1–5 Aug 2010

14. Krajnović, S., Basara, B.: Numerical simulation of the flow around a tall finite cylinder using LES and PANS. In: Progress in Turbulence III. Springer Proceedings in Physics, vol. 131, pp. 115–118 (2010)
15. Krajnović, S., Lárusson, R.: Large eddy simulations of the flow around pyramids. In: CFD Society of Canada Conference, Canmore, Canada (2012)
16. Serre, E., Minguez, M., Pasquetti, R., Guilmineau, E., Deng, G.B., Komhaas, M., Schäfer, M., Frölich, J., Hinterberger, C., Rodi, W.: On simulating the turbulent flow around the Ahmed body: a French German collaborative evaluation of LES and DES. Comput. Fluids **78**, 10–23 (2013)
17. Spalart, P.R., Jou, W.H., Strelets, M., Allmaras, S.R.: Detached-eddy simulation of an airfoil at high angle of attack. In: Advances in DNS/LES, Proceedings of the First AFOSR International Conference on DNS/LES (1997)

Simulation of Smooth Surface Separation Using the Partially Averaged Navier-Stokes Method

Pooyan Razi and Sharath S. Girimaji

Abstract The objective of this study is to investigate the separation of a turbulent flow over smooth surfaces using the Partially Averaged Navier Stokes (PANS) Methodology. The degree of resolution in PANS, which can range from fully resolved to fully modeled turbulence, is parametrized by the ratio of modeled to resolved kinetic energy and dissipation. The flow geometry considered in this study consists of a channel constricted by a series of periodic hills which are spaced 9 hill heights apart from each other. This configuration yields flow separation over a curved surface with well defined flow conditions which makes it ideal as a benchmark case for turbulence closure model validation. Flow Reynolds number based on the bulk velocity above the crest of the hill and hill height is 37,000. Several PANS simulations are performed to study the effect of cut-off length scale and grid size. Turbulence statistics such as averaged mean velocity and Reynolds stresses are compared against well documented experimental and numerical data.

1 Introduction

Flow separation over smooth curved surfaces occurs in many engineering applications such as flow over wings and airfoils, turbine blades, ships, automobile bodies and curved obstructions in pipes. Reliable and accurate modeling of this phenomenon is important for effective and safe design of the aforementioned industrial components. The prediction of flow separation over curved and continuous surface is challenging due to the spatial and temporal fluctuations of the separation line and failure of the law of the wall assumption for separated shear layer regions. Separation from curved

P. Razi (✉) and S.S. Girimaji
Texas A&M University, College Station, TX, USA
e-mail: razi.pooyan@gmail.com

S.S. Girimaji
e-mail: girimaji@tamu.edu

S. Girimaji et al. (eds.), *Progress in Hybrid RANS-LES Modelling*,
Notes on Numerical Fluid Mechanics and Multidisciplinary Design 130,
DOI 10.1007/978-3-319-15141-0_33

surfaces differs from the one from sharp edges in that the point or line of separation is not fixed in space and is very sensitive to external flow properties, turbulence level and development of streamwise pressure gradient [1].

Rapp and Manhart [2] carried out an experimental study for the flow over periodic hills at four Reynolds numbers in the range of 5,600–37,000 using particle image velocimetry and laser doppler anemometry measurements. The experiment was performed in a water channel with 10 hills in the streamwise direction to ensure periodicity for the range of Reynolds numbers under investigation. In order to minimize the effect of side walls on the flow statistics, the spanwise direction was extended to 18 hill heights. They observed formation of secondary vortex structures due to the destabilizing curvature of streamlines on the windward side of the hills particularly at low Reynolds numbers. They also noticed frequent break up of the separation bubble and strong intermittency in the location of the reattachment line. Finally, it was seen out that the reattachment length decreases with increasing Reynolds number.

Computations of flow over periodic hills have also been performed by several researchers [3, 4]. Chaouat and Schiestel [4] simulated the flow over periodic hills using a hybrid RANS/LES method referred as partially integrated transport modeling (PITM) for Re = 37,000 and compared their results against experimental data [2]. Simulations were performed on coarse and medium grid sizes and aimed to achieve reasonable agreement with experimental data and LES calculation [3] at low computational cost. At the finest grid size of 0.9 million cells, the mean turbulence quantities are well predicted. It is noteworthy that the LES calculation was performed on a 4.7 million grid nodes at lower Reynolds number of 10,590. Chaouat and Schiestel [4] demonstrated the failure of Reynolds stress transport model to predict the correct behavior of mean velocity and stress components at different streamwise locations. It was also observed that PITM method may not be accurate if the grid is too coarse especially in the spanwise direction.

This study is aimed at evaluating PANS method [5, 6] employing an open source CFD solver referred as OpenFOAM. The goal is to demonstrate the ability of PANS in predicting the mean turbulence quantities associated with flows with highly separated region and non-trivial periodic boundary condition. In Sect. 2, the governing equations for PANS k-ω model are presented and the length scale cut off parameters, f_k and f_ω are introduced. In Sect. 3, details of the computational domain, simulation procedure and specification of the test cases studied here are provided. The effect of cut-off length scale and grid resolution on flow statistics are discussed in Sect. 4. Section 5 concludes this paper with a discussion of overall accuracy and computational cost of PANS method to investigate a separated turbulent flow field.

2 Governing Equations for PANS

In the PANS methodology, the resolved flow field evolves according to the following mass and momentum conservation equations [5, 6]

$$\frac{\partial \langle u_i \rangle}{\partial x_i} = 0, \tag{1}$$

$$\frac{\partial \langle u_i \rangle}{\partial t} + \langle u_j \rangle \frac{\partial \langle u_i \rangle}{\partial x_j} + \frac{\partial}{\partial x_j} \left(\tau \left(u_i, u_j \right) \right) = -\frac{\partial \langle p \rangle}{\partial x_i} + \nu \frac{\partial^2 \langle u_i \rangle}{\partial x_j x_j}, \tag{2}$$

where, $\langle x \rangle$ of an instantaneous quantity x represents the resolved part. Here, u is the velocity field, p is the pressure, ν is the kinematic viscosity and $\tau(u_i, u_j)$ is the generalized second moment ($\tau_{ij} = \langle u_i u_j \rangle - \langle u_i \rangle \langle u_j \rangle$). The closure for sub filter stress term, τ_{ij} can be obtained by using the Boussinesq approximation,

$$\tau \left(u_i, u_j \right) = -\nu_u \left(\frac{\partial \langle u_i \rangle}{\partial x_j} + \frac{\partial \langle u_j \rangle}{\partial x_i} \right) + \frac{2}{3} k_u \delta_{ij}, \tag{3}$$

where k is the kinetic energy, $\nu_u = k_u / \omega_u$ is the eddy viscosity and ω is the turbulence frequency parameter. Here, the subscript u denotes the unresolved field. The PANS filter is defined as,

$$f_k = \frac{k_u}{k}, \quad f_\varepsilon = \frac{\varepsilon_u}{\varepsilon}, \tag{4}$$

which gives $f_\omega = f_\varepsilon / f_k$. The unresolved kinetic energy and dissipation are defined as,

$$k_u = \frac{1}{2} \tau(u_i, u_j), \quad \varepsilon_u = \nu \tau \left(\frac{\partial u_i}{\partial x_j}, \frac{\partial u_j}{\partial x_i} \right) \tag{5}$$

The parameters f_k and f_ε represent the ratio of unresolved to resolved kinetic energy and dissipation, respectively. When $f_k = 1$, the formulation takes the form of RANS since all the turbulent kinetic energy is unresolved. On the other hand, for $f_k = 0$, the unresolved component goes to zero and hence the simulation behaves like a DNS. PANS $k - \omega$ equations [7] are given by

$$\frac{\partial k_u}{\partial t} + \langle u_j \rangle \frac{\partial k_u}{\partial x_j} = P_u - \beta^* k_u \omega_u + \frac{\partial}{\partial x_j} \left[\left(\nu + \frac{\nu_u}{\sigma_{ku}} \right) \frac{\partial k_u}{\partial x_j} \right] \tag{6}$$

$$\frac{\partial \omega_u}{\partial t} + \langle u_j \rangle \frac{\partial \omega_u}{\partial x_j} = \alpha \frac{P_u \omega_u}{k_u} - \beta' \omega_u^2 + \frac{\partial}{\partial x_j} \left[\left(\nu + \frac{\nu_u}{\sigma_{\omega u}} \right) \frac{\partial \omega_u}{\partial x_j} \right] \tag{7}$$

where the coefficients are related by,

$$\beta' = \alpha \beta^* - \frac{\alpha \beta^*}{f_\omega} + \frac{\beta}{f_\omega}, \quad \sigma_{ku} = \sigma_k \frac{f_k}{f_\omega}, \quad \sigma_{\omega u} = \sigma_\omega \frac{f_k}{f_\omega} \tag{8}$$

and their values given as, $\beta^* = 0.09, \beta = 0.075, \alpha = 5/9, \sigma_k = 2.0, \sigma_\omega = 2.0$. Here, P_u is the unresolved production.

OpenFoam [8], an open source finite volume code written in C++, is used to solve the equations. OpenFoam is easily parallelizable and highly flexible for turbulence model development. Out of the many available solvers in OpenFOAM, an incompressible transient solver, ChannelFOAM, was used with a second order accurate spatial discretization scheme. The Backward time scheme, which is second order accurate, was applied for time integration.

3 Simulation Procedure

The computational domain and flow configuration are summarized in Fig. 1. The geometry and domain dimensions are consistent with [3, 4]. A body fitted, curvilinear grid very similar to the one studied in [3] is generated for the flow geometry. The hills constrict the channel by almost one third of its height. In the streamwise direction, the domain extends from one crest to the next for a total of 9 h. In the spanwise direction, the domain size is 4.5 h. The flow is periodic in both streamwise and spanwise directions and no slip boundary condition is used at the bottom and top walls. Flow is driven by pressure gradient which is added as a source term to the momentum equation. The flow Reynolds number is calculated based on the following equation

$$Re = \frac{U_b h}{\nu} \tag{9}$$

where, U_b is the bulk velocity at the inlet and h is the hill height.

The flow statistics are studied for several PANS simulations over a range of cut-off length scales and grid resolutions. It is important to investigate the mean flow properties after flow has reached statistically steady state condition. For LES calculation [3], mean quantities were collected after 23 flow-through times and over a time period of 55T. The time to start averaging and the averaging period for PANS calculations depends on the cut-off length scale; the lower the cut-off ratio, the higher averaging period is required to obtain the steady flow statistics. It is also important to note that the mean flow properties are also averaged in the spanwise direction.

Fig. 1 Flow configuration

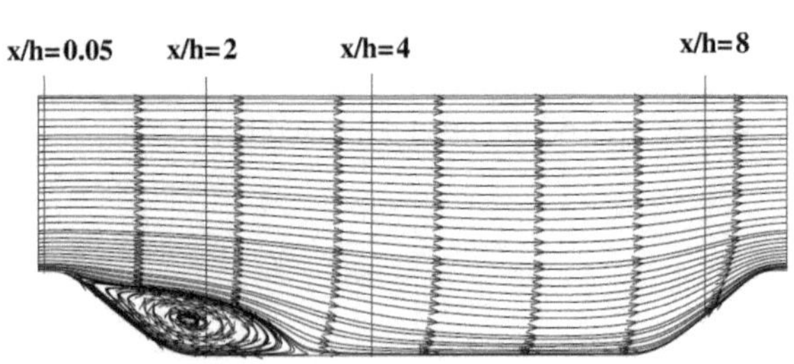

Table 1 Details of the test cases simulated

Study	f_k	f_ε	Grid	Averaging period
f_k study	1	1	$150 \times 100 \times 60$	10–15T
f_k study	0.35	1	$150 \times 100 \times 60$	10–15T
f_k study	0.25	1	$150 \times 100 \times 60$	10–24T
f_k study	0.15	1	$150 \times 100 \times 60$	18–36T
Resolution study	0.35	1	$150 \times 100 \times 60$	10–15T
Resolution study	0.35	1	$100 \times 100 \times 30$	10–15T
Resolution study	0.15	1	$150 \times 100 \times 60$	18–36T
Resolution study	0.15	1	$100 \times 100 \times 30$	18–36T

A summary of the various test cases simulated in this study are given in Table 1 which details the range of f_k values along with various grid resolutions investigated in this study.

4 Results

For each simulation, we examine mean flow and turbulence statistics. Two categories of simulations are performed to investigate the effects of physical resolution and computational resolution. In PANS, the physical resolution of the model is controlled by f_k. The physical resolution must be commensurate with computational resolution as given by Girimaji and Abdol-Hamid [9]

$$f_k \geq 3\left(\frac{\delta}{\lambda}\right)^{\frac{2}{3}}, \quad \lambda = \frac{k^{3/2}}{\varepsilon} \tag{10}$$

where λ is the local Taylor length-scale and δ is the grid size. At high enough computational resolution, it is very illustrative to compare simulations of different f_k values to demonstrate the capabilities of PANS closure model. Further, it is important to establish that for high f_k values, good results can be obtained on coarse grids. Thus, both f_k and grid resolution studies yield important insight into the closure model performance.

4.1 Variation of f_k Study

We first investigate the effect of varying f_k on flow statistics. In this section, all the studies are performed on the finest grid of about 0.9 million grid nodes. This grid permits computations of $f_k > 0.15$. Figure 1 shows the four streamwise locations at which the mean flow properties are investigated in this paper. Based on the experimental study [2], the selected positions are associated with the most important physics occurring in this flow configuration. The physical features are discussed in detail later on in this paper. Figure 2 shows the streamwise velocity profiles at four locations of x/h = 0.05, 2, 4 and 8. The results are shown for resolutions: $f_k = 0.35$, 0.25 and 0.15. The PANS results are compared against RANS $k - \omega$ model and experimental data [2].

At x/h = 0.05, there is a peak in near wall streamwise velocity which is attributed to the flow acceleration towards the windward slope of the hill. As can be seen from Fig. 2a, RANS model completely fails to predict the flow acceleration near the wall, whereas PANS results at all of the selected f_k values are in good agreement with experimental data. The next location, x/h = 2, is in the middle of recirculation zone where the boundary layer is detached and there is an interaction between the free shear layer separating from the hill crest and the reverse flow below that. Although the near wall velocity is accurately recovered by all turbulence models, poor prediction of the RANS simulation for streamwise velocity near the top boundary is seen in Fig. 2b. The flow is expected to be reattached at x/h = 4 which is well predicted by PANS method. However, a more elongated separation bubble is predicted by the

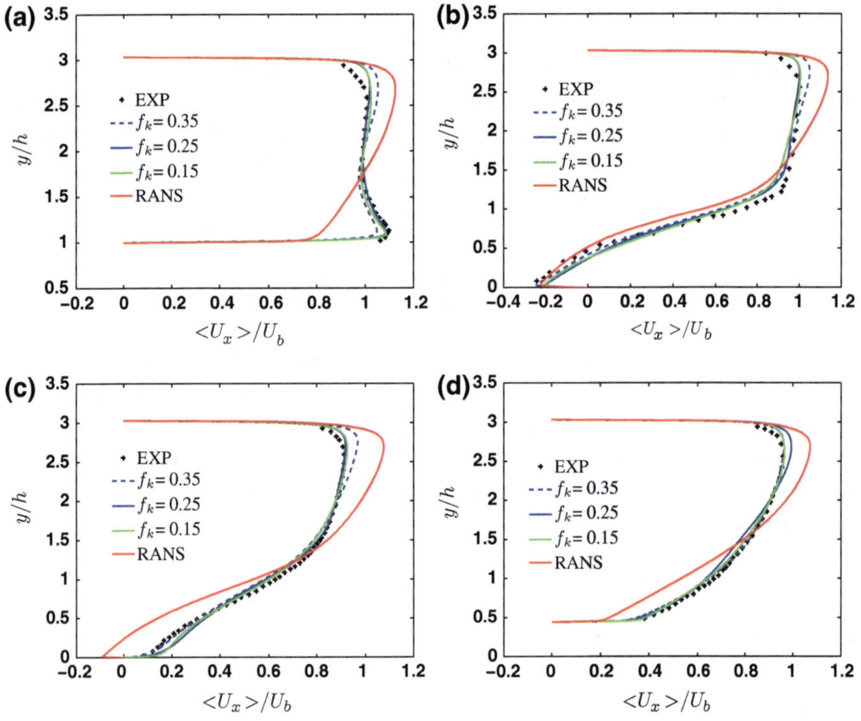

Fig. 2 Streamwise velocity profiles at different locations. **a** $x/h = 0.05$, **b** $x/h = 2$, **c** $x/h = 4$, **d** $x/h = 8$

RANS simulation. At x/h = 8, the flow is accelerated on the windward slope of the hill. This feature is again well captured by PANS simulation. As can be seen from Fig. 2d, the near wall velocity for RANS simulation is noticeably lower than the corresponding PANS values.

An important observation from Fig. 2 is that the agreement on streamwise velocity profile with experimental data improves progressively with lower f_k values. Among the RANS and PANS data, the best agreement is achieved for $f_k = 0.15$ wherein a wider range of turbulence length scales are resolved in the highly unsteady regions.

Figure 3 shows the shear stress profiles at the four streamwise locations. The total shear stress is computed as the sum of resolved and modeled stresses. At x/h = 0.05, the maximum value for $< u'v' >$ at $y/h \approx 1.6$ is related to the local minimum of streamwise velocity. As can be seen, this peak value is well captured by PANS simulation for $f_k = 0.15$. At the next two locations, x/h = 2 and x/h = 4, the peak value of shear stress predicted by PANS is in close agreement with experimental data. Location at x/h = 8 is regarded as post reattachment region where the flow is recovering from upstream separation with flow components of different history. As can be seen from Fig. 3a, b, inability of RANS to capture separation results in over-prediction of shear stress at this region. Overall, Figs. 2 and 3 indicate that the

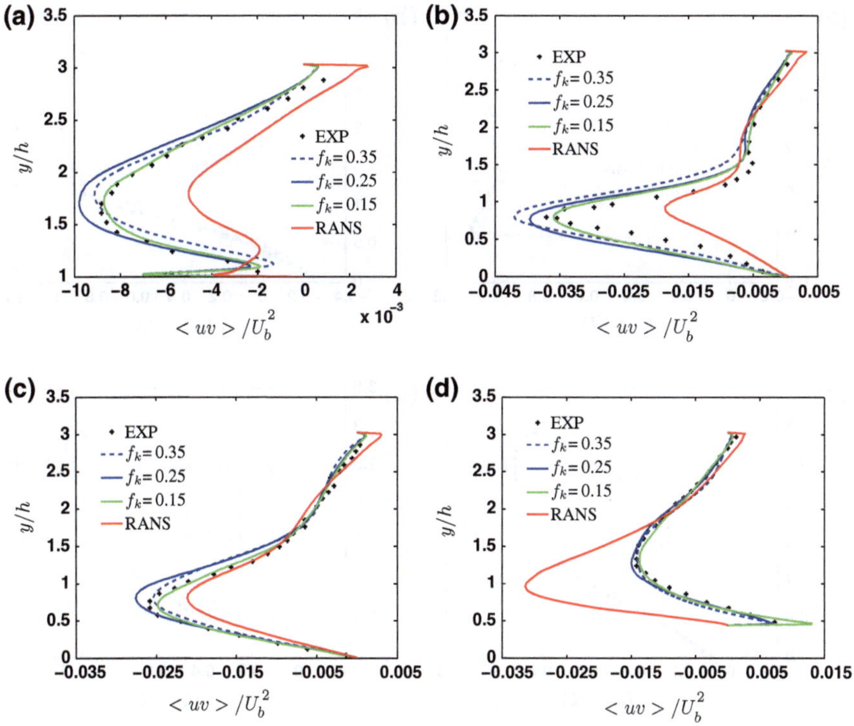

Fig. 3 Shear stress profiles at different locations. **a** $x/h = 0.05$, **b** $x/h = 2$, **c** $x/h = 4$, **d** $x/h = 8$

mean flow quantities are well predicted by PANS simulation at all the cut-off ratios investigated here. However, the results for $f_k = 0.15$ are much closer than RANS computation.

4.2 Resolution Study

In this section, the grid sensitivity of the PANS simulations with cut-off ratios of $f_k = 0.15$ and $f_k = 0.35$ is investigated for two grid resolutions.

It must be noted that the coarser grid still supports $f_k = 0.35$ while is marginally adequate for $f_k = 0.15$. The detail on the grid sizes is summarized in Table 1. As seen in Table 1, the coarse grid is generated by reducing the grid nodes in the streamwise and spanwise directions while keeping the resolution fixed in the normal direction.

Figures 4 and 5 show the mean streamwise velocity and shear stress for PANS calculations on the coarse and fine grid sizes. As seen in these plots, at almost all the locations, the mean quantities are hardly distinguishable for the PANS simulations on the both grid sizes. Regarding the PANS calculations, the only noticeable difference

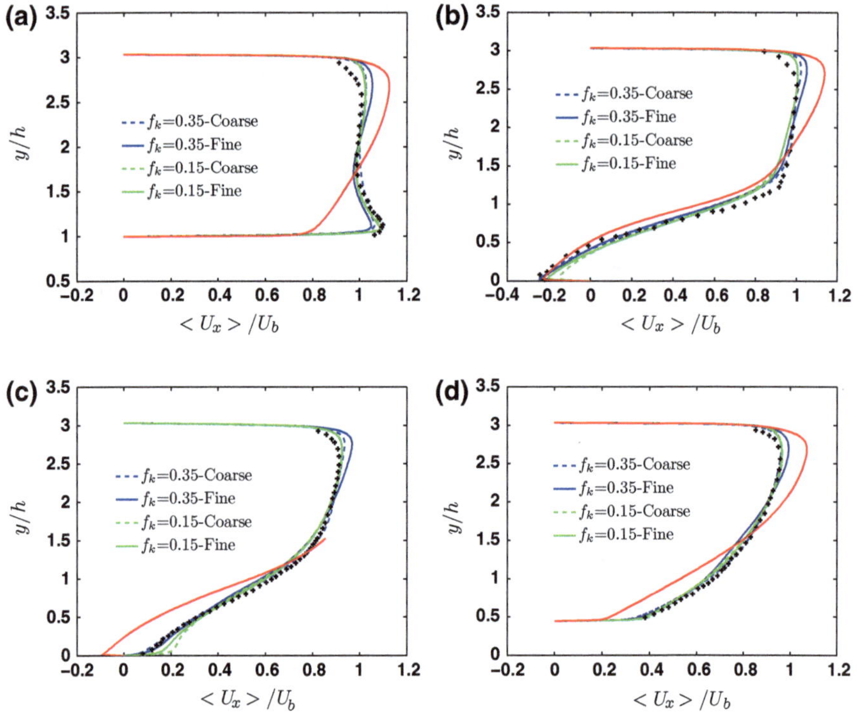

Fig. 4 Streamwise velocity (symbols for Exp. and RANS are consistent with Fig. 2). **a** $x/h = 0.05$, **b** $x/h = 2$, **c** $x/h = 4$, **d** $x/h = 8$

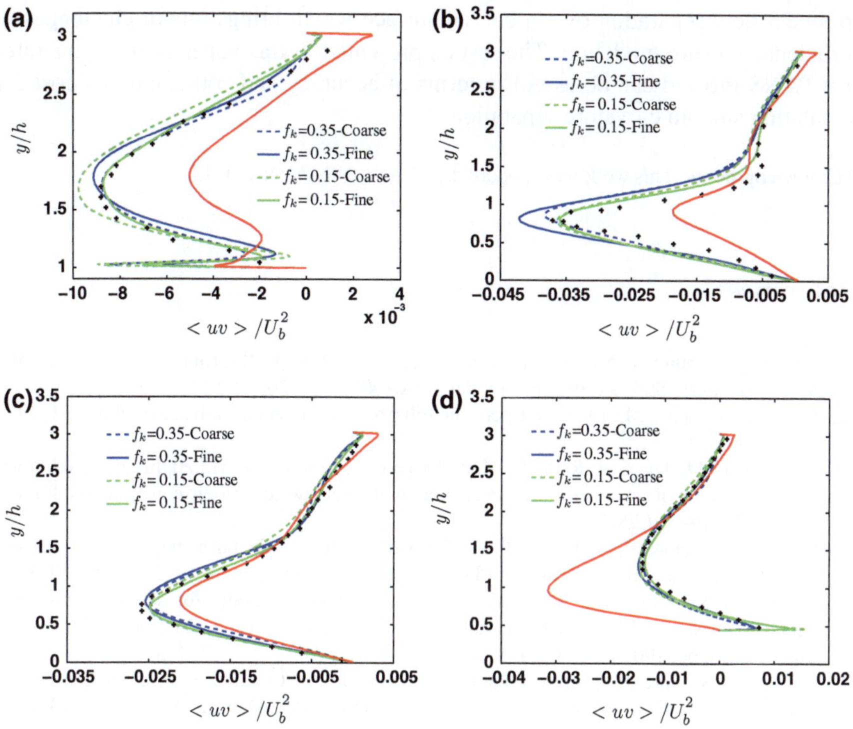

Fig. 5 Shear stress (symbols for Exp. and RANS are consistent with Fig. 3). **a** $x/h = 0.05$, **b** $x/h = 2$, **c** $x/h = 4$, **d** $x/h = 8$

between the results of the two grid resolutions is observed at x/h = 2 and x/h = 4 for $f_k = 0.15$. This is expected since as f_k is reduced, the grid should be fine enough to capture the small turbulence length scales.

5 Conclusion

Variable resolution simulations of flow over a periodic hill are performed using PANS implemented in OpenFoam. The results for the flow Reynolds number of 37,000 are compared against an experimental study and RANS. Two studies are performed here to investigate the effect of reducing cut-off ratio, f_k and grid resolution on the accuracy of the PANS results. For the finest grid investigated in this work, reducing f_k value results in improved agreement with experimental data regarding the size of separation bubble and flow statistics. For the two grid resolutions studied here, PANS method did not show much sensitivity to grid resolution. This paper demonstrates the adaptability of PANS to OpenFoam solver and further highlights its capability to model flow separation over a curved surface which brings lots of challenges to turbulence closure modeling. The results presented in this paper further illustrates that PANS method can be useful in terms of accuracy and computational cost for simulating smooth curvature separation.

Acknowledgments This work was supported by NASA NRA#NNXAI61A.

References

1. Basara, B., Krajnovic, S., Girimaji, S.S., Pavlovic, Z.: Near-wall formulation of the partially averaged Navier-Stokes turbulence model. AIAA **49**, 2627–2636 (2011)
2. Rapp, C., Manhart, M.: Flow over periodic hills: an experimental study. Exp. Fluids **51**, 247–269 (2011)
3. Frohlich, J., Mellen, C.P., Rodi, W., Temmerman, L., Leschziner, M.: Highly resolved large-eddy simulation of separated flow in a channel with streamwise periodic constrictions. J. Fluid Mech. **526**, 19–66 (2005)
4. Chaouat, B., Schiestel, R.: Hybrid RANS/LES simulations of the turbulent flow over periodic hills at high Reynolds number using PITM method. J. Comput. Fluids **84**, 279–300 (2013)
5. Girimaji, S.S., Srinivasan, R., Jeong, E.: PANS turbulence model for seamless transition between RANS and LES: fixed-point analysis and preliminary results. In: ASME Conference Proceedings, pp. 1901–1909 (2003)
6. Girimaji, S.S.: Partially-averaged Navier-Stokes model for turbulence: a Reynolds-averaged navier-stokes to direct numerical simulation bridging method. ASME J. Appl. Mech. **73**, 413–421 (2006)
7. Lakshmipathy, S., Girimaji, S.S.: Partially avaraged Navier-Stokes method for turbulent flows: $k - \omega$ model implementation. AIAA (2006)
8. OpenCFD: OpenFOAM—The Open Source CFD Toolbox—User Guide, 1.4 edn. OpenCFD Ltd., Bracknell (2007)
9. Girimaji, S.S., Abdol-Hamid, K.S.: Partially-averaged Navier Stokes model for turbulence: implementation and validation. AIAA (2005)

Partially-Averaged Navier-Stokes (PANS) Simulations of Lid-Driven Cavity Flow—Part II: Flow Structures

Pooyan Razi, Vishnu Venugopal, Shriram Jagannathan and Sharath Girimaji

Abstract The vortical flow structures in low and high Reynolds number lid-driven cavity flows are examined using Partially-averaged Navier-Stokes (PANS) and unsteady Reynolds-averaged Navier-Stokes (URANS) simulations. The spanwise aspect ratio (SAR) of the cavity is 3:1:1 and the Reynolds numbers based on cavity height and lid velocity are 10^4 and 10^6. It is demonstrated that, while the mean flow statistics are nearly the same for URANS and PANS, the complex vortex structures are captured much better by PANS. The difference between the URANS and PANS structures are even more distinct at higher Reynolds numbers. Furthermore, it is shown that the PANS small-scale statistics at different levels of resolution are self-similar and scale according to established turbulence theory.

1 Introduction

Numerical simulation of turbulent flows is always challenging, especially with increasing Reynolds number, due to the presence of wide range of length and time scales and the necessity to resolve at least the large scales. High fidelity computational approaches like Direct Numerical Simulations (DNS) and Large Eddy Simulations (LES) provide the most accurate solutions, but its their application is limited for high Reynolds number flows due to the extraordinary computational resource required. Reynolds averaged Navier-Stokes (RANS) simulations, on the other hand, are computationally inexpensive but resolve only the time-averaged mean velocity field, while the fluctuating fields are modeled. Since the accuracy of RANS for engineering applications is sub-optimal and DNS/LES is enormously expensive, it has been suggested [1–3] that a variable resolution method would be computationally more appropriate as they can provide any intermediate level of resolution. Partially averaged Navier-Stokes (PANS) method, while embodying much of the RANS paradigm, is a variable resolution method that can take the form of RANS or LES

P. Razi · V. Venugopal · S. Jagannathan · S. Girimaji (✉)
Texas A&M University, College Station, TX, USA
e-mail: girimaji@tamu.edu

S. Girimaji et al. (eds.), *Progress in Hybrid RANS-LES Modelling*,
Notes on Numerical Fluid Mechanics and Multidisciplinary Design 130,
DOI 10.1007/978-3-319-15141-0_34

method depending upon the specification of appropriate filter parameters. The turbulence field is filtered into a resolved and unresolved field by choosing a cut-off position in the energy spectrum such that the ratio of unresolved to total kinetic energy (f_k) and dissipation (f_ε) are at desired levels. While $f_k = 1$ would indicate that the entire field is modeled and thus a RANS simulation, a value of $f_k = 0$ would correspond to DNS where all the spatial and temporal scales are resolved without any model component.

One of the crucial steps towards successfully modeling turbulent flows is to validate the model under different conditions and flow types. The lid driven cavity flow problem is widely considered as one of the benchmark problems for code validation in computational fluid dynamics. The complex flow structures that appear in a simple geometry have attracted much work on this problem in the past few decades. Fluid is contained in a rectangular domain, with three stationary sides and one moving side (with velocity tangent to the side). The spanwise aspect ratio (SAR) of a cuboid type cavity is defined as *spanwise-length:height:width* of the cavity (see Fig. 1a). The Reynolds number is defined as $\mathrm{Re} = Uh/\nu$, where h is the cavity height, U is the lid velocity and ν is the kinematic viscosity of the fluid. For a low Re, the flow is two-dimensional and quickly reaches a steady state. However, if Re is high, the flow becomes turbulent and three-dimensional.

In Part-I [4] of the study, the large scale turbulence statistics have been examined using PANS, URANS and LES. The objective of this study is to investigate whether PANS is capable of capturing the vortical flow structures and important features of intermediate-scale fluctuations. Such structures and features are very important for high fidelity simulations of mixing processes which are dominated by small scale fluctuations.

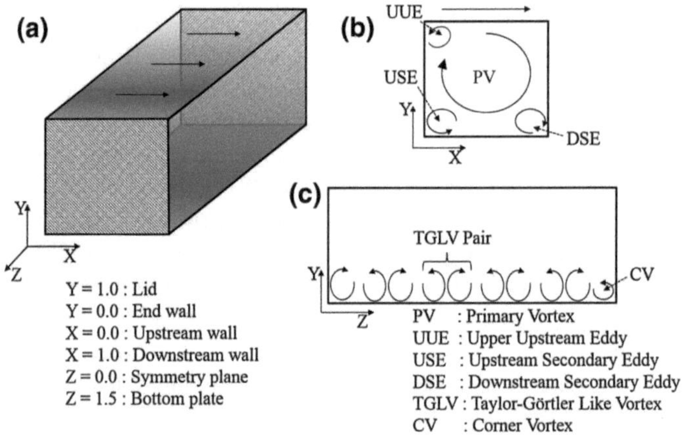

Fig. 1 **a** Computational domain for lid-driven cavity with SAR 3:1:1. **b** Vorical structures near the symmetry plane. **c** Vortical structures near the downstream wall

2 Governing Equations for PANS

The governing equation for PANS is obtained by filtering [1] the Navier-Stokes equation into resolved and unresolved parts. Though the filter width can be arbitrary, it is assumed to be commutative and invariant in space and time. The resolved field is described by

$$\frac{\partial \bar{u_i}}{\partial x_i} = 0, \tag{1}$$

$$\frac{\partial \bar{u_i}}{\partial t} + \bar{u_j}\frac{\partial \bar{u_i}}{\partial x_j} + \frac{\partial}{\partial x_j}\left(\tau\left(u_i, u_j\right)\right) = -\frac{\partial \bar{p}}{\partial x_i} + \nu\frac{\partial^2 \bar{u_i}}{\partial x_j x_j}, \tag{2}$$

where, $\bar{f}$ and f' represent the resolved and unresolved parts of any quantity f. Here u is the velocity field, p is the pressure, ν is the kinematic viscosity and $\tau(u_i, u_j)$ is the generalized second moment (also known as the sub-filter stress (SFS), $\tau_{ij} = \overline{u_i u}_j - \bar{u_i}\bar{u_j}$ [5]), and represents the interaction between the resolved and unresolved field. We use here the Boussinesq approximation to relate the stress to the strain as,

$$\tau\left(u_i, u_j\right) = -\nu_u\left(\frac{\partial \bar{u_i}}{\partial x_j} + \frac{\partial \bar{u_j}}{\partial x_i}\right) + \frac{2}{3}k_u\delta_{ij}, \tag{3}$$

where k is the kinetic energy (the subscript u denotes the unresolved field), $\nu_u = k_u/\omega_u$ is the eddy viscosity, and ω is the turbulence frequency parameter. We can now define the PANS filter as,

$$f_k = \frac{k_u}{k}, \quad f_\omega = \frac{\varepsilon_u}{\varepsilon}, \tag{4}$$

which yields $f_\omega = f_\varepsilon/f_k$. The unresolved kinetic energy and dissipation are defined as,

$$k_u = \frac{1}{2}\tau(u_i, u_j), \quad \varepsilon_u = \nu\tau\left(\frac{\partial u_i}{\partial x_j}, \frac{\partial u_j}{\partial x_i}\right). \tag{5}$$

The parameters f_k and f_ε are given by the ratio of unresolved to resolved kinetic energy and dissipation respectively and denotes a cut-off in the energy spectrum. When $f_k = 1$, the formulation takes the form of RANS since all the turbulent kinetic energy is unresolved. On the other hand for $f_k = 0$, the unresolved component goes to zero and hence the simulation behaves like a DNS. It is important to note here that the value of f_k, f_ε should be consistent with the grid resolution—for instance, it is not appropriate to use a small f_k for really coarse grid sizes. The smallest f_k supported by a given grid size (Δ) is given by

$$f_k \geq 3\left(\frac{\Delta}{\lambda}\right)^{2/3}, \tag{6}$$

where $\lambda = k^{3/2}/\varepsilon$ [2].

For problems of current interest, the cut-off is in the so-called inertial range of the energy spectrum where the dissipation scales are unresolved, allowing us to choose $f_\varepsilon = 1$. We use the PANS $k - \omega$ implementation, the details of which are given in [6], and only the equations are mentioned here,

$$\frac{\partial k_u}{\partial t} + \bar{u_j}\frac{\partial k_u}{\partial x_j} = P_u - \beta^* k_u \omega_u + \frac{\partial}{\partial x_j}\left(\frac{\nu_u}{\sigma_{ku}}\frac{\partial k_u}{\partial x_j}\right), \tag{7}$$

$$\frac{\partial \omega_u}{\partial t} + \bar{u_j}\frac{\partial \omega_u}{\partial x_j} = \alpha\frac{P_u \omega_u}{k_u} - \beta' \omega_u^2 + \frac{\partial}{\partial x_j}\left(\frac{\nu_u}{\sigma_{ku}}\frac{\partial \omega_u}{\partial x_j}\right), \tag{8}$$

where the coefficients are related by,

$$\beta' = \alpha\beta^* - \frac{\alpha\beta^*}{f_\omega} + \frac{\beta}{f_\omega} \tag{9}$$

and their values given as, $\beta^* = 0.09$, $\alpha = 5/9$, $\sigma_k = 2.0$, $\sigma_\omega = 2.0$. Here, P_u is the unresolved production.

3 Simulation Details and Flow Structures

Simulation Details: The open source *finite volume method* based code OpenFOAM is used for the simulations. The simulations are second order accurate in space and first order accurate in time. The RANS $k-\omega$ model constants are changed accordingly to simulate the required PANS cases. The computational domain with SAR 3:1:1 (see Fig. 1a) is divided into non-uniform structured brick type cells. Grid is maintained fine enough near the walls to predict the right near wall behaviour. Ensemble averaging for turbulence statistics is done after allowing the initial transients in the flow to pass. The flow is let to evolve for $150T$ time units, where $T = L_x/U$ is the turbulence time scale, and statistics are subsequently sampled for $150T$. Different cases selected for the present study are listed in Table 1. It should be noted that f_ε is set to unity in all the cases.

Flow structures: Extensive literature is available on the lid-driven cavity problem. Shankar and Deshpande [7] note that in the "streamwise" (X-Y) plane, there are three eddies at three different corners apart from the primary eddy (PE) in the center (see Fig. 1b). The eddy at the bottom downstream corner is called downstream secondary eddy (DSE) and those at the bottom and top upstream corners are called as upstream secondary eddy (USE) and upper upstream eddy (UUE) respectively. The

Table 1 Simulation parameters

Case	Grid	Type	f_k	f_ε	Re
1	85^3	URANS	–	–	10^4
2	85^3	PANS	0.60	1.00	10^4
3	85^3	PANS	0.35	1.00	10^4
4	85^3	PANS	0.25	1.00	10^4
5	85^3	URANS	–	–	10^6
6	85^3	PANS	0.25	1.00	10^6

particle movement taken from Chiang et al. [8] shows how a particle covers almost all spanwise direction at high Re, indicating the highly three-dimensional nature of the flow. The velocity profiles comparison shows that the flow is 2-D in a 3-D cavity up to very small Re (~20–30). As Re increases, flow tends to be three-dimensional due to presence of end wall. At higher Re, the cavity flow is unsteady and contains complex flow structures in the spanwise direction (see Fig. 1c). These are corner vortex (CV) and meandering Taylor-Görtler like vortex (TGLV) pairs [9]. Solution multiplicity of the lid-driven cavity problem is also another interesting issue. In three similar repeated experiments, Aidun et al. [10] changed Re from 2,000 to 5,000 and observed three different states, each state having different number of TGL vortex pairs. While the flow structures may change with the Reynolds number, qualitatively they have been observed for Re $= 10^4$ [7, 11], which is the Reynolds number for most of the simulations in current investigation.

4 Results

The effect of f_k on turbulence statistics is presented for different grid resolutions in the companion study [4]. Simulations are carried out for both low and high Reynolds number (10^4 and 10^6 respectively). It is shown that URANS, PANS and LES reasonably capture the mean flow in the low Reynolds number case. The focus of the present study is on the ability of PANS to simulate vortical structure and other small scale features of the flow.

In the first part of this section, it is demonstrated that the complex vortex structures are captured much better by PANS. The difference between structures that are captured by URANS and PANS are even more distinct at high Reynolds numbers. In the second part, it is shown that the PANS small-scale statistics at different levels of resolution are self-similar and scale according to established turbulence theory.

Vortical structures: We compare the flow structures observed from URANS and PANS simulation for both low and high Reynolds number. These are educed by computing the instantaneous vorticity vectors. Since the secondary flow structures (DSE, USE, UUE, TGLV and CV) are highly unsteady, it may not be completely descriptive to use instantaneous time snapshots for a one-to-one comparison.

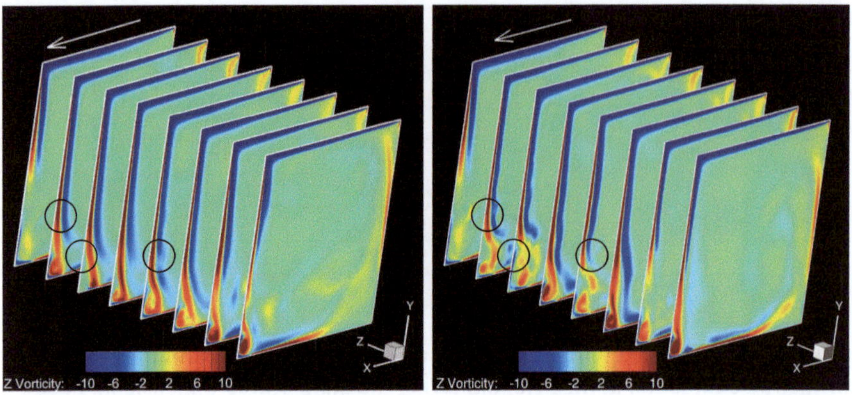

Fig. 2 Z-Vorticity contours on XY planes with URANS (*left*) and PANS (*right*) for Re $= 10^4$

Nevertheless, they provide a good qualitative picture on the ability of URANS and PANS methods in capturing the flow features. Figure 2 shows the Z-vorticity contours at different X-Y planes from symmetry plane till the end wall for URANS and a representative PANS case ($f_k = 0.35$, $f_\varepsilon = 1.00$) for Re $= 10^4$. First, the extent of primary vortex is very well resolved in both the cases and qualitatively agrees with that of [9]. Since this is a large scale phenomenon and less unsteady as compared to other flow structures, it is expected that URANS and PANS would behave identically. Second, the DSE and USE (which is of a lower strength compared to that of DSE) are well represented by both URANS and PANS simulations. It is also worth mentioning here that since we are performing an unsteady URANS, a smaller fraction of the fluctuations in the flow variables (and thus the structures) are already captured. A steady RANS, on the other-hand, would not be able to resolve any of these unsteady statistics. However, the unsteadiness associated with the formation of these vortices and the primary vortex are captured much better in PANS (cf. circles in Fig. 2), but are smoothened out in URANS (spatially along the Z direction) which is purported only for very large scale unsteadiness. It is well known that lid driven cavity flow has significant three-dimensional effects owing to the presence of the end wall [9]. It can be clearly seen that the instabilities near the planes close to end wall is very well captured in the PANS case (see Fig. 2). However, the corresponding URANS simulation fails to capture these effects. Successfully modeling these unsteady and three-dimensional characteristics of a flow are critical to any turbulence model since they may correspond to accurately predicting, for instance, instabilities, extent and structure of vortices, separation, reattachment etc. In order to ensure that this observation is consistent in time, a series of similar time snapshots were compared and the results were similar.

Finer flow features that appear due to the end wall are the meandering Taylor-Görtler like vortex pairs and corner vortex [9]. These can be visualized using the X-vorticity contours near the bottom plate and end wall respectively as shown in Fig. 3. The number of TGLV pairs have been reported to be approximately 11 at

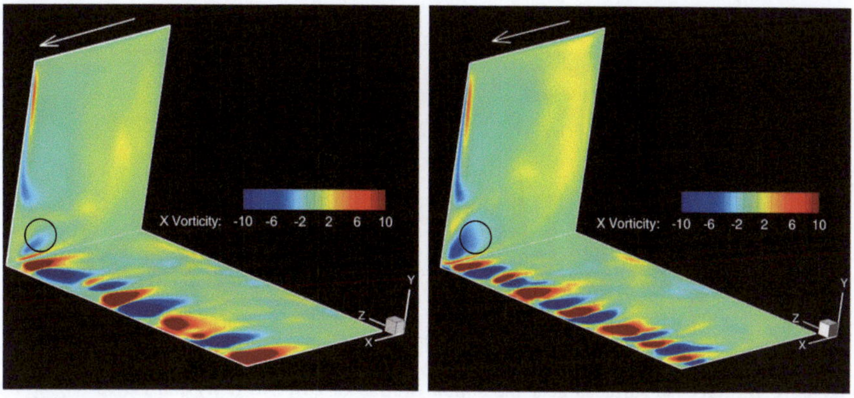

Fig. 3 X-Vorticity contours near the *bottom* plate and end wall with URANS (*left*) and PANS (*right*) for $\mathrm{Re} = 10^4$

$\mathrm{Re} \approx 6000$ [7]. We observe approximately 5 and 7 vortex pairs for URANS and PANS respectively at $\mathrm{Re} = 10^4$. As the flow has been simulated with symmetry boundary condition for the spanwise wall (Z = 0), this would correspond to 10 and 14 TGL vortex pairs in the full cavity. Since the simulated Re is higher than what is reported, it is natural to expect the flow to be more unsteady with a higher number of TGL pairs. Though there is very little data in the literature to conclusively state the number of vortex pairs at high Reynolds number, one could speculate that PANS does predict the overall number of TGLV pairs better than URANS. Owing to the unsteadiness of these meandering structures, a series of time snapshots were compared and results were identical. Due to the meandering nature of TGLV pairs, it is also essential to investigate the degree of their unsteadiness. The wavy tails that extend towards upstream wall in Fig. 3 are well represented by PANS, while they are smoothened out in URANS. This further demonstrates the ability of PANS method in successfully capturing the smooth transitions between successive stable configurations. Experiments [9] also reveal the presence of a highly unsteady corner vortex close to the end wall that is well captured by PANS (cf. circles in Fig. 3).

Since engineering applications are typically at high Reynolds number, it is essential that the model performs equally well at high Reynolds numbers. Hence, we now observe the flow structures at a Reynolds number of $\mathrm{Re} = 10^6$ with a very coarse grid of 85^3 (refer cases 5 and 6 in Table 1). Figure 4 illustrates iso-surfaces of x-vorticity near the bottom plate for the high Reynolds number case. It could be clearly seen that the URANS model completely fails in predicting the flow structures (TGLV pairs) at high Reynolds number. A corresponding PANS simulation, on the other hand, captures the TGLV pairs as well as the CV. It should also be noted that the PANS resolves these flow structures with a very coarse grid of 85^3. An LES simulation, on the other hand, would require a much larger grid for the same Reynolds number. Therefore, for a high Reynolds number turbulent flow, PANS method predicts the right flow features, at a lesser computational cost compared to LES.

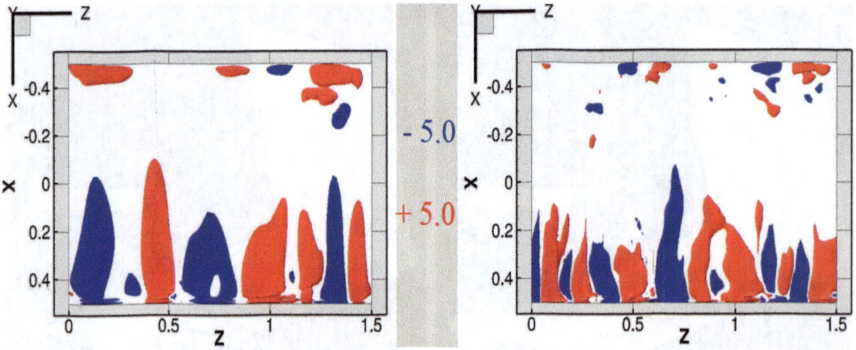

Fig. 4 X-Vorticity iso-surfaces near the *bottom* wall with URANS (*left*) and PANS (*right*) for Re $= 10^6$

PANS self-similarity: While capturing the fluctuations, and thus the structures, it is of paramount importance to ensure that these are indeed physical. This can be accomplished by extending the Kolmogorov Similarity Hypotheses (KSH) [12] to PANS and has been done in [13]. The first similarity hypothesis states that the statistics of smallest scales of motion are uniquely determined by the mean dissipation rate and viscosity. In this section, we investigate the PDF (probability density function) of the mean dissipation rate and unresolved viscosity separately. For a decaying homogeneous isotropic turbulent flow, it has been shown that the unresolved fluctuations that are modeled in PANS adhere to the first KSH [14]. However, most engineering flows exhibit inhomogeneity, anisotropy, separation or reattachment and are more difficult to model and hence it is imperative for a robust model to perform well under these flow conditions. As a first step, we investigate how the unresolved fields behave for an inhomogeneos and anisotropic flow. It should also be noted that the flow here is not turbulent throughout the domain, but only near regions of DSE.

The PDF of unresolved dissipation ($\varepsilon_u = \beta^* \omega_u k u$) normalized by its time mean is shown in Fig. 5a for different filter cut-offs. If the fluctuations behave according

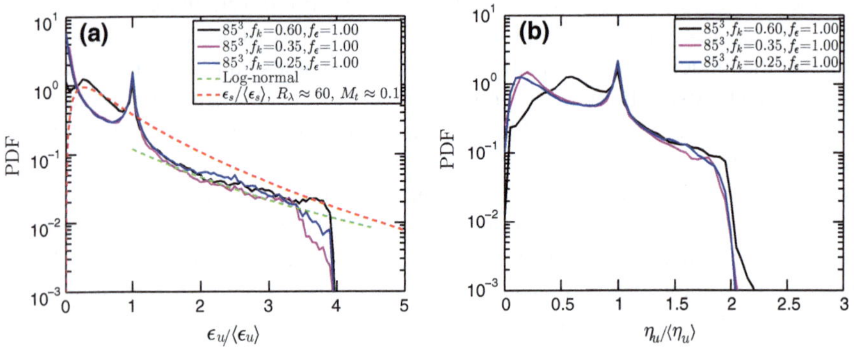

Fig. 5 **a** PDF of ε. **b** PDF of η

to KSH, we would expect the curves to fall on top of each other in regions where the flow is turbulent. Since the flow is turbulent only in certain regions in the flow, it is natural expect a collapse only in those regions. As seen in Fig. 5a, for $\varepsilon_u / \langle \varepsilon_u \rangle > 1$, we see a very good collapse. It has also been argued in the literature [15] that for a narrow range of fluctuations close to the mean, the PDF of dissipation tend to follow log-normal behaviour and is also included in Fig. 5a. We observe that for a range of fluctuations ($1 < \varepsilon_u / \langle \varepsilon_u \rangle < 5$), the PDF of dissipation does follow a log-normal behaviour. To further corroborate the result, we have included the PDF of solenoidal dissipation from DNS of forced stationary compressible turbulence at low turbulent Mach number (M_t) where compressibility effects are negligible [16]. Overall, the agreement between PANS and DNS, is good for the range of fluctuations indicated. For $\varepsilon_u / \langle \varepsilon_u \rangle < 1$, though the PDFs from PANS simulations collapse fairly well, they are different from that of the DNS. It is possible that regions may correspond to the flow that is not fully turbulent at this Reynolds number.

The PDF of $\eta_u / \langle \eta_u \rangle$, which is shown in Fig. 5b, also demonstrates self-similarity. The peak in the PDF for both unresolved dissipation and Kolmogorov scale occur around one when they are normalized by their respective time mean values. This suggests that the fluctuations do not vary in time and tend to stay close to the mean, which indicates that this should represent the region of flow where the unsteadiness in the flow is very small (since they are normalized by the time mean values). Considering that most of the structures like DSE, USE, and TGLV are highly unsteady, it is thus conceivable that this region may actually correspond to the primary vortex which doesn't exhibit fully developed turbulent behaviour at this Reynolds number [9].

5 Conclusions

In hybrid simulations of turbulence, it is important to establish the fidelity of flow structures and small-scale fluctuations. Such an investigation is performed in this paper for the case of cavity flow. Despite the simple geometry, cavity flow exhibits complex vortical flows and small-scale behaviour. It is clearly demonstrated that the PANS simulations capture these features at reasonable computational effort. In particular, the small-scale statistics are shown to be self-similar at different resolutions and are consistent with turbulence scaling laws.

Acknowledgments The computational time for the simulations presented in this work was awarded by the Texas A&M Supercomputering Facility and the Texas Advanced Center for Computations (TACC).

References

1. Girimaji, S.S.: J. Appl. Mech. **73**(3), 413 (2005)
2. Girimaji, S.S., Abdol-Hamid, K.: Aerospace Sciences Meetings, vol. 502. American Institute of Aeronautics and Astronautics, Dallas (2005)
3. Girimaji, S.S., Suman, S.: Progress in Hybrid RANS-LES Modelling, pp. 29–43. Springer, New York (2012)
4. Akula, B., Roy, P., Razi, P., Anderson, S., Girimaji, S.S.: HRLM (2014)
5. Germano, M.: J. Fluid Mech. **238**, 325 (1992)
6. Lakshmipathy, S., Girimaji, S.S.: Aerospace Sciences Meetings, vol. 10. American Institute of Aeronautics and Astronautics, Dallas (2006)
7. Shankar, P.N., Deshpande, M.D.: Annu. Rev. Fluid Mech. **32**(1), 93 (2000)
8. Chiang, T., Sheu, W., Hwang, R.R.: Int. J. Numer. Methods Fluids **26**(5), 557 (1998)
9. Koseff, J.R., Street, R.L.: J. Fluids Eng. **106**(4), 390 (1984)
10. Aidun, C.K., Triantafillopoulos, N.G., Benson, J.D.: Phys. Fluids A: Fluid Dyn. (1989–1993) **3**(9), 2081 (1991)
11. Prasad, A.K., Koseff, J.R.: Phys. Fluids A: Fluid Dyn. (1989–1993) **1**(2), 208 (1989)
12. Kolmogorov, A.N.: Dokl. Akad. Nauk. SSSR **30**, 299 (1941)
13. Reyes, D.: Advancing the theoritical foundation of the partially-averaged navier-stokes approach. Ph.d. thesis, Texas A&M University, Dallas (2013)
14. Reyes, D., Girimaji, S.: Bull. Am. Phys. Soc. 57 (2012)
15. Wang, L.P., Chen, S., Brasseur, J.G., Wyngaard, J.C.: J. Fluid Mech. **309**, 113 (1996)
16. Jagannathan, S., Donzis, D.A.: Proceedings of the 1st Conference of the Extreme Science and Engineering Discovery Environment: Bridging from the eXtreme to the Campus and Beyond, XSEDE '12, pp. 23:1–23:8. ACM, New York, NY, USA (2012)

Part VI
Applications

Evaluation of Scale-Adaptive Simulations for Transonic Cavity Flows

S.V. Babu, G. Zografakis and G.N. Barakos

Abstract This paper demonstrates the Scale-Adaptive Simulation approach for the computation of flows around transonic weapon bays idealised as rectangular cavities. Results are also compared with Detached-Eddy Simulations for the M219 cavity with and without doors. The Mach and Reynolds numbers (based on the cavity length) are 0.85 and 6.5×10^6 respectively, with a grid size of 5.0 million for the cavity with doors-off and 5.5 million for the cavity with doors-on. Instantaneous Numerical schlieren contours made it possible to visualise the propagation of pressure waves in and around the cavities and also showed the high level of unsteadiness and breakdown of the shear layer for both doors on and doors off cases. Both cavities were seen to have similar acoustic signatures reaching maximum sound pressure levels of 170 dB. Spectral analyses revealed that the addition of the doors caused the second Rossiter mode to dominate along the length of the cavity. Scale-Adaptive Simulation results showed good agreement with experimental data for the M219 cavity at a tenth of the time required for Detached-Eddy Simulations.

1 Introduction

Store carriage and separation from weapon bays has received large interest in recent years. Early works have focused on weapon bays idealised as rectangular cavities but the focus is shifting to weapon bays installed in Unmanned Combat Air Vehicles (UCAVs) and complex bays with that carry stores. Barakos et al. [1] and Lawson and Barakos [2] give a detailed account of the work relating to UCAVs and transonic cavities. Computational Fluid Dynamics (CFD) simulations of store separation from

S.V. Babu · G. Zografakis · G.N. Barakos (✉)
CFD Laboratory, School of Engineering, University of Liverpool, Liverpool L69 3GH, UK
e-mail: s.babu@liverpool.ac.uk

G. Zografakis
e-mail: g.zografakis@liverpool.ac.uk

G.N. Barakos
e-mail: g.barakos@liverpool.ac.uk

S. Girimaji et al. (eds.), *Progress in Hybrid RANS-LES Modelling*,
Notes on Numerical Fluid Mechanics and Multidisciplinary Design 130,
DOI 10.1007/978-3-319-15141-0_35

cavities, for store clearance, are becoming expensive with increasing mesh sizes that account for geometric complexities. This requires faster calculations to allow for numerous simulations under varying conditions. While DES is capable of accurately predicting these flows, it still takes a considerable amount of time on a large number of processors. Since its introduction by Menter et al. [3–5] in 2003, the Scale-Adaptive Simulation approach gained popularity due to its LES-like behaviour in highly separated flow regions and found place in several studies [6–10]. A detailed explanation of the theory and description of the model was given by Menter and Egorov [11] following which Egorov et al. [12] presented the application of the SAS model, implemented in ANSYS-FLUENT and ANSYS-CFX, for a range of complex flows. Other than the demonstration of SAS for the M219 cavity case without doors by Egorov et al. [12], no other cavity work using SAS is available in the open literature. The evaluation of SAS for cavity flows and later for cavities with stores may encourage its use in store separation simulations.

In view of the above, this paper presents results from ongoing studies of the SAS approach for cavity flows. The widely used M219 cavity is used for numerical computations using DES, SAS and URANS using the Helicopter Multi-Block (HMB2) flow solver of Liverpool [13]. The M219 clean cavity with and without doors is used here with experimental data obtained from Nightingale et al. [14]. DES using the Spalart-Allmaras (S-A) turbulence model [15], SAS using the SST k-ω turbulence model and URANS using the SST k-ω turbulence model were employed for the computations.

2 Flow Solver and Turbulence Modelling

HMB2 [13], used for the flow computations presented in this paper, is a parallel, cell-centred finite volume flow solver for the three-dimensional, unsteady, compressible Navier-Stokes equations on multi-block structured grids and was developed at the University of Liverpool. More details of the employed CFD solver and turbulence models are given in Barakos et al. [1]. For the work presented in this paper, DES was employed along with the Spalart-Allmaras turbulence model [15].

Direct Numerical Simulation (DNS) was not practical for this flow, due to computational resources, and so early studies of these flows used CFD approaches that relied on the use of Reynolds-Averaged Navier-Stokes (RANS) equations together with a turbulence model. Unsteady RANS (URANS), however, was unable to predict the full spectrum of turbulent scales and with Large-Eddy Simulation (LES) being too expensive near wall regions, Detached-Eddy Simulation (DES) became the popular choice for these flows. The original idea of DES was postulated by Spalart et al. [15]. Its underlying principle involved using RANS for the near-wall and boundary layer and LES everywhere outside. Spalart et al. [15] modified the S-A model to achieve a DES equivalent. The wall distance (d) is now recomputed according to the DES principle and represented by $\tilde{d}$. In the pure one-equation S-A turbulence model [15], the terms $\tilde{d}$ and d are identical. However, the DES formulation of $\tilde{d}$ is given by:

$$\tilde{d} = C_{DES}\Delta \tag{1}$$

where C_{DES} is a constant and Δ is the metric of the grid size. In practice, the distance to the wall in the DES formulation of the one-equation S-A model is expressed as a comparison between the actual distance to the wall and that calculated by $C_{DES}\Delta$, which essentially computes the size of the maximum cell length.

$$\tilde{d} = \min\left(d, C_{DES}\Delta\right) \tag{2}$$

$$\Delta = \max\left(\Delta_x, \Delta_y, \Delta_z\right) \qquad \forall \ \text{cell}. \tag{3}$$

When the cell length ($C_{DES}\Delta$) is less than the actual distance to the nearest wall (d), LES is triggered. RANS is activated when the converse occurs. This boundary between LES and RANS is therefore completely dependent on the geometry and on the density of the computational mesh.

The governing equations of the SST-SAS model differ from those of the SST-RANS model by the additional SAS source term Q_{SAS} in the transport equation for the turbulence eddy frequency ω given by

$$Q_{SAS} = max\left[\rho\zeta_2\kappa S^2\left(\frac{L}{L_{vK}}\right)^2 - C\frac{2\rho k}{\sigma_\Phi}max\left(\frac{1}{\omega}\frac{\partial\omega}{\partial x_j}\frac{\partial\omega}{\partial x_j}, \frac{1}{k}\frac{\partial k}{\partial x_j}\frac{\partial k}{\partial x_j}\right), 0\right] \tag{4}$$

where $\zeta_2 = 3.51$, $\sigma_\Phi = 2/3$ and $C = 2$.

The length scales of the modelled turbulence L and the von Karman length scale L_{vK} are defined as:

$$L = \frac{\sqrt{k}}{c_\mu^{1/4}\omega}, \quad L_{vK} = \frac{\kappa S}{|U''|} \tag{5}$$

where $\kappa = 0.41$ is the von Karman constant.

To provide proper damping of resolved turbulence at high wave numbers, the SST-SAS model requires a lower constrain on the L_{vK} given by:

$$L_{vK} = max\left(\frac{\kappa S}{|U''|}, C_s\sqrt{\frac{\kappa\zeta_2}{(\frac{\beta}{c_\mu}) - \alpha}} \cdot \Delta\right) \tag{6}$$

3 Model Geometry and Mesh Generation

For all configurations, the geometry, the structured multi-block topologies and the mesh (collectively are know as the grid) were generated using ICEMCFD v13. The experimental Mach and Reynolds number (based on the cavity length), for doors-on and doors-off cavities, were 0.85 and 6.78 million respectively. Unsteady pressure measurements were taken inside the cavity via 10 pressure transducers, on the cavity floor [14, 16] for which data was sampled at a rate of 6 kHz for approximately 3.5 s. The geometry used to create the grids was modelled on the wind-tunnel experiments that measured 20 in. in length and 4 in. in width and depth giving an L/D of 5 and W/D of 2. Barakos et al. [1] provide a detailed description of multi-block topologies employed in the M219 clean cavity. CFD grids of approximately 5.0 million cells (doors-off) and 5.5 million (doors-on) were used to perform the computations, where the grid densities were based on previous experience with the numerical method [1]. For all computed cases, the free-stream Mach number was kept at 0.85 and the Reynolds number based on the length of the cavity was 6 million (based on the cavity length).

4 Results and Discussion

From past experience with cavity flows [2], DES computations were run at a non-dimensional time-step of 0.001, for results presented in this paper, while SAS computations used a larger time-step of 0.01, similar to that used by Egorov et al. [12]. Figure 1 shows the comparison between DES and SAS computations run at the same non-dimensional time-step of 0.01 for the clean cavity with doors on. Evidently the SAS model can run at a larger time-step which was adopted for the rest of the paper.

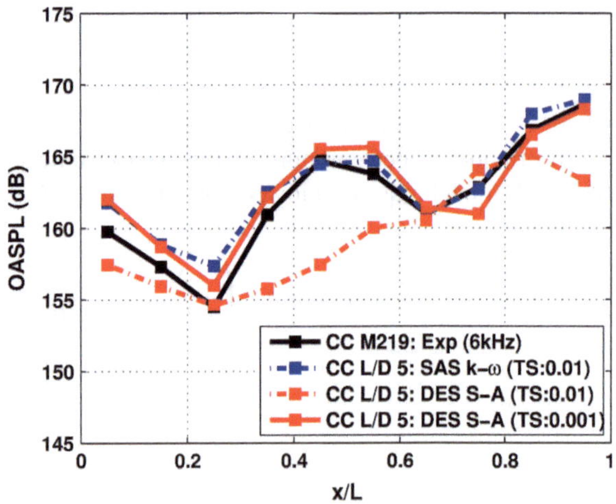

Fig. 1 OASPL for the clean cavity with doors on comparing DES and SAS results with experimental results from Nightingale et al. [14]. DES results are shown for two non-dimensional time-steps of 0.01 and 0.001, while SAS results are shown for a time-step of 0.01

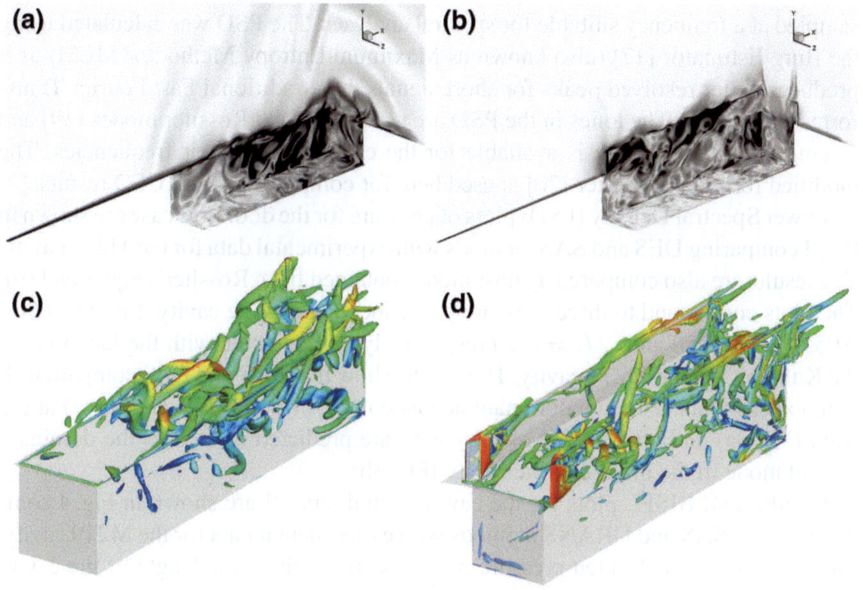

Fig. 2 Instantaneous contours of numerical schlieren (**a**, **b**) and iso-surface of Q-Criteria (**c**, **d**) for the cavity with doors-on and doors-off as predicted using SAS. Planes are located at x/L = 0.99, y/L = −0.19 and z/L = −0.04. Iso-surfaces at Q = 2000 are shown and coloured with Mach number ranging between 0.0 (*blue*) and 1.0 (*red*). Numerical Schlieren is calculated as: $NS = c_1 \exp[-c_2(x - x_{min})/(x_{max} - x_{min})]$, where $x = |\nabla\rho|$, c_1 and c_2 are constants, $c_1 = 0.8$ and $c_2 = 10$

Instantaneous contours of numerical schlieren and iso-surfaces of Q-Criteria as predicted using SAS, for the clean cavity with doors-off and-on are shown in Fig. 2 with one slice in each direction to make up a three-dimensional view inside the cavity. The unsteadiness and breakdown of the shear layer is seen as it separates from the leading edge of the cavity at about a third of its length. The doors-off case showed 'spillages' over the edges of the cavity. The addition of the doors, however, caused the early breakdown of the shear layer and restricted the flow along the spanwise direction. The propagation of noise out of the cavity is visualised along three directions. Strong acoustic waves are observed for the cavity with doors-off. Pressure waves were seen to travel towards the aft wall, and are then reflected out and towards the front wall of the cavity. Iso-surface of Q-Criteria showed small and long structures along the shear layer and by the doors of the cavity. While the doors-off case showed structures spilling out of the cavity towards the aft wall, the doors-on case had longer structures along the length of the cavity originating at the doors.

HMB2 outputs flow-field files, that are written at specific instances of time as specified by the user before the computation is started. In addition to this, HMB2 also outputs data from specific probes that are placed in the flow. The locations of the probes are defined at the beginning of the computation and are then written at every time step. For the cavity flow computations, these probes were defined in the same locations as the Kulite^{TM} pressure transducers in experiments and are therefore sampled at a frequency suitable for spectral analyses. The PSD was calculated using the Burg Estimator [17] (also known as Maximum Entropy Method or MEM) as it produces better resolved peaks for short signals than traditional Fast Fourier Transforms (FFT) [18]. The tones in the PSD are usually termed Rossiter modes [19] and a semi-empirical formula is available for the estimation of their frequencies. The modified formula by Heller [20] is used here for comparisons with CFD results.

Power Spectral Density (PSD) plots of pressure for the doors-off case are shown in Fig. 3 comparing DES and SAS methods with experimental data for the M219 cavity. The results are also compared against modes obtained from Rossiter's equation [19]. The plots correspond to three pressure probe locations on the cavity floor at $x/L = 0.05$, $y/L = 0.50$ and $z/L = 0.90$ respectively that coincide with the locations of the Kulites^{TM} in the M219 cavity. The results show that SAS and DES compare well with experiments. The two dominant acoustic modes (modes two and three) at the front (Fig. 3a) and aft (Fig. 3c) of the cavity are predicted as well as the dominant second mode in the middle of the cavity (Fig. 3b).

OASPL and BISPL plots for the cavity with doors-off are shown in Fig. 4 comparing DES, SAS and URANS methods with experimental data for the M219 cavity. The plots correspond to ten pressure probe locations along the length of the cavity

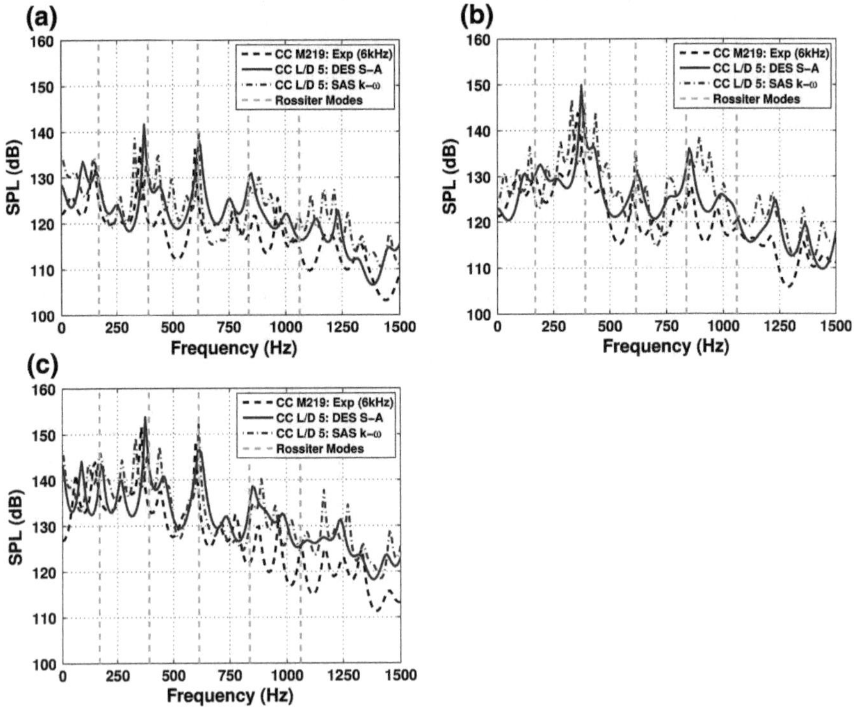

Fig. 3 PSD plots for the clean cavity with doors-off comparing results from DES and SAS methods to experimental data for the M219 cavity from Nightingale et al. [14]. Plots are for the front (**a**), middle (**b**) and rear (**c**) transducers on the cavity floor and presented in terms of SPL. *CC* Clean Cavity, *SA* Spalart Allmaras. **a** $x/L = 0.05$, **b** $x/L = 0.50$, **c** $x/L = 0.95$

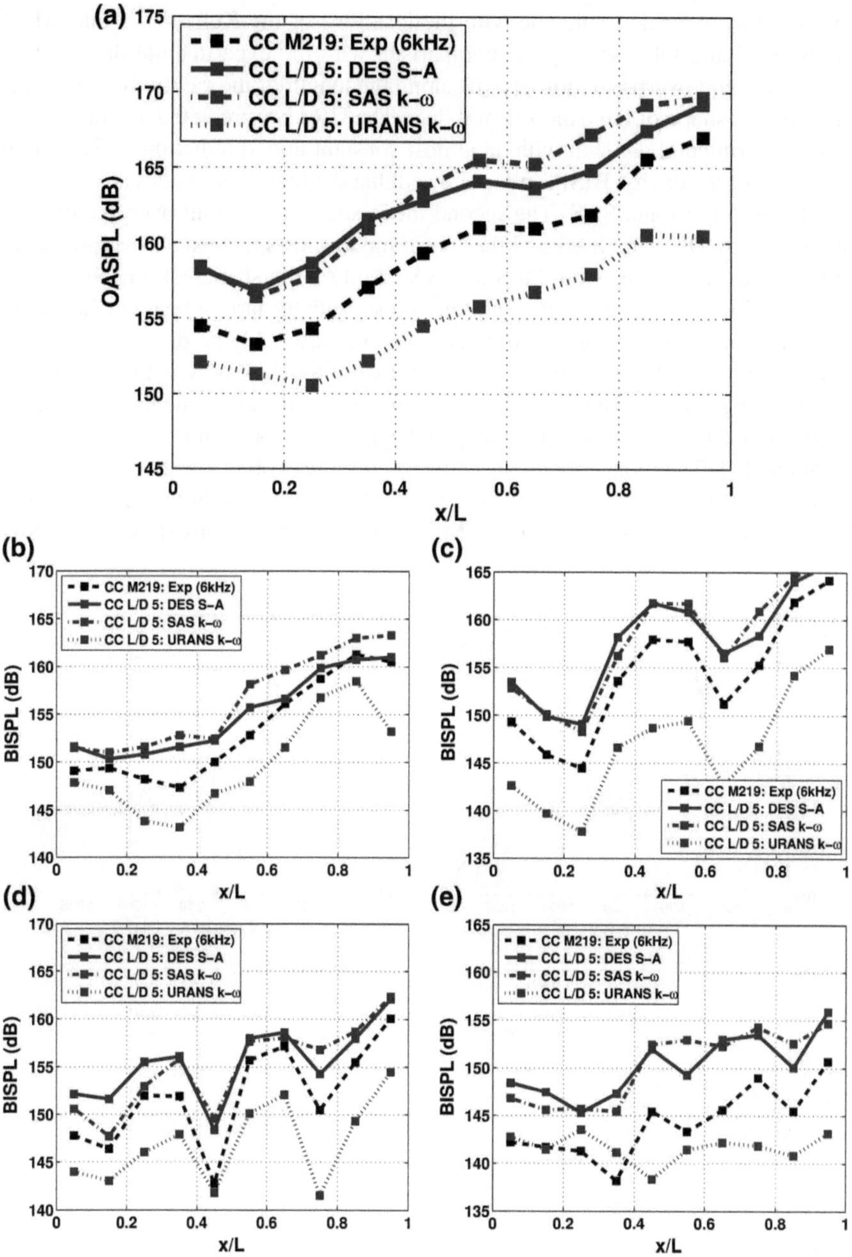

Fig. 4 OASPL (**a**) and BISPL (**b**–**e**) along the cavity floor for the clean cavity with doors-off. Plots compare results from DES, SAS and URANS to experimental data from Nightingale et al. [14]. **a** OASPL, **b** Mode 1: $50 \leq f \leq 250$ HZ, **c** Mode 2: $250 \leq f \leq 450$ HZ, **d** Mode 3: $500 \leq f \leq 700$ HZ, **e** Mode 4: $700 \leq f \leq 900$ HZ

on the cavity floor that coincide with the locations of the Kulites^{TM} in the M219 cavity. SAS and DES show good comparison with the experimental data with an almost constant overprediction of 5 dB along the length of the cavity. Both methods captured the shape of the experimental data. The URANS predicted a similar shape to the experimental data but with an almost constant underprediction of 5 dB along the length of the cavity. BISPL plots showed that the first three modes are predicted well by both DES and SAS. The second mode shows a constant overprediction of about 4 dB by DES and SAS and the fourth mode shows a constant overprediction between 4 dB and 6 dB. Both DES and SAS capture the shapes of all four modes. URANS shows underprediction in all four modes with the largest being in the second mode of about 5 dB. The shapes of the curves are reasonably predicted.

Power Spectral Density (PSD) plots of pressure for the cavity with doors-on are shown in Fig. 5 comparing DES and SAS methods with experimental data for the M219 cavity. The results are also compared against modes obtained from Rossiter's equation [19]. The plots correspond to three pressure probe locations on the cavity floor at $x/L = 0.05$, $y/L = 0.50$ and $z/L = 0.90$ respectively that coincide with the locations of the Kulites^{TM} in the M219 cavity. Comparisons showed that SAS

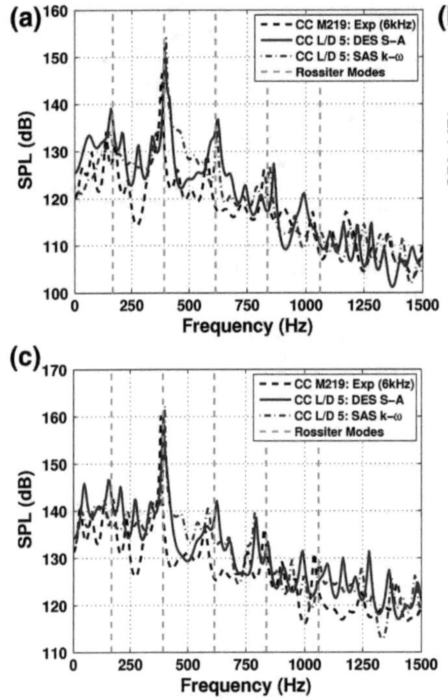

Fig. 5 PSD plots for the clean cavity with doors-on comparing results from DES and SAS methods to experimental data for the M219 cavity from Nightingale et al. [14]. Plots are for the front (**a**), middle (**b**) and rear (**c**) transducers on the cavity floor and presented in terms of SPL. *CC* Clean Cavity, *SA* Spalart Allmaras **a** $x/L = 0.05$, **b** $x/L = 0.50$, **c** $x/L = 0.95$

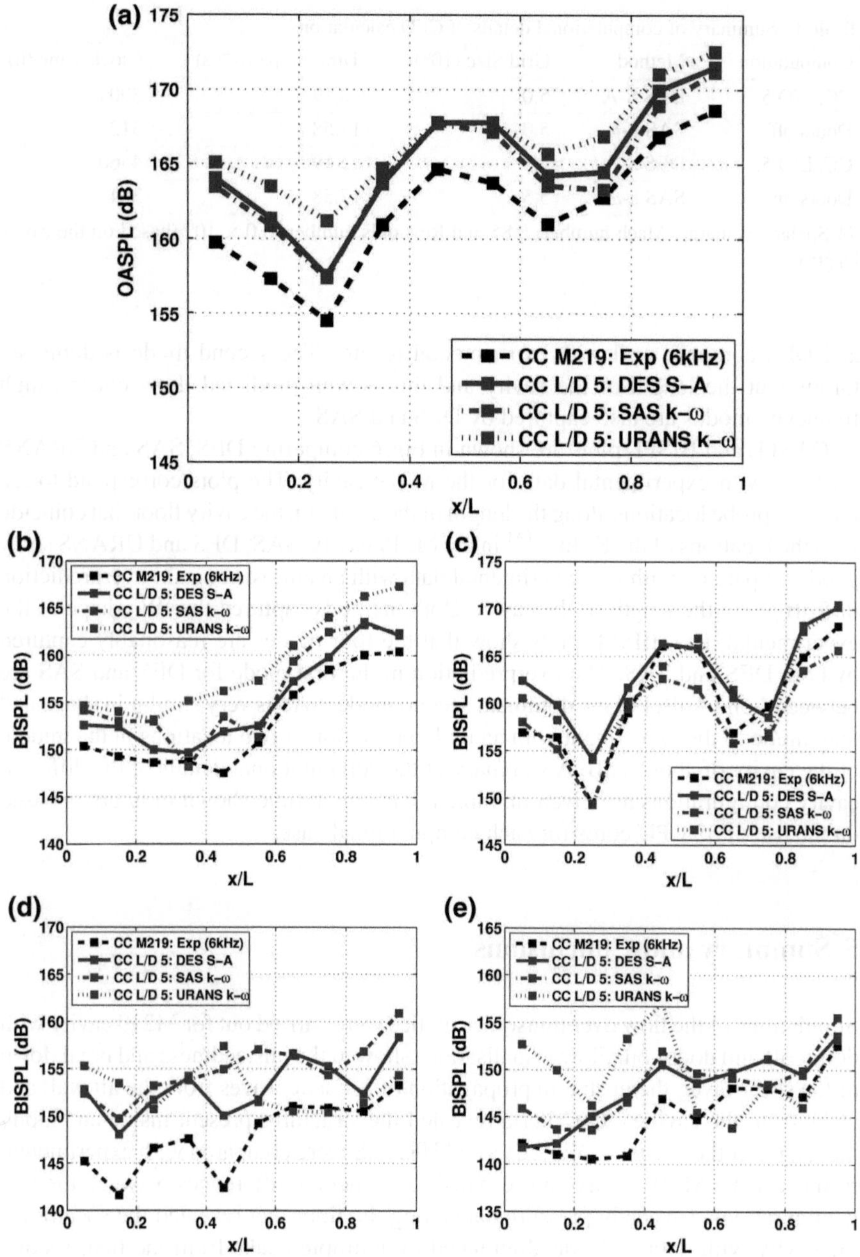

Fig. 6 OASPL (**a**) and BISPL (**b–e**) along the cavity floor for the clean cavity with doors-on. Plots compare results from DES, SAS and URANS to experimental data from Nightingale et al. [14]. **a** OASPL, **b** Mode 1: $50 \leq f \leq 260$ HZ, **c** Mode 2: $250 \leq f \leq 450$ HZ, **d** Mode 3: $500 \leq f \leq 700$ HZ, **e** Mode 4: $700 \leq f \leq 900$ HZ

Table 1 Summary of computational details of CFD calculations

Computation	Method	Grid Size (10^6)	Time-Step (10^{-5}s)	Clock Time (h)
CC, L/D 5,	DES S-A	5.0	2.19	3909
Doors off	SAS k-ω	5.0	17.58	312
CC, L/D 5,	DES S-A	5.5	2.19	4560
Doors on	SAS k-ω	5.5	17.58	364

SA Spalart Allmaras, Mach number: 0.85 and Reynolds number: 6.0×10^6 (based on the cavity length)

and DES compare well with experimental results. The second mode is dominant throughout the length of the cavity and this is well predicted. Low energy, high frequency modes are also captured by DES and SAS.

OASPL and BISPL plots are shown in Fig. 6 comparing DES, SAS and URANS methods with experimental data for the M219 cavity. The plots correspond to ten pressure probe locations along the length of the cavity on the cavity floor that coincide with the locations of the Kulites^{TM} in the M219 cavity. SAS, DES and URANS show good comparison with the experimental data with an almost constant overprediction of 5 dB along the length of the cavity. Both methods captured the 'W' shape of the experimental data. BISPL plots show that the four modes are reasonably captured by both DES and SAS. The overprediction in the first mode for DES and SAS are between 1 and 4 dB. For the dominant second mode, SAS is very similar in shape and magnitude to the experiment with an underprediction at two locations in the middle of the cavity of about 5 dB. A summary of the computational details of the different cavity configurations are given in Table 1. The clock time shown in hours is based on the use of 32 CPU cores for each computational case.

5 Summary and Conclusions

Simulations of the flow over transonic cavities were carried out for M219 cavity with doors-off and doors-on. Flow visualisations showed the unsteadiness and breakdown of the shear layer, the upstream propagation of acoustic waves from the aft wall and away from the cavities. Q-Criteria revealed the structures present inside and along the length of the cavities. DES, SAS and URANS were compared with experimental results for the M219 cavity and revealed fundamental differences between the two configurations. Unsteady pressure data along the floor that revealed the spectra for the cavity with doors-off was dominated by multiple peaks from the first, second and third modes, however, the cavity with doors-on was dominated by the second mode only. SAS showed good agreement with experimental results for the cavity with doors-off and doors-on. SAS produced these results at about a 10th of the time of DES. Future work will focus on simulations of a store in the cavity to further demonstrate the use of SAS and the possibility to extend it to the problem of store release from cavities.

Acknowledgments The financial support of the Engineering and Physical Sciences Research Council and MBDA through Industrial CASE: 09000510 is gratefully acknowledged. The authors would also like to thank Nigel Taylor of MBDA for his support in this work. The use of the POLARIS HPC cluster of N8 and the Chadwick HPC cluster of the University of Liverpool are also gratefully acknowledged.

References

1. Barakos, G.N., Lawson, S.J., Steijl, R., Nayyar, P.: Numerical simulations of high speed turbulent cavity flows. Flow Turbul. Combus. **83**(4), 569–585 (2009)
2. Lawson, S.J., Barakos, G.N.: Evaluation of DES for weapons bays in UCAVs. Aerosp. Sci. Technol. (2010, in Press). doi:10.1016/j.ast.2010.04.006
3. Menter, F.R., Kuntz, M., Bender, R.: A scale-adaptive simulation model for turbulent flow predictions. In: 41st aerospace sciences meeting and exhibit (2003). doi:10.2514/6.2003-767. Accessed 06–09 Jan 2003
4. Menter, F.R., Egorov, Y.: Revisiting the turbulent length scale equation. One hundred years of boundary layer research, Gottingen. In: IUTAM symposium (2004)
5. Menter, F.R., Egorov, Y.: A scale-adaptive simulation model using two-equation models. In: AIAA paper 20051095, Reno, NV (2005)
6. Davidson, L.: Evaluation of the SST-SAS model: channel flow, asymmetric diffuser and axi-symmetric hill, ECCOMAS CFD 2006. Egmond aan Zee, The Netherlands (2006). Accessed 5–8 Sept 2006
7. Davidson, L.: The SAS model: a turbulence model with controlled modelled dissipation. In: 20th Nordic Seminar on Computational Mechanics, Gotenburg, 20–23 Nov 2007
8. Egorov, Y., Menter, F.R.: Development and application of SST-SAS turbulence model in the DESIDER project. In: Advances in hybrid RANS-LES modelling. Notes on numerical fluid mechanics and multidisciplinary design vol. 97, p. 261 (2008)
9. Smirnov, P.E., Kapetanovic, S., Braaten, M.E., Egorov, Y., Menter, F.R.: Application of the SAS turbulence model to buoyancy driven cavity flows. In: ASME Paper GT2009-59621 (2009)
10. Menter, F.R., Schutze, J., Kurbatskii, K.A., Gritskevich, M., Garbaruk, A.: Scale-resolving simulation techniques in industrial CFD. In: 6th AIAA Theoretical Fluid Mechanics Conference, Honolulu, Hawaii, 27–30 June 2011
11. Menter, F.R., Egorov, Y.: The scale-adaptive simulation method for unsteady turbulent flow predictions. Part 1 Theory Model Descr. Flow Turbul. Combust. **85**, 113 (2010)
12. Egorov, Y., Menter, F., Lechner, R., Cokljat, D.: The scale-adaptive simulation method for unsteady turbulent flow predictions. Part 2 Appl. Complex Flows Flow Turbul. Combust. **85**, 139 (2010)
13. Barakos, G., Steijl, R., Badcock, K., Brocklehurst, A.: Development of CFD capability for full helicopter engineering analysis. In: 31st European Rotorcraft Forum, vol. 2005, pp. 91.1–91.15 (2005)
14. Nightingale, D.A., Ross J.A., Foster, G.W.: Cavity unsteady pressure measurements—examples from wind-tunnel tests. Technical Report Version 3, Aerodynamics and Aeromechanics Systems Group, QinetiQ (2005)
15. Spalart, P.R., Allmaras, S.R.: A one-equation turbulence model for aerodynamic flows. La Recherche Aerospatiale **1**, 1–5 (1994)
16. Ross, J.A.: Cavity acoustic measurements at high speeds. Technical Report DERA/MSS/MSFC2/ TR000173, QinetiQ (2000)

17. Childers, D.G. (ed.): Modern Spectrum Analysis. Chapter 2, pp. 23148. IEEE Press, New York (1978)
18. Larcheveque, L., Sagaut, P., Le, T.-H., Comte, P.: Large-eddy simulation of a compressible flow in a three-dimensional open cavity at high Reynolds number. J. Fluid Mech. **516**, 265–301 (2004)
19. Rossiter, J.E.: Wind tunnel experiments on the flow over rectangular cavities at subsonic and transonic speeds. Technical Report 64037, Royal Aircraft Establishment (1964)
20. Heller, H.H., Holmes, D.G., Covert, E.: Flow induced pressure oscillations in shallow cavities. J. Sound Vib. **18**, 545 (1971)

Challenges in Variable Resolution Simulations of Separated Flow Over Delta Wings

Jacob M. Cooper and Sharath S. Girimaji

Abstract Variable-resolution simulations of a 50° swept, sharp leading edge, delta wing are performed at low Reynolds number using the Unsteady Reynolds Averaged Navier Stokes (URANS) and Partially Averaged Navier Stokes (PANS) turbulence models. The study focuses on the ability of the models to capture the vortex structure both forward and aft of vortex breakdown. A challenging aspect of the flow is that the pre-breakdown vortex is nearly laminar, and turbulent flow appears only after breakdown occurs near mid-chord. If the upstream resolution is adequate to resolve all laminar features, then both URANS and PANS reasonably capture the large scale features of the flow in the downstream region. It is found that inadequate resolution in the mostly laminar forward vortex leads to poor simulation of the aft vortex and aft flow features irrespective of the model fidelity.

1 Introduction

Future unmanned combat aerial vehicles (UCAVs) will employ delta wing geometries due to their ability to maintain their stability, control, and lift at high incidence angles. Delta wing aerodynamics at large angles of attack is characterized by large vortices which separate from the leading edge of the wing creating a complex separated flow field. There have been numerous computational and experimental studies performed on delta wings with various leading edge sweep angles and leading edge curvature, [1–4].

The delta wings which are currently being proposed for use in UCAVs such as the Northrop Grumman X-47A will have moderate leading edge sweep angles (45°–60°). A comprehensive high-fidelity delta wing data set was generated by the direct numerical simulation (DNS) study of Gordnier and Visbal [5]. Their DNS computes a sharp leading edge delta wing with 50° sweep angle at $Re = 26{,}000$ at 5°, 10°, and 15° angle of attack using a 6th order compact difference scheme, with

J.M. Cooper (✉) · S.S. Girimaji
Texas A&M University, College Station, TX 77843-3141, USA
e-mail: Jacob.Cooper.06@gmail.com

S. Girimaji et al. (eds.), *Progress in Hybrid RANS-LES Modelling*,
Notes on Numerical Fluid Mechanics and Multidisciplinary Design 130,
DOI 10.1007/978-3-319-15141-0_36

an 8th order low pass spatial filter. Although the Reynolds number for their study is significantly lower than the Reynolds numbers expected for flight, these simulations are useful for understanding fundamental flow physics of delta wings and provide an important low Reynolds number benchmark for validating computational tools.

The principal objective of this study is to determine the degree of physical/numerical resolution required to capture different aspects of the delta wing flow features. In this study we will employ URANS and PANS to compute the low Reynolds number case of Gordnier and Visbal [5]. The main challenges to simulating delta wing separated flows will be identified.

2 PANS Closure Modeling

The PANS model is derived by decomposing the flow into resolved and unresolved parts, followed by applying an arbitrary filter which commutes with temporal and spatial differentiation to the Navier Stokes equations [6].

$$\frac{\partial \langle u_i \rangle}{\partial t} + \frac{\partial \langle u_i \rangle \langle u_k \rangle}{\partial x_k} = -\frac{\partial \langle p \rangle}{\partial x_i} + 2\nu \frac{\partial \langle s_{ik} \rangle}{\partial x_k} - \frac{\partial \tau (u_i, u_k)}{\partial x_k} \tag{1}$$

The resolution of the PANS simulation is determined via the filter control parameters f_k and f_ε which are the ratios of unresolved-to-total turbulence kinetic energy and unresolved-to-total turbulence dissipation rate. In the $k-\omega$ context, f_ε is transformed into f_ω via $f_\omega = f_\varepsilon / f_k$.

$$f_k = \frac{k_u}{k}; \qquad f_\omega = \frac{\omega_u}{\omega} \tag{2}$$

$$\tau (u_i, u_j) = -\nu_u \left(\frac{\partial \langle u_i \rangle}{\partial x_j} + \frac{\partial \langle u_j \rangle}{\partial x_i} \right) + \frac{2}{3} k_u \delta_{ij}; \quad \nu_u = \frac{k_u}{\omega_u} = \frac{f_k k}{f_\omega \omega} \tag{3}$$

The final form of the PANS $k - \omega$ model equations can be summarized as:

$$\frac{\partial k_u}{\partial t} + \langle u_j \rangle \frac{\partial k_u}{\partial x_j} = P_u - \beta^* k_u \omega_u + \frac{\partial}{\partial x_j} \left[\left(\nu + \frac{\nu_u}{\sigma_{ku}} \right) \frac{\partial k_u}{\partial x_j} \right] \tag{4}$$

$$\frac{\partial \omega_u}{\partial t} + \langle u_j \rangle \frac{\partial \omega_u}{\partial x_j} = \alpha \frac{P_u \omega_u}{k_u} - \beta' \omega_u^2 + \frac{\partial}{\partial x_j} \left[\left(\nu + \frac{\nu_u}{\sigma_{\omega_u}} \right) \frac{\partial \omega_u}{\partial x_j} \right] \tag{5}$$

The RANS closure coefficients are unchanged for α and β^*, while the remaining closure coefficients are modified as given below:

$$\sigma_{ku} \equiv \sigma_k \frac{f_k}{f_\omega}; \qquad \beta' \equiv \alpha\beta^* - \frac{\alpha\beta^*}{f_\omega} + \frac{\beta}{f_\omega}; \qquad \sigma_{\omega u} \equiv \sigma_\omega \frac{f_k}{f_\omega} \tag{6}$$

The user can achieve accuracy-on-demand by varying the filter control parameter from a coarse RANS simulation to a fully-resolved DNS simulation depending upon the grid size: $f_k \geq 3\,(\Delta/\Lambda)^{(2/3)}$ [7] where Δ is the cube-root of the cell volume and $\Lambda = k^{(3/2)}/\varepsilon$. The PANS model is uniquely suited to simulate the separated delta wing flow which is comprised of regions which require high fidelity resolution, along with regions where a low fidelity model may be applied. The details of the original derivation of the PANS model can be found in [8], and its validity has been established in various works: [9–13].

3 Simulation and Results

This section presents results for the 50° leading edge sweep, flat plate delta wing of [5] simulated at $\alpha = 15°$ with URANS and PANS turbulence models. Comparisons are made against DNS data when possible. In the first subsection, we briefly identify various vortical flow features which should be found in our simulations. In the second subsection, we compute the delta wing flow by performing simulations on the fine DNS grid at flow conditions (Re = 26,000) identical to the DNS study. The third subsection provides URANS and PANS results at the same flow conditions using a significantly coarsened grid to determine whether the important flow features can be computed at lower resolutions.

3.1 Flow Features and Challenges

According to [5], there are several flow features which one would expect to encounter when simulating such a case. At $\alpha = 15°$, there should be a distinct primary and secondary vortex, and a subtle tertiary vortex present in the pre-breakdown region. Instabilities should be present in the separated shear layer and within the primary vortex by $x/c = 0.3$. Vortex breakdown should occur in the range of $x/c = 0.40 - 0.54$. After breakdown, the dominant primary and secondary vortices are disintegrated into fine scale structures with no semblance of a dominant vortex. A plot of stream-wise vorticity iso-surfaces from Gordnier and Visbal [5] is provided in Fig. 1 to orient the reader.

The flow is nearly laminar before vortex breakdown. However, the laminar vortex is complex featuring steep gradients and separation. The shear layer in this region harbors the instabilities that lead to vortex breakdown and ultimately turbulence. Thus, despite being laminar, the forward vortex region needs high numerical resolution. After vortex breakdown, turbulence develops rapidly in the aft half of the wing. These complex flow features present several challenges to hybrid computations.

To understand the importance of high fidelity computations of the laminar region we perform two sets of hybrid simulations. The first set employs the high resolution DNS grid ensuring accurate simulations of the laminar region. The second set uses

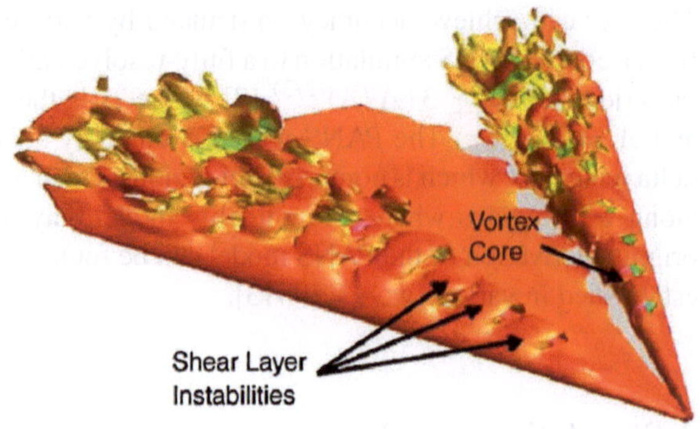

Fig. 1 Isosurface of stream-wise vorticity [5]

a coarse grid in the laminar region and beyond. Contrasting the results can lead to a clearer understanding of the resolution needs of this complex flow.

3.2 Low Reynolds Number Fine Grid Simulations

The first simulations are performed using nearly identical test conditions and an identical grid to the DNS study of Gordnier and Visbal [5]. The only difference between the two simulations is that the present study employs a 4th order spatial discretization scheme while the DNS study utilizes a 6th order scheme. The purpose of this study is to examine the closure model capability in resolving the various flow features. Despite the fine grid, URANS is not expected to capture much of the turbulent scales of motion. Similarly, PANS is expected to capture the range of scales permitted by f_k specification. By ensuring that the grid resolution is adequate, the ability of the model to resolve flow features can be isolated.

Before starting a PANS simulation, it is useful to complete a URANS simulation to determine the appropriate values for f_k and f_ε. The results shown in Fig. 2 are contours of $f_k = 3\,(\Delta/\Lambda)^{(2/3)}$ and turbulent Reynolds number $R_t = k/(\omega\nu)$ for the URANS simulation at 8 stream-wise stations along the chord ranging from $x/c = 0.2$ to $x/c = 0.9$. The contours of R_t clearly indicate that the turbulence levels are substantial only in the aft regions after the vortex breakdown. In the forward region, the flow is nearly laminar as expected. This clearly reveals that in the forward region all closure models and DNS should provide similar results as the flow is nearly laminar. Even more importantly, the R_t contours imply that coarsening the grid in the forward region will lead to poor resolution of the laminar vortex evolution.

The other half of Fig. 2 shows contours of f_k. These results indicate that the only location where one might expect to see improvement with a PANS simulation is on the aft portion of the wing after breakdown has occurred. The contours indicate that, in the aft region, the grid should support a very low f_k near zero. The implication is clear: in this region, the grid is suitable for DNS. One weakness of this analysis is

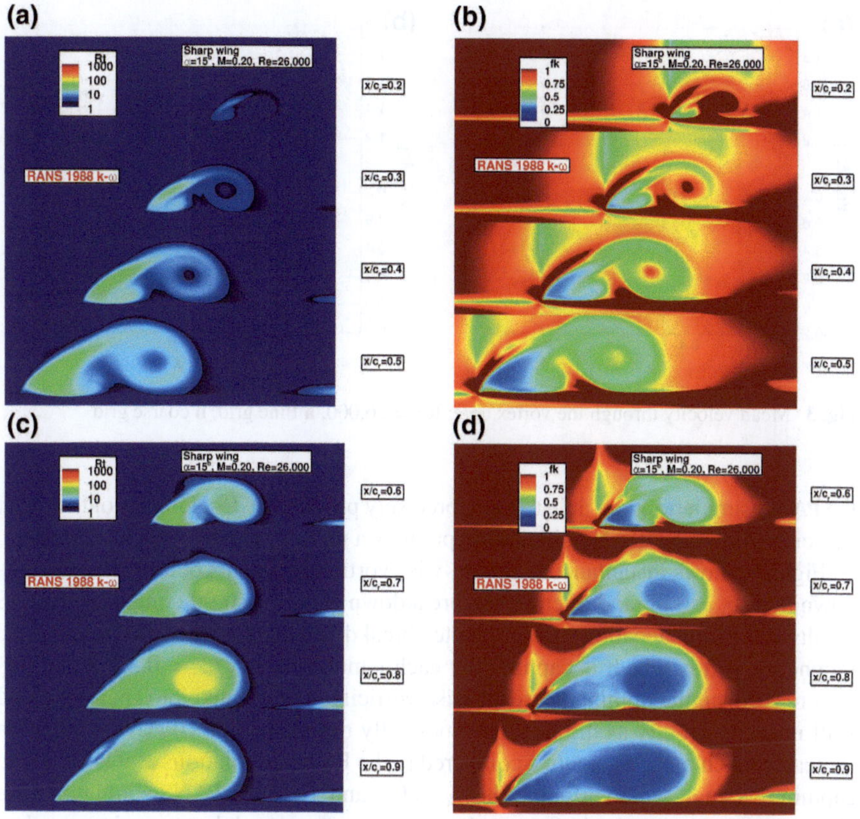

Fig. 2 URANS results; Re = 26,000; fine grid. **a** Forward R_t contours, **b** forward f_k contours, **c** aft R_t contours **d** aft f_k contours

the use of turbulence quantities in an essentially laminar flow regions. However, we only perform this type of analysis in order to gauge the overall levels of turbulence in the flow and to locate regions where more flow structures can be liberated via a PANS simulation.

Following the URANS simulations, we perform PANS simulations at $f_k = 0.5$ and $f_k = 0.1$. One of the few quantitative results presented in [5] is the mean velocity magnitude along a straight ray which starts at the wing apex and passes through the core of the primary vortex. The results for the mean velocity magnitude along this line are presented in Fig. 3a. The URANS and PANS simulations are all able to capture the mean velocity magnitude to a reasonable degree. It is well established that the strong primary vortex acts in a jet-like manner accelerating the flow up until vortex breakdown where the vortex acts in a wake-like manner decelerating the flow. This trend is captured by the present simulations. The location of vortex breakdown correlates with the velocity peak, and the present simulations predict an early vortex breakdown compared to the DNS. As the primary vortex undergoes breakdown,

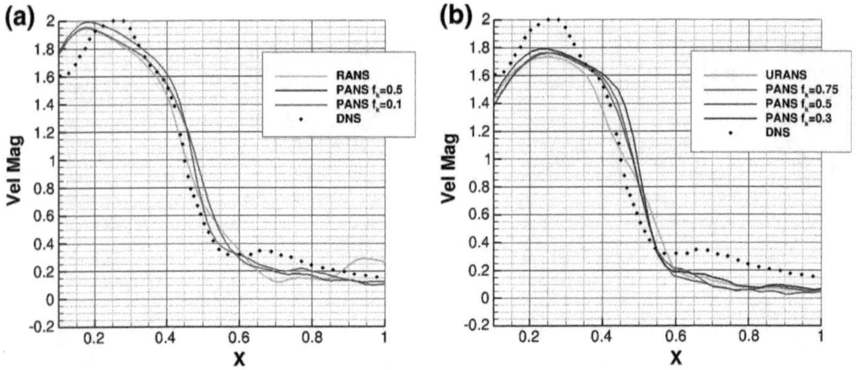

Fig. 3 Mean velocity through the vortex core; Re = 26,000. **a** Fine grid, **b** coarse grid

the PANS simulations are able to more precisely predict the deceleration of the mean velocity while the URANS simulation predicts a more gradual vortex breakdown.

Figure 4 shows instantaneous stream-wise vorticity contours before vortex breakdown at $x/c = 0.32$ and after vortex breakdown at $x/c = 0.99$ compared to DNS results at the same locations. Before vortex breakdown there is a distinct primary and secondary vortex which is captured by each simulation at $x/c = 0.32$. After vortex breakdown, the dominant stream-wise vorticity is reoriented into span-wise and wall-normal vorticity as the flow becomes fully turbulent. The URANS simulation is clearly much more dissipative compared to the PANS simulations as it is unable to capture the small scale vorticity. The $f_k = 0.5$ and $f_k = 0.1$ PANS simulations are able to resolve increasingly finer scale structures after breakdown much more than the URANS simulation, as expected. The difference between the $f_k = 0.1$ and DNS can be attributed to two reasons: (i) the finer structures are due to fluctuations that carry less than 10 % of the turbulence kinetic energy, and (ii) the numerical scheme of PANS is only 4th order compared to the 6th order DNS scheme.

The results for the mean stream-wise vorticity are presented in Fig. 5. There are very few discernible differences in the mean vorticity between the four simulations. The URANS simulation dissipates the strength of the vorticity in the separated shear layer, while the PANS simulations preserve the intensity of the shear layer vorticity. In addition, the tertiary vortex is resolved slightly better by the PANS simulations compared to the URANS simulation. However, in general, the URANS simulation performs reasonably well in predicting the mean flow structure which is not entirely unexpected given the low Reynolds number and fine grid resolution. With the mean flow velocity and vorticity reasonably well captured, we proceed to perform further PANS calculations.

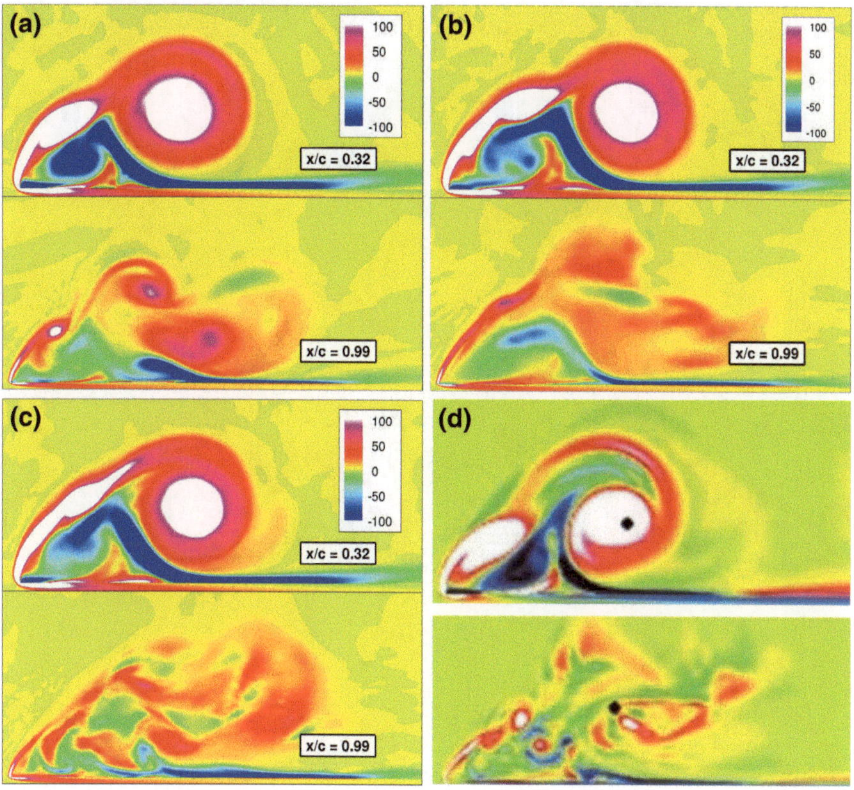

Fig. 4 Instantaneous contours of stream-wise vorticity; Re = 26,000; fine grid. **a** URANS, **b** $f_k = 0.5$, **c** $f_k = 0.1$, **d** DNS [5]

3.3 Low Reynolds Number Coarse Grid Simulations

It is evident from Fig. 2 that the coarser grids may not be adequate in the forward region of the wing. The complex vortical flow here is nearly laminar and must be adequately resolved. In the aft region, coarser grids may be used with appropriate closure models. To examine the effect of under-resolving the laminar portion, we perform computations on a coarser grid. To this end, the original 4.5 million cell DNS grid was coarsened to approximately 0.9 million cells by removing every other grid point in the stream-wise and span-wise directions, and removing a small number of grid points in the wall-normal direction while keeping the wall spacing constant.

As in the previous section, we begin with a URANS simulation to determine the turbulence levels in the flow, and the appropriate f_k for the subsequent PANS simulations. Figure 6 presents results for the URANS simulation using the coarsened grid at the same flow conditions presented in the previous section. As mentioned before, the contours of R_t demonstrate that the Reynolds number of the flow is too

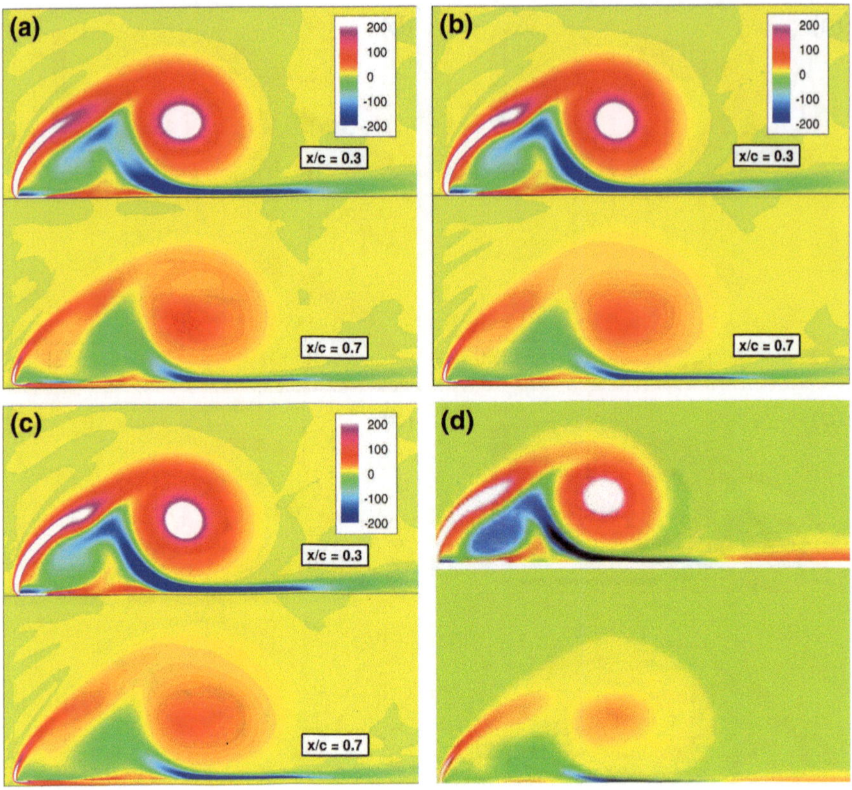

Fig. 5 Mean contours of stream-wise vorticity; Re = 26,000; fine grid. **a** URANS, **b** $f_k = 0.5$, **c** $f_k = 0.1$, **d** DNS [5]

low for any appreciable amount of turbulence to be generated until the most aft areas of the wing. Essentially, the separated vortices remain laminar until breakdown near $x/c = 0.5$. The contours of f_k in Fig. 6 reiterate that point. At $x/c = 0.2$, 0.3, and 0.4, the f_k contours in the core of the primary vortex remain at 1.0. This is an indication that there is essentially no turbulence in this solid body rotation vortex, and the vortex simply must be resolved with a finer grid. It is only well after vortex breakdown that enough turbulence is generated to employ any f_k reduction with a PANS simulation.

Before proceeding with the PANS simulations, it must be stated that Fig. 6 indicates that the grid could be quite inadequate for the pre-breakdown region $x/c < 0.5$. As seen in Fig. 6, in this region the flow is nearly laminar and the complex features include abrupt flow separation and strong, tight vortices. If these laminar aspects are not adequately resolved over the forward half of the wing, many subsequent turbulent features may be lost despite the fidelity of the closure model.

The mean velocity magnitude along the line which passes through the core of the primary vortex is shown in Fig. 3b beside the fine grid result in Fig. 3a. Neither the

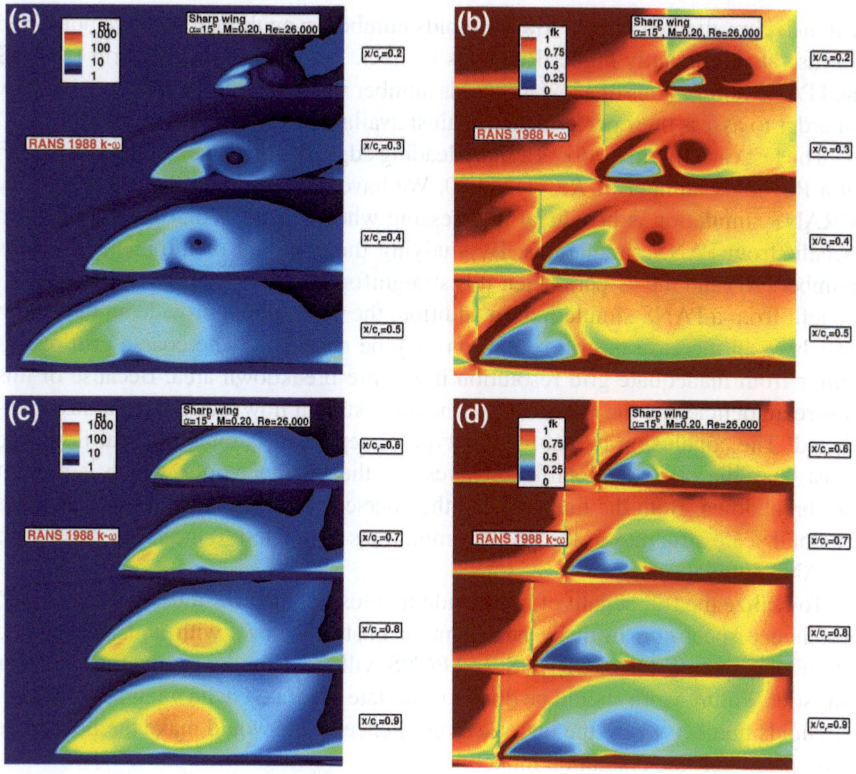

Fig. 6 URANS results; Re = 26,000; coarse grid. **a** Forward R_t contours, **b** forward f_k contours, **c** aft R_t contours **d** aft f_k contours

URANS simulation nor the PANS simulations are able to accurately capture the peak velocity in the vortex core. However, the PANS simulations predict the correct slope of the velocity in the core, while the URANS simulation predicts a more shallow slope indicating a slower, more dissipative breakdown. The velocity in the furthest aft portion of the wing is under-predicted by all simulations. Referring back to Fig. 6 which indicates that the laminar vortex requires an adequate grid resolution, the result in Fig. 3b demonstrates that the downstream flow features cannot possibly be modeled correctly unless the upstream region is properly resolved.

4 Conclusions

While most aerodynamic flows of practical relevance involve high Reynolds numbers, many of the high fidelity experimental and numerical studies are carried out at a much lower Reynolds number. Thus the development of a practical CFD tool can be extensively validated only at low Reynolds number, even though they are purported for use at significantly higher Reynolds number. In this work we perform URANS and PANS simulations of a low Reynolds number sharp leading edge delta wing flow in order to assess their performance against available DNS data.

The test case for this work is a sharp leading edge, flat plate delta wing at $\alpha = 15°$ at a Reynolds number of $Re = 26{,}000$. We have demonstrated that a preliminary URANS simulation is beneficial in assessing whether a particular flow is likely to benefit from PANS simulations. By studying the contours of turbulent Reynolds number (R_t) and the f_k parameter, it is straightforward to determine if the case will benefit from a PANS simulation. In addition, the f_k contours provide an indication of where an increase in grid resolution may be necessary. The coarse grid results suffer from inadequate grid resolution in the pre-breakdown area. Because of this upstream deficiency, the remainder of the downstream flow was not accurately predicted. The conclusion is that for this type of separated laminar vortical flow, one must "pay the price" and sufficiently resolve the laminar vortex which is present pre-breakdown. Both the fine grid and the coarse grid cases showed some increase in small scale structure in the PANS simulations compared to the more dissipative URANS simulations.

To realize the full benefit of higher fidelity closure, the flow must exhibit a broad turbulence spectrum. In many near-laminar unsteady flows with under-developed turbulence spectra, DNS, URANS, and PANS will yield similar results for low order statistics. Future work should be done to simulate separated delta wing flows at high Reynolds number where the increased turbulence levels would make the flow more appropriate for fine resolution modeling.

References

1. Hummel, D.: On the vortex formation over a slender delta wing at large angles of incidence. Technical Report 15, Technische Universitat Braunschweig (1978)
2. Levin, D., Katz, J.: Dynamic load measurements with delta wings undergoing self-induced roll oscillations. J. Aircr. **21**(1), 30–36 (1983)
3. Soemarwoto, B.I., Boelens, O.J.: Simulation of vortical flow over a slender delta wing experiencing vortex breakdown. Technical report NLR-TP-2003-396, National Aerospace Laboratory NLR (2003)
4. Luckring, J.M.: Reynolds number, compressibility, and leading-edge bluntness effects on delta-wing aerodynamics. In: 24th International Congress of the Aeronautical Sciences, Hampton, VA (2004)
5. Gordiner, R.E., Visbal, M.R.: Higher-order compact difference scheme applied to the simulation of a low sweep delta wing flow. In: 41st Aerospace Sciences Meeting and Exhibit, Reno, NV (2003)
6. Germano, M.: Turbulence: The Filtering Approach. Technical report, The Massachusetts Institute of Technology (1972)
7. Girimaji, S.S., Abdol-Hamid, K.S.: Partially-averaged Navier Stokes model for turbulence: implementation and validation. Number AIAA 2005–502. In: 43rd AIAA Aerospace Sciences Meeting and Exhibit (2005)

8. Girimaji, S.S.: Partially-averaged Navier-Stokes method for turbulence: a reynolds-averaged Navier-Stokes to direct numerical simulation bridging method. J. Appl. Mech. **73**, 413–421 (2006)
9. Lakshmipathy, S.: PANS method for turbulence: simulations of high and low reynolds number flows past a circular cylinder. Master's thesis, Texas A&M University (2004)
10. Murthi, A.: Effect of turbulent transport models and grid spacing on PANS calculations of a id-driven Cavity. Master's thesis, Texas A&M University (2004)
11. Song, C., Park, S.: Numerical simulation of flow past a square cylinder using partially-averaged Navier-Stokes model. J. Wind Eng. Ind. Aerodyn. **97**, 37–47 (2009)
12. Girimaji, S.S., Jeong, E., Srinivasan, R.: Partially averaged Navier-Stokes method for turbulence: fixed point analysis and comparison with unsteady partially averaged Navier-Stokes. J. Appl. Mech. **73**, 422–429 (2006)
13. Lakshmipathy, S., Girimaji, S.S.: Partially averaged Navier-Stokes: PANS method for turbulence simulations: flow past a circular cylinder. J. Fluids Eng. **132**, 121202 (2010)

LES and DES of Swirling Flow with Rotor-Stator Interaction

Ardalan Javadi and Håkan Nilsson

Abstract A highly swirling turbulent flow engendered by the rotor-stator interaction of a swirl generator is investigated using LES and DES. The delayed DES Spalart-Allmaras (DDES-SA), improved DDES-SA, shear stress transport DDES (DDES-SST) and a dynamic *k*-equation LES are studied. A mesh sensitivity study is performed on the hybrid methods, including the ability to capture the details of the flow field. It is shown that all the methods are capable of predicting the large-scale flow features, e.g. the vortex breakdown and the corresponding on-axis recirculation region. It is also shown that all the hybrid methods capture most of the small-scale coherent structures, even with a relatively coarse mesh resolution. The various shielding functions of the hybrid methods are analyzed, distinguishing the location of the transition between RANS and LES mode.

1 Introduction

The swirling flow is found in numerous technical applications. The effects of the swirl may be favourable, e.g. when the turbulent mixing is increased by vortex breakdown in combustors [5]. On the other hand, the vortex breakdown may cause severe pressure fluctuations that may damage the construction. In hydraulic turbines, such pressure pulsations appear in the draft tube at off-design operation [8]. The occurrence of the pressure pulsations in hydraulic turbines depends on a complex interaction between several flow features, which is not yet fully understood. Running a hydraulic turbine at an off-design operating condition yields a remaining swirl after the runner. The mean swirl profiles develop through the diffusing draft tube, and may approach unstable conditions that cause a vortex breakdown [4]. The flow separation in the

A. Javadi (✉) · H. Nilsson
Division of Fluid Dynamics, Chalmers University of Technology,
412 96 Gothenburg, Sweden
e-mail: ardalan.javadi@chalmers.se

H. Nilsson
e-mail: hani@chalmers.se

S. Girimaji et al. (eds.), *Progress in Hybrid RANS-LES Modelling*,
Notes on Numerical Fluid Mechanics and Multidisciplinary Design 130,
DOI 10.1007/978-3-319-15141-0_37

boundary layers of the runner blades and the draft tube walls, and massive separations at the runner blade suction sides, may further trigger such instabilities. As the large-scale fluctuation has occurred, the dynamic response of the whole hydraulic circuit may further amplify the effect.

To supplement experimental studies, it is desirable to increase the knowledge of swirling flow and vortex breakdown by using computational fluid dynamics (CFD). The unsteady Reynolds-averaged Navier-Stokes (URANS) simulations are sometimes used in the literature, despite their limitations and inabilities to capture the physics of separation, transition, reattachment and pulsation of the flow. The use of such turbulence models in unsteady flows is justified only if the scales associated with the resolved unsteady motion are substantially larger than the scales of the modelled turbulence. This condition may be satisfied in the large-scale and low-frequency dynamics of draft tube surge, but not necessarily in the vortex shedding at the trailing edges and wakes of the runner blades. Although the large-scale structure of the vortex breakdown may be captured by URANS methods, there is a need to gain a better knowledge of the interaction between the vortex breakdown and the small-scale structures. In particular, this may be used to develop passive or active flow control devices that mitigate the unwanted effects of the vortex breakdown, using a minimum of energy. The large-eddy simulation (LES) methodology is a well-known accurate approach for resolving all the unsteadiness of the flow, down to the scales where the energy transfer to the dissipating scales is well-defined. However, LES requires that those scales are resolved by the numerical mesh. That is certainly impossible in the boundary layers of high-Reynolds flow, such as that in hydraulic turbines. A practical solution to capture more unsteadiness, without the excessive resolution requirements of the LES model in the boundary layers, is the hybrid RANS-LES methods. Those methods use the benefits of LES outside the boundary layers and model the turbulence in the boundary layers using RANS. Detached-eddy simulation (DES) is a promising hybrid RANS-LES strategy capable of simulating internal flows dominated by large-scale detached eddies at practical Reynolds numbers. The method aims at treating the boundary layers with URANS while the detached eddies in separated regions or outside the boundary layers are resolved using LES.

The present work evaluates LES and DES methods in turbulent swirling flow at an intermediate industrial Reynolds number. The case used in the study is a swirl generator [2, 20] that is designed to generate a draft tube flow field similar to that of a Francis hydro turbine running at off-design. The simulations involve rotor-stator interaction, using rotating parts of the mesh and sliding interfaces. The numerical results are compared with detailed LDA measurements. The study is a step towards the application of DES methods in model-scale hydraulic turbines, at higher Reynolds numbers. The simulations are performed using the OpenFOAM open source software, and the full URANS case and post-processing routines have been made available through the OpenFOAM Turbomachinery Working Group [8, 12].

2 Flow Configuration

The simulations are done for a swirl generator for which detailed LDA measurements are available. The experimental test rig was developed to analyze the decelerated swirling flow in a conical diffuser with a 2 × 8.6° cone angle [13]. Figure 1 shows the test rig. It is located at the Hydraulic Machinery Laboratory at the Politehnica, University of Timisoara, Romania. The test rig is employed to generate a draft tube flow similar to the one encountered at a 70 % partial discharge of a Francis turbine [14]. At this regime, the vortex breakdown, also referred to as the *vortex rope* in the hydraulic turbine field, is well developed and generates the largest pressure pulsations.
The swirling flow apparatus is installed with two main parts, the swirl generator and the convergent-divergent test section, see Fig. 1. The swirl generator has an upstream section with stationary and rotating blades for generating the swirling flow. It has three components: the supporting struts, the guide vanes and the free runner. The guide vanes and the free runner are mounted in the annular cylindrical section, with diameters $D_{shroud} = 0.15$ m and $D_{hub} = 0.09$ m. The runner rotates freely, without shaft torque. The runner is designed to act as a turbine at the inner part of the blades

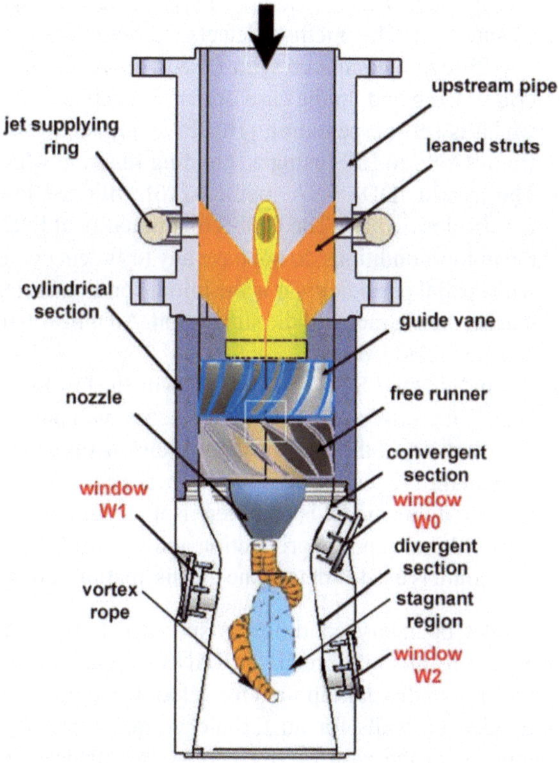

Fig. 1 The test rig geometry, strut, guide vane, runner and draft tube. Experimental velocity profiles are available at sections W0, W1 and W2

and as a pump at the outer part of the blades, yielding a finite rotational speed of 920 rpm at a water flow of 30 l/s. This gives an excess in the axial velocity near the shroud and a deficit of axial velocity near the hub, and the desired draft tube inlet flow profile that corresponds to that of a Francis turbine at off-design operation. The Reynolds number based on the bulk velocity and the throat diameter is 3.81×10^5. The experimental mean velocity is available at cross-sections W0, W1 and W2, see Fig. 1. The velocity profiles normal to the traversing axis, in the meridional plane, and in the tangential direction were measured with a two-component LDA system.

3 Computational Framework

The calculations reported herein are performed using the finite-volume method in the OpenFOAM open source CFD code. The second-order linear central difference is used to discretize the diffusion terms, and the second-order linear upwind scheme is adopted to approximate the convection term [7]. Time marching is performed with an implicit second-order accurate backward differentiation scheme.

The turbulence is modelled using three hybrid DES methods and a k-equation LES:

- The hybrid DDES-SA method [18] with the low Reynolds Spalart-Allmaras RANS closure [17]. This method detects the boundary layer thickness and prolongs the RANS region compared with DES. This method is less sensitive to wall parallel grid spacing and, in the case of massive separation, LES mode takes over RANS, which is self-perpetuating [18]. In a simpler word, the method obliges a switch from RANS to LES using a shielding function which defined in Eq. (2).
- The hybrid IDDES-SA method [16] with the low Reynolds Spalart-Allmaras RANS closure [17]. The DDES functionality of IDDES becomes active only when the inflow conditions do not have any turbulent content. In the case of a simulation with spatial periodicity, it is the initial conditions rather than the inflow conditions that set the character of the simulation. An inflow with a turbulent content activates wall-modeled LES.
- The hybrid DDES-SST method [3] with the low Reynolds $k-\omega$ SST RANS closure [11]. This is the same as DDES-SA, but with another RANS model closure and a recalibration of the empirical constants involved in the delay function.
- A dynamic k-equation LES model, here referred to as dynOneEq [9]. An important feature of this model is that there is no assumption of a local balance between the sub-grid scale energy production and the dissipation rate. A dynamic localization procedure is used to determine the dissipation and the diffusion model coefficients.

The inlet boundary condition is specified between the strut and the guide vane, using the results of a precursor RANS simulation. A homogenous Neumann outlet boundary condition is applied for velocity and pressure. The homogenous Neumann is applied on walls for all turbulence quantities, $\tilde{\upsilon}$ and k_{sgs}. Regarding the wall treatment, in the case of IDDES-SA, we made sure that the average y^+ was of order one, since a Spalding's law [19] based on unified wall functions was applied. The major advantage of using such unified wall functions is that, the first grid point close to the wall can be placed into the buffer or viscous region, $y^+ < 30$, without the loss of accuracy associated with the logarithmic profiles. In the case of other turbulence models, no wall treatment is used.

The equations are discretized on the computational domain using two different structured resolutions. The domain was realized in ICEM Hexa, and is divided in three different parts. Those different parts are coupled in OpenFOAM. The General Grid Interface (GGI) [1] is used at the interfaces between the rotor and the stator. The main advantage of the GGI is that it allows non-conformal meshes at the interface. It makes mesh generation easier for complex geometries and facilitates a sliding grid approach. It has been shown to give a close agreement for the velocity results between non-conformal and conformal meshes. However, the present study required a proper circumferential mesh resolution on each side of the GGI to yield good results for the turbulent kinetic energy. Nilsson et al. [12] validated the use of non-conformal meshes arguing that the spacing should be comparable in the radial and axial directions.

To achieve a near-optimal parallel load balancing, the computational meshes are subdivided into blocks of equal size, submitted to individual cores of a Linux cluster. The CFD code is parallelized using domain decomposition and the Message Passing Interface (MPI) library. The simulation is performed using an AMD Opteron 6220 super computer. The finest mesh was run on 12 nodes with 16 cores each (192 cores) for a period of 98 days which two third of this period is initial transient flow. The maximum *CFL* number in the entire domain is on average 0.06, with a local maximum of 4.

4 Mesh Sensitivity Study

A mesh sensitivity study conducted for the hybrid methods using DDES-SA. Table 1 shows the details of the meshes for the coarse (total of 13.25×10^6 cells) and the fine (total of 19.27×10^6 cells) resolutions. Resolving the boundary layers in the runner at this high Reynolds number flow is a challenging task. The maximum dimensionless wall units given in Table 1 are defined as:

Table 1 Mesh resolution details

Mesh	Guide vane	Runner	Draft tube	$\Delta z^+{}_{max}$	$\Delta(r\theta)^+{}_{max}$
Coarse	2.45×10^6 $y^+{}_{max} = 2.2$	2.4×10^6, $y^+{}_{max} = 8.1$	8.4×10^6, $y^+{}_{max} = 2.5$	23.82	39.2
Fine	2.45×10^6	3.2×10^6	13.62×10^6	18.27	41.23

$$\Delta z^+ = \sqrt{\frac{\partial W}{\partial z}\frac{1}{\upsilon}}\Delta z, \quad \Delta(r\theta)^+ = \sqrt{\frac{\partial W}{\partial(r\theta)}\frac{1}{\upsilon}}\Delta(r\theta), \tag{1}$$

where υ is the kinematic viscosity, r is the radial position, θ is the tangential position and z is the axial direction.
Figure 2 shows that the mean axial and tangential velocity distributions at both W0 and W1 are very similar for the two mesh resolutions. The velocity field is reasonably mesh independent. The maximum wall units for different parts are given in Table 1. The stagnation region close to the leading edge has a locally higher y^+ value; furthermore, close to the hub, y^+ is larger due to the special shape of the blade.

5 Results and Discussion

The flow downstream of the runner blade is highly swirling and turbulent. The mean velocity profiles below the runner are determined by a time averaged over at least five complete revolutions of the runner to filter out all unsteadiness. The quantities are normalized by the bulk velocity and the diameter of the throat. The swirl number is found by integration of the mean velocity profiles. Equation (2) shows the swirl:

$$Sr = \frac{1}{R}\frac{\int_{R1}^{R2} r^2 U W dr}{\int_{R1}^{R2} r U^2 dr}\quad, \tag{2}$$

here, R is the radius, W is the mean axial velocity and U is the mean tangential velocity. The numerical results show that the swirl number at W0, W1 and W2 is 2.5, 1 and 0.6, respectively. These values correspond well with experimental results. Figure 3 shows the mean axial and mean tangential velocity distributions of the different turbulence models, compared with the experimental results at W0 and W1. The velocity components at W0 increase linearly ($\partial P/\partial r = \rho V_\theta^2/r$) from hub to shroud due to the centrifugal force and the wake generated from the trailing edge of the runner blades. The axial velocity is predicted by hybrid methods better than dynOneEq. This can be related to the coarse used resolution for such a high Reynolds number flow. The flow is more influenced by wall effects at W0, therefore the predicted results by dynOneEq at W0 is more deviated from experimental results than other method. For the same reason the tangential velocity is underestimated by all turbulence methods close to the wall at W0. Figure 3b shows the velocity components at section W1. For high swirl flows, a central recirculation region may develop and vortex breakdown occurs. Vortex breakdown is an abrupt change in the core of a slender vortex and typically develops downstream into a recirculatory "bubble" or a helical pattern [10, 15]. In the present flow field, a small recirculation region develops as a result of the separation from the runner hub, forming a helical vortex that immediately deflects outwards due to the swirl. A central recirculation region thus occurs which ends at the throat, yielding an on-axis stagnation region at W1.

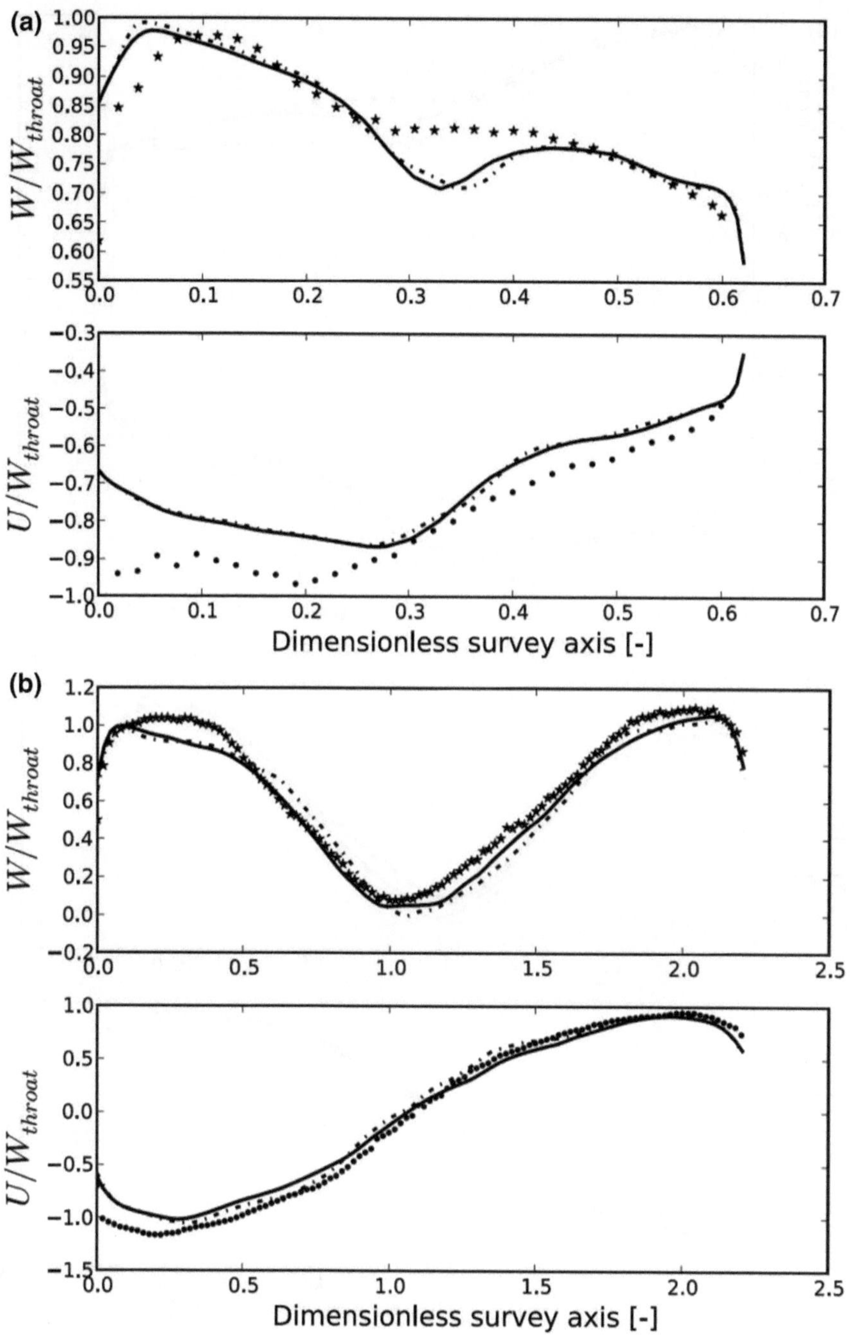

Fig. 2 Mesh refinement study of mean axial (*W*) and tangential (*U*) velocity, compared to experimental data (markers). *Solid curves* coarse mesh. *Dash-dotted curves* fine mesh. Cross-section W0 (**a**) and W1 (**b**). Survey axis measured from the wall normalized by the throat diameter

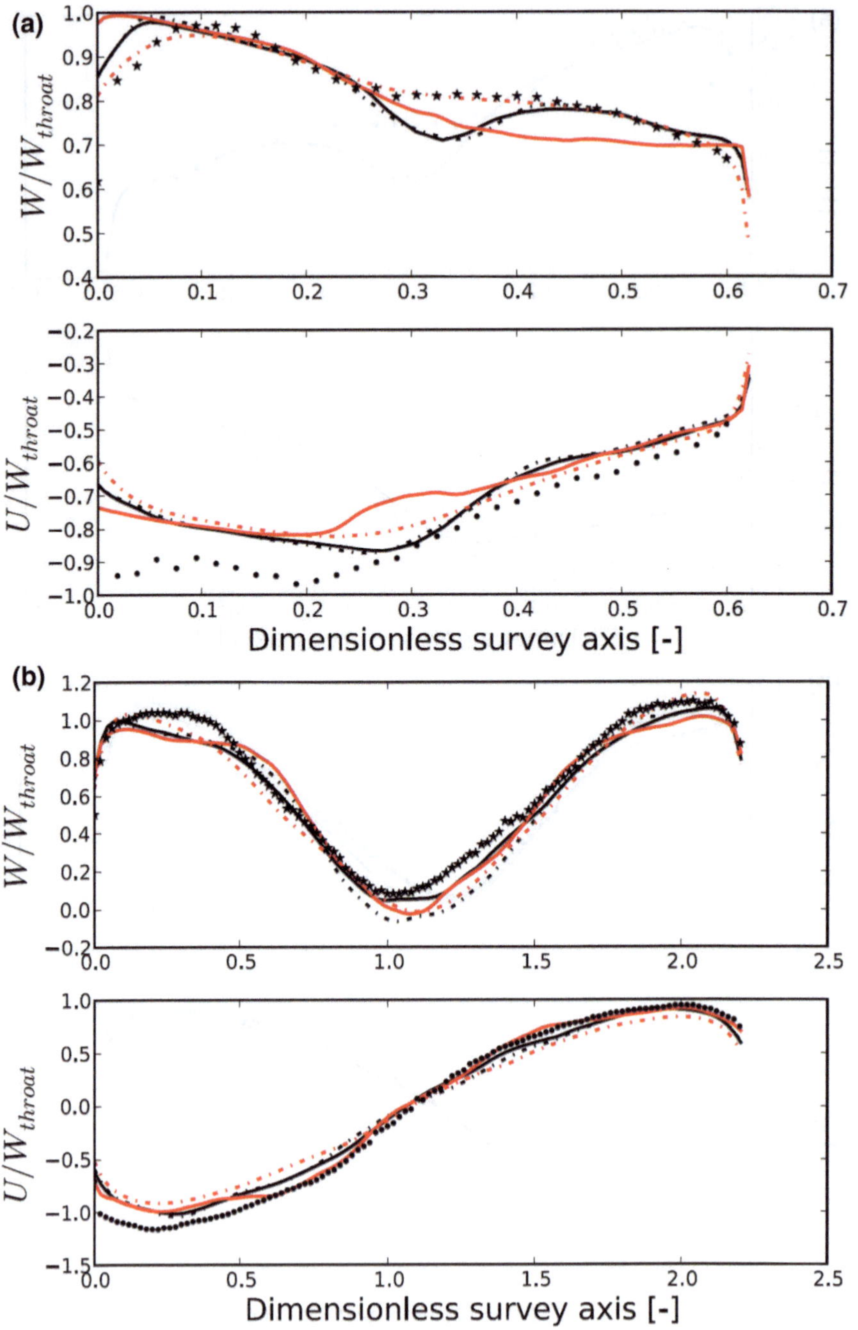

Fig. 3 Axial and tangential velocity at **a** W0; **b** W1; *black solid* DDES-SA; *black dash-dot* IDDES-SA; *solid plus* DDES-SST; *red solid* dynOneEq; markers experiment

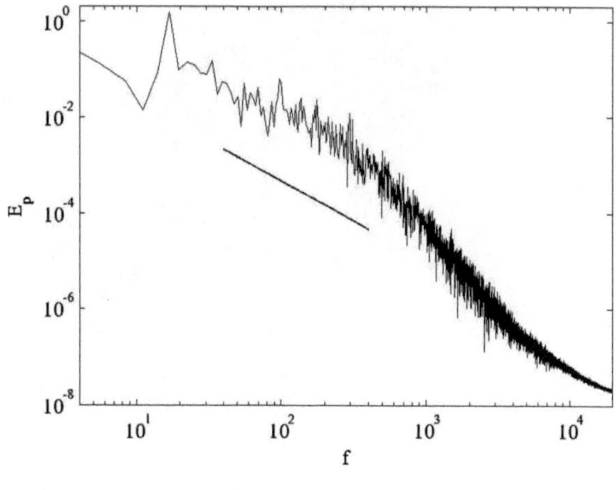

Fig. 4 Energy spectrum at a point downstream of the draft tube hub: *solid line* −5/3 slope

The experimental results show $W/W_{throat} = 0.08$ on-axis at W1 while numerical results predict $W/W_{throat} = -0.010, -0.026, -0.041$ and -0.187 with DDES-SA, dynOneEq, DDES-SST and IDDES-SA, respectively. All models predict somewhat larger on-axis recirculation region while DDES-SA presents better agreement with experimental results. Around the vortex core, the tangential velocity is linearly proportional to the radius, a characteristic of a rigid body rotation. Outside the vortex core, the tangential velocity gradually becomes inversely proportional to the radius, thus exhibiting a free vortex tail.

Figure 4 shows energy the spectrum captured at a point in 0.1 diameter of the throat downstream of the draft tube nozzle with fine resolution. The fact that the slope of the spectrum is reasonable indicates that the resolution is fine enough to resolve most of the unsteadiness of this complex flow field. The energy spectrum shows the dominant frequency of the flow and other periodicities, which mostly depend on the rotational speed of the runner. The power spectrum shows that the flow is characterized by the existence of the vortex breakdown with a dominant frequency, which captured 7 % less than the runner rotation. Figure 5 shows the Q-criterion [6] in the draft tube. With too a high swirl ratio, the vortex rope disintegrates. A large number of unorganized smaller vortices replace the single corkscrew vortex.

Figure 6 shows the of the viscosity ratio, υ_t/υ, simulated by DDES-SA in the center plane and a plane at the downstream of the trailing edge. The high ratio region corresponds to URANS activity and the low ratio region corresponds to LES activity. The current switching behavior is not only a function of the resolution but also the solution. The boundary layer flow at the walls and the shear layers of the runner blade wake are modelled mostly by URANS.

Figure 7 shows the shielding function, f_d, for DDES at W1 and W2 normalized by its own radios. The shielding function is given by Eq. (3). The DES limiter is deactivated if the function $f_d = 0$, i.e. URANS is manipulated. All methods switch reasonably smooth between URANS and LES. Figure 7a shows the f_d of DDES-SA for the coarse mesh. Figure 7b shows the f_d function for DDES-SST, where the URANS

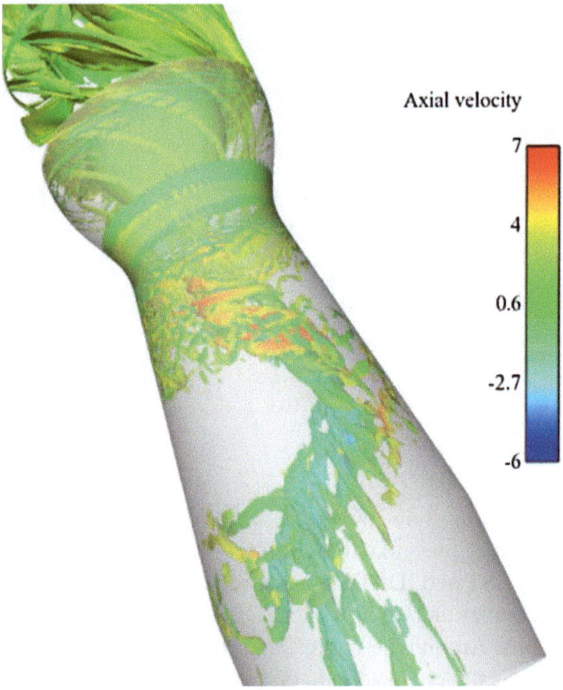

Fig. 5 Iso-surface of Q colored by axial velocity

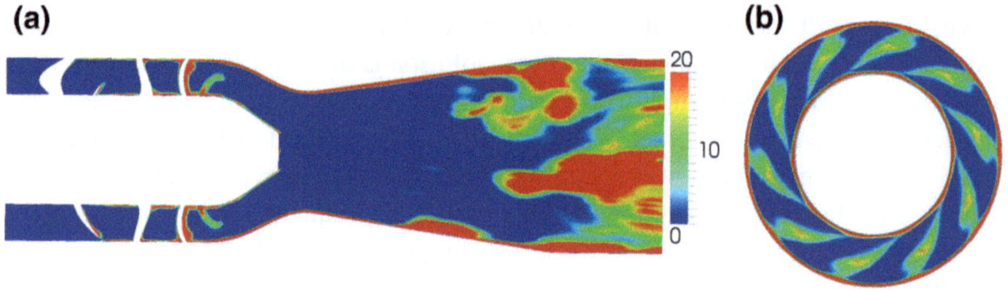

Fig. 6 Viscosity ratio, υ_t/υ, for DDES-SA; **a** centre plane; **b** downstream of the trailing edge

area is wider and a larger part of the flow is modelled.

$$f_d = 1 - \tanh\left[(C_d r_d)^3\right], \qquad r_d = \frac{\upsilon_t + \upsilon}{\kappa^2 d_w^2 \sqrt{0.5\left(S^2 + \Omega^2\right)}}, \tag{3}$$

here, υ_t and υ are the eddy and molecular viscosities, respectively, S and Ω are the strain rate and vorticity tensor invariants, respectively, $\kappa = 0.41$ is the von Karman constant and d_w is the distance to the wall. The constant C_d for DDES-SST and

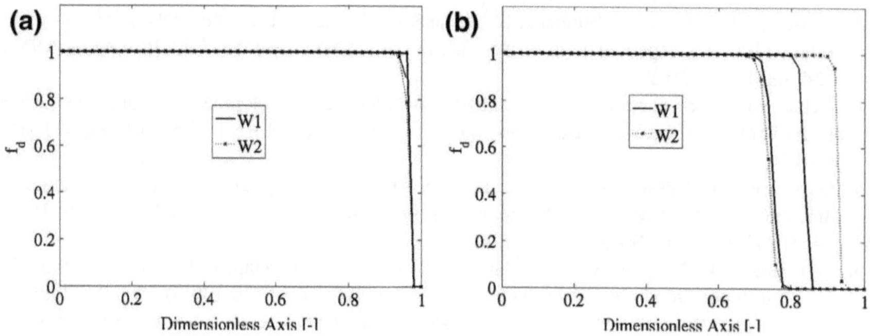

Fig. 7 Shielding function (fd) at W1 and W2; **a** DDES-SA coarse mesh; **b** DDES-SST coarse mesh. W1 and W2 are normalized by their own radios

DDES-SA is 20 and 8, respectively. The turbulent viscosity is determined according to each underlying turbulence model.

6 Conclusion

The highly turbulent swirling flow emanating from a rotor-stator is investigated using DDES-SA, IDDES-SA and DDES-SST hybrid RANS-LES methods and a *k*-equation LES (dynOneEq). The high Reynolds number of the flow makes LES too expensive in capturing a detailed unsteadiness, although the main feature of the flow (precessing vortex and vortex breakdown) are simulated reasonably. The precessing vortex is far from wall effects; accordingly it is possible to mimic with a coarse resolution close to the wall. In terms of smaller coherent structures such as the wake of the trailing edge, LES needs a finer resolution. The hybrid methods are capable of reproducing an accurate flow field, even with the coarser resolution than LES

Acknowledgments The research presented was carried out as a part of the "Swedish Hydropower Centre—SVC". SVC is established by the Swedish Energy Agency, Elforsk and Svenska Kraftnät together with Luleå University of Technology, The Royal Institute of Technology, Chalmers University of Technology and Uppsala University, www.svc.nu. The computational facilities are provided by C^{3}SE, the center for scientific and technical computing at Chalmers University of Technology, and SNIC, the Swedish National Infrastructure for Computing.

References

1. Beaudoin, M., Jasak, H.: Development of a generalized grid interface for turbomachinery simulation with OpenFOAM. In: Open Source CFD International Conference, Berlin, Germany (2008)

2. Bosioc, A.I., Resiga, R., Muntean, S., Tănasă, C.: Unsteady pressure analysis of a swirling flow with vortex rope and axial water injection in a discharge cone. J. Fluid. Eng.-T ASME **134**, 081104-1 (2012)
3. Gritskevich, M.S., Garbaruk, A.V., Schütze, J., Menter, F.L.: Development of DDES and IDDES formulations for the k-ω shear stress transport model. Flow Turbul. Combust. **88**, 431–449 (2012)
4. Gyllenram, W., Nilsson, H., Davidson, L.: On the failure of the quasi-cylindrical approximation and the connection to the vortex breakdown in turbulent swirling flow. Phys. Fluids **19**(4), 045108 (2007). doi:10.1063/1.2717724
5. Gyllenram, W., Nilsson, H.: Design and validation of a scale-adaptive filtering technique for LRN turbulence modeling of unsteady flow. J. Fluid Eng.-T ASME **130**(5), 051401 (2008)
6. Hunt, J.C.R., Wray, A., Moin, P.: Eddies, stream, and convergence zones in turbulent flows. In: Center for Turbulence Research Report CTR-S88 (1988)
7. Jasak, H.: Error analysis and estimation for the finite volume method with applications to fluid flows. Ph.D. thesis, Imperial College, University of London (1996)
8. Javadi, A., Bosioc, A., Nilsson, H., Muntean, S., Resiga, S.R.: Velocity and pressure fluctuations induced by the precessing helical vortex in a conical diffuser. In: 27th IAHR Symposium on Hydraulic Machinery and Systems, Montreal, Canada (2014)
9. Kim, W.W., Menon, S.: A new dynamic one-equation subgrid-scale model for large eddy simulations. In: AIAA Paper, Reno, NV (1995)
10. Leibovich, S.: Vortex stability and breakdown: survey and extension. AIAA J. **22**, 1192–1206 (1984)
11. Menter, F.R.: Two-equation eddy-viscosity turbulence models for engineering applications. AIAA J. **32**(8), 1598–1605 (1994)
12. Nilsson, H., Page, M., Beaudoin, M., Gschaider, B., Jasak, H.: The OpenFOAM turbomachinery working-group and conclusion from the turbomachinery session of the 3rd OpenFOAM workshop. In: IAHR, 24th Symposium on Hydraulic Machinery and Systems, Foz do Iguassu, Brazil (2008)
13. Resiga S.R., Muntean, S., Bosioc, A.: Blade design for swirling flow generator. In: Proceedings 4th German-Romanian Workshop on Turbomachinery Hydrodynamics. GROWTH-4, Stuttgart, Germany (2008)
14. Resiga, S.R., Muntean, S.: Decelerated swirling flow control in the discharge cone of Francis turbine. In: 4th International Symposium on Fluid Machinery and Fluid Engineering, Beijing, China (2008)
15. Shtern, V., Hussain, F.: Collapse, symmetry, breaking, and hysteresis in swirling flow. Annu. Rev. Fluid Mech. **31**, 537–566 (1999)
16. Shur, M.L., Spalart, P.R., Strelets, M.K., Travin, A.K.: A hybrid RANS-LES approach with delayed-DES and wall-modelled LES capabilities. Int. J. Heat Fluid Flow **29**, 1638–1649 (2008)
17. Spalart, P.R., Allmaras, S.R.: A one-equation turbulence model for aerodynamic flows. AIAA Paper, 92-0439 (1992)
18. Spalart, P.R., Deck, S., Shur, M.L., Squires, K.D., Strelets, M.K., Travin, A.K.: A new version of detached-eddy simulation, resistant to ambiguous grid densities. Theoret. Comput. Fluid Dyn. **20**(3), 181–195 (2006)
19. Spalding, D.B.: A single formula for the law of the wall. J. Appl. Mech. **28**(3), 455–458 (1961)
20. Tănasă, C., Resiga, R.S., Muntean, S., Bosioc, A.: Flow-feedback method for mitigating the vortex rope in decelerated swirling flows. J. Fluids Eng. **135**(6), 061304 (2013)

Experimental and Numerical Studies of Flow in a Duct with a Diaphragm

A. Prieto, P. Spalart, M. Shur, M. Strelets and A. Travin

Abstract Results of an experimental study and numerical simulations are presented for the flow in a circular duct with an obstruction plate and an orifice (diaphragm). This is a model of components of an airplane's Environmental Control System, or ECS. The Reynolds number of the flow is close to that in actual applications. The simulations rest on versions of the Detached-Eddy Simulation (DES) technique, with some differences in the definition of the Sub-Grid-Scale modelling length-scale. Two different CFD codes are used, the unstructured-grid BCFD code of Boeing and the high-order structured-grid NTS code. The number of grid points is around 8 million. The paper focuses on a comparison of computed and measured primary flow quantities (mean and fluctuating wall pressures). The ultimate objective is to evaluate the capabilities of the simulation approaches in predicting noise radiated into the airplane's cabin. The agreement with experiment is generally encouraging, including spectra, with the high-order code reaching higher frequencies.

1 Introduction

Recent regulations limit the noise exposure of airplane crews and ground personnel, and are tightening the interior noise levels requested by the aircraft operators. This brought forth a new wave of attention to the airport and aircraft noise problem as the quest for silent aircraft continues. The impact of cabin noise on passenger comfort is also substantial, so that the commercial incentives are also strong. Multiple sources are responsible for the noise signature inside an airplane's cabin. One of the more noticeable sources is the Environmental Control System (ECS) which may be easier

A. Prieto (✉) · P. Spalart
Boeing Commercial Airplanes, PO Box 3707, Seattle 98124, USA
e-mail: Alvaro.F.Prieto@boeing.com

M. Shur · M. Strelets · A. Travin
New Technologies and Services, Saint Petersburg 197198, Russia

M. Shur · M. Strelets · A. Travin
Saint Petersburg State Polytechnic University, Saint Petersburg 195220, Russia

S. Girimaji et al. (eds.), *Progress in Hybrid RANS-LES Modelling*,
Notes on Numerical Fluid Mechanics and Multidisciplinary Design 130,
DOI 10.1007/978-3-319-15141-0_38

to mitigate (if only by reducing the velocity levels, using larger ducts) than the noise of the turbulent boundary-layer on the outside of the airplane, for which heavy insulation is the primary solution.

The propagation of sound within ducts and the noise generated by the interaction of the flow with the different components of the duct system such as elbows, bifurcations and diaphragms are primary sources of ECS noise. Obstructions such as the one studied here are needed to finely control the pressure within the ECS duct system, which is fairly complex, and naturally create flow separation, strong turbulence, and noise. Therefore, in order to improve the acoustic performance of aircraft, as it pertains to interior noise, there is a need for a better understanding and a more accurate prediction of the noise-generation mechanisms of flow-induced noise inside duct systems.

In order to achieve this goal, an experimental study was conducted of a simple ducted flow with a circular plate obstruction. In the experiment, the ducts upstream and downstream of the orifice plate were suited with an adequate amount of instrumentation to characterize the behaviour of the flow, as well as the generation and propagation of acoustic waves within the duct. In the present paper this data set is used to validate capabilities of two unsteady flow solvers, BCFD and NTS. These use recent versions of the DES technique, which is well established for unsteady simulations in aerodynamics including engine and airframe noise. These flows have physics similar to those in the obstructed duct flow, and in particular, massive separation.

2 Experimental Setup

The experiments (see Fig. 1) were conducted for the configuration including circular ducts with an inner diameter of 15.21 cm and obstruction plates of thickness 0.1 cm with circular orifices of different diameters, with particular emphasis given to the source region just downstream of the obstruction.

The duct downstream of the orifice is instrumented with microphones arranged in three circular arrays and one line array. Each circular array consists of six microphones evenly distributed along the perimeter of the duct's cross-section. The

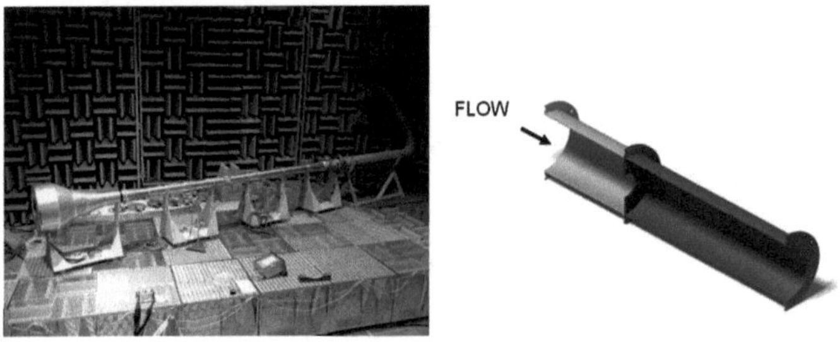

Fig. 1 Test layout and duct geometry

circular arrays have 30 cm of separation between them along the axis. The first array is located 50 cm downstream of the orifice plate. Successive arrays have an offset of 30° azimuthally with respect to each other. The main purpose of these three circular arrays is to characterize the noise source in the near-field, which in turn helps validate CFD solutions. Eight more microphones were introduced on a line along the length of the duct to monitor the evolution of the propagating waves. The first microphone from this set was located at approximately 3 cm from the orifice plate, and the rest followed a geometric spacing distribution that covers the entire length of the duct. Two microphones in the circular arrays were on the line, thus completing a ten microphone line array.

An anechoic termination was also introduced at the end of the duct, as seen in Fig. 1, which helped to reduce the flow velocity before it entered the anechoic chamber, and also provided the acoustic duct with a matched impedance termination thus minimizing reflections of the outgoing acoustic waves.

The last set of instruments added to the experimental setup focuses on tracking static and total pressure at different stations. A total of nine static pressure taps were installed along the length of the setup with two upstream of the orifice and seven downstream of the orifice. Also, an extra pair of static and total pressure ports was added on each side of the diaphragm to measure radial profiles of total pressure upstream and downstream of the orifice, by traversing the probes. The first pair of total/static pressure ports was placed about 60 cm upstream of the diaphragm measured from the orifice's normal plane to the incoming flow. The second pair of ports was located approximately 107 cm downstream of the test article. Adding up the instrumentation of all regions yields a total of twenty six flush mounted 1/2 in. microphones using the standard configuration, eleven static pressure ports, and two total pressure ports.

3 Simulation Approaches and Numerical Details

Simulations were carried out in the framework of the standard Delayed DES (DDES) [1] and Improved DDES (IDDES) [2] hybrid RANS-LES approaches with the use of the Spalart-Allmaras RANS model [3] as a background model. In addition, IDDES of one of the considered flow regimes was carried out with the use of a new definition of the subgrid length-scale (SGS). Recall that on typical (strongly anisotropic) grids used in the initial region of the separated shear layers, where the flow is locally two-dimensional and does not have any turbulent content, the use of the standard DES definition of SGS length-scale, $\Delta = \Delta_{max}$ (where Δ_{max} is the maximum local grid-spacing in the three directions), results in blocking the Kelvin-Helmholtz instability and in a tangible delay of transition to turbulence farther downstream (this is part of the so called "grey area issue" [4]). This has motivated proposals of alternative, solution-dependent, definitions of the length-scale aimed at a mitigation of this delay.

A recent example is the definition proposed by Deck et al. [5], which accounts for the abovementioned peculiarity of the grid and flow in the initial part of the separated shear layers by detecting the alignment of the vorticity with the spanwise direction, and ensures a considerable shortening of the grey area in the presented numerical examples. However, this definition is considered by the present authors as somewhat "non-physical", since it still allows the minimum grid-spacing to control the resulting length-scale (because it reduces to the square root of the product of the streamwise and lateral grid steps, $\sqrt{\Delta x \Delta y}$). The definition employed in the present study, which does not suffer from this deficiency, is as follows (see also the paper of Mockett et al. [6] in this book).

Let us consider a cell with the center at point $\mathbf{r}$ and vertices at points $\mathbf{r}_n$ ($n = 1$–8). Then the proposed definition reads:

$$\tilde{\Delta}_\omega = \frac{1}{\sqrt{3}} \max_{n,m=1,8} |(\mathbf{l}_n - \mathbf{l}_m)|\,, \tag{1}$$

where we have introduced the vector $\mathbf{l}_n = \mathbf{n}_\omega \times (\mathbf{r}_n - \mathbf{r})$, and $\mathbf{n}_\omega$ is the unit vector aligned with the vorticity vector.

Thus, the quantity $\tilde{\Delta}_\omega$ is the diameter of the set of cross-product points $\mathbf{l}_n$ divided by $\sqrt{3}$ (so that it agrees with $\Delta_{\max}$ in a cubic grid cell. In the 2D flow regions with $\mathbf{n}_\omega$ aligned with the z-axis it reduces to $\frac{1}{\sqrt{3}}(\Delta x^2 + \Delta y^2)^{1/2}$, i.e., it is $O(\max\{\Delta x,\ \Delta y\})$. In 3D cases, $\tilde{\Delta}_\omega$ is of the order of $\Delta_{\max}$, except for the situation when the vorticity vector is aligned with one of coordinate-directions of the grid, in which case it reduces to $O(\max\{\Delta x_i,\ \Delta x_j\})$ where i and j are the two directions orthogonal to the vorticity. Therefore, the smallest grid-spacing never sets the resulting SGS length-scale, but the roll-up of the vortex sheet is facilitated when the grid is favourable to it. This is helpful for mixing layers, including in jets, and here for the shear layer which just flowed off the orifice.

As mentioned in the Introduction, simulations were carried out with the use of two codes, BCFD of Boeing [7] and the in-house NTS code [8, 9].

BCFD uses an implicit first-order time integration scheme. Unstructured grids are solved using a point-implicit technique whereas structured grids are solved implicitly on a computational plane. For steady-state flows, variable time steps based on local eigenvalues are used to speed convergence and a multigrid algorithm is available for accelerating the convergence on unstructured grids. For time-accurate calculation, BCFD offers a Global Newton/dual-time algorithm which provides second-order accurate updates in the physical-time domain across all zones. Additionally, a Runge-Kutta method is available for high temporal accuracy with structured grids.

BCFD has a number of numerical schemes including the Roe, HLLE, and Lax-Friedrichs/Rusanov schemes. The user is limited to 1st and 2nd order spatial accuracy for unstructured grids but can choose from 1st, 2nd, 3rd, 4th, or 5th order accuracy on structured grids. The default explicit spatial operator is a second-order HLLE scheme for the unstructured grid solver while the structured solver uses Roe's scheme in a second-order accurate implementation in the physical domain.

NTS is a structured finite-volume CFD code accepting multi-block overset grids of Chimera type. The compressible branch of the code used in the present work employs the Roe scheme [10]. The viscous terms are approximated with second-order central differences, whereas the spatial approximation of the inviscid fluxes is different in the RANS and LES zones. In particular, in the RANS zone a 3rd order upwind-biased scheme is used, and in the LES zone a 4th -order central scheme is employed. For the time integration, implicit 2nd order backward Euler scheme with sub-iterations is used (a typical number of sub-iterations, ensuring 2 orders of magnitude drop of the residuals, is 10).

4 Results and Discussion

Table 1 presents a matrix of the simulations we performed, which were carried out at conditions corresponding to two orifice diameters and two bulk velocities. Case 1 (simulations 1a–d) was computed with the use of the two codes and the three turbulence simulation approaches. A comparison of the results of these simulations with each other and with the data allows an assessment of the standard DDES code-sensitivity and of the performance of all the considered simulation approaches at experimental conditions 1, whereas a comparison of the BCFD DDES solutions 1a, 2, and 3 with the data gives an idea about the capability of DDES to capture the effect of the orifice diameter and bulk velocity on the major mean and unsteady flow characteristics.

Note that the computational grids used in the simulations carried out with the two codes were somewhat different (see Fig. 2). However both grids have similar topology and a size of about 8 million cells. As a result, the comparison of solutions obtained with the two codes within the same DDES simulation approach seems justified.

As mentioned in the Introduction, a primary objective of this work is the evaluation of the capability of DDES and IDDES to predict the major characteristics of the flow with an emphasis on the unsteady flow features which are essential for predicting the noise radiated by the ECS into the airplane's cabin.

Table 1 Summary of simulations

Case #	Open area of diaphragm (%)	Bulk velocity (m/s)	Simulation approach	Code
1a	25	35.6	DDES, Δ_{max}	BCFD
1b	25	35.6	DDES, Δ_{max}	NTS
1c	25	35.6	IDDES, Δ_{max}	NTS
1d	25	35.6	IDDES, $\tilde{\Delta}_{\omega}$	NTS
2	25	25.4	DDES, Δ_{max}	BCFD
3	35	35.6	DDES, Δ_{max}	BCFD

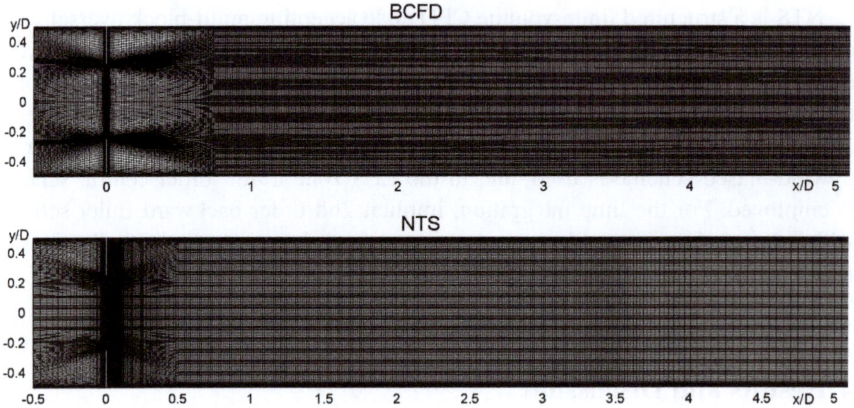

Fig. 2 Computational grids used in the simulations

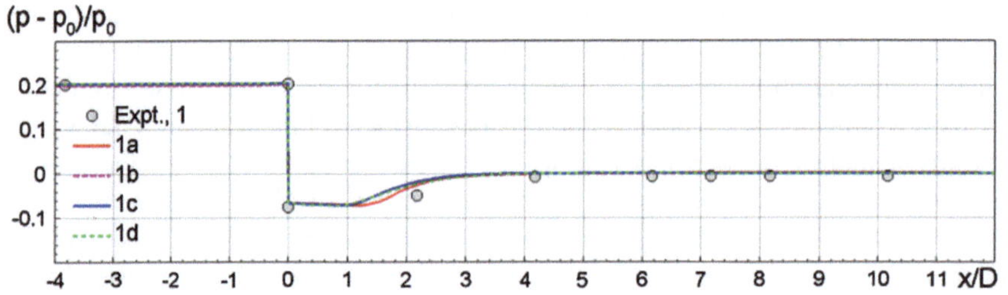

Fig. 3 Comparison of wall pressure distributions along the duct wall predicted by different codes and simulation approaches with experimental data for case 1

As far as the mean flow is concerned, the results of the simulations suggest that both codes and all the approaches to turbulence representation used in this study are capable of predicting them with high accuracy. As an example, Fig. 3 compares the mean wall pressure distributions computed based on the simulations 1a–d with the corresponding experimental distribution. One can see that all the computed distributions are virtually identical and agree well with the experimental data, although there is a noticeable difference near $x/D = 2$. The reattachment appears to take place a little too early.

For the unsteady flow characteristics, the situation is not quite as simple. Particularly, as seen in Fig. 4 which presents flow visualizations from the simulations 1a-1d, the resolution of fine-grained turbulence ensured by the BCFD flow solver within the standard DDES approach (frame 1a) is tangibly worse than that of NTS. Also, the figure suggests that the standard IDDES predicts somewhat earlier shear layer roll-up than the standard DDES, and that the use of the new definition of the SGS length-scale (1) within IDDES, in line with our expectations, results in some additional acceleration of this process. Note also that, despite these differences, all four

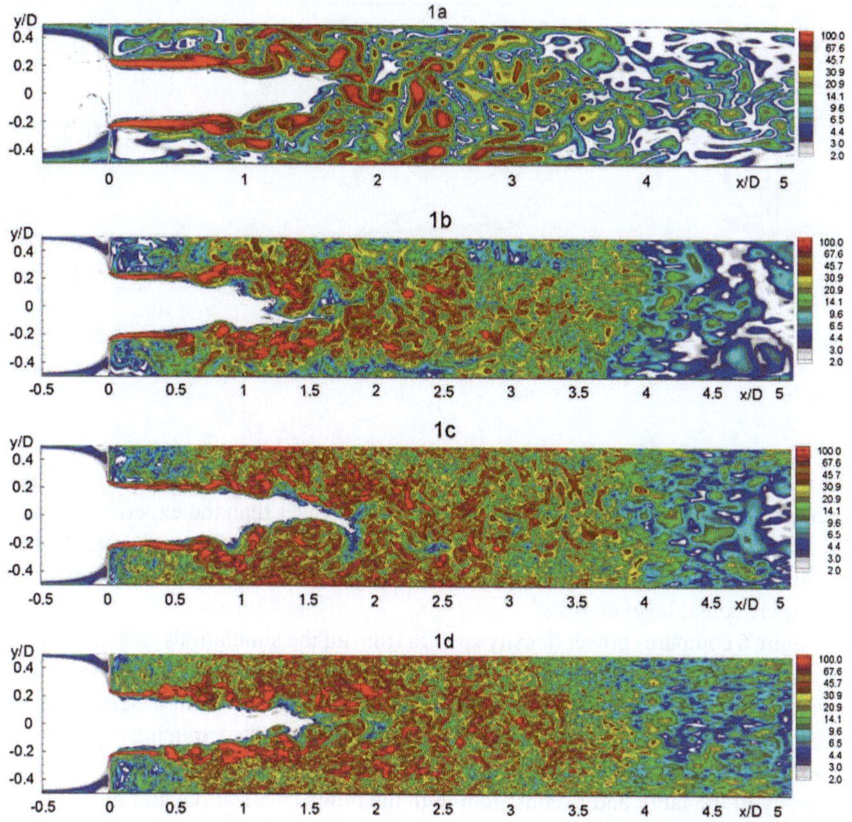

Fig. 4 Snapshots of contours of vorticity magnitude in a meridian *plane* of the duct from simulations 1a-1d in Table 1

simulations predict roughly the same length of the potential core of the jet issuing from the diaphragm orifice, namely a little over $1.5D$.

These peculiarities of the different DES versions are reflected quantitatively in Fig. 5 where we present distributions of the RMS of the wall pressure fluctuations from different simulations of case 1 and compare them with the corresponding experimental data. Namely, all the simulations predict almost identical position of the maximum of $p_{\mathrm{rms}}(x/D)$, which is consistent with the close predictions of the length of the jet potential core, but this position turns out to be shifted upstream compared to the experiment by around one duct diameter, not unlike the apparent reattachment. As far as the maximum intensity of the fluctuations is concerned, all the simulations overestimate it, and the scatter of its prediction by different approaches turns out to be rather wide, from around 1,100 Pa in the standard DDES and IDDES of NTS down to 900 Pa in the standard DDES of BCFD and 850 Pa in the IDDES of NTS with new subgrid length-scale, respectively. The last two values are close to the experimental value of the maximum $p_{\mathrm{rms}} \approx 800$ Pa. At the same time, downstream of

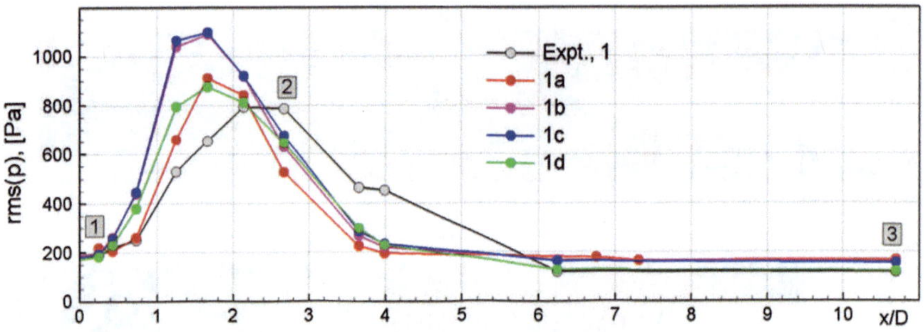

Fig. 5 Computed and measured distributions of RMS of wall pressure for case 1 in Table 1. Marks 1, 2, and 3 in *grey squares* refer to points at which the PSD of wall-pressure are plotted in Figs. 6, 8 and 9

the maximum position (at $3 < x/D < 5$), the predicted distributions of $p_{\rm rms}$ again become close to each other, but are around 200 Pa lower than the experimental one. Finally, at $x/D > 6$, all the versions of DES, except for IDDES employing the new definition of the subgrid length-scale, again somewhat (nearly by 50 Pa) overestimate the experimental level of $p_{\rm rms}$.

Figure 6 compares power density spectra from all the simulations with the experimental ones. It suggests that all the DES versions are capable of capturing the strong variation of the spectral shape and level along the duct observed in the experiments. Other than that, it shows that the BCFD flow-solver predicts a much earlier cut–off of the spectra than NTS, which is not surprising considering the higher order of accuracy of the latter and is consistent with the flow visualizations in Fig. 4.

Let us finally move to Figs. 7, 8 and 9, which illustrate the capability of capturing the effects of the diaphragm orifice diameter and duct flow-rate on the unsteady wall pressure provided by the standard DDES technique, carried out with the use of BCFD. One can see that the model does reproduce the trends observed in the experiment, namely a rather strong drop of the wall pressure unsteadiness following an increase of the open area of the diaphragm from 25 up to 35 %, and a relatively small drop with the decrease of the bulk velocity from 35.6 down to 25.4 m/s (see Fig. 7). The variation of the level and shape of the spectra along the duct with the increase of the orifice open area and bulk velocity of the flow are also captured correctly (Figs. 8 and 9). However the accuracy of predictions of the $p_{\rm rms}$ level and wall pressure spectra is nearly the same or even lower than that for the case 1 considered above. It probably could be improved by the use of IDDES with the new definition of the SGS length-scale and the less-dissipative NTS flow-solver, but based on the results obtained with this approach for case 1 (see Figs. 4, 5 and 6 above), this will still not ensure a perfect agreement with the data. The most probable reason of the remaining discrepancy between the CFD and experiment is insufficiently fine grid. This conjecture will be checked in the near future.

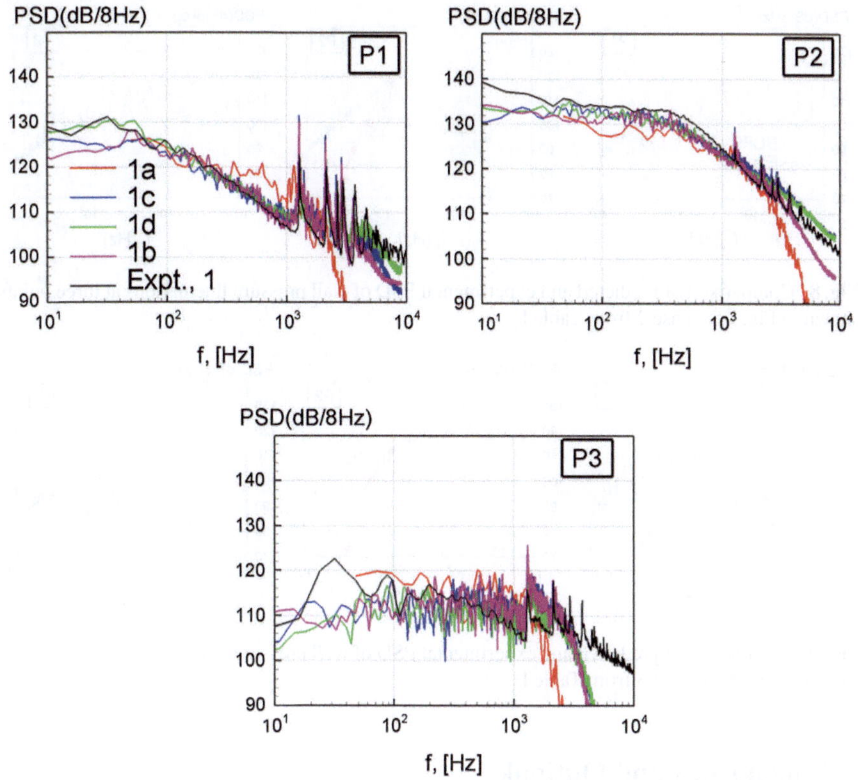

Fig. 6 Computed and experimental PSD of wall pressure for case 1 from Table 1 at points shown in Fig. 5

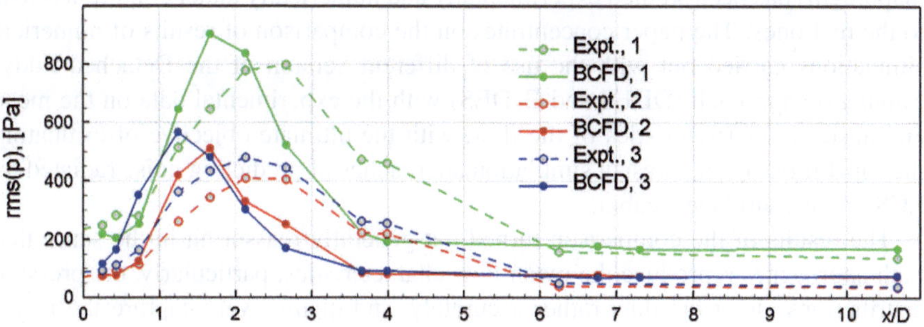

Fig. 7 Comparison of computed and measured distributions of RMS of wall pressure for cases 1, 2, and 3 from Table 1

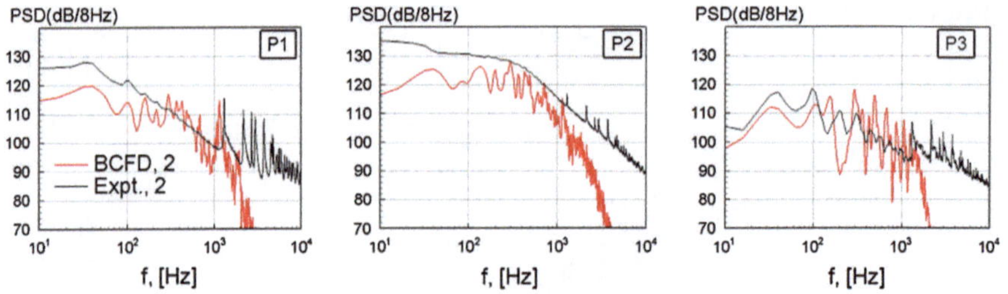

Fig. 8 Comparison of predicted and experimental PSD of wall pressure fluctuations at three points shown in Fig. 5 for case 2 from Table 1

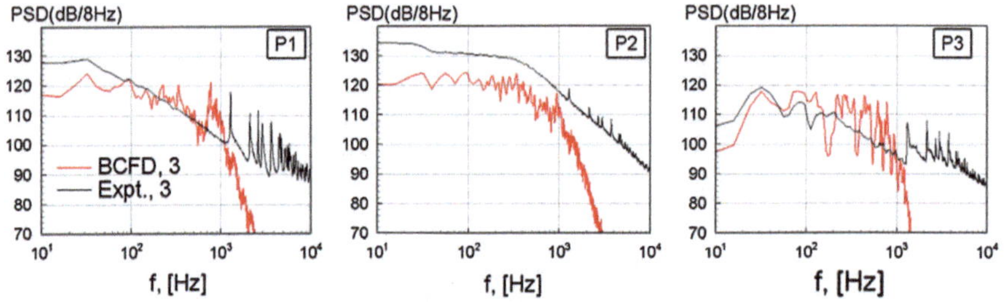

Fig. 9 Comparison of predicted and experimental PSD of wall pressure fluctuations at three points shown in Fig. 5 for case 3 from Table 1

5 Conclusions and Outlook

The flow in a model of components of an airplane's ECS (a circular duct with diaphragm) has been studied experimentally and numerically under conditions close to the real ones. The paper concentrates on the comparison of results of numerical simulations carried out with the use of different version of the Detached-Eddy-Simulation approach (DDES and IDDES) with the experimental data on the mean and unsteady characteristics of the flow, with the ultimate objective of evaluating the capabilities of the present simulation approaches of predicting noise radiated by ECS into the airplane's cabin.

The results of the comparison turned out generally reassuring in the sense that both approaches reproduce the mean flow characteristics, particularly, the pressure distributions along the duct rather accurately, and qualitatively capture the major unsteady pressure characteristics and their response to variations of diameter of the diaphragm orifice and duct flow-rate. However, these characteristics turned out to be sensitive to both the numerics and the simulation approaches used, although not to a disturbing degree. In terms of numerics, the results obtained with the use of the high-order NTS code reveal considerably better resolution of the fine-grained turbulent structures on comparable grids, which ensures a considerably wider range of resolved frequencies of the pressure fluctuations.

In terms of the simulation approaches, the IDDES version with the newly proposed definition of the SGS length scale is shown to be preferable, but the relatively coarse grid used in the simulations may be insufficient to provide a very accurate prediction of the evolution of the separated shear layer downstream of the orifice and hence the streamwise location of the maximum intensity of the wall pressure fluctuations. This issue will be addressed in future work.

Acknowledgments The study was partially supported by the Russian Scientific Foundation (Project no. 14-11-00060) and by the EU within the Project Go4Hybrid (under grant agreement no. 605361).

References

1. Spalart, P.R., Deck, S., Shur, M.L., Squires, K.D., Travin, A.K.: A new version of detached-eddy simulation, resistant to ambiguous grid densities. Theoret. Comput. Fluid Dyn. **20**, 181–195 (2006)
2. Shur, M.L., Spalart, P.R., Strelets, MKh., Travin, A.K.: Improvement of delayed detached-eddy simulation for LES with wall modeling. Int. J. Heat Fluid Flow **29**, 1638–1649 (2008)
3. Spalart, P.R., Allmaras, S.R.: A one-equation turbulence model for aerodynamic flows. In: AIAA Paper, 92–0439 (1992)
4. Spalart, P.R.: Detached-eddy simulation. Annu. Rev. Fluid Mech. **41**, 181–202 (2009)
5. Deck, S.: Recent improvements of the zonal detached eddy simulation (ZDES) formulation. Theoret. Comput. Fluid Dyn. **26**, 523–550 (2012)
6. Mockett, C., Fuchs, M., Garbaruk, A., Shur, M., Spalart, P., Strelets, M., Thiele, F., Travin, A.: Two non-zonal approaches to accelerate RANS to LES transition of free shear layers in DES. This book
7. Cary, A., Dorgan, A., Mani, M.: Toward accurate flow predictions using unstructured meshes. In: AIAA, 2009–3650 (2009)
8. Strelets, M.: Detached-eddy simulation of massively separated flows. In: AIAA Paper, 2001–0879 (2001)
9. Shur, M., Strelets, M., Travin, A.: High-order implicit multi-block Navier-Stokes code: ten-years experience of application to RANS/DES/LES/DNS of turbulent flows. In: 7th Symposium on Overset Composite Grids & Solution Technology, Huntington Beach, California. http://agarbaruk.professorjournal.ru/c/document_library/get_file?uuid=a8f303a6-ddf5-4f03-9b30-fe11c95b1655&groupId=199655 (2004)
10. Roe, P.L.: Approximate Riemann solvers, parameter vectors and difference schemes. J. Comput. Phys. **46**, 357–378 (1981)

Scale-Resolving Simulations of Wall-Bounded Flows with an Unstructured Compressible Flow Solver

Axel Probst and Silvia Reuß

Abstract The fully-developed channel flow at $Re_\tau \approx 395$ is used to validate scale-resolving simulations with the unstructured compressible DLR-TAU code. In a sensitivity study based on wall-resolved LES a low-dissipative spatial scheme is derived, which allows to predict the channel flow in fair agreement with DNS. Then the scheme is used in Improved Delayed DES computations in order to assess TAU's capabilities for wall-modelled LES. As pointed out in a grid study, a tangential resolution of about $\Delta x^+ \approx 40$, $\Delta z^+ \approx 20$ is required to obtain acceptable mean-flow results. Besides, the combination of IDDES with a vorticity-dependent subgrid filter width is shown to yield consistent results, and the effect of the underlying RANS approach up to Reynolds-stress modelling is analysed.

1 Introduction

In the recent years the application range of hybrid RANS/LES methods (HRLM) has been expanded from massive separation to weakly separated or wall-bounded flows [10]. One example is the flow around multi-element airfoils, where resolved turbulence from the separated cove regions is convected into the attached boundary layers on the downstream elements, cf. [2]. The requirement to resolve parts of the small near-wall structures poses high demands on the numerical method and the scale-resolving simulation (SRS) approach. This is particularly true for unstructured compressible solvers such as DLR-TAU [9], which has been primarily focused on stability in complex flow problems. Besides, due to the unstructured finite-volume approach, an extension towards higher-order accuracy, as e.g. [4], is difficult.

In order to qualify the DLR-TAU code for wall-bounded SRS, basic numerical studies of the fully-developed plane channel flow at $Re_\tau \approx 395$ are conducted in this work. The first part focuses on wall-resolved LES computations which ensure minimal modelling influences, thus allowing to specifically analyse the numerical

A. Probst (✉) · S. Reuß
DLR (German Aerospace Center), Bunsenstr. 10, 37073 Göttingen, Germany
e-mail: axel.probst@dlr.de

S. Girimaji et al. (eds.), *Progress in Hybrid RANS-LES Modelling*,
Notes on Numerical Fluid Mechanics and Multidisciplinary Design 130,
DOI 10.1007/978-3-319-15141-0_39

scheme. By identifying relevant parameters of the spatial discretization, an optimized 'low-dissipation' scheme is derived. The second part focuses on more relevant wall-modelled LES computations which are based on Improved Delayed DES [10] and the newly derived numerical settings. Besides a basic validation of the wall-modelled approach in TAU, its sensitivities w.r.t. grid resolution and modelling aspects are pointed out. Concerning the latter, novel variants of IDDES using a vorticity-dependent filter width and a Reynolds-stress background model are tested.

2 Basic Flow Setup and Numerical Method

The qualification of the DLR-TAU code for SRS of wall-bounded flows is based on the fully-developed turbulent channel flow with half-height δ at a nominal friction-velocity Reynolds number of $Re_\tau = u_\tau \cdot \delta/\nu = 395$. The case is selected for its clear focus on the relevant wall phenomena, which are well documented by DNS data [5].

Following common practice [5], the channel geometry is set up as a rectangular box of height 2δ, a length of $2\pi \cdot \delta$ and a width of $\pi \cdot \delta$. For all flow variables periodic boundary conditions are applied in stream- and spanwise directions. To compensate the neglected pressure gradient along the channel, a uniform driving force term is applied which ensures constant bulk velocity U_{bulk} corresponding to $Re_\delta = U_{bulk} \cdot \delta/\nu = 6875$ from DNS [5]. This allows to fix one global flow parameter during the computation, while the degree of deviation in the other parameter, Re_τ, can be regarded a quality measure of the simulation approach. A sufficiently low bulk Mach number of $Ma = 0.15$ is chosen to avoid any compressibility effects.

Overall simulation times vary around 30 convective time units based on the channel length and U_{bulk}, with about 20 time units for obtaining temporal statistics.

2.1 *Reference Numerical Scheme in DLR-TAU*

The DLR-TAU code [9] solves the compressible flow equations based on an unstructured finite volume discretization, which provides wide flexibility w.r.t. different cell types. Generally speaking, compared to common RANS approaches the demands on the numerical scheme for SRS are shifted from *stability* towards *accuracy*. Based on the methods available in TAU a preliminary reference scheme for the spatial discretization in SRS applications has been defined:

- A 2nd-order non-dissipative central scheme for the convective fluxes is applied. To prevent spurious kinetic-energy production, the compressible fluxes follow the energy-conserving skew-symmetric form proposed by Kok [4].
- Since strict energy conservation by the skew-symmetric form can only be proven for incompressible flows and curvilinear grids [4], a certain amount of 4th-order matrix-valued artificial dissipation (MD) [11] is added to the compressible fluxes.

The dissipative flux at a face between two grid points i and j is defined as:

$$\mathbf{d}_{ij}^{(4)} = k^{(4)} \cdot |\mathbf{PA}|_{ij} \cdot \phi_{ij} \cdot \left\{\nabla^2 \mathbf{w}_i - \nabla^2 \mathbf{w}_j\right\}. \quad (1)$$

For the reference scheme the global scaling factor $k^{(4)}$ is set to $k^{(4)} = 1/128$, based on experience from previous DES of separated flows [6]. The term $|\mathbf{PA}|_{ij}$ is the matrix-dissipation operator including the preconditioning matrix $\mathbf{P}$ (see below), and $\mathbf{w}_i$, $\mathbf{w}_j$ are the vectors of conservative variables. The "cell-stretching coefficient" (CSC) ϕ_{ij} increases the dissipation in the direction of local cell stretching in order to stabilize computations on grid cells with high aspect ratios [1].

- To handle incompressible flow regions the compressible DLR-TAU code applies low-Mach preconditioning (LMP) which alters both convergence and accuracy properties. For the latter, the artificial dissipation in Eq. (1) is modified by a matrix $\mathbf{P}$ which is designed to reduce the disparity of eigenvalues of the compressible equations, thus providing almost Ma-independent accuracy [7].
  To prevent singularities in $\mathbf{P}$, a lower bound for the so-called 'artificial speed of sound' β is introduced, $\beta^2 = \min\left[\max\left(|\mathbf{u}|^2, (K \cdot U_{bulk})^2\right), a^2\right]$, where $\mathbf{u}$ is the velocity vector, U_{bulk} is the channel bulk velocity, and a is the speed of sound. K is a global cut-off which controls the level of preconditioning close to walls. The reference value, $K = 1$, originates from earlier Euler/RANS studies.

3 Wall-Resolved LES of Plane Channel Flow

Wall-resolved LES computations of the fully-developed channel flow at $Re_\delta = 6{,}875$ (corresponding to $Re_\tau \approx 395$) are conducted with the compressible DLR-TAU solver. By using established subgrid-scale models for wall-bounded flows and a well-resolved grid the modelling uncertainties are minimized, thus allowing to focus on the capabilities and sensitivities of the numerical scheme.

3.1 Subgrid-Scale Modelling and Numerical Setup

In LES only the large scales of the turbulent energy spectrum are directly resolved, whereas the smaller structures below the local grid resolution require appropriate modelling. Two subgrid-scale models suited for wall-resolved LES (WR-LES) are considered: The classic Smagorinsky model (denoted 'SMAG') with near-wall damping and the Wall-Adapting Local Eddy-viscosity model ('WALE').

Both models apply the Boussinesq approach to compute the subgrid stresses from a scalar eddy viscosity, which is related to the local grid spacing according to:

$$\mu_{t,SGS} = \rho \cdot (f_{vD} \cdot C \cdot \Delta_{vol})^2 \cdot \left|\tilde{S}\right| \quad \text{with:} \quad \Delta_{vol} = (\Delta x \cdot \Delta y \cdot \Delta z)^{1/3}. \quad (2)$$

The van-Driest damping function f_{vD} and the general gradient term $\left|\tilde{S}\right|$ are model-dependent: As the Smagorinsky model employs the strain-rate magnitude, i.e. $\left|\tilde{S}\right| = |S| = \sqrt{2S_{ij}S_{ij}}$, near-wall damping is required to correctly capture the vanishing eddy viscosity at solid walls. In this work, $f_{vD} = \sqrt{1 - e^{(-y^+/25)^3}}$.

In contrast, the WALE model employs a more complex expression for $\left|\tilde{S}\right|$ (see [3] for details) which automatically fulfils the asymptotic near-wall requirements for $\mu_{t,SGS}$, so that $f_{vD} = 1$. For the general model constant C the commonly accepted values for wall-bounded turbulence are adopted, i.e. $C_{SMAG} = 0.1$ and $C_{WALE} = 0.325$.

The numerical discretization of the simulation setup described in Sect. 2 follows common practice for wall-resolved LES. The hexahedral grid (cf. Table 2 for more details) provides a normalized spacing in wall units of: $\Delta x^+ = 31$, $\Delta y(1)^+ = 1$, $\Delta z^+ = 13$. For time discretization a non-dimensional time step of $\Delta t^+ = 0.4$ is used.

3.1.1 Assessment and Optimization of the Numerical Scheme

The strategy to optimize the numerical scheme for wall-bounded SRS can be summarized as follows: Starting from simulations with the reference scheme, a sensitivity study based on the SMAG model is conducted, in which each of the available numerical parameters, cf. Sect. 2, is varied individually. The parameter variations during the analysis and their reference values ('Ref-scheme') are listed in Table 1.

As mentioned above, the deviation of the resulting Reynolds number Re_τ from the DNS is considered a meaningful accuracy measure. Thus, Fig. 1 (left) presents a sensitivity map showing Re_τ for all performed WR-LES computations in dependency of the numerical parameters. Note that for a fair assessment, the line indicating the DNS corresponds to the *actual* computed value $Re_\tau = 392.24$ [5].

Starting with the reference numerics, the relative deviation with SMAG from DNS data amounts to about 10 %. The effect of switching to WALE is notable, but smaller than the numerical sensitivities. This is also reflected in the wall-normal profiles of mean velocity and the relevant Reynolds-stress components in Figs. 1 (right) and 2 (left): Although qualitative agreement with DNS data can be stated, both models overpredict both u^+ and $\overline{u'^2}$ to a very similar extent.

Table 1 Numerical parameters used in sensitivity study and reference/low-dissipation settings

	MD: $k^{(4)}$	CSC: ϕ_{ij}	LMP: K
Sensitivity analysis	[1/128; 1/256; 1/512; 1/1024; 0]	[ϕ_{ij}; 1]	["∞"; 1; 0.3; 0.1]
Ref-scheme	1/128	ϕ_{ij}	1
LD-scheme	1/1024	1	0.3

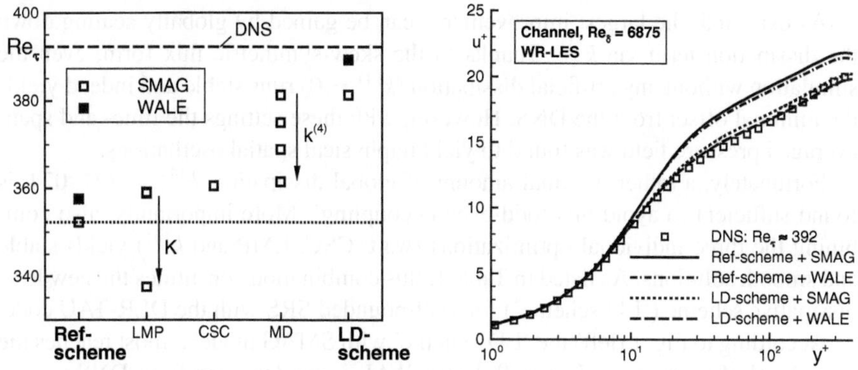

Fig. 1 *Left* Sensitivity map of the computed friction Reynolds number in WR-LES of the *plane* channel as function of different numerical settings. *Right* Non-dimensional mean velocity profiles computed with different numerical schemes and subgrid-scale models in comparison with DNS

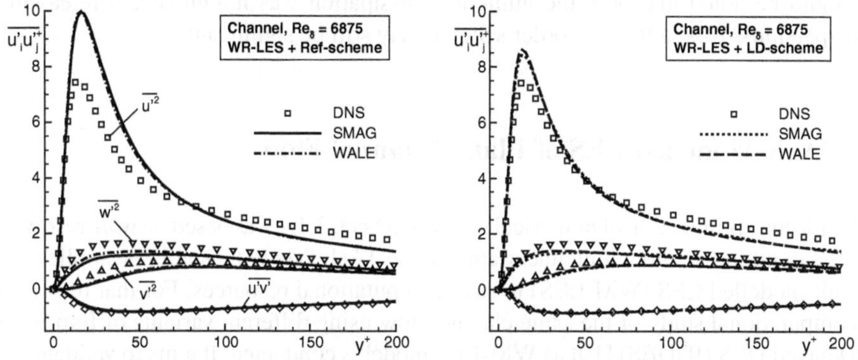

Fig. 2 Non-dimensional resolved Reynolds-stress profiles computed with different subgrid-scale models in comparison with DNS. *Left* Reference numerical scheme. *Right* Optimized LD scheme

According to Fig. 1 (left) all numerical parameters clearly affect Re_τ. The cell-stretching coefficient (CSC), which was initially introduced to stabilize RANS computations on anisotropic near-wall cells, can be switched off here (i.e. set to 1) without causing notable stability issues. Its effect on $Re_\tau = 395$, though, is comparable to globally reducing the dissipation by a factor of 2 (i.e. $k^{(4)} = 1/256$).

Similar improvements can be achieved by reducing the cut-off factor K in the preconditioning (LMP) matrix, but the effect levels out at $K = 0.3$ (note that the Re_τ values for $K = 0.3$ and $K = 0.1$ in Fig. 1 (left) virtually coincide). As a further reduction may potentially cause stability issues, the value $K = 0.3$ appears to be the optimal compromise. On a side note, the effect of deactivating LMP completely is analysed as well (i.e. $K =$ "∞"). Without LMP the deviation from DNS data increases to about 14 % which clearly illustrates its importance in SRS at low Mach numbers.

As expected, the largest improvements can be gained by globally scaling down the dissipation term via $k^{(4)}$. Thanks to the skew-symmetric flux form, even the simulation without any artificial dissipation ($k^{(4)} = 0$) runs stable and indeed yields the minimal offset from the DNS. However, with these settings the time- and span-averaged pressure field was found to yield unphysical spatial oscillations.

Fortunately, a rather minimal amount of global dissipation, $k^{(4)} = 1/1,024$, is found sufficient to avoid this 'odd-even decoupling'. More importantly, also combining the three individual optimizations (w.r.t. CSC, LMP and $k^{(4)}$) yields stable and smooth solutions. As listed in Table 1, this combination constitutes the new low-dissipative scheme ('LD-scheme') for wall-bounded SRS with the DLR-TAU code.

According to Fig. 1 (left) the 'LD-scheme' with SMAG model almost reaches the same level of accuracy as $k^{(4)} = 0$. Using WALE, the deviation from DNS is even further reduced to only about 0.7 %. The mean velocity in Fig. 1 (right) and Reynolds stresses in Fig. 2 (right) confirm these clear improvements over the 'Ref-scheme', although the DNS data for the stresses are not perfectly matched. In this regard it should be noted that only the numerical dissipation was minimized, whereas the dispersion errors in the 2nd-order scheme may still be significant [4].

4 Wall-Modelled LES of Plane Channel Flow

While the optimization of numerical settings in Sect. 3.1.1 was based on wall-resolved LES to minimize modelling uncertainties, the TAU code will usually be applied for wall-modelled LES (WM-LES) to save computational resources. For that reason, a computational study of the same channel flow using different variants of Improved Delayed DES (IDDES) [10] as WM-LES model is conducted. It aims to validate the 'LD-scheme' for WM-LES, and to identify grid and modelling sensitivities.

4.1 The Improved Delayed DES in Wall-Modelled LES Mode

Based on the Delayed DES, which discerns between RANS in attached flow and LES in separation, the Improved DDES (IDDES) [10] adds another branch for wall-modelled LES. Thus, for resolved turbulence in boundary layers, IDDES provides a consistent switch from near-wall RANS modelling to LES in the outer region.

In common with any DES approach, IDDES replaces the integral length scale in the underlying RANS model by a hybrid length scale l_{hyb}. In the periodic channel, l_{hyb} defined by IDDES roughly corresponds to its WM-LES part, i.e.: $l_{hyb} \approx l_{\text{WMLES}} = f_B\,(1 + f_e)\,l_{\text{RANS}} + (1 - f_B)\,l_{\text{LES}}$. Here, the function f_B provides a purely grid-dependent blending from RANS to LES, while f_e increases the modelled turbulence at the interface to correct the so-called 'log-layer mismatch' [10].

As second key element, the filter width in the LES length scale $l_{\text{LES}} = \Psi C_{\text{DES}} \Delta$ (where Ψ is a low-Re correction) is expressed as a function of the wall distance d_w, the

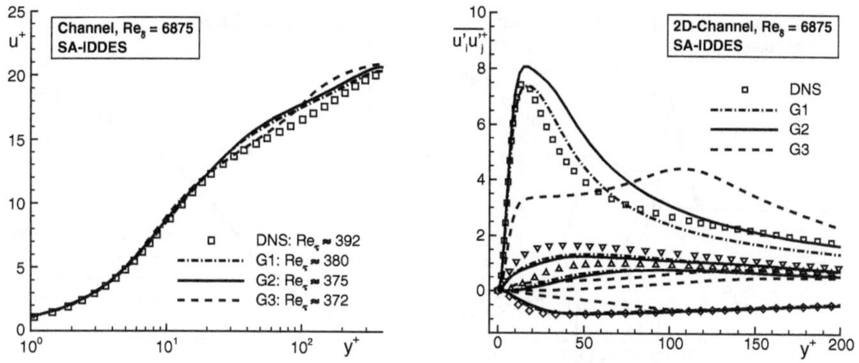

Fig. 3 Influence of the grid resolution on the (non-dimensional) results of SA-IDDES computations of the plane channel flow. *Left* Mean velocity. *Right* Resolved Reynolds stresses

wall-normal spacing h_{wn}, as well as the classic $h_{\max} = \Delta_{\text{DES}} = \max(\Delta x, \Delta y, \Delta z)$:

$$\Delta = \Delta_{\text{IDDES}} = \min\left\{\max\left[C_w \cdot d_w, C_w \cdot h_{\max}, h_{wn}\right], \Delta_{\text{DES}}\right\}, \quad (3)$$

where $C_w = 0.15$. In essence, this definition allows to use a unique calibration constant C_{DES} for both wall-bounded and free turbulent flow. For details refer to [10].

4.1.1 Grid Sensitivity

By design, IDDES should provide consistent wall-modelling over a certain range of grid resolutions. To assess this feature in combination with DLR-TAU, three different grids as listed in Table 2 are considered. 'G1' corresponds to the mesh used in Sect. 3 for WR-LES and should yield only a minimal near-wall RANS layer. 'G2' mimics the channel discretization used in [10] for basic validation, thus it is considered a typical IDDES grid. Finally, 'G3' is coarsened in stream- and spanwise directions in order to shift the RANS/LES interface farther away from the wall.

The simulations apply Spalart-Allmaras as RANS background model (SA-IDDES [10]). Figure 4 (left) shows the effect of the grid resolution on the location of the

Table 2 Properties of grids used in numerical sensitivity study with IDDES

Grid	G1	G2	G3
$n_x \times n_y \times n_z$	$81 \times 97 \times 97$	$62 \times 65 \times 65$	$21 \times 65 \times 34$
	$\approx$ 762.000	$\approx$ 262.000	$\approx$ 46.000
$\Delta x^+ \times \Delta y^+ \times \Delta z^+$	$31 \times 1 \times 13$	$41 \times 0.8 \times 19.5$	$120 \times 0.8 \times 37$
Geometric stretching (Δy)	1.07	1.14	1.14

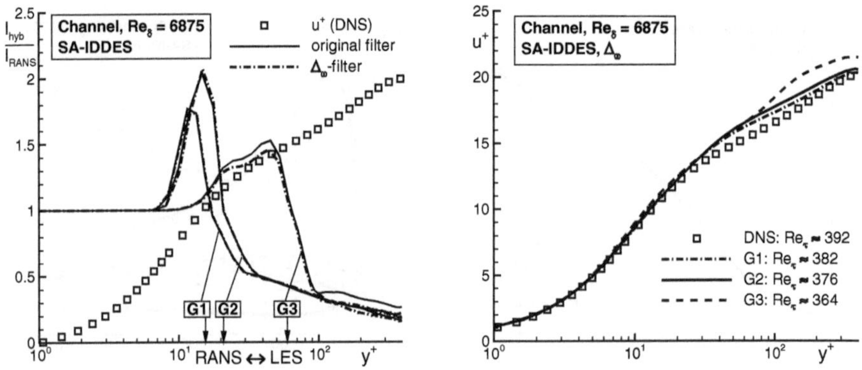

Fig. 4 *Left* Hybrid length-scale ratio and locations of the RANS/LES interface (indicated by arrows) from SA-IDDES on different grids and for different filter scales. *Right* Non-dimensional mean velocity profiles on different grids for SA-IDDES with vorticity-dependent filter width

formal RANS/LES interface (i.e. where the function f_B starts to depart from $f_B = 1$) and on the resulting length-scale ratio $l_{hyb}/l_{\mathrm{RANS}}$ (where values < 1 indicate LES). As intended, 'G3' places the RANS/LES interface well within the logarithmic region of the boundary layer, whereas 'G1' and 'G2' already switch in the buffer layer.

Comparing the velocity profiles of 'G1' and 'G2' in Fig. 3 (left), only very little grid influence is observed. Apart from the lower logarithmic layer, the curves agree reasonably with DNS.

However, the coarse grid 'G3' shows a variable slope in the logarithmic layer which resembles a 'log-layer mismatch'. As depicted in Fig. 3 (right), this phenomenon which is localized around $y^+ = 100$, is accompanied by an overpredicted resolved stress component $\overline{u'^2}$. Thus, at least for the present DLR-TAU approach, the resolution provided by the grid 'G3' appears infeasible for IDDES.

4.1.2 Modelling Variants

In hybrid RANS/LES the actual modelling (both RANS and subgrid) can become crucial when tackling more complex problems, such as flows with incipient separation or delayed formation of resolved turbulence ("grey area"). Thus, compatibility of IDDES with different model variants and extensions has to be ensured. Two aspects, namely a modified LES filter and the underlying RANS model, are addressed.

4.2 LES Filter Definition

To address the grey-area problem in free shear/mixing layers, Deck [2] proposed to replace the grid-dependent filter in Zonal-DES by a flow-dependent filter:

$$\Delta_\omega = \sqrt{N_x^2 \Delta y_{max} \Delta z_{max} + N_y^2 \Delta x_{max} \Delta z_{max} + N_z^2 \Delta x_{max} \Delta y_{max}}. \quad (4)$$

By incorporating the orientation of the vorticity vector, $\mathbf{N} = \{N_x, N_y, N_z\}^T$, it only accounts for the relevant grid-spacing directions to resolve vortical structures.

For a combination with IDDES, Δ_ω needs to be sensibly adopted in the IDDES filter without altering its near-wall functionality. For this it was found crucial to only replace the term Δ_{DES} in Eq. (3) by Δ_ω, while leaving other instances of $h_{\max} = \Delta_{\text{DES}}$ unchanged. Indeed, as depicted in Fig. 4 (left), the modification retains the RANS/LES transition on the grids 'G1' and 'G2', along with satisfying mean-flow predictions, cf. Fig. 4 (right). Only on 'G3', which was found unsuitable for IDDES anyway, the ratio l_{hyb}/l_{RANS} in the outer boundary layer is notably changed, which leads to a larger velocity offset from DNS. Note that the potential of IDDES+Δ_ω in reducing the 'grey area' was demonstrated for a backward-facing step flow in [8].

4.3 Underlying RANS Model

The quality of RANS modelling in HRLM is of particular importance in separating flow, where the RANS/LES interface is not fixed by the geometry, but rather dependent on local boundary-layer characteristics. However, for WM-LES of the present channel-flow setup, we expect rather similar results irrespective of the RANS model. Apart from SA-IDDES, two more RANS approaches are tested against this requirement: The SST-based IDDES as presented in [10], and a preliminary combination with an ε^h-based Reynolds-stress model (RSM), which has already been coupled with DDES [6]. The definition of RSM-IDDES combines that RSM-DDES with the procedures in [10], whereby two RANS-model-dependent coefficients in the function f_e are simply adopted from SST-IDDES for now.

Simulation results for different RANS models were obtained on the IDDES grid 'G2' only. As depicted in Fig. 5, both mean velocity and resolved stresses are almost identical for SA- and SST-IDDES, clearly in line with expectations [10]. Remarkably, the preliminary RSM-IDDES matches the DNS in the log-law region and w.r.t. Re_τ even better, but in turn underpredicts the velocity and the resolved stresses $\overline{u'^2}$ closer to the wall ($y^+ \approx 10-30$). A possible explanation is found in the 'elevating' function f_e depicted in Fig. 5 (left): In RSM-IDDES f_e acts closer to the wall than in SA- and SST-IDDES, thus causing additional damping of resolved near-wall turbulence. Thus, despite the promising first results for the present test case, further adjustments and more detailed validation of RSM-IDDES seems advisable.

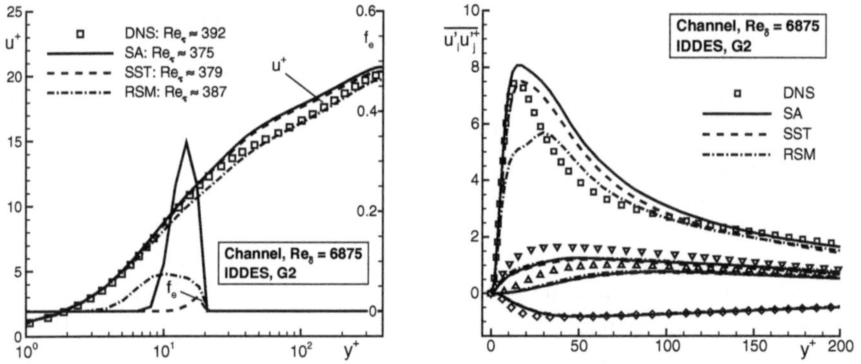

Fig. 5 Influence of the underlying RANS model in IDDES on the channel results for the medium grid 'G2'. *Left* Mean velocity and 'elevating' function. *Right* Resolved Reynolds stresses

5 Conclusion

The fully-developed turbulent channel flow was used as validation test case for the compressible unstructured DLR-TAU solver in order to assess and improve its scale-resolving simulation capabilities for wall-bounded flows.

In a sensitivity study based on wall-resolved LES, the relevant parameters of the spatial discretization scheme were identified and carefully adjusted to derive a stable 'low-dissipation' (LD) scheme for SRS. Applying this scheme with two different subgrid-scale models, the TAU code was shown to predict the important characteristics of the channel flow in reasonable agreement with DNS.

The scheme was then adopted in wall-modelled LES computations using the IDDES hybrid RANS/LES approach. For grid resolutions that provide a RANS/LES interface sufficiently close to the wall, satisfying and consistent mean-flow predictions for Spalart-Allmaras- and Menter-SST-based IDDES were obtained. Moreover, a novel IDDES variant, which is equipped with a vorticity-dependent filter scale for reducing 'grey areas' in detached flow, was shown to retain the desired wall-modelling behaviour in the channel flow. Finally, first promising results of a preliminary RSM-based IDDES were presented which call for further assessment.

References

1. Blazek, J.: Computational Fluid Dynamics: Principles and Applications, 2nd edn. Elsevier, Amsterdam (2005)
2. Deck, S.: Recent improvements in the zonal detached eddy simulation (ZDES) formulation. Theoret. Comput. Fluid Dyn. **26**(6), 523–550 (2012)
3. Ducros, F., Nicoud, F., Poinsot, T.: Wall-adapting local eddy-viscosity models for simulations in complex geometries. In: Proceedings Conference on Numerical Methods Fluid Dynamics, Oxford, UK (1998)

4. Kok, J.: A high-order low-dispersion symmetry-preserving finite-volume method for compressible flow on curvilinear grids. J. Comput. Phys. **228**(18), 6811–6832 (2009)
5. Moser, R., Kim, J., Mansour, N.: Direct numerical simulation of turbulent channel flow up to Re = 590. Phys. Fluids **11**(4), 11–13 (1999)
6. Probst, A., Radespiel, R., Knopp, T.: Detached-eddy simulation of aerodynamic flows using a Reynolds-stress background model and algebraic RANS/LES sensors. In: AIAA Paper 2011–3206 (2011)
7. Radespiel, R., Turkel, E., Kroll, N.: Assessment of preconditioning methods. Technical Report, DLR-FB 95–29 (1995)
8. Reuß, S., Knopp, T., Probst, A., Orlt, M.: Assessment of local LES-resolution sensors for hybrid RANS/LES simulations. In: 5th Symposium on Hybrid RANS-LES Methods TEXAS A&M University, College Station, Houston, USA, 19–21 March 2014
9. Schwamborn, D., Gerhold, T., Heinrich, R.: The DLR TAU-code: recent applications in research and industry. In: Wesseling, P., Oñate, E., Périaux, J. (eds.) ECCOMAS CFD. TU Delft, The Netherlands (2006)
10. Shur, M.L., Spalart, P.R., Strelets, M.K., Travin, A.K.: A hybrid RANS-LES approach with delayed-DES and wall-modelled LES capabilities. Int. J. Heat Fluid Flow **29**(6), 406–417 (2008)
11. Swanson, R., Turkel, E.: On central-difference and upwind schemes. J. Comput. Phys. **306**, 292–306 (1992)

Noise Generated by an Airfoil Located in the Wake of a Circular Cylinder

Man Zhang and Abdelkader Frendi

Abstract In this paper, a problem involving noise radiation from a bluff body is solved numerically using a hybrid RANS-LES method. In particular the problem of noise radiated by an airfoil leading edge located in the wake of a circular cylinder is addressed. Our results compare well to experimental measurements and other CFD computations. It is found that the hybrid RANS-LES method is able to resolve enough turbulent scales to compute the nearfield noise spectra and the directivity pattern. Our CFD results indicate that the coherent structures are responsible for the peak Strouhal number in the spectra.

1 Introduction

One of the biggest challenges facing aerospace engineers is the problem of noise pollution. As air traffic increases in coming years, the noise level in communities surrounding airports will increase leading to a deterioration of the quality of life. This has prompted the Federal Aviation Administration (FAA) to set a goal of reducing noise emission from aircrafts by 10 dB which is a very aggressive target [1]. This in turn has resulted in a renewed research effort to identify the sources of noise and to develop noise control strategies; active and/or passive.

From a computational view point, in order to identify the sources of noise, one needs to perform unsteady turbulent flow computations. In the past, these computations were carried-out using either direct numerical simulation (DNS) or large eddy simulation (LES), which are computationally intensive and not practical at high Reynolds numbers. However, in the last two decades, a new class of methods known as bridging methods have been developed. These methods take advantage of the large experiences gained in the RANS methods to bridge the gap to the LES methods [2–6] and offer a cost-effective way of resolving more scales of turbulence based on available computational resources.

M. Zhang · A. Frendi (✉)
MAE Department, Uahuntsville, Huntsville, AL 35899, USA
e-mail: kader.frendi@uah.edu

S. Girimaji et al. (eds.), *Progress in Hybrid RANS-LES Modelling*,
Notes on Numerical Fluid Mechanics and Multidisciplinary Design 130,
DOI 10.1007/978-3-319-15141-0_40

2 Mathematical Model and Method of Solution

The mathematical model is composed of the standard unsteady Reynolds Averaged Navier-Stokes equations with a two-equation eddy viscosity model based on Mentor shear stress transport model [7]. This model has been modified to accommodate a hybrid RANS-LES method (HRLES) [8–10]. The LES turbulent kinetic energy equation, k^{sgs}, is used to obtain the subgrid scale quantities

$$\frac{\partial}{\partial t}\left(\overline{\rho}k^{sgs}\right)+\frac{\partial}{\partial x_i}\left(\bar{\rho}\tilde{u}_j k^{sgs}\right)=\bar{\rho}\tau_{ij}^{sgs}\frac{\partial\tilde{u}_i}{\partial x_j}-C_\varepsilon\bar{\rho}\frac{\left(k^{sgs}\right)^{3/2}}{\Delta}+\frac{\partial}{\partial x_j}\left[\left(\frac{\tilde{\mu}}{\mathrm{Pr}}+\frac{\mu^{sgs}}{\mathrm{Pr}_t}\right)\frac{\partial k^{sgs}}{\partial x_j}\right] \quad (1)$$

where

$$\mu^{sgs}=C_\nu\bar{\rho}\Delta\sqrt{k^{sgs}} \quad (2)$$

is the subgrid eddy viscosity. In the above equations, $\bar{\rho}$ is the time averaged velocity, $\tilde{u}$ and $\tilde{\mu}$ the mass-averaged fluid velocity and viscosity, Δ the subgrid scale filter width, $\bar{\rho}\tau_{ij}$ the unresolved stress and Pr and Pr_t are the molecular and turbulent Prandtl numbers, respectively. The coefficients C_ν and C_ε are obtained dynamically as part of the solution. The "sgs" superscript refers to subgrid scale quantity.

The hybrid RANS-LES governing equations are written in a generic from as

$$\frac{\partial}{\partial t}\left(\vec{E}\right)+\frac{\partial}{\partial x_j}\left(\tilde{u}_j\vec{E}\right)=\frac{\partial}{\partial x_j}\left(\vec{G}_{trans}\right)+\vec{G}_{src}+\frac{\partial}{\partial x_j}\left(\vec{G}_{Ttrans}^{hybrid}\right)+\vec{G}_{Tsrc}^{hybrid} \quad (3)$$

where $\vec{E}=\left\{\bar{\rho},\bar{\rho}\tilde{u}_j,E,\bar{\rho}k\right\}$ and the right hand side of the above equation consists of the original transport $\left(\vec{G}_{trans}\right)$ and source $\left(\vec{G}_{src}\right)$ vectors without the fluctuating turbulence terms. These later terms are written in hybrid vectors as;

$$\vec{G}_{Ttrans}^{hybrid}=F\vec{G}_{Ttrans}^{rans}+(1-F)\,\vec{G}_{Ttrans}^{sgs} \quad (4)$$

$$\vec{G}_{Tsrc}^{hybrid}=F\vec{G}_{Tsrc}^{rans}+(1-F)\,\vec{G}_{Tsrc}^{sgs}. \quad (5)$$

The blending function is defined as

$$F=\tanh(\arg_1^4) \quad (6)$$

with

$$\arg_1=\max\left(2\frac{\sqrt{k}}{0.09\omega d},\frac{500\bar{\nu}}{d^2\omega}\right) \quad (7)$$

where d is the distance to the wall. In the near wall region, the HRLES model becomes a RANS model using $(k - \omega)$ equations for closure [7]. Away from the wall, the HRLES model becomes an LES that uses the k-equation to obtain the sub-grid viscosity.

All the computational results presented in this paper are obtained using a finite volume flow solver developed at NASA Langley and known as FUN3D which stands for Fully Unstructured Navier-Stokes [11]. The code has been used extensively on a wide range of applications and its user base has grown beyond NASA Langley to include industry and academic institutions. An unstructured grid generation software known as VGRID [12] is used to generate our 3D grids. To generate an unsteady solution using FUN3D, we first obtain a steady base flow using the RANS model, then switch to the unsteady model that uses the described hybrid RANS-LES model. Five sub-iterations per time-step are used, and an optimized second-order backward differencing scheme is chosen to obtain at least two orders of magnitude reduction in the residuals of governing equations.

3 Results and Discussions

In order to validate our mathematical model and numerical approach we use the experimental setup given in Jacob et al. [13]. Figure 1 shows the geometric arrangement with all the distances given in mm. There is a 2 mm off-set in the vertical direction between the cylinder axis and the airfoil leading edge this was balanced

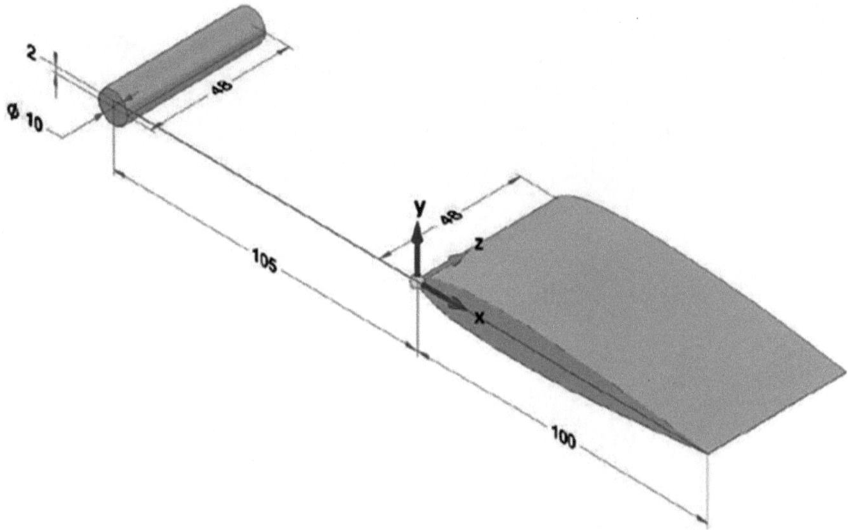

Fig. 1 Rod-Airfoil geometric configuration

by a 2° angle of attack. A conventional NACA0012 airfoil with a 100 mm cord and 12 mm thickness was used in the test and hence in our computations. The cylinder diameter is 10 mm, and the distance between the cylinder and the airfoil leading edge is also one cord length. The spanwise extent of the cylinder and airfoil is 48 mm. Air at a pressure of 98.9 kPa and temperature of 293 K is used. The flow condition used are; mean flow Mach number $M_0 = 0.21$, mean flow velocity $U_0 = 72\,\mathrm{m/s}$, Reynolds number based on cylinder diameter $\mathrm{Re}_d = 48{,}000$. The computational domain includes all the geometric features shown on Fig. 1 and extends $(-150, 300)$, $(-100, 100)$ and $(0, 48)$ mm in the x, y and z directions, respectively. It is important to note that the leading edge of the airfoil serves as the origin of our coordinate system and therefore the cylinder center is at -105 mm in x-direction. Figure 2 shows a cross-section of the grid in the streamwise and vertical directions with Fig. 2a showing the grid between the cylinder and the airfoil, Fig. 2b showing a zoomed-in view of the grid around the cylinder and Fig. 2c that around the airfoil leading

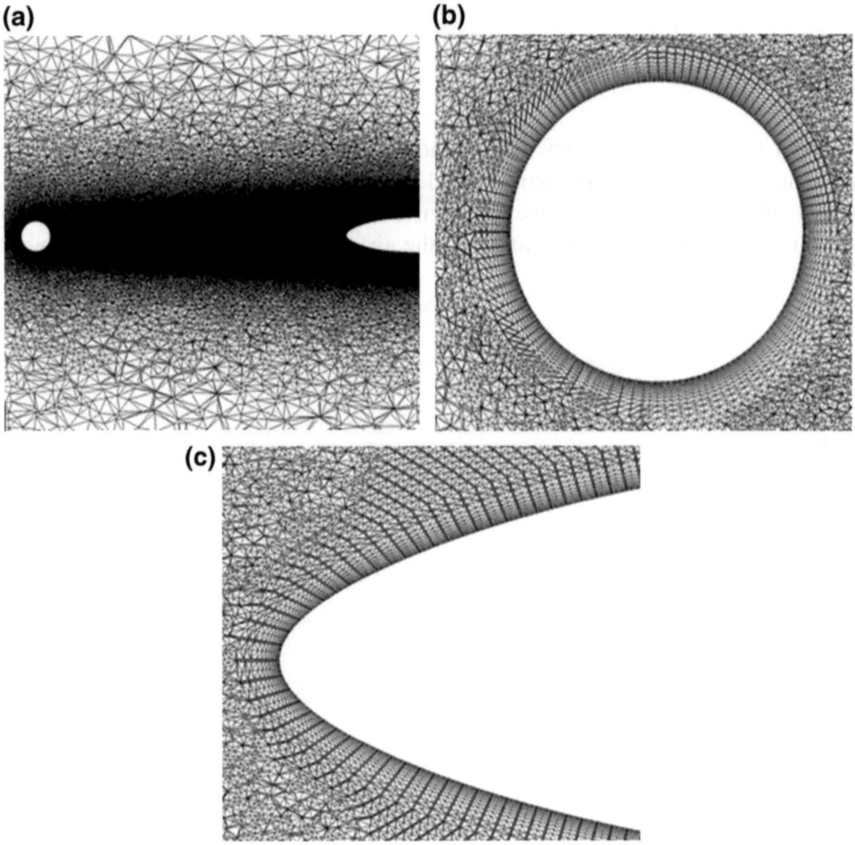

Fig. 2 Cross-section of the grid used; **a** grid between cylinder and airfoil, **b** grid around cylinder and **c** grid around airfoil

edge. Our resolution is such that $y^+ < 1$in the direction normal to solid wall and x^+ and z^+ ~50. These numbers are widely used in the literature and are adequate for a hybrid RANS-LES computation. Given these parameters, our overall grid was composed of 64 million cells. Grid refinement studies are not possible when using hybrid RANS-LES modelling as the space filtering operation is grid size dependent. None the less, it is critical to use a grid that captures the physics of the problem as accurately as possible. To this end, comparisons to experimental data are even more critical when using hybrid RANS-LES.

In the discussion that follows, the lengths are scaled by the cord and the velocities by U_0. In addition to comparing our results to the experimental data of Jacob et al. [13], we also compare to the CFD results of Greschner et al. [14] who used a different hybrid RANS-LES model in their computations. Figure 3 shows the nondimensional mean velocity profile and the nondimensional root-mean-square (rms) of the velocity fluctuations at two different nondimensional downstream locations; (a) −0.255 and (b) 0.25. Our results compare favorably to the experimental data and previous CFD results especially those of the rms velocity. It is important to point out that in the CFD

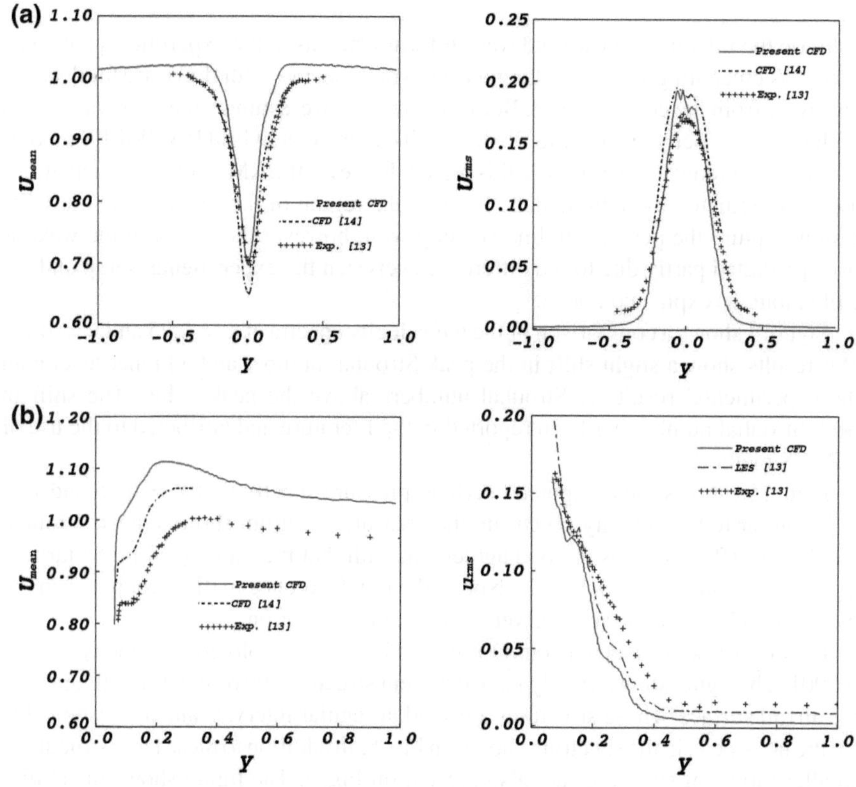

Fig. 3 Comparison of the mean velocity profile (*left*) and rms velocity fluctuation profile (*right*) to experimental data [13] and prior CFD results [14]. **a** x = −0.255 and **b** x = 0.25

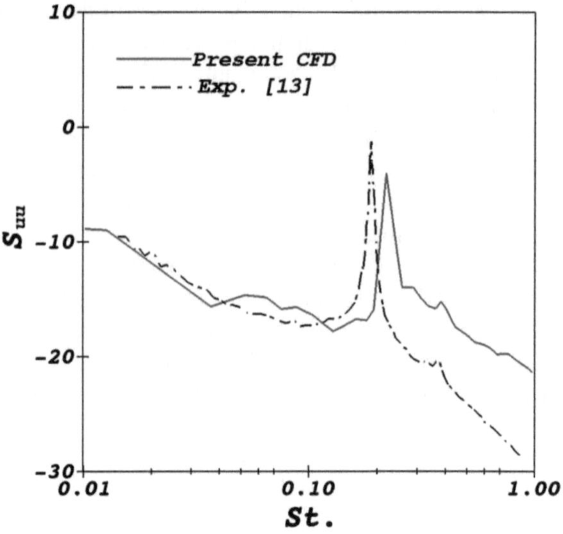

Fig. 4 Comparions of the predicted velocity spectra to that measured for x = 0.25 and y = 0.08

solution the inlet mean flow used was uniform whereas in the experiments the mean flow was emanating from a rectangular jet which decays with downstream distance and away from the core of the jet. Both CFD results overestimate the near wall region of the mean velocity profile, Fig. 3b. It is well-known that hybrid RANS-LES models tend to over predict the velocity in this region, however the RMS velocity fluctuations are well predicted by both models. In the wake region of the rod, Fig. 3a, our CFD results capture the peak centreline velocity well, however the extent of the wake is over-predicted partly due to the differences between the experimental setup and the CFD input, as explained above.

Figure 4 shows a comparison of the the velocity spectra at x = 0.25 and y = 0.08. Our results show a slight shift in the peak Strouhal number and a higher level than the experimental results at Strouhal numbers above the peak value. The shift in peak Strouhal number has been reported in the literature and attributed to the use of HRLES method.

Figure 5a shows the computed nearfield pressure spectrum at x = 0.75 and y = 0.7. Similar to the velocity spectrum, the pressure spectrum shows a peak Strouhal number of 0.24, which is in good agreement with that measured [13] in the farfield at x = 0.75 and y = 15.0, Fig. 5b. Notice the big difference in dB-level between the two spectra because of the observer location, i.e. y-location.

Figure 6 show an instantaneous Q-criterion iso-surface colored by velocity for q = 0.0001. The figure shows clearly large coherent structures traveling downstream over the airfoil surface. These structures are shed at regular intervals and are responsible for the large peak in the spectrum shown on Fig. 5. In addition to these large structures, smaller turbulent structures are also present on Fig. 6. The figure shows the ability of HRLES to capture various size turbulent strcutures.

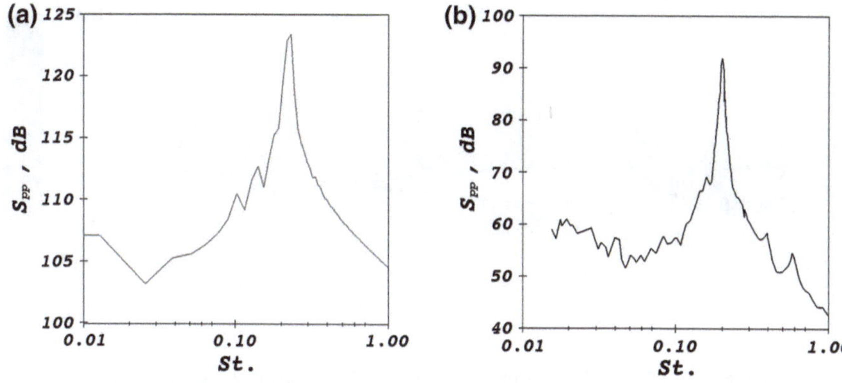

Fig. 5 **a** Computed nearfield pressure spectrum, x = 0.75 and y = 0.7; **b** measured farfield pressure spectrum [13], x = 0.75 and y = 15.0

Fig. 6 Instantaneous Q-criterion iso-surface colored by velocity

Figure 7 shows instantaneous CFD schlieren, which shows the log of the density gradient contours. The figure clearly shows the location of the coherent structures represented by the darker spots over the airfoil. This is an indication that these structures are indeed responsible for large density gradients in the flowfield and hence are the source of noise both in the nearfield and farfield.

Figure 8 shows the noise radiation directivity around the leading, Fig. 8a, and trailing, Fig. 8b, edges of the airfoil on a circle of radius 0.7. Both figures show that the highest noise level is in the area near the surface and directly above and below the airfoil.

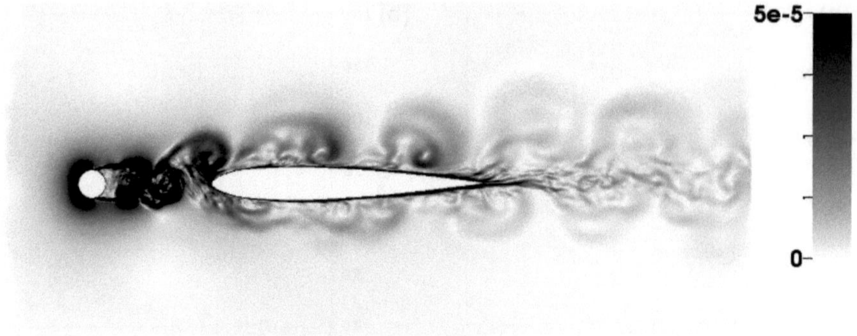

Fig. 7 Instantaneous CFD schlieren showing the locus of the density gradient

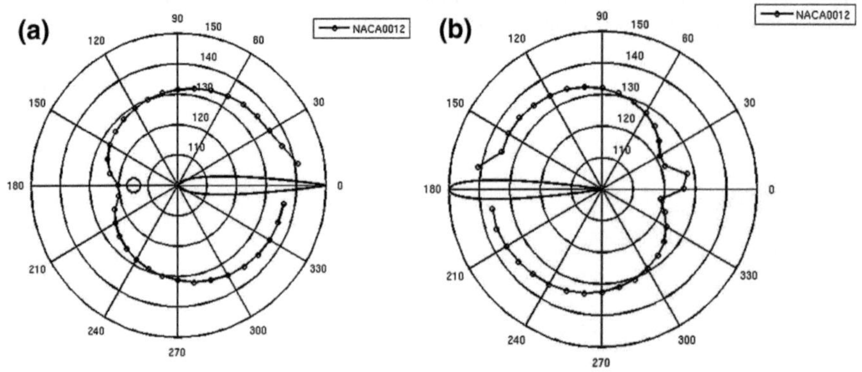

Fig. 8 Noise radiation directivity around the airfoil **a** leading edge **b** trailing edge

4 Conclusions

In this paper a hybrid RANS-LES method has been used to compute the noise radiated by the leading and trailing edges of a NACA0012 airfoil. Comparison of our computational results to experimental data showed good overall agreement. Our computational results show the ability of HRLES to capture turbulent structures relevant to noise radiation. Our computational results revealed that large coherent structures are responsible for the peak Strouhal number. The noise directivity is directly above and below the airfoil surface.

References

1. NASA Aeronautics Blueprint: Toward a bold new era in aviation. http://www.aerospace.nasa.gov/aero_blueprint/cover.html

2. Girimaji, S.S.: Partially-averaged Navier-Stokes for turbulence: a Reynolds averaged Navier-Stokes to direct numerical simulation bridging method. J. Appl. Mech. **73**(3), 413–421
3. Squires, K.D., Frosythe, J.R., Spalart, P.R.: Detached-eddy simulation of the separated flow around a forebody cross-section. In: Friedrich, G.R., Metais, O. (eds.) Direct and Large-Eddy Simulation IV, ERCOFTAC series, vol. 8, pp. 481–500. Kluwer Academic Press (2001)
4. Frendi, A., Tosh, A., Girimaji, S.: Flow past a backward facing step: comparison of PANS, DES and URANS results with experiments. Int. J. Comput. Methods Eng. Sci. Mech. **8**(1), 23–38 (2006)
5. Hahn, P.V., Frendi, A.: Interaction of three-dimensional protuberances with a supersonic turbulent boundary layer. AIAA J. **51**(7), 1657–1666 (2013)
6. Peugeot, J.W., Frendi, A.: Toward the understanding of flow-induced vibrations in rocket-engine manifold. J. Propul. Power **29**(6), 1468–1477 (2013)
7. Menter, F.R.: Two-equation eddy-viscosity turbulence models for engineering applications. AIAA J. **32**(8), 1598–1605 (1994)
8. Vatsa V.N., Lockard, D.P.: Assessment of hybrid RANS/LES turbulence models for aeroacoustics applications. In: AIAA-2010-4001 (2010)
9. Lynch, C.E., Smith, M.J.: Extension and exploration of a hybrid turbulence model on unstructured grids. AIAA J. **49**, 2585–2591 (2011)
10. Baurle, R.A., Tam, C.J., Edwards, J.R., Hassan, H.A.: Hybrid simulation approach for cavity flows: blending, algorithm, and boundary treatment issues. AIAA J. **41**, 1463–1480 (2003)
11. FUN3D: http://fun3d.larc.nasa.gov
12. VGRID: http://tetruss.larc.nasa.gov/vgrid/
13. Jacob, M.C., Boudet, J., Casalino, D., Michard, M.: A rod-airfoil exper iment as benchmark for broadband noise modeling. Theoret. Comput. Fluid Dyn. **19**, 171–196 (2005)
14. Greschner, B., Thiele F., Jacob, M.C., Casalino, D.: Prediction of sound generated by a rod-airfoil configuration using EASM DES and the generalized Lighthill/FW-H analogy. Comput. Fluids **37**(2008), 402–413 (2008)

[illegible]

MIX
Papier aus verantwortungsvollen Quellen
Paper from responsible sources
FSC® C105338
FSC
www.fsc.org

If you have any concerns about our products,
you can contact us on
ProductSafety@springernature.com

In case Publisher is established outside the EU,
the EU authorized representative is:
Springer Nature Customer Service Center GmbH
Europaplatz 3, 69115 Heidelberg, Germany

Printed by Libri Plureos GmbH
in Hamburg, Germany